D0151565

Circuit Design and Simulation with VHDL

second edition

Circuit Design and Simulation with VHDL

second edition

Volnei A. Pedroni

The MIT Press
Cambridge, Massachusetts
London, England

© 2010 Massachusetts Institute of Technology

All rights reserved. No part of this book may be reproduced in any form by any electronic or mechanical means (including photocopying, recording, or information storage and retrieval) without permission in writing from the publisher.

For information about special quantity discounts, please email special_sales@mitpress.mit.edu

This book was set in Times New Roman on 3B2 by Asco Typesetters, Hong Kong.
Printed and bound in the United States of America.

Library of Congress Cataloging-in-Publication Data

Pedroni, Volnei A.
Circuit design and simulation with VHDL / Volnei A. Pedroni. — 2nd ed.
    p.   cm.
Rev. ed. of: Circuit design with VHDL / Volnei A. Pedroni. 2004.
Includes bibliographical references and index.
ISBN 978-0-262-01433-5 (hardcover : alk. paper) 1. VHDL (Computer hardware description language).
2. Electronic circuit design. 3. System design. I. Pedroni, Volnei A. II. Title.
TK7885.7.P43   2010
621.39′5—dc22   2009045909

10   9   8   7   6   5   4   3

This second edition is dedicated to the many people, in so many countries, who made the first edition of the book so successful.

Contents

Preface

The book presents one of the most comprehensive coverages so far of VHDL and its applications to the design and simulation of real, industry-standard circuits. It does not focus only on the VHDL language, but also on its use in building and testing digital circuits. In other words, besides explaining VHDL in detail, it also shows why, how, and which types of circuits are inferred from the language constructs, and how any of the four simulation categories can be implemented, all demonstrated by means of numerous examples. A rigorous distinction is made between VHDL for synthesis and VHDL for simulation. In both cases, the VHDL codes are always complete, not just partial sketches, and are accompanied by circuit theory, code comments, and simulation results whenever applicable. The book also reviews fundamental concepts of digital electronics and digital design, resulting in a very practical, self-contained approach. A series of modern extended and advanced designs are also presented, covering state machines, memory implementations, serial data communications circuits, video interfaces, and more.

Main Features

- The book focuses on the *use* of VHDL rather than solely on the language itself. In other words, besides explaining VHDL in detail, it also shows why, how, and which types of circuits are inferred from the language constructs.

- The book makes a clear distinction between the parts of VHDL that are for *synthesis* versus those that are for *simulation* (contrary to other books, which usually mix up all VHDL constructs).

- The VHDL codes in all design examples are complete, not just partial sketches. Circuit diagrams, physical synthesis in Field Programmable Gate Arrays (FPGAs), simulation results, and explanatory comments are also included in the designs.

- It teaches all indispensable features of VHDL in a very concise format.

- It is the first text to also include a detailed analysis of circuit simulation with VHDL testbenches in all four categories (nonautomated and fully automated, functional and timing

simulations), accompanied also by related tutorials (like ModelSim), which allow complete
end-to-end practical examples to be presented.

· The book also reviews fundamental concepts of digital electronics and digital design,
resulting in a very practical, self-contained approach.

· To ease the understanding of the language and its applications, the review and the examples are separated into *combinational* and *sequential* circuits. Further distinction is made
between *logical* versus *arithmetic* combinational circuits, as well as between *regular* versus
state-machine-based sequential circuits.

· The book is divided into three parts, with circuit-level VHDL in part 1 (chapters 1–8),
system-level VHDL and simulation in part 2 (chapters 8–10), and finally extended and
advanced designs in part 3 (chapters 11–17). In summary, chapters 1–10 teach VHDL,
while chapters 11–17 show a series of extended and advanced designs using VHDL.

· Inclusion of new, modern digital circuits, like advanced state machines, serial data communications circuits, and video interfaces, all with theory, complete VHDL codes, simulation, and explanatory comments, makes the lab sections much more productive.

· All examples and exercises are named to ease the identification of the circuit/design
under analysis.

· Finally, a series of 15 appendices show tutorials on very important design tools, such as
ISE, Quartus II, and ModelSim, plus descriptions of programmable logic devices (CPLDs/
FPGAs, in which the designs are implemented), of the DE2 development board, of stan-
dard VHDL packages, and more.

Main Differences Relative to the First Edition

The book was updated, extended, and immensely improved. The VHDL language is now
covered in chapters 1–10 (including fundamental designs and simulation), while chapters
11–17 present extended and advanced designs. Below is a summary of the main improve-
ments with respect to the first edition, preceded by the total number of examples, exercises,
and figures in both editions.

Enumerated design examples: 79 (first edition); 94 (second edition)

Exercises: 96 (first edition); 231 (second edition)

Figures: 145 (first edition); 278 (second edition)

With Respect to the Language

The study of VHDL (chapters 1–10) was updated, extended, and deepened. The syntax
was improved, with better coverage and a simplified representation adapted from the
Backus-Naur Form. Features of VHDL 2008 were also included. New theoretical details

were included in the descriptions of basically all circuits. Additionally, the number of examples and of exercises grew substantially.

In chapters 2–4, the study of VHDL libraries/packages was extended. Numerous new details, particularly for synthesis, were described and included in the examples and exercises. The description of the code structure, in chapter 2, was modernized, including additional details about the syntax and synthesis packages as well as new introductory design examples.

In chapter 3, the description of data types was updated and immensely expanded. A more rigorous distinction between the several data type families was provided, including several type classifications. A successful technique introduced in the first version, which bases any data type on its number of bits, was again employed and extensively used in the examples. Type conversion and the analysis of unsigned versus signed types were also deepened.

In chapter 4, the description of operators and attributes was updated and expanded. All predefined options are now present in the text. A series of synthesis attributes, to prevent logic or register simplifications or for automated pin assignments, were also included.

The study of concurrent code, in chapter 5, received new examples and new topics, including recommendations for the implementation of signed systems, followed by respective examples. The use of special synthesis attributes, described in chapter 4, was also illustrated. Additionally, the study of sequential code, in chapter 6, was modernized. The examples were reorganized and new examples and new exercises were provided.

In chapter 7, the study of signal versus variable was modernized, with six rules introduced for the proper understanding of their differences. A new topic, which introduces a technique to allow multiple signal assignments, was also included.

Chapter 8's description of packages and components was also updated. New examples and exercises were included, as well as configuration declarations and multiple component instantiations.

The study of functions and procedures, in chapter 9, was reorganized and expanded. Additional details were presented, with more information on overloading and the inclusion of new function implementations.

Contrary to all other chapters, which deal exclusively with synthesis, chapter 10 is completely dedicated to simulation (this did not exist in the previous edition). All four categories of VHDL simulation with testbenches are described, and several practical examples are given. The use of text files, which are very helpful in simulations, is also described. This chapter is a crucial distinguishing feature with respect to the previous edition (in fact, with respect to any other VHDL book).

With Respect to the Extended and Advanced Design Examples

Chapters 11–17 are dedicated exclusively to the presentation of extended and advanced designs, constituting another major addition to the book. Another difference is that these designs are not done without first going through the theory on the involved circuits, including the analysis of official standards, when applicable. This part of the book replaces chapters 8, 9, and 12 of the previous edition. The main differences are summarized below.

Chapter 11 presents detailed design techniques for finite state machines (FSMs) using VHDL. A new technique for complex, timed machines is introduced in this new edition. Another new topic discusses the state-bypass problem in FSMs. The result is the most comprehensive study of FSM design using VHDL so far.

Chapter 12 is another completely new addition to the book. It presents a series of designs involving basic displays (LEDs, SSDs, and LCDs). Because these devices provide physical (visual) feedback to the students, lab exercises involving them are very motivating. As in all other chapters, each design is preceded by theoretical information on the circuits to be designed.

Chapter 13, also new, shows a study of memory implementations. This is important because basically any modern digital system requires some sort of memory.

Chapter 14 is the longest of the new chapters. It deals with a very modern topic, present in almost all large designs, which consists of serial data communications circuits. Several modern interfaces are described (I^2C, SPI, TMDS, PS2) and subsequently used in actual applications, driving actual ICs.

Chapters 15–17 are also new. All three deal with video circuits, again providing interesting, advanced circuits for lab experiments. Chapter 15 deals with the traditional VGA interface, used to connect computers to analog video monitors. Chapter 16 describes the DVI interface, fully digital and much more complex, used to connect desktop computers to LCD monitors. Finally, chapter 17 deals with the FPD-Link interface, which is a modern video interface for industrial and hand-held applications.

With Respect to the Exercises

The exercise sections were modernized and greatly expanded (there were 96 exercises in the previous edition; there are now 231). A broader coverage is achieved, with numerous interesting exercises for implementation and testing in FPGA boards during the lab sections.

With Respect to the Overall Presentation

The text was fully revised and expanded, with a smoother presentation and the inclusion of many additional details, both theoretical and practical. The same occurred with the figures, with the inclusion of many new ones and the complete reconstruction of those brought over from the first edition.

Audience

The book is mainly intended for the following:

· electrical engineering undergraduate and graduate students,
· computer engineering undergraduate and graduate students,

- computer science undergraduate and graduate students,
- digital design professors and instructors,
- VHDL professors and instructors,
- digital design engineers and practitioners in the industry, and
- digital design consultants and other practitioners at all levels.

Companion Books

The following two references are highly recommended:

- IEEE 2008 for additional details on the VHDL language.
- Pedroni 2008 for theoretical background on digital concepts and digital circuits.

Circuit Design and Simulation with VHDL

second edition

I CIRCUIT-LEVEL VHDL

1 Introduction

1.1 About VHDL

This chapter concisely describes what VHDL is and what it is used for. Popular synthesis and simulation tools are also listed, and a typical design flow is summarized.

VHDL is a *hardware description language*. The code describes the behavior or structure of an electronic circuit, from which a compliant physical circuit can be inferred by a compiler. Its main applications include synthesis of digital circuits onto CPLD/FPGA (Complex Programmable Logic Device/Field Programmable Gate Array) chips and layout/mask generation for ASIC (Application-Specific Integrated Circuit) fabrication.

VHDL stands for VHSIC (Very High Speed Integrated Circuits) Hardware Description Language, and resulted from an initiative funded by the U.S. Department of Defense in the 1980s. Its first version was VHDL 87, later upgraded by VHDL 93, then VHDL 2002, and finally VHDL 2008. It was the first hardware description language standardized by the IEEE, through the 1076 and 1164 standards. VHDL is technology/vendor independent, so VHDL codes are portable and reusable.

VHDL allows circuit *synthesis* as well as circuit *simulation* (both are covered in the book). The former is the translation of a source code into a hardware structure that implements the intended functionality, while the latter is a testing procedure to ensure that such functionality is indeed achieved by the synthesized circuit. In all chapters we will concentrate on VHDL constructs that are synthesizable, except for chapter 10, which deals exclusively with VHDL for simulation.

1.2 VHDL Versions

The first VHDL version was released in 1987, through the *IEEE 1076-1987 Standard VHDL Language Reference Manual*. Several versions followed (all listed below).

- *IEEE 1076-1987 Standard VHDL Language Reference Manual* (archived)
- *IEEE 1076-1993 Standard VHDL Language Reference Manual* (archived)

- *IEEE 1076-2000 Standard VHDL Language Reference Manual* (archived)

- *IEEE 1076-2002 Standard VHDL Language Reference Manual* (archived)

- *IEEE 1076-2008 Standard VHDL Language Reference Manual* (active: released in February 2009).

Other related documents, mainly with synthesis packages, are listed below.

- *IEEE 1164-1993 Standard Multivalue Logic System for VHDL Model Interoperability* (active)

- *IEEE 1076.3-1997 Standard VHDL Synthesis Packages* (active)

- *IEEE 1076.6 Standard for VHDL Register Transfer Level (RTL) Synthesis* (active).

Because the 2000 document was rapidly replaced with the 2002 version, it is usually considered that four VHDL versions exist: VHDL 87, VHDL 93, VHDL 2002, and VHDL 2008.

The expansions introduced in VHDL 2008 will be explained in the corresponding chapters. The package-related additions will also be mentioned in the corresponding data packages of appendices H–N.

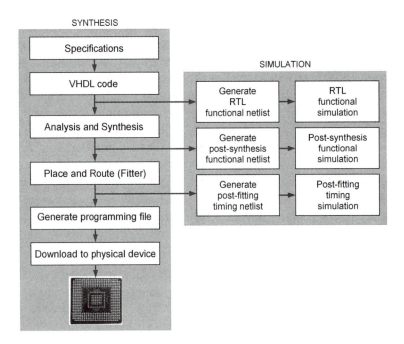

Figure 1.1
Simplified VHDL design flow.

1.3 Design Flow

A simplified view of the design flow is presented in figure 1.1. We assume that the designer already has a set of specifications for which a compliant circuit should be generated. The first step is to write a VHDL code that fulfills such specifications. The code must be saved in a text file with the extension *.vhd* and the same name as that of its main entity. Next, the code is compiled using a synthesis tool (a list of such tools is given in section 1.4). Several files are generated during the compilation process. The synthesizer breaks down the code into the hardware structures that are available within the chosen device, so during fitting (place and route) each structure inferred by the synthesizer is assigned a specific place inside the device. This positional information is important because it greatly influences the resulting circuit's timing behavior. With the timing information generated by the fitting process, the software allows the circuit to be fully simulated. Once the specifications have been met, the designer can proceed to the final step (implementation), during which a programming file for the device (when using a CPLD or FPGA) or for the masks (for ASICs) is generated. In the case of CPLDs/FPGAs, the design is concluded by downloading the programming file from the computer to the target device.

1.4 EDA Tools

There are several EDA (Electronic Design Automation) tools available for circuit synthesis and simulation using VHDL. Some tools are offered by CPLD/FPGA companies (Altera, Xilinx, etc.), while others are offered by third-party software companies (Mentor Graphics, Synopsys, Cadence, etc.). Some examples are listed below.

- From Altera: *Quartus II* (for synthesis and graphical simulation)
- From Xilinx: *ISE* (*XST* for synthesis, *ISE Simulator* for simulation)
- From Mentor Graphics: *Precision RTL* and *Leonardo Spectrum* (synthesis), *ModelSim* (simulation)
- From Synopsys/Synplicity: *Design Compiler Ultra* and *Synplify Pro/Premier* (synthesis), *VCS* (simulation)
- From Cadence: *NC-Sim* (simulation)
- From Aldec: *Active-HDL* (simulation).

The designs presented in the book were compiled and simulated using *Quartus II Web Edition 8.1* or newer, from Altera, and occasionally *ISE WebPack 10.1* or newer, from Xilinx. In the simulations with testbenches of chapter 10, *ModelSim 6.3g* was also employed.

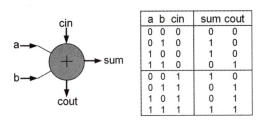

Figure 1.2
Full-adder diagram and truth table.

```
ENTITY full_adder IS
   PORT (a, b, cin: IN BIT;
         sum, cout: OUT BIT);
END full_adder;
-----------------------------------------------------
ARCHITECTURE dataflow OF full_adder IS
BEGIN
   sum <= a XOR b XOR cin;
   cout <= (a AND b) OR (a AND cin) OR (b AND cin);
END dataflow;
```

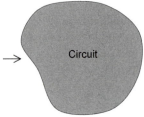

Figure 1.3
VHDL code for the full-adder unit of figure 1.2.

In the tutorials, the following EDAs are described:

- Quartus II 9.0 sp1 Web Edition (appendix B)
- ISE 11.1 WebPack (appendix C)
- ModelSim 6.3g Web Edition for Altera (appendix D).

1.5   Translation of VHDL Code into a Circuit

A full-adder unit is depicted in figure 1.2. In it, *a* and *b* represent the input bits to be added, *cin* is the carry-in bit, *sum* is the sum bit, and *cout* the carry-out bit. As shown in the truth table, *sum* must be high whenever the number of inputs that are high is odd (odd parity function), while *cout* must be high when two or more inputs are high (majority function).

A VHDL code for the full-adder of figure 1.2 is shown in figure 1.3. As can be seen, it consists of an ENTITY, which contains a description of the circuit ports (pins), and of an ARCHITECTURE, which describes how the circuit must function. We see in the latter that the outputs are computed by $sum = a \oplus b \oplus cin$ and $cout = a.b + a.cin + b.cin$.

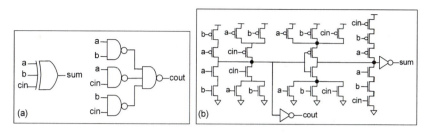

Figure 1.4
Implementation examples for the full-adder circuit of figure 1.2: (a) With conventional gates; (b) At transistor level (with CMOS logic).

Note: VHDL is not case sensitive, except possibly for the STD_ULOGIC symbols ('Z', 'X', etc.). Just to make the code easier to read, throughout the book capital letters are employed for the reserved words, while lowercase is used for the words that can be chosen by the user.

From the VHDL code shown on the left of figure 1.3, a physical circuit is inferred. There are, however, several ways of implementing the equations described above, so the actual circuit will depend on the target technology as well as on the compilation setup. Just to illustrate such possibilities, two examples are depicted in figure 1.4. In (a), *sum* and *cout* are computed using conventional gates (this might be the case when the target is a CPLD or FPGA), while in (b) a transistor-level implementation, utilizing CMOS logic, is shown (proper for ASICs). Moreover, the synthesis tool can be set to optimize the layout for area or for speed, which might also affect the final circuit.

1.6 Circuit Simulation

Whatever the final circuit inferred from the code is, its operation should always be verified by means of simulations still at the design level (after synthesis), as indicated in figure 1.1. Of course, the physical system should eventually also be verified. However, contrary to basically any other kind of electronic design, VHDL designs, once they pass careful EDA simulations, are basically guaranteed to pass physical evaluation as well.

Testing using VHDL will be seen in chapter 10. For this brief introduction, let us consider just simulations with graphical inputs (done directly in the Quartus II or ISE tool). In this case, waveforms similar to those depicted in figure 1.5 will be displayed by the simulator. Indeed, figure 1.5 contains the simulation results from the circuit synthesized with the VHDL code of figure 1.3, which implements the full-adder unit of figure 1.2. In figure 1.5a, a *functional* simulation is shown (propagation delays through the circuit are neglected), while figure 1.5b shows a *timing* simulation (propagation delays are taken into account— see highlighted area). Nearly all simulations presented in the book are in the latter category.

Figure 1.5
(a) Functional versus (b) timing simulation results obtained from the VHDL code of figure 1.3.

The input ports (characterized by an inward arrow with an "I" marked inside) and the output ports (characterized by an outward arrow with an "O" marked inside) are those listed in the ENTITY of figure 1.3. We can freely choose the values (waveforms) for the input signals (*a*, *b*, and *cin* in this case), and the simulator will compute and plot the output signals (*sum* and *cout*). As can be observed in figure 1.5, the outputs do behave as expected.

1.7 VHDL Syntax

In the description of VHDL, great care was taken with respect to the syntax. In order to make it easier to understand, some simplifications were made with respect to that in the *IEEE 1076-2008 Standard VHDL Language Reference Manual*, which follows approximately the EBNF (Extended Backus-Naur Form) style. The simplifications were made without compromising the quality of the information; indeed, additional comments and examples were included whenever necessary to clarify the syntax as much as possible, always emphasizing items that are synthesizable.

1.8 Number and Character Representations in VHDL

Integers

Integers are normally represented with base-10 (decimal) numbers. Their default range in VHDL is from $-(2^{31} - 1)$ to $+(2^{31} - 1)$. The underscore character (_) can be used anywhere in the number to help visualize it, with no effect on the synthesized value. Exponents are also accepted. Though unusual, any other base from 2 to 16 can also be employed, in which case the base value must precede the number, and the number must be surrounded by the pound (sharp) symbol (#). Examples are shown below.

- Base 10 (decimals): 5, 32, 3250, 3_250, 3E2 ($= 3 \cdot 10^2 = 300$)
- Other bases (from 2 to 16):

2#0111# (this is the integer 7, because $0 \cdot 2^3 + 1 \cdot 2^2 + 1 \cdot 2^1 + 1 \cdot 2^0 = 7$)

5#320# ($3 \cdot 5^2 + 2 \cdot 5^1 + 0 \cdot 5^0 = 85$)

16#9F# ($9 \cdot 16^1 + 15 \cdot 16^0 = 159$)

3#201#E4 ($(2 \cdot 3^2 + 0 \cdot 3^1 + 1 \cdot 3^0) \cdot 3^4 = 1539$).

Binary Values

Binary values are surrounded by either single quotes (single bit) or double quotes (multi bits). Besides the regular binary representation, multi-bit words can also be expressed in *octal* or *hexadecimal* form. In such cases, an 'O' (for octal) or 'X' (for hexadecimal) must precede the bit vector. For binary, an optional 'B' can also be used. Since VHDL is not case sensitive, lowercase letters are fine too. Examples are shown below, with the equivalent decimal value included between parentheses.

- Regular binary form:

'0' ($= 0$), "0111" ($= 7$), b"0111" ($= 7$), B"11110000" ($= 240$)

- Octal and hexadecimal forms:

O"54" ($5 \cdot 8^1 + 4 \cdot 8^0 = 44$), o"0" ($0 \cdot 8^0 = 0$), X"C2F" ($12 \cdot 16^2 + 2 \cdot 16^1 + 15 \cdot 16^0 = 3119$), x"D" ($13 \cdot 16^0 = 13$)

Unsigned Values

In *unsigned* systems, all numbers are non-negative, hence ranging from 0 to $2^N - 1$, where N is the number of bits. For example, with 8 bits, values from 0 ("00000000") up to 255 ("11111111") can be encoded.

Signed Values

On the other hand, in *signed* systems, the numbers can be negative. With N bits, the range of integers from -2^{N-1} to $2^{N-1} - 1$ is covered. The usual representation for negative numbers is *two's complement*. If the MSB (Most Significant Bit; by convention, the leftmost) is '0', then the number is positive; if it is '1', it is negative. To obtain such a representation, we start with the positive value and complement (reverse) all bits, then add '1' to the result. Examples are shown below (note that $b^+ + |b^-| = 2^N$, where b^+ and b^- are the positive and negative representations for the binary number b).

"0111" = +7

"1001" = −7 (complement of +7 is "1000"; adding '1', "1001" results)

"010000" = +16

"110000" = −16 (complement of +16 is "101111"; adding '1', "110000" results).

Characters

Characters from an extended ASCII table (appendix H) are synthesizable. Following the same style of bit and bit vector, a single character is represented surrounded by a pair of single quotes, while a string of characters (also synthesizable) is surrounded by a pair of double quotes. Examples are shown below.

'A', 'a', '$', "VHDL", "mp3".

2 Code Structure

2.1 Fundamental VHDL Units

This chapter describes the fundamental sections that comprise a piece of regular VHDL code: Library declarations, ENTITY, and ARCHITECTURE. A few introductory design examples are also included.

As depicted in figure 2.1a, a basic VHDL code is composed of three sections:

- Library/package declarations: Contains a list of all libraries and respective packages needed in the design. The most commonly used libraries are *ieee*, *std*, and *work* (the last two are made visible by default).
- ENTITY: Specifies mainly the circuit's I/O ports, plus (optional) generic constants.
- ARCHITECTURE: Contains the VHDL code proper, which describes how the circuit should function, from which a compliant hardware is inferred.

LIBRARY is a collection of commonly used pieces of code. Placing them inside a library allows the code to be reused and also shared by other designs. The typical structure of a library is illustrated in figure 2.1b. Any previously designed circuit can be part of a library. Such circuits can then be used (instantiated) in other designs using the COMPONENT keyword. Another popular option is to write commonly used pieces of code in the form of FUNCTION or PROCEDURE (called *subprograms*), then place them inside a PACKAGE, which is located in a library. General declarations of data types are also usually located in libraries.

The fundamental units of VHDL (figure 2.1a) are studied in the first part of the book (*circuit-level VHDL*, chapters 1–7), while the library-related units (PACKAGE, COMPONENT, FUNCTION, and PROCEDURE—figure 2.1b) plus simulation are seen in the second part (*system-level VHDL*, chapters 8–10).

2.2 VHDL Libraries and Packages

The standard VHDL libraries are *std* and *ieee*, and their main packages are listed below. Popular nonstandard packages (sharewares) are also listed.

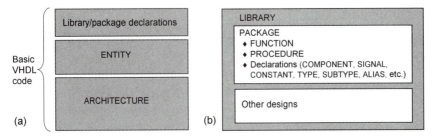

Figure 2.1
(a) Fundamental sections of a VHDL code; (b) Fundamental parts of a library.

Library *std*

▪ Package *standard* (appendix H): This package, specified in the IEEE 1076 standard, is part of VHDL since its first version (1987). Among other things, it contains several data type definitions (BIT, INTEGER, BOOLEAN, CHARACTER, etc.) and respective logic, arithmetic, comparison, shift, and concatenation operators. This package was expanded in VHDL 2008.

▪ Package *textio* (appendix M): A resource package for text and files, also specified in the IEEE 1076 standard and also expanded in VHDL 2008.

Library *ieee*

▪ Package *std_logic_1164* (appendix I): Defines the 9-value data types STD_ULOGIC and STD_LOGIC, whose main feature, compared to the original type BIT, is the existence of the additional synthesizable values *don't care* ('-') and *high-impedance* ('Z') (BIT only allows '0' and '1'). This package is specified in the IEEE 1164 standard.

▪ Package *numeric_std* (appendix J): Introduces the types SIGNED and UNSIGNED and corresponding operators, having STD_LOGIC as the base type. This package is specified in the IEEE 1076.3 standard.

▪ Package *numeric_bit*: Same as above, but with BIT as the base type.

▪ Package *numeric_std_unsigned* (appendix N): Introduced in VHDL 2008, this package is expected to replace the nonstandard package *std_logic_unsigned*.

▪ Package *numeric_bit_unsigned*: Also introduced in VHDL 2008, this package is similar to that above, but operates with the type BIT_VECTOR instead of STD_LOGIC_VECTOR.

▪ Package *env*: Introduced in VHDL 2008, it includes stop and finish procedures for communication with the simulation environment.

▪ Package *fixed_pkg* (plus associated packages): Developed by Kodak and introduced in VHDL 2008, it defines the unsigned and signed fixed-point types UFIXED and SFIXED and related operators.

• Package *float_pkg* (plus associated packages): Also from Kodak and VHDL 2008, it defines the floating-point type FLOAT and related operators.

Nonstandard Packages

• Package *std_logic_arith* (appendix K): Defines the types SIGNED and UNSIGNED and corresponding operators. This package is *partially* equivalent to *numeric_std*.

• Package *std_logic_unsigned*: Introduces functions that allow arithmetic, comparison, and some shift operations with signals of type STD_LOGIC_VECTOR operating as unsigned numbers.

• Package *std_logic_signed* (appendix L): Same as above, but operating as signed numbers.

The last two packages can be considered as complements to the package *std_logic_1164*, because the latter does not contain arithmetic or comparison operators for the type STD_LOGIC_VECTOR, while the former two do.

The packages listed will be studied in detail in chapters 3 and 4, when dealing with data types and operators, respectively.

2.3   Library/Package Declarations

To make a package visible to the design, two declarations are needed, one for the *library* where the package is located, the other a use clause pointing to the specific *package*. The corresponding syntax is shown below.

```
LIBRARY library_name;
USE library_name.package_name.all;
```

The most frequently used packages are:

• Package *standard*, from the library *std* (visible by default).
• Library *work* (folder where the project files are saved; also visible by default).
• Package *std_logic_1164*, from the library *ieee* (when needed, must be explicitly declared).

Corresponding declarations are shown below.

```
1   -------------------------------------------
2   LIBRARY std;              --optional declaration
3   USE std.standard.all;     --optional declaration
4   LIBRARY work;             --optional declaration
5   USE work.all;             --optional declaration
```

```
6   LIBRARY ieee;
7   USE ieee.std_logic_1164.all;
8   USE work.my_package.all;
9   -------------------------------------------
```

The package *standard* (lines 2–3) is made visible by default, so there is no need to explicitly declare it. The same is true for the *work* library (lines 4–5). On the other hand, the package *std_logic_1164* (lines 6–7) needs to be declared when the STD_(U)LOGIC type is used in the project. If an extra, user-made package is also needed, then it too must be declared, as shown in line 8 above (line 8 does not need line 4 because the latter is already in the default list).

Note in the declarations above that a semi-colon (;) indicates the end of a declaration or statement, while a double dash (--) indicates a comment. VHDL code is not case sensitive (except, possibly, for the STD_(U)LOGIC symbols). As already mentioned, other details about libraries/packages will be given in the next two chapters.

2.4   ENTITY

The main part of an ENTITY is PORT, which is a list with specifications of all input and output ports (pins) of the circuit. A simplified syntax is shown below (the complete syntax will be shown shortly).

```
ENTITY entity_name IS
    PORT (
        port_name: port_mode signal_type;
        port_name: port_mode signal_type;
        ...);
END [ENTITY] [entity_name];
```

The entity's name can be basically any word, except VHDL (and a few other) reserved words (appendix G). The same is true for the port names.

All members of the PORT field in the syntax above are *signals* (in contrast with variables); that is, wires that go in and out of the circuit. Their *mode* can be IN, OUT, INOUT, or BUFFER. As illustrated in figure 2.2a, IN and OUT are truly unidirectional wires, while INOUT is bidirectional and BUFFER is employed when a signal is sent out but it must also be used (read) internally. Finally, the *type* can be BIT, INTEGER, STD_LOGIC, and so on (data types will be studied in chapter 3).

The use of BUFFER can be avoided by creating internal auxiliary signals. Moreover, the mode OUT will be available for internal reading when VHDL 2008 is implemented.

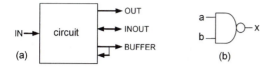

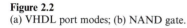

Figure 2.2
(a) VHDL port modes; (b) NAND gate.

The use of INOUT is particularly important when implementing memories, which often employ the same data bus for writing and reading (this will be seen in chapter 13).

Example   Below is a possible ENTITY for the NAND gate of figure 2.2b, under the name *nand_gate*. Its meaning is the following: the circuit has three I/O ports, of which two are inputs (*a* and *b*, mode IN) and the other is an output (*x*, mode OUT). The type of all three signals is BIT.

```
----------------------
ENTITY nand_gate IS
   PORT (a, b: IN BIT;
         x: OUT BIT);
END ENTITY;
----------------------
```

In the previous syntax, only the PORT field was shown. However, as shown next, an entity can contain three other fields, which are a GENERIC declarations section (before PORT), a general declarative part (after PORT), and finally a section with passive calls or processes (also after PORT).

```
ENTITY entity_name IS
   [GENERIC (
      const_name: const_type const_value;
      ...);]
   [PORT (
      signal_name: mode signal_type;
      ...);]
   [entity_declarative_part]
[BEGIN
   entity_statement_part]
END [ENTITY] [entity_name];
```

Only the PORT part of an entity is mandatory (for synthesis; for simulation, as will be shown in chapter 10, an empty entity can be employed). The optional GENERIC part (described in section 2.6) is for declaring constants that are globally visible to the design, including to PORT.

The optional *declarative* part (after PORT), though rarely used, can contain the following: subprogram declaration, subprogram body, type declaration, subtype declaration, constant declaration, signal declaration, shared variable declaration, file declaration, alias declaration, attribute declaration, attribute specification, disconnection specification, use clause, group template declaration, and group declaration.

Finally, the optional *statements* section, also rarely used, can contain passive calls and/or passive processes (that is, those that do not involve any signal assignments. They can be used, for example, to test PORT values). While the GENERIC field is often used, the other two optional sections are seldom employed (general system declarations are normally placed in the declarative part of the ARCHITECTURE or in a separate PACKAGE).

Example   The ENTITY below contains the first three of the four sections mentioned above.

```
-------------------------------------------------
ENTITY controller IS
    GENERIC (N: INTEGER := 8);
    PORT (a, b: IN INTEGER RANGE 0 TO 2**N-1;
          x: OUT STD_LOGIC_VECTOR(N-1 DOWNTO 0));
    TYPE byte IS ARRAY (7 DOWNTO 0) OF STD_LOGIC;
    CONSTANT mask: byte "00001111";
END ENTITY;
-------------------------------------------------
```

In VHDL 2008, the declarative part of the entity can also contain the following: subprogram instantiation declaration, package declaration, package body, package instantiation declaration, and PSL declarations. See other features in section 2.9.

2.5 ARCHITECTURE

ARCHITECTURE contains a description of how the circuit should function, from which the actual circuit is inferred. A simplified syntax is shown below.

```
ARCHITECTURE architecture_name OF entity_name IS
    [architecture_declarative_part]
BEGIN
    architecture_statements_part
END [ARCHITECTURE] [architecture_name];
```

As shown, an architecture has two parts: a *declarative* (optional) part, and the *statements* (code) part (from BEGIN down). The former can contain the same items as the declarative part of an entity, plus component declarations and configuration specifications (in either VHDL 2002 or VHDL 2008). The latter is where the VHDL statements are placed. As in the case of an entity, the name of an architecture can be basically any word, including the same name as the entity's.

Example Below is a possible ARCHITECTURE for the NAND gate of figure 2.2b, under the name *arch*. Its meaning is the following: the circuit must perform the NAND operation between *a* and *b*, assigning the result to *x*. In this example, there are no declarations in the declarative part, and the code contains just a single logical statement.

```
---------------------------------
ARCHITECTURE arch OF nand_gate IS
BEGIN
   x <= a NAND b;
END ARCHITECTURE;
---------------------------------
```

2.6 GENERIC

GENERIC declarations allow the specification of *generic* parameters (that is, generic constants, which can be easily modified or adapted to different applications). Their purpose is to parameterize a design, conferring the code more flexibility and reusability.

As seen in the syntax for ENTITY in section 2.4, GENERIC is the only declaration allowed before the PORT clause, which causes such constants to be truly global because they can be used even in the PORT specifications. A simplified syntax for GENERIC declarations is shown below.

```
GENERIC (constant_name: constant_type := constant_value;
         constant_name: constant_type := constant_value;
         ... );
```

Example The GENERIC declaration in the entity below specifies two parameters, called *m* and *n*. The first is of type INTEGER and has value 8, while the second is of type BIT_VECTOR and has value "0101". Therefore, whenever *m* and *n* are encountered in the code (including in the ENTITY itself), the values 8 and "0101" are automatically assigned to them.

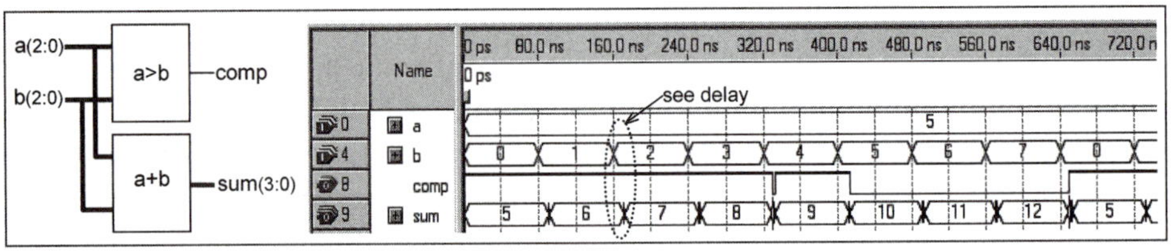

Figure 2.3
Circuit of example 2.1 and respective simulation results.

```
--------------------------------------------------
ENTITY my_entity IS
   GENERIC (m: INTEGER := 8;
            n: BIT_VECTOR(3 DOWNTO 0) := "0101");
   PORT (...);
END my_entity;
--------------------------------------------------
```

GENERIC MAP: If a COMPONENT containing a GENERIC declaration (like the one above) is instantiated in another design, the values of the generic constants that appear in the component being instantiated can be overwritten by the main design. This is done with a GENERIC MAP declaration, which will be seen in chapter 8 while studying the instantiation of components (section 8.4). An example illustrating the usefulness of GENERIC is presented in the next section (example 2.4).

In VHDL 2008, besides the traditional generic constants, generic types and generic subprograms are also supported. A generic constant can be used in the specification of other generic constants in the same generic list. The places where generics can be declared were also expanded; besides ENTITY and BLOCK headers, it can also be done in PACKAGE (chapter 8) and subprogram (chapter 9) headers.

2.7 Introductory VHDL Examples

In this section we present several introductory examples of VHDL code. Though we have not yet studied the constructs that appear in the examples, they will help illustrate fundamental aspects regarding the overall code structure. Each example is accompanied by explanatory comments and simulation results.

Example 2.1: Compare-Add Circuit

On the left of figure 2.3, a two-block circuit is shown. The inputs are two unsigned 3-bit values (*a* and *b*, ranging from 0 to 7), while the outputs are *comp* (single bit) and *sum* (to avoid overflow, 4 bits are needed, hence ranging from 0 to 15). The upper part must compare a to b, producing a '1' when $a > b$ or '0' otherwise. The lower part must add a and b, producing sum.

A VHDL code for this circuit is shown below. Note that dashed lines (lines 1, 4, 10, 16) were used to better organize the code (separating it into the three fundamental sections mentioned earlier). A library declaration appears in lines 2–3. The entity, named *comp_add*, is in lines 5–9. Finally, the architecture, called *circuit*, appears in lines 11–15.

```
1   -----------------------------------------
2   LIBRARY ieee;
3   USE ieee.std_logic_1164.all;
4   -----------------------------------------
5   ENTITY comp_add IS
6      PORT (a, b: IN INTEGER RANGE 0 TO 7;
7             comp: OUT STD_LOGIC;
8             sum: OUT INTEGER RANGE 0 TO 15);
9   END ENTITY;
10  -----------------------------------------
11  ARCHITECTURE circuit OF comp_add IS
12  BEGIN
13     comp <= '1' WHEN a>b ELSE '0';
14     sum <= a + b;
15  END ARCHITECTURE;
16  -----------------------------------------
```

Note that the entity contains all I/O ports. The inputs are a and b (mode IN, line 6), both of type INTEGER and ranging from 0 to 7 (3-bit unsigned values). The outputs are *comp* (line 7) and *sum* (line 8), the former of type STD_LOGIC (single bit), the latter of type INTEGER, ranging from 0 to 15 (4-bit unsigned value).

The architecture contains only two statements, with the first (line 13) making the comparison (by means of the WHEN statement), while the second (line 14) computes the sum (by means of the "+" operator). In this example, there are no declarations in the architecture's declarative part.

Simulation results are included in figure 2.3. Note that any input signal is preceded by an arrow with an "I" written inside, while each output shows an arrow with an "O" inside. A fixed value (5) was assigned to a, while b varies over the whole 3-bit range (0 to 7). The results are *comp* = '1' when $a > b$ and *sum* = $a + b$ (without overflow). Observe that it is a *timing* simulation because internal propagation delays were taken into consideration.

Note the glitch that occurs on *comp* when b changes from 3 to 4. It is because in this transition all bits of b change ("011" → "100"), so since the bits do not all change at exactly the same time, and moreover the actual transitions are not instantaneous (it is rather like a ramp instead of a vertical step), for a brief moment $b \geq a$ might occur, so this type of glitch is absolutely normal.

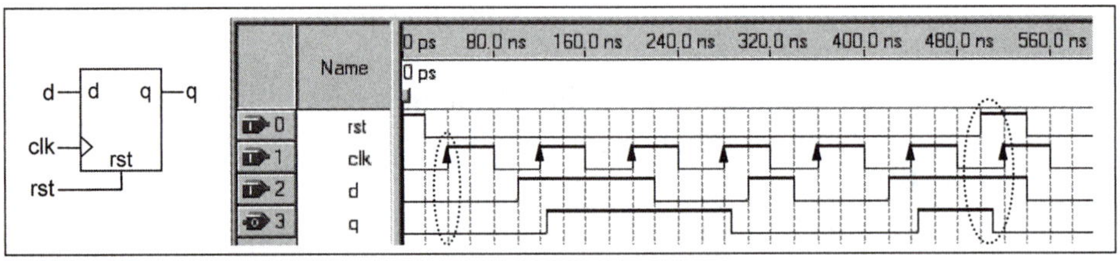

Figure 2.4
Circuit of example 2.2 and respective simulation results.

Example 2.2: D-type Flip-Flop (DFF)

Figure 2.4 shows a DFF (Pedroni 2008), which is one of the most fundamental storage circuits (there are thousands of them in FPGAs). Its inputs are *d* (data), *clk* (clock), and *rst* (reset), while *q* (stored data) is its output. In this case, the DFF is triggered at the positive (upward) clock transition, but the opposite is also possible. The output copies the input ($q <= d$) at the moment when *clk* changes from '0' to '1', remaining so until a new upward clock edge happens. Reset is asynchronous (that is, it does not depend on *clk*), so the output is immediately zeroed if *rst* = '1' occurs.

There are several ways of implementing a DFF, one being the solution presented below. One must remember, however, that VHDL code is inherently concurrent (contrary to regular computer programs, which are sequential), so to implement any clocked circuit (flip-flops, for example) we have to "force" VHDL to be sequential, which can be done with a PROCESS, as shown below.

```
1    -----------------------------------------
2    LIBRARY ieee;
3    USE ieee.std_logic_1164.all;
4    -----------------------------------------
5    ENTITY flip_flop IS
6        PORT (d, clk, rst: IN STD_LOGIC;
7              q: OUT STD_LOGIC);
8    END ENTITY;
9    -----------------------------------------
10   ARCHITECTURE flip_flop OF flip_flop IS
11   BEGIN
12     PROCESS (clk, rst)
13     BEGIN
14       IF (rst='1') THEN
15         q <= '0';
16       ELSIF (clk'EVENT AND clk='1') THEN
17         q <= d;
```

```
18        END IF;
19      END PROCESS;
20   END ARCHITECTURE;
21   ----------------------------------------
```

Comments about the code above follow.

Lines 2–3: First part (library declarations) of the code. Recall that this type of declaration consists of a library *name* followed by a library *use* clause. Because the data type STD_LOGIC is employed in this design, the package *std_logic_1164* must be included. The other two indispensable libraries (*std* and *work*) are made visible by default.

Lines 5–8: Second part (ENTITY) of the code, in this example named *flip-flop*.

Lines 10–20: Third part (ARCHITECTURE) of the code, here with the same name as the entity.

Line 6: Input ports, all of type STD_LOGIC.

Line 7: Output port, also of type STD_LOGIC.

Lines 12–19: Code part of the architecture (starts after the word BEGIN). In this case, the code contains just a PROCESS, needed because we want to implement a sequential (clocked) circuit (code inside a process is executed sequentially).

Line 12: Note that two signals (*clk*, *rst*) are included in the process's sensitivity list (the process is run whenever any of these signals change).

Lines 14–15: If *rst* goes to '1', the flip-flop is reset, regardless of *clk*.

Lines 16–17: If *rst* is not active, plus *clk* has changed (an EVENT occurred on *clk*), and such an event was a rising edge (*clk* = '1'), then the input signal (*d*) is stored into the flip-flop ($q <= d$).

Lines 15 and 17: The operator "$<=$" is used to assign a value to a SIGNAL (all ports are signals by default). In contrast, "$:=$" would be used for a VARIABLE.

Lines 1, 4, 9, and 21: Employed to better organize the code.

Simulation results from this code are included in figure 2.4 (note that it is again a *timing* simulation). The reader is invited to check them to confirm the DFF functionality. Arrows were included in the clock waveform to highlight the (only) points where the circuit is "transparent" (that is, when the output copies the input). Observe also that *rst* is indeed asynchronous.

Example 2.3: Registered Comp-Add Circuit

Figure 2.5 shows a circuit that combines those seen in the previous two examples; that is, DFFs are added at the outputs of the *comp_add* circuit in order to "register" (store) *comp* and *sum* (then called *reg_comp* and *reg_sum*).

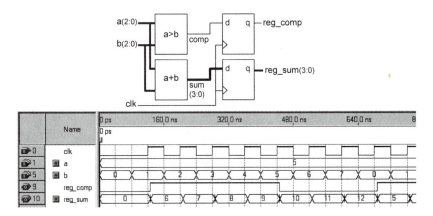

Figure 2.5
Circuit of example 2.3 and respective simulation results.

A VHDL code for this circuit is presented below. The original signals (*comp*, *sum*) are computed in the initial part of the architecture (lines 16–17). A process (lines 18–24) then follows, needed for flip-flop inference (sequential circuit). Note that a total of five DFFs are needed. Observe also that because *comp* and *sum* are now internal signals, they were specified in the declarative part of the architecture (lines 13–14).

```
1    -----------------------------------------------
2    LIBRARY ieee;
3    USE ieee.std_logic_1164.all;
4    -----------------------------------------------
5    ENTITY registered_comp_add IS
6       PORT (clk: IN STD_LOGIC;
7             a, b: IN INTEGER RANGE 0 TO 7;
8             reg_comp: OUT STD_LOGIC;
9             reg_sum: OUT INTEGER RANGE 0 TO 15);
10   END ENTITY;
11   -----------------------------------------------
12   ARCHITECTURE circuit OF registered_comp_add IS
13      SIGNAL comp: STD_LOGIC;
14      SIGNAL sum: INTEGER RANGE 0 TO 15;
15   BEGIN
16      comp <= '1' WHEN a>b ELSE '0';
17      sum <= a + b;
18      PROCESS (clk)
19      BEGIN
20         IF (clk'EVENT AND clk='1') THEN
21            reg_comp <= comp;
```

ena	address	word_line
0	X X	1 1 1 1
1	0 0	1 1 1 0
	0 1	1 1 0 1
	1 0	1 0 1 1
	1 1	0 1 1 1

Figure 2.6
Address decoder of example 2.4.

```
22              reg_sum <= sum;
23           END IF;
24      END PROCESS;
25   END ARCHITECTURE;
26   -----------------------------------------------
```

Simulation results are included in figure 2.5. Observe that now, contrary to example 2.1, the outputs are only updated when positive clock edges occur.

Example 2.4: Generic Address Decoder

A top-level diagram for a *generic N*-bit address decoder is depicted in figure 2.6. The circuit has two inputs, called *address* (N bits) and *ena* (enable, one bit), and one output, called *word_line* (2^N bits). As shown in the truth table (for $N = 2$), the output has only one bit dissimilar from all the others, located in the position determined by the input value. Note that when *ena* = '0' all output bits must be high.

Below is a VHDL code for this circuit. Library declarations are not needed because only data types from the package *standard* (visible by default) are employed in this example. The ENTITY is in lines 2–7, containing GENERIC and PORT declarations. N is entered as a *generic* parameter (line 3), so the code can be easily adapted to any address decoder size. The input and output signals (PORT declarations, lines 4–6) are from figure 2.6.

The ARCHITECTURE is in lines 9–16. It is totally generic because no changes are required when the size (N) of the circuit is modified (the only change needed is in line 3). The GENERATE statement (studied in chapter 5) is used to create a loop, which causes all output bits to be '1' when *ena* = '0', or produces just one bit equal to '0' (whose position coincides with the value represented by *address*) when *ena* = '1'.

```
1   --------------------------------------------------
2   ENTITY address_decoder IS
3      GENERIC (N: NATURAL := 3);
4      PORT (address: IN NATURAL RANGE 0 TO 2**N-1;
5            ena: BIT;
6            word_line: OUT BIT_VECTOR(2**N-1 DOWNTO 0));
7   END address_decoder;
8   --------------------------------------------------
```

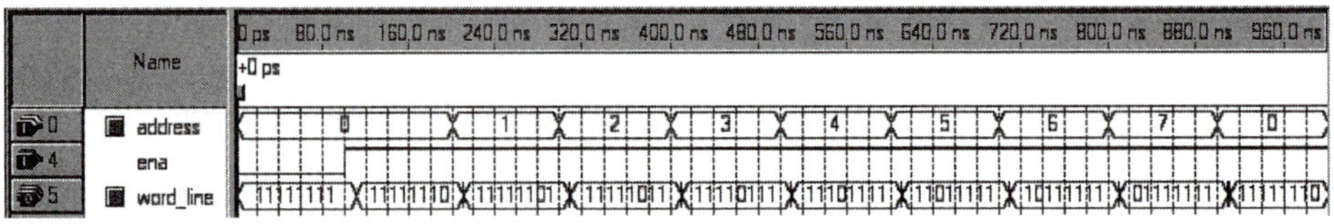

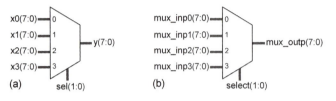

Figure 2.7
Simulation results from the address decoder of example 2.4.

Figure 2.8
Multiplexer represented with (a) short and (b) more meaningful signal names.

```
 9   ARCHITECTURE address_decoder OF address_decoder IS
10   BEGIN
11      gen: FOR i IN address'RANGE GENERATE
12         word_line(i) <= '1' WHEN ena='0' ELSE
13                         '0' WHEN i=address ELSE
14                         '1';
15      END GENERATE;
16   END address_decoder;
17   -----------------------------------------------------
```

Simulation results, for $N = 3$, are displayed in figure 2.7. As can be seen, all outputs are high when *ena* = '0'. After *ena* is asserted, one output bit is turned low, in the position defined by *address*.

2.8   Coding Guidelines

In order to save book space, the VHDL codes are generally presented with short signal names (so statements can fit in one line) and multiple signal definitions are made in the same line whenever possible. Additionally, in order to illustrate the usage of the different data types, a variety of types are used in the examples. However, when writing VHDL code for a large project, especially when more than one design team is involved, it is important to standardize the coding style (especially for the interfacing signals) as much as possible and use more meaningful signal names. Taking figure 2.8 as an example, a space-saving code could be the following (figure 2.8a):

```
1   ------------------------------------------------------------
2   LIBRARY ieee;
3   USE ieee.std_logic_1164.all;
4   ------------------------------------------------------------
5   ENTITY multiplexer IS
6      PORT (x0, x1, x2, x3: IN STD_LOGIC_VECTOR(7 DOWNTO 0);
7            sel: IN NATURAL RANGE 0 TO 3;
8            y: OUT STD_LOGIC_VECTOR(7 DOWNTO 0));
9   END ENTITY;
10  ------------------------------------------------------------
11  ARCHITECTURE multiplexer OF multiplexer IS
12  BEGIN
13     y <= x0 WHEN sel=0 ELSE
14          x1 WHEN sel=1 ELSE
15          x2 WHEN sel=2 ELSE
16          x3;
17  END ARCHITECTURE;
18  ------------------------------------------------------------
```

On the other hand, a more spread-out code could be as follows (figure 2.8b):

```
1   ----------------------------------------------------
2   LIBRARY ieee;
3   USE ieee.std_logic_1164.all;
4   ----------------------------------------------------
5   ENTITY multiplexer_4x8 IS
6      GENERIC (
7         N: NATURAL := 8;      --bits in in/out signals
8         M: NATURAL := 2);     --bits in select
9      PORT (
10        mux_inp0:    IN STD_LOGIC_VECTOR(N-1 DOWNTO 0);
11        mux_inp1:    IN STD_LOGIC_VECTOR(N-1 DOWNTO 0);
12        mux_inp2:    IN STD_LOGIC_VECTOR(N-1 DOWNTO 0);
13        mux_inp3:    IN STD_LOGIC_VECTOR(N-1 DOWNTO 0);
14        select:      IN STD_LOGIC_VECTOR(M-1 DOWNTO 0);
15        mux_outp:    OUT STD_LOGIC_VECTOR(N-1 DOWNTO 0));
16  END ENTITY;
17  ----------------------------------------------------
19  ARCHITECTURE multiplexer_4x8 OF multiplexer_4x8 IS
20  BEGIN
21     mux_outp <= mux_inp0 WHEN select="00" ELSE
22                 mux_inp1 WHEN select="01" ELSE
23                 mux_inp2 WHEN select="10" ELSE
24                 mux_inp3;
```

```
25  END ARCHITECTURE;
26  --------------------------------------------------------
```

Observe the following in the above code:

1) More meaningful (normally much longer) signal names are employed (lines 5 and 10–15.

2) Each signal in the entity is specified in a separate line (lines 6–15).

3) Values that appear often and can make the code more "generic" (that is, valid for other data buses sizes) are declared using GENERIC declarations (lines 7–8)

4) STD_LOGIC(_VECTOR) is used in all ports (in/out signals, lines 10–15), which is the usual interface in multiteam designs (industry standard).

5) Only descending indexes are used to specify the data ranges, and the final value is always zero (little endian, lines 10–15). Hence the MSB is always on the left and has the highest index value, which is always "number of bits – 1."

6) Reserved words are typed using capital letters, while the other words (chosen by the user) employ lowercase. This helps understand and debug the code (recall that VHDL is not case sensitive). Another (but less visible) option is to employ boldface for reserved words.

7) The use of separating lines (lines 1, 4, 17, 26) between the three fundamental code sections (library declarations, entity, and architecture) helps organize the code. The use of additional (but shorter) separating lines within the architecture is also helpful (others might prefer to leave a blank line instead). The use of optional labels (for PROCESS, for example, seen later) might also be helpful.

Additional useful practices include:

8) Adopting the same name for the project, the main file, and the main entity.

9) Avoiding using the mode BUFFER. To do so, employ auxiliary signals, specified in the architecture's declarative part.

10) Avoiding using more than one ENTITY-ARCHITECTURE pair in the same code, which would then require the use of CONFIGURATION or part of the code to be commented out.

11) For proper results, be rigorous (obey all formalities described in chapter 11) when designing finite state machines (FSMs). Be aware of their enormous potential, but keep in mind the problems that a bad design might entail.

12) Finally, try to always practice with the following questions:

—Is the circuit that I am designing combinational or sequential?

—If combinational, is it logical or arithmetic?

—If sequential, is it a regular or an FSM-based design?

—If sequential, how many flip-flops should I expect the compiler to infer? (Remember that this can always be determined exactly.)

2.9 VHDL 2008

With respect to the material covered in this chapter, the main additions specified in VHDL 2008 are those listed below. VHDL 2008 is backward compatible with VHDL 2002.

1) The packages *standard*, *textio*, *std_logic_1164*, and *numeric_std* were expanded (details will be given in chapters 3 and 4; see also the corresponding appendices).

2) The packages *numeric_std_unsigned*, *numeric_bit_unsigned*, *env*, *fixed_pkg*, and *float_pkg* are new (details will be given in chapters 3 and 4; see also a corresponding appendix).

3) The list of items that can be included in the declarative part of an ENTITY or ARCHITECTURE was expanded, also allowing the following: subprogram instantiation declaration, package declaration, package body, package instantiation declaration, and PSL declarations.

4) The options for GENERIC declarations were also expanded. Besides the traditional generic constants, unspecified generic types and generic subprograms can also be declared, as indicated in the simplified syntax below.

```
GENERIC (CONSTANT const_name: const_type := const_value;
         TYPE type_name;
         FUNCTION function_name (parameter_declarations);
         PROCEDURE procedure_name (parameter_declarations);
         ... );
```

5) Moreover, a generic constant can be used in the specification of other generic constants in the same generic list, as illustrated in the example below.

```
GENERIC (
   CONSTANT N: NATURAL := 8;
   CONSTANT M: NATURAL := 2**N;
   TYPE decoder_type;
   FUNCTION decode (i: decoder_type) RETURN o: BIT_VECTOR(N-1 DOWNTO 0));
```

6) The places where generics can be declared were also expanded, including PACKAGE and subprogram headers. A PACKAGE and a FUNCTION with generics are depicted below.

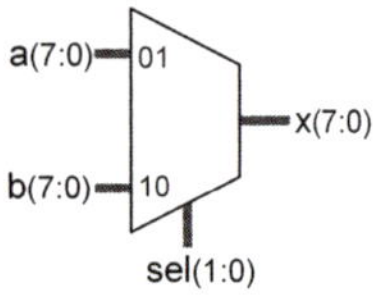

sel	x
"00"	"00000000"
"01"	a
"10"	b
"11"	"ZZZZZZZZ"

Figure 2.9

```
PACKAGE generic_type IS
   GENERIC (CONSTANT words: NATURAL;
            TYPE: word_type);
   TYPE gen_type IS ARRAY 1 TO words OF word_type;
END PACKAGE;

FUNCTION my_function IS
   GENERIC (VARIABLE word: BIT_VECTOR(15 DOWNTO 0);
BEGIN
   ...
END FUNCTION;
```

7) A package with a generic list is called an *uninstantiated package*, which must be instantiated with a *package instantiation* declaration, shown in the simplified syntax below. An example of instantiation for the package *generic_type* above is presented subsequently.

```
PACKAGE package_name IS NEW uninstant_package_name GENERIC MAP (inst_list);
```

```
LIBRARY ieee;
USE ieee.std_logic_1164;
PACKAGE memory_array IS NEW work.generic_type
   GENERIC MAP (words => 256, word_type => STD_LOGIC_VECTOR(15 DOWNTO 0));
```

2.10 Exercises

Exercise 2.1: Multiplexer

A multiplexer is depicted in figure 2.9. According to the truth table, the output should be equal to one of the inputs if *sel* = "01" ($x = a$) or *sel* = "10" ($x = b$), but should be zero or high impedance if *sel* = "00" or *sel* = "11", respectively.

a) Complete the VHDL code below.

b) Write relevant comments regarding your solution (as in example 2.2).

c) Compile and simulate the code, checking whether it works as expected.

Note: A solution using IF was employed in the code below, because it is more intuitive. However, a multiplexer can also be implemented with other statements, like WHEN or CASE (see example in section 2.8).

```
1   -----------------------------------------------
2   LIBRARY ieee;
3   USE _____;
4   -----------------------------------------------
5   ENTITY mux IS
6      PORT (___, ___: ___ STD_LOGIC_VECTOR(7 DOWNTO 0);
7             sel: IN _____;
8             ___: OUT STD_LOGIC_VECTOR(__ DOWNTO 0));
9   END _____;
10  -----------------------------------------------
11  ARCHITECTURE example OF _____ IS
12  BEGIN
13     PROCESS (a, b, ____)
14     BEGIN
15        IF (sel="00") THEN
16           x <= "00000000";
17        ELSIF (_____) THEN
19           x <= a;
20        _____ (sel="10") THEN
21           x <= __;
22        ELSE
23           x <= "_____";
24        END ___;
25     END _____;
26  END _____;
27  -----------------------------------------------
```

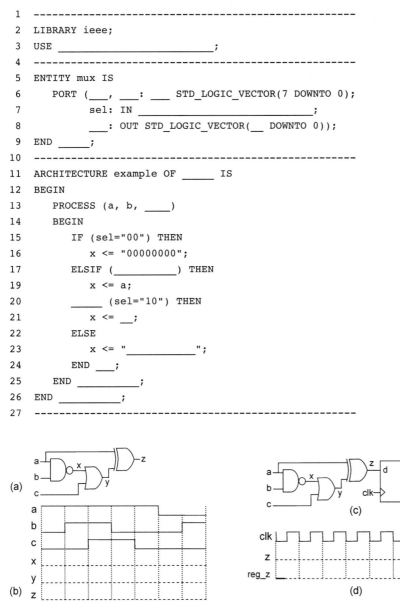

Figure 2.10

Exercise 2.2: Logic Gates

a) Show that the expression for z in the circuit of figure 2.10a is $z = abc' + a'$.

b) Given the waveforms for a, b, and c in figure 2.10b, draw the waveforms for x, y, and z (consider that the propagation delay is negligible → *functional* analysis).

c) Write a VHDL code for this circuit. Note that only logical operators (AND, OR, NAND, NOR, XOR, etc.) are needed.

d) Compile your code, then simulate it with the same waveforms given in figure 2.10b and check whether the resulting waveforms coincide with your answers above.

Exercise 2.3: Registered Logic

Consider now the inclusion of a DFF at the output of the circuit designed above, producing a registered signal (*reg_z*), as shown in figure 2.10c.

a) Copy the waveform for z from the previous exercise to figure 2.10d, then draw the waveform for *reg_z*.

b) Write a VHDL code for this circuit. Note that now a process is needed (as in example 2.3).

c) Compile your code, then simulate it with the same waveforms given for a, b, and c in figure 2.10b and check whether the resulting waveform for *reg_z* matches yours.

Exercise 2.4: Basic VHDL Data Types

(Even though data types will be discussed in detail in the next chapter, introductory analysis is proposed in this and in the next exercises.)

As mentioned in section 2.2, the library *std* contains a package called *standard* that defines the basic VHDL data types (see appendix H). Examine it and list all the numeric types and subtypes that it contains (do not include those that are not intended for synthesis, like TIME, DELAY_LENGTH, etc., or that have reduced synthesis support, like REAL). For each type in your list, give the values or range of values that it can assume.

Exercise 2.5: Type STD_LOGIC

As mentioned in section 2.2, a very popular data type (industry standard) that is not part of the package *standard* (examined in the exercise above) is STD_LOGIC. This type is defined in the package *std_logic_1164* (appendix I). Examine it and list all the values that a signal of type STD_LOGIC can assume.

3 Data Types

3.1 Introduction

In order to write VHDL code efficiently, it is indispensable to know which data types are allowed and how to specify and use them.

VHDL types can be divided into two categories: *predefined* and *user-defined*. The former is standardized and is available in the libraries that accompany the VHDL compiler, while the latter can be created by the user to handle special situations.

Both categories are described in detail in this chapter. Special emphasis is given to the types that are synthesizable (VHDL for simulation is discussed separately in chapter 10). Discussions on data compatibility and type conversion are also included.

3.2 VHDL Objects

Before we start describing VHDL types, let us briefly describe the VHDL objects, because it is for them that type specifications are intended (other details about VHDL objects will be seen in chapters 6, 7, and 10).

An object is a named item of a specific type that has a value. In other words, objects constitute the means through which values are passed around. Such VHDL objects are CONSTANT, SIGNAL, VARIABLE, and FILE. For example, all GENERIC and PORT items are objects because the former are constants and the latter are signals.

SIGNAL and VARIABLE are more complex than CONSTANT, so greater attention is needed. Their study is covered in three parts, as follows.

• Section 3.2 (present): Concept, syntax, and examples concerning object declarations (needed in the discussions on data types that follow in this chapter).

• Section 6.1: Summary of their main properties (needed to explain sequential code, which is the subject of that chapter).

• Chapter 7: SIGNAL versus VARIABLE comparison and details regarding the main properties, usage, and flip-flop inference.

The FILE object is also described in this section, but only briefly. Because files are particularly important for simulation, details will be given in chapter 10, which deals specifically with the subject (simulation with testbenches).

CONSTANT

As the name says, it is an object whose value cannot be changed. A simplified syntax for the declaration of constants is shown below.

```
CONSTANT constant_name: constant_type := constant_value;
```

The *name* can be essentially any word, except reserved words. The *type* can be any VHDL type, either predefined or user-defined (seen in the coming sections of this chapter). Finally, the *value* can be a constant or an expression involving constants.

Examples The name of the first constant below is *bits*, its type is INTEGER, and its value is 16. The second constant is called *words*, its type is also INTEGER, and its value is $2^{16} = 65536$. The third constant is called *flag*, its type is BIT, and its value is '1' (a single bit must be surrounded by a pair of single quotes). Finally, the fourth constant is named *mask*, its type is BIT_VECTOR, with a total of 8 bits, indexed in ascending order from 1 to 8, and its value is "00001111" (a pair of double quotes is used with multiple bits).

```
------------------------------------------------
CONSTANT bits: INTEGER := 16;
CONSTANT words: INTEGER := 2**bits;
CONSTANT flag: BIT := '1';
CONSTANT mask: BIT_VECTOR(1 TO 8) := "00001111";
------------------------------------------------
```

CONSTANT can be declared in the declarative part of ENTITY, ARCHITECTURE, PACKAGE, PACKAGE BODY, BLOCK, GENERATE, PROCESS, FUNCTION, and PROCEDURE (the last two are called *subprograms*). When declared in a package, for example, it is truly global because a package can be used by any design file. When declared in an entity (after PORT), it is global only to the architectures that follow that entity. When declared in an architecture (in its declarative part), it is global only to that particular architecture.

Deferred constant
A constant declared without its value is said to be a *deferred* constant. Such a declaration is allowed in a PACKAGE, but then the complete specification (including the value) must appear in the corresponding PACKAGE BODY (packages will be studied in chapter 8).

Keyword OTHERS

OTHERS is a useful keyword for making assignments. It represents all index values that were left unspecified.

Examples

The constant below is $a =$ "000000".

```
CONSTANT a: BIT_VECTOR(5 DOWNTO 0) := (OTHERS=>'0');
```

The next constant is $b =$ "01111111" (index 7 gets '0', the others, '1').

```
CONSTANT b: BIT_VECTOR(7 DOWNTO 0) := (7=>'0', OTHERS=>'1');
```

The signal below is $c =$ "01100000" ("|" means "or").

```
SIGNAL c: STD_LOGIC_VECTOR(1 TO 8) := (2|3=>'1', OTHERS=>'0');
```

The variable below is $d =$ "1111111100000000".

```
VARIABLE d: BIT_VECTOR(1 TO 16) := (1 TO 8=>'1', OTHERS=>'0');
```

SIGNAL

SIGNAL serves to pass values in and out of the circuit, as well as between its internal units. In other words, a signal represents circuit interconnects (wires). All ports of an entity are signals by default.

Signal declarations can be made in the declarative part of ENTITY, ARCHITEC-TURE, PACKAGE, BLOCK, and GENERATE. Signal declarations are not allowed in sequential code (i.e., PROCESS and subprograms), but signals can be *used* there. A simplified syntax (without a resolution specification, seen in chapter 7) for signal declarations is shown below.

```
SIGNAL signal_name: signal_type [range] [:= default_value];
```

Examples The name of the first signal below is *enable*, its type is BIT, and its default (initial) value is '0'. The second signal is called *temp* and its type is BIT_VECTOR, with a total of 4 bits, indexed in descending order. The third signal is called *byte* and its type is

STD_LOGIC_VECTOR, with a total of 8 bits, also indexed in descending order. Finally, the fourth signal is called *count* and its type is NATURAL, ranging from 0 to 255. Default values were not specified for the last three signals.

```
-------------------------------------------
SIGNAL enable: BIT := '0';
SIGNAL temp: BIT_VECTOR(3 DOWNTO 0);
SIGNAL byte: STD_LOGIC_VECTOR(7 DOWNTO 0);
SIGNAL count: NATURAL RANGE 0 TO 255;
-------------------------------------------
```

A very important aspect related to SIGNAL, when used inside a section of sequential code (PROCESS or subprogram), is that its update is *not* immediate. Instead, the new value is only expected to be ready after the conclusion of the current run of the PROCESS or subprogram.

Another important aspect concerns the case when multiple assignments are made to a signal. In concurrent code, the compiler will issue an error message and quit compilation. In sequential code, only the last assignment will be considered. In summary, multiple signal assignments must not be made.

To assign a value to a SIGNAL, the proper operator is "<=", while for CONSTANT or VARIABLE (or for default values) it is ":=". For example, "enable <= '1';". As seen, the keyword OTHERS can be helpful to make signal assignments.

VARIABLE

Contrary to CONSTANT and SIGNAL, VARIABLE represents only *local* information because it can only be seen and modified inside the sequential unit (i.e., PROCESS or subprogram) where it was created (it is slightly different for SHARED VARIABLE, explained in section 7.3). On the other hand, its update is immediate, so the new value can be promptly used in the next line of code. Also, because the update is immediate, multiple assignments to the same variable are fine. A simplified syntax for variable declarations is shown below.

```
VARIABLE variable_name: variable_type [range] [:= default_value];
```

Examples The name of the first variable below is *flip*, its type is STD_LOGIC, and its default (initial) value is '1'. The second variable is called *address* and its type is STD_LOGIC_VECTOR, with a total of 16 bits, indexed in ascending order from 0 to 15. Finally, the third variable is called *counter* and its type is INTEGER, ranging from 0 to 127. Default values were not specified for the last two variables.

```
-------------------------------------------
VARIABLE flip: STD_LOGIC := '1';
VARIABLE address: STD_LOGIC_VECTOR(0 TO 15);
VARIABLE counter: INTEGER RANGE 0 TO 127;
-------------------------------------------
```

To assign a value to a VARIABLE, the proper operator is ":=". For example, "flip := '0';". The keyword OTHERS can be helpful to make variable assignments.

FILE

The fourth and last VHDL object is FILE. However, to declare an *object* of that kind, a FILE *type* must first be created.

FILE type
A simplified syntax for file types is presented below. It contains the name chosen to represent the type and the type of the data contained in the file (only one type is allowed), which can be any VHDL type, either predefined or user-defined.

```
TYPE type_name IS FILE OF type_in_file;
```

FILE object
A simplified syntax for the declaration of a file object is shown below. It contains an identifier (name) chosen to represent that object, followed by the type name (seen above), then the optional keyword OPEN with the corresponding file-open mode (*read_mode*, *write_mode*, or *append_mode*, defined in the package *standard* of the library *std*). The optional *expression* at the end of the syntax can be, for example, a file name (between double quotes).

```
FILE file_identifier: type_name [[OPEN open_mode] IS expression];
```

Example   A file type, followed by a file object, are declared below.

```
TYPE bit_file IS FILE OF BIT;
FILE file01: bit_file IS "my_file.txt";
```

As already mentioned, details about the use of files will be seen in chapter 10.

SIGNAL versus VARIABLE

In chapter 7, a detailed discussion on the differences between SIGNAL and VARIABLE, and on the consequences of using one or the other, will be presented.

3.3   Data-Type Libraries and Packages

As mentioned in section 2.2, VHDL contains a series of predefined data types, specified in different packages. The fundamental packages for dealing with binary logic and with integer numbers are:

- Package *standard* (expanded in VHDL 2008)
- Package *std_logic_1164* (expanded in VHDL 2008)
- Package *numeric_bit* (expanded in VHDL 2008)
- Package *numeric_std* (expanded in VHDL 2008)
- Package *std_logic_arith* (shareware, nonstandard)
- Package *std_logic_unsigned* (shareware, nonstandard)
- Package *std_logic_signed* (shareware, nonstandard)
- Package *textio* (expanded in VHDL 2008)
- Package *numeric_bit_unsigned* (introduced in VHDL 2008)
- Package *numeric_std_unsigned* (introduced in VHDL 2008)

There are also several new packages, introduced in VHDL 2008, for dealing with fixed- and floating-point numbers. The main ones are (see section 3.8 for details about compatibility with previous VHDL versions):

- Package *fixed_pkg*
- Package *fixed_generic_pkg*
- Package *float_pkg*
- Package *float_generic_pkg*
- Package *fixed_float_types*

A brief description for each package listed above is presented next.

Package *standard* (See Appendix H)

This package is specified in the *IEEE 1076-2008 Standard VHDL Language Reference Manual* and is a member of the *std* library. It defines the following data types:

- Bit-related (synthesizable): BIT, BIT_VECTOR, BOOLEAN
- Integer-related (synthesizable): INTEGER, NATURAL, POSITIVE
- Character-related (synthesizable): CHARACTER, STRING
- Floating-point (limited synthesis support): REAL
- Time-related (not for synthesis): TIME, DELAY_LENGTH

- File-related (not for synthesis): FILE_OPEN_KIND, FILE_OPEN_STATUS
- Communication with the compiler: SEVERITY_LEVEL

It also includes definitions for logical, arithmetic, shift, comparison, and concatenation operators for the types above.

In VHDL 2008 (to be implemented), the following was added to the package *standard*:

1) New types: BOOLEAN_VECTOR, INTEGER_VECTOR, REAL_VECTOR, TIME_VECTOR.

2) Matching operators: ?=, ?/=, ?<, ?<=, ?>, ?>=.

3) Other functions: ??, MINIMUM, MAXIMUM, RISING_EDGE, FALLING_EDGE, TO_STRING, TO_OSTRING, TO_HSTRING.

Package *std_logic_1164* (See Appendix I)

This package, which is a member of the *ieee* library, is specified in the IEEE 1164 standard. It was introduced along with VHDL 93, and received additions in VHDL 2008. Both versions (93 and 2008) are shown in appendix I. The main types defined in that package are:

- STD_ULOGIC, STD_ULOGIC_VECTOR
- STD_LOGIC, STD_LOGIC_VECTOR (industry standard)

It specifies also logical operators (only logical) for the types above, and includes several type-conversion functions, such as TO_BIT, TO_BITVECTOR, TO_STDLOGICVECTOR, and so on.

In VHDL 2008, the following was added to the package *std_logic_1164* (compare part II to part I of appendix I):

1) STD_LOGIC_VECTOR is now a subtype of STD_ULOGIC_VECTOR, hence operators defined for the latter are automatically overloaded to the former.

2) Inclusion of more logical operator options, with the XNOR operator uncommented.

3) Inclusion of some shift operators.

4) Inclusion of the matching comparison operators (?=, ?/=, ?<, ?<=, ?>, ?>=).

5) Inclusion of condition (??), to-string conversion (TO_STRING, TO_OSTRING, TO_HSTRING) and READ/WRITE operators.

6) Inclusion of several short-hand aliases, like TO_SLV for TO_STDLOGICVECTOR.

Package *numeric_bit*

This package, which is also a member of the *ieee* library, is similar to the next package, but has BIT as the base type instead of STD_LOGIC.

Package *numeric_std* **(See Appendix J)**

This package, which is also a member of the *ieee* library, was specified in 1997, having received additions in VHDL 2008. Both versions are shown in appendix J. The types below are defined in it:

- UNSIGNED (based on STD_LOGIC)
- SIGNED (also based on STD_LOGIC)

It includes also definitions for logical, arithmetic, shift, and comparison operators for the types above. Several type-conversion functions are included in this package, like TO_INTEGER, TO_UNSIGNED, etc.

In VHDL 2008, the following was added to the package *numeric_std* (compare part II to part I of appendix J):

1) The definitions of UNSIGNED and SIGNED were slightly modified.

2) More arithmetic operator options were included.

3) More logical operator options were included.

4) More shift operator options were included.

5) Inclusion of the matching comparison operators ?=, ?/=, ?<, ?<=, ?>, ?>=.

6) Inclusion of MINIMUM, MAXIMUM, TO_STRING, TO_OSTRING, TO_HSTRING, READ, and WRITE functions.

Package *std_logic_arith* **(See Appendix K)**

A shareware (from Synopsis) whose new types are:

- UNSIGNED (based on STD_LOGIC)
- SIGNED (also based on STD_LOGIC)

It includes also arithmetic, shift, and comparison operators (logical not included) for the types above. Several type-conversion functions are included, such as CONV_INTEGER and CONV_STD_LOGIC_VECTOR. This package is only *partially* equivalent to *numeric_std*.

Package *std_logic_unsigned*

This package is similar to the next package, but with unsigned operators instead of signed.

Package *std_logic_signed* **(See Appendix L)**

Another shareware (from Synopsis) that defines some arithmetic, comparison, and shift operators with signals of type STD_LOGIC_VECTOR operating as *signed* numbers. A type-conversion function, called CONV_INTEGER, is also included. No new data types are defined here.

Package *textio* (See Appendix M)

This package is specified in the *IEEE 1076 Standard VHDL Language Reference Manual*, so it is a member of the *std* library. It defines types (LINE, TEXT, etc.) and procedures (READ, WRITE) for dealing with text and files.

   In VHDL 2008, the following functions were added to this package: FLUSH, MINIMUM, MAXIMUN, TO_STRING, JUSTIFY, TEE, and additional READ and WRITE options.

Package *numeric_bit_unsigned*

This package was introduced in VHDL 2008. It is essentially similar to the following package, but has BIT and BIT_VECTOR as the base types instead of STD_LOGIC and STD_LOGIC_VECTOR.

Package *numeric_std_unsigned*

This package was also introduced in VHDL 2008. It is expected to replace the non-standard package *std_logic_unsigned* in the future. It contains a large set of arithmetic, comparison, matching, and shift operators, plus type-conversion functions for the types STD_LOGIC and STD_LOGIC_VECTOR (see appendix N).

Packages *fixed_pkg*, *float_pkg*, and associated packages

These packages were introduced in VHDL 2008. However, compatibility files were also provided to allow operation with VHDL 93 and 2002. The main types specified in these packages are:

```
TYPE UFIXED IS ARRAY (INTEGER RANGE <>) OF STD_LOGIC;
TYPE SFIXED IS ARRAY (INTEGER RANGE <>) OF STD_LOGIC;
TYPE FLOAT IS ARRAY (INTEGER RANGE <>) OF STD_LOGIC;
```

Details will be given in section 3.8.

3.4   Type Classifications

Before we start describing the synthesizable data types, several classifications are presented in this section in order to help the reader visualize their nature and extent.

Classification according to the source of the declaration

Predefined data types: Defined in the provided VHDL packages (section 3.3).

User-defined data types: Defined by the user to handle specific situations.

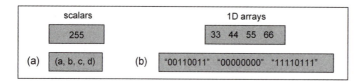

Figure 3.1
Value-based classification: (a) Scalar (single integer or enumerated) and (b) 1D (vector of values).

Classification according to the nature of the elements

Numeric types: Consist of *integer*, *fixed-point* and *floating-point* types.

Examples

```
-----------------------------------------------------------------
TYPE INTEGER IS RANGE -2147483647 TO 2147483647; --default range
TYPE grade IS RANGE 0 TO 10; --user-defined integer
-----------------------------------------------------------------
```

Enumerated types: Values are represented by symbols, all explicitly listed.

Examples

```
-----------------------------------------------------------------
TYPE color IS (red, green, blue); --3 values
TYPE BIT IS ('0', '1'); --2 values
TYPE STD_ULOGIC IS ('U','X','0','1','Z','W','L','H','X'); --9 values
-----------------------------------------------------------------
```

Classification according to the number of values (figure 3.1)

Scalar types: Types with a single value. VHDL defines as *scalar* the numeric (integer, floating-point, etc.), enumerated, and physical types. Examples of scalar types are given in figure 3.1(a).

Composite types: This category includes two subgroups, called *array* (collection of elements of the same type) and *record* (collections of scalar and/or array elements that can be of different types). Examples of *array* types are given in figure 3.1(b).

Classification according to the number of bits (figure 3.2)

This is not a usual classification, but it is introduced here because a fundamental parameter from a *design* (hardware) perspective is the *number of bits*, not the number of values. As shown later, this classification will help understand the construction of data arrays. Its six main cases are described below.

• *Scalar*: Single bit (figure 3.2(a)). Examples: '1', 'Z', FALSE.

• *1D array*: A bit vector (figure 3.2(b)). Examples: "01000", "111100ZZ", 255, 'A'.

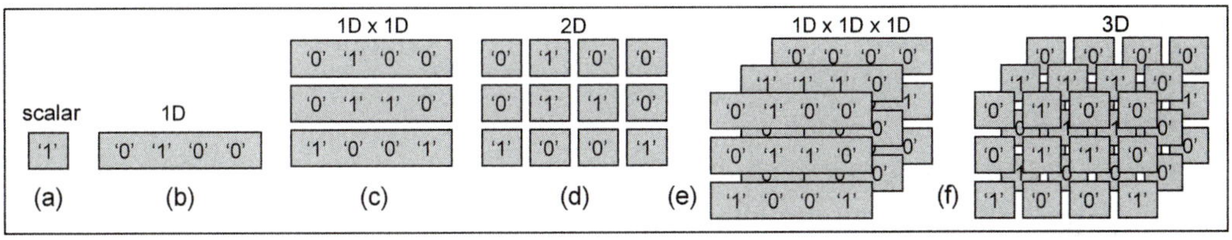

Figure 3.2
Bit-based arrays: (a) Scalar (single bit), (b) 1D (vector of bits), (c) 1D × 1D (pile of vectors), (d) 2D (matrix of bits), (e) 1D × 1D × 1D (block of vectors), and (f) 3D (block of bits).

- *1D × 1D array*: A pile of bit vectors (figure 3.2(c)). Examples: ("0000", "0111"), (43, 5, 25, 0).
- *2D array*: A matrix of bits (figure 3.2(d)). Example: (('0', '1', '0'), ('0', '0', '0'), ('Z', '1', 'Z')).
- *1D × 1D × 1D array*: A block of vectors. See example in figure 3.2(e).
- *3D array*: A block of bits. See example in figure 3.2(f).

Classification according to the package of origin

This is another unusual classification, again introduced here because it helps in understanding the purpose and structure of the several data types.

- *Standard types*: From the package *standard* of the *std* library.
- *Standard-logic types*: From the package *std_logic_1164* of the *ieee* library.
- *Unsigned/Signed types*: From the package *numeric_std* of the *ieee* library or package *std_logic_arith* (shareware).
- *Fixed/Floating-point types*: From the packages *fixed_pkg* and *float_pkg*.

The classifications presented above will help as follows.

1) The classification according to the *package of origin* will help describe the predefined data types. *Standard* types will be seen in section 3.5, *standard-logic* types in section 3.6, *unsigned/signed* types in section 3.7, and *fixed/floating-point* types in section 3.8.

2) The classification according to the *nature of the elements* will help describe the *scalar* (section 3.10), the *array* (section 3.11), and the *record* (section 3.14) user-defined data types.

3) Finally, the classification according to the *number of bits* will help understand and deal with all sorts of data types, particularly *arrays* (section 3.11). Even more important, it will help understand the hardware.

3.5   Standard Data Types

This section describes the *synthesizable* data types from the package *standard* (appendix H). Recall that those introduced in VHDL 2008 might still be undergoing implementation.

- BIT
- BIT_VECTOR
- BOOLEAN
- BOOLEAN_VECTOR (2008)
- INTEGER
- NATURAL
- POSITIVE
- INTEGER_VECTOR (2008)
- CHARACTER
- STRING

As seen in section 3.3, data-type packages define essentially three things:

1) New data types;

2) Operators for them (logical, arithmetic, shift, comparison, and concatenation);

3) Type-conversion functions involving the new and other data types.

Because operators are part of the present discussion, but will only be studied in the next chapter, a list of them is presented below.

Logical operators: NOT, AND, NAND, OR, NOR, XOR, XNOR (section 4.2.2)

Arithmetic operators: +, −, *, /, **, ABS, REM, MOD (section 4.2.3)

Comparison (relational) operators: =, /=, >, >=, <, <= (section 4.2.4)

Shift operators: SLL, SRL, SLA, SRA, ROR, ROL (section 4.2.5)

Concatenation operator: & (section 4.2.6)

Matching comparison operators: ?=, ?/=, ?>, ?>=, ?<, ?<= (section 4.2.7, from VHDL 2008).

BIT

Its definition is presented below, showing a two-value enumerated type. It supports logical and comparison operations (see appendix H). In terms of the number of bits, it is a scalar type (figure 3.2a).

```
TYPE BIT IS ('0', '1');
```

Example Below, the objects *a*, *x*, and *y* are declared as SIGNAL, all of type BIT, then the value '1' is assigned to *x*, while the value "NOT *a*" is assigned to *y*.

```
--------------------
SIGNAL a, x, y: BIT;
x <= '1';
y <= NOT a;
--------------------
```

In VHDL 2008, the following operators were added: ?=, ?/=, ?<, ?<=, ?>, ?>=, ??, MINIMUM, MAXIMUM, RISING_EDGE, FALLING_EDGE, TO_STRING.

BIT_VECTOR

This is the vector form of BIT (1D array of figure 3.2b), as shown in the definition below. It supports logical, comparison, shift, and concatenation operations (see appendix H).

```
TYPE BIT_VECTOR IS ARRAY (NATURAL RANGE <>) OF BIT;
```

The "NATURAL RANGE <>" specification (the symbol "<>" is called a "box") indicates that the range is unconstrained, with its only limitation that it must fall within the NATURAL range (default = 0 to $2^{31} - 1$). For logical and shift operations, the vectors are required to have the same length, which is not necessary for comparison.

Example In the code below, the objects a, b, x, and y are declared as 8-bit vectors with descending index range, while v is an 8-bit vector with ascending index range and w is a single bit. Subsequently, the value "11110000" is assigned to x, so $x(7) = x(6) = x(5) = x(4) = $ '1' and $x(3) = x(2) = x(1) = x(0) = $ '0'. y receives the result of a logical operation (XOR) between a and b, while v receives the result of a shift left logical (SLL) operation on a. Finally, w is asserted whenever $a > b$ (comparison operation).

```
-----------------------------------------------
SIGNAL a, b: BIT_VECTOR(7 DOWNTO 0);   --8 bits
SIGNAL x, y: BIT_VECTOR(7 DOWNTO 0);   --8 bits
SIGNAL v: BIT_VECTOR(1 TO 8);          --8 bits
SIGNAL w: BIT;                         --1 bit
x <= "11110000";
y <= a XOR b;
v <= a SLL 2;
w <= '1' WHEN a>b ELSE '0';
-----------------------------------------------
```

In VHDL 2008, the following operators were added: ?=, ?/=, MINIMUM, MAXIMUM, TO_STRING, TO_OSTRING, TO_HSTRING.

BOOLEAN

Its definition is presented below, showing another two-value enumerated type. It supports logical and comparison operations (see appendix H). In terms of the number of bits, it is a scalar type (figure 3.2a).

```
TYPE BOOLEAN IS (FALSE, TRUE);
```

Example   The value of x below changes from "000" to "111" when *ready* is TRUE (= '1').

```
x<="111" WHEN ready ELSE "000";
```

In VHDL 2008, the following functions were added: MINIMUM, MAXIMUM, RISING_EDGE, FALLING_EDGE, TO_STRING.

BOOLEAN_VECTOR

Introduced in VHDL 2008, this is the vector form of BOOLEAN (1D array of figure 3.2b). Must support logical, comparison, shift, and concatenation operations (see appendix H), plus the operators ?=, ?/=, MINIMUM, and MAXIMUM. Its definition is shown below.

```
TYPE BOOLEAN_VECTOR IS ARRAY (NATURAL RANGE <>) OF BOOLEAN;
```

INTEGER

Its definition is shown below. It supports arithmetic and comparison operations (see appendix H).

```
TYPE INTEGER IS RANGE implementation_defined;
TYPE INTEGER IS RANGE -2147483647 TO 2147483647;
```

The last line above shows the default range of INTEGER, which consists of a 32-bit representation, from $-(2^{31}-1)$ to $+(2^{31}-1)$. The actual bounds are referred to as INTEGER'LOW (on the left) and INTEGER'HIGH (on the right).

In VHDL codes for synthesis, it is important to always specify the range (as in the examples below) for objects of type INTEGER (or its subtypes), because otherwise the compiler will employ 32 bits to represent them.

Example   In the code below, first four signals (a, b, x, y) are declared, then the first two are added and also compared. The number of bits of the signals involved in either operation are not required to be equal.

```
--------------------------------------------
SIGNAL a: INTEGER RANGE 0 TO 15;     --4 bits
SIGNAL b: INTEGER RANGE -15 TO 15;   --5 bits
SIGNAL x: INTEGER RANGE -31 TO 31;   --6 bits
SIGNAL y: BIT;
x <= a + b;
y <= '1' WHEN a>=b ELSE '0';
--------------------------------------------
```

In VHDL 2008, the following functions were added: MINIMUM, MAXIMUM, TO_STRING.

NATURAL

Non-negative integers. As shown in the definition below, it is a *subtype* of INTEGER (has the same dimensionality and supports the same operators).

```
SUBTYPE NATURAL IS INTEGER RANGE 0 TO INTEGER'HIGH;
```

POSITIVE

Positive integers. As shown in the definition below, it is also a *subtype* of INTEGER (has the same dimensionality and supports the same operators).

```
SUBTYPE POSITIVE IS INTEGER RANGE 1 TO INTEGER'HIGH;
```

INTEGER_VECTOR

Introduced in VHDL 2008, it is the vector form of INTEGER (1D × 1D array of figure 3.2c). Must support comparison and concatenation operations (see appendix H), plus the functions MINIMUM and MAXIMUM. Its definition is shown below.

```
TYPE INTEGER_VECTOR IS ARRAY (NATURAL RANGE <>) OF INTEGER;
```

CHARACTER

A 256-symbol enumerated type. Its (partial) definition is shown below (it goes from $nul =$ "00000000" up to $ÿ =$ "11111111"). This type supports only comparison operations (see appendix H).

```
TYPE CHARACTER IS (NUL, SOH, ... , '0', '1', '2', ..., 'ÿ');
```

The symbols are from the ISO 8859-1 character set, with the first 128 symbols comprising the regular ASCII code. Because each symbol is represented by 8 bits, this type falls in the 1D category of figure 3.2b.

Example   In the code below, two signals (*char1*, *char2*) are declared of type CHARACTER. Then two outputs (*outp1*, *outp2*), of type BIT, are used to compare *char1* and *char2* (*outp2* is '1' when *char1* appears in the ASCII table earlier than *char2*).

```
-------------------------------------------------
SIGNAL char1, char2: CHARACTER;
SIGNAL outp1, outp2: BIT;
outp1 <= '1' WHEN char1='a' OR char1='A' ELSE '0';
outp2 <= '1' WHEN char1<char2 ELSE '0';
-------------------------------------------------
```

In VHDL 2008, the following functions were added: MINIMUM and MAXIMUM TO_STRING.

STRING

This is the vector form of CHARACTER ($1D \times 1D$ array of figure 3.2c). Its definition is shown below. This type supports comparison and concatenation operations (see appendix H).

```
TYPE STRING IS ARRAY (POSITIVE RANGE <>) OF CHARACTER;
```

Example   In the code below, *str* is a string with four characters (hence 4-by-8 bits).

```
---------------------------------------
SIGNAL str: STRING(1 TO 4);
SIGNAL output: BIT;
output <= '1' WHEN str="VHDL" ELSE '0';
---------------------------------------
```

In VHDL 2008, the following functions were added: MINIMUM and MAXIMUM.

Even though this section focuses on types that are fully synthesizable, REAL- and TIME-based types are briefly described below. The first has limited synthesis support; the second is for simulations (so it will be used in chapter 10).

REAL

Floating-point numbers, with support for arithmetic and comparison operators. Its definition is shown below.

```
TYPE real IS RANGE implementation_defined;
```

In VHDL 2008, the following functions were added: MINIMUM, MAXIMUM, TO_STRING.

REAL_VECTOR

Introduced in VHDL 2008, this is the vector form of REAL (see appendix H). Its definition is shown below.

```
TYPE REAL_VECTOR IS ARRAY (NATURAL RANGE <>) OF REAL;
```

TIME

Represented by integers with the same range as INTEGER, with support for arithmetic (most of them) and comparison operators. Its definition is shown below (intended for simulation).

```
TYPE time IS RANGE implementation_defined;
```

In VHDL 2008, the following functions were added: MINIMUM, MAXIMUM, TO_STRING.

TIME_VECTOR

Introduced in VHDL 2008, this is the vector form of TIME (see appendix H). Its definition is shown below.

```
TYPE TIME_VECTOR IS ARRAY (NATURAL RANGE <>) OF TIME;
```

See other details about VHDL 2008 in section 3.22. See also the data-type summary in figure 3.6.

3.6 Standard-Logic Data Types

We describe now the types STD_LOGIC and STD_LOGIC_VECTOR, which are the industry standard. They are defined in the *std_logic_1164* package, introduced along with VHDL 93, with several new features added in VHDL 2008. Both versions (1993 and 2008) of this package are shown in appendix I.

In fact, the type defined in that package is STD_ULOGIC (the "U" stands for unresolved), of which STD_LOGIC is a "resolved" subtype. Their definitions are as follows (note that, like BIT, BOOLEAN, and CHARACTER, they too are *enumerated* data types):

```
TYPE STD_ULOGIC IS ('U','X','0','1','Z','W','L','H','-');
TYPE STD_LOGIC IS resolved STD_ULOGIC;
```

The meanings (and possible usage) for the nine STD_(U)LOGIC symbols are listed below.

'U' Uninitialized

'X' Forcing unknown

'0' Forcing low

'1' Forcing high

'Z' High impedance

'W' Weak unknown

'L' Weak low

'H' Weak high

'-' Don't care

The vector forms (1D bit arrays) of STD_ULOGIC and STD_LOGIC are the following:

```
TYPE STD_ULOGIC_VECTOR IS ARRAY (NATURAL RANGE <>) OF STD_ULOGIC;
TYPE STD_LOGIC_VECTOR IS ARRAY (NATURAL RANGE <>) OF STD_LOGIC;
```

The main feature of the STD_LOGIC type, compared to the original BIT type, is the inclusion of the *high-impedance* ('Z') and *don't care* ('-') values, which allow the construction of tri-state buffers and a better hardware optimization for lookup tables, respectively (illustrated ahead in examples 3.1 and 3.2).

The package *std_logic_1164* defines only *logical* operators for the types above. However, if the package *std_logic(un)signed* is also declared in the code, then some *arithmetic, comparison*, and *shift* operations will also be allowed.

STD_LOGIC is said to be a *resolved* subtype because if more than one source drives a common node the resulting logic level is determined by a predefined *resolution* function. The resolution function for STD_LOGIC (copied from the PACKAGE BODY of the *std_logic_1164* package) is shown below.

```
 1   -------------------------------------------------------------
 2   CONSTANT resolution_table : stdlogic_table := (
 3        -----------------------------------------------------
 4        -- U    X    0    1    Z    W    L    H    -      |   |
 5        -----------------------------------------------------
 6        ( 'U', 'U', 'U', 'U', 'U', 'U', 'U', 'U', 'U' ), -- | U |
 7        ( 'U', 'X', 'X', 'X', 'X', 'X', 'X', 'X', 'X' ), -- | X |
 8        ( 'U', 'X', '0', 'X', '0', '0', '0', '0', 'X' ), -- | 0 |
 9        ( 'U', 'X', 'X', '1', '1', '1', '1', '1', 'X' ), -- | 1 |
10        ( 'U', 'X', '0', '1', 'Z', 'W', 'L', 'H', 'X' ), -- | Z |
11        ( 'U', 'X', '0', '1', 'W', 'W', 'W', 'W', 'X' ), -- | W |
12        ( 'U', 'X', '0', '1', 'L', 'W', 'L', 'W', 'X' ), -- | L |
13        ( 'U', 'X', '0', '1', 'H', 'W', 'W', 'H', 'X' ), -- | H |
14        ( 'U', 'X', 'X', 'X', 'X', 'X', 'X', 'X', 'X' )  -- | - |);
15        -----------------------------------------------------
16   FUNCTION resolved (s: STD_ULOGIC_VECTOR) RETURN STD_ULOGIC IS
17      VARIABLE result: STD_ULOGIC_ := 'Z';  --weakest state default
18      ATTRIBUTE synthesis_return OF result: VARIABLE IS "WIRED_THREE_STATE";
19   BEGIN
20      IF (s'LENGTH=1) THEN RETURN s(s'LOW);
21      ELSE
22        FOR i IN s'RANGE LOOP
23           result := resolution_table(result, s(i));
24        END LOOP;
25      END IF;
26      RETURN result;
27   END resolved;
28   -------------------------------------------------------------
```

Note that the code above contains a resolution table (lines 2–14), entered as a CON-STANT. The name of the function is *resolved* (line 16); it receives a parameter called *s*, of type STD_ULOGIC_VECTOR (hence with any number of bits), returning a single bit of type STD_ULOGIC. In the declarative part of the function, a VARIABLE named *result* is declared (line 17), having 'Z' as its initial value (note in the resolution table that 'Z' is the weakest value, because "anything" versus 'Z' is "anything"). In the first part of the code proper (line 20), the size of *s* is checked; if it is a single bit, then *s* itself is returned (there is nothing to be resolved). However, if *s* contains more than one digit, then in part two of the code (lines 21–25) the table is used to determine the winning value. For exam-ple, if the competing values are '0', '1', and 'Z', then the result (according to the expression in line 23) is 'X' (unknown) (this is because 'Z', the initial value of *result*, versus '0' is '0'; '0' versus '1' is 'X'; and finally 'X' versus 'Z' is again 'X').

From a design perspective, what is essential to know is how this data type (STD_LOGIC) is synthesized. In summary, the following occurs (just as a remark, note that the choice of symbol for "don't care" was unfortunate, because the traditional repre-sentation for "don't care" in digital design is 'X'):

• '0' and 'L' are both synthesized as '0' (for inputs and outputs);

• '1' and 'H' are both synthesized as '1' (for inputs and outputs);

• 'Z' is synthesized as 'Z' (for outputs);

• The others as synthesized as '-' (*don't care*—for outputs).

Because in regular digital designs the input specifications of interest are '0', '1', and '-' (don't care), while the output values of interest are '0', '1', 'Z', and '-', all values are covered by STD_LOGIC, hence its importance. Unfortunately, even though VHDL 2002 allows "don't care" outputs, it does not support "don't care" inputs, a limitation that was resolved in VHDL 2008.

Another important feature in VHDL 2008 is that STD_LOGIC_VECTOR is a subtype of STD_ULOGIC_VECTOR (compare part I to part II in appendix I), that is:

```
TYPE STD_ULOGIC_VECTOR IS array (NATURAL RANGE <>) of STD_ULOGIC;
SUBTYPE STD_LOGIC_VECTOR IS (resolved) STD_ULOGIC_VECTOR;
```

Consequently, operators defined for the latter are automatically overloaded to the former.

Another addition made in VHDL 2008 is the package *numeric_std_unsigned*, which defines unsigned operators (no new types) for the types STD_LOGIC and STD_LOGIC_VECTOR (there is also a package called *numeric_bit_unsigned*, which specifies similar operators for the types BIT and BIT_VECTOR).

See other details about VHDL 2008 in section 3.22. See also the summary in figure 3.6.

Examples illustrating its usefulness of STD_LOGIC are presented next.

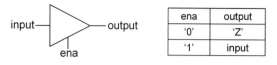

ena	output
'0'	'Z'
'1'	input

Figure 3.3
Tri-state buffer of example 3.1.

Example 3.1: Tri-state Buffer

As already mentioned, a fundamental synthesizable value is 'Z' (high impedance), which is needed to create tri-state buffers, like that depicted in figure 3.3 (see its truth table). Write a VHDL code from which this circuit can be inferred.

Solution   A corresponding VHDL code is shown below, with 'Z' employed in the WHEN statement of line 12. While the enable port is asserted (*ena* = '1'), the input is copied to the output. However, if *ena* = '0', the buffer is physically disconnected from the output node (high-impedance state). Note that the package *std_logic_1164* (lines 2–3) is needed because it is in that package that 'Z' is defined (BIT would not do because it can only be '0' or '1').

```
1    ----------------------------------------
2    LIBRARY ieee;
3    USE ieee.std_logic_1164.all;
4    ----------------------------------------
5    ENTITY tri_state IS
6    PORT (input, ena: IN STD_LOGIC;
7          output: OUT STD_LOGIC);
8    END ENTITY;
9    ----------------------------------------
10   ARCHITECTURE tri_state OF tri_state IS
11   BEGIN
12      output <= input WHEN ena='1' ELSE 'Z';
13   END ARCHITECTURE;
14   ----------------------------------------
```

Example 3.2: Circuit with 'Don't Care' Outputs

Figure 3.4a depicts a circuit whose input (*x*) and output (*y*) are 2-bit signals for which two sets of specifications are given in the truth tables of figures 3.4b–c. In the former, all outputs are specified with '0's and '1's, while in the latter there is a "don't care" output (*y* = "--"). Design this circuit by hand, then write a VHDL code for each truth table and observe the consequences of employing the logic value '-' in the code.

Solution   Using Karnaugh maps (Pedroni 2008), we obtain the following optimal equations for *y* in figure 3.4b: $y_1 = x_1' \cdot x_0$, $y_0 = x_1 \cdot x_0'$. On the other hand, for figure 3.4c, the

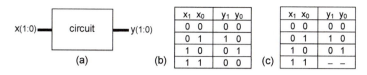

Figure 3.4
Circuit with "don't care" outputs of example 3.2.

result is simply $y_1 = x_0$ and $y_0 = x_1$, so the circuit requires less hardware than the previous one (lower cost, less power consumption, higher speed).

A corresponding VHDL code is presented below (for the case of figure 3.4b, the only change is obviously in line 15), from which similar conclusions are obtained. (*Note:* The compiler does not always reach the very optimal expressions for a given combinational circuit, so one must be careful when making this type of comparison.)

```
1    ---------------------------------------------
2    LIBRARY ieee;
3    USE ieee.std_logic_1164.all;
4    ---------------------------------------------
5    ENTITY circuit IS
6        PORT (x: IN STD_LOGIC_VECTOR(1 DOWNTO 0);
7             y: OUT STD_LOGIC_VECTOR(1 DOWNTO 0));
8    END ENTITY;
9    ---------------------------------------------
10   ARCHITECTURE circuit OF circuit IS
11   BEGIN
12       y <= "00" WHEN x="00" ELSE
13            "01" WHEN x="10" ELSE
14            "10" WHEN x="01" ELSE
15            "--";
16   END ARCHITECTURE;
17   ---------------------------------------------
```

3.7   Unsigned and Signed Data Types

As seen in section 3.3, UNSIGNED and SIGNED data types are defined in two (competing) packages called *numeric_std* (from the *ieee* library, appendix J) and *std_logic_arith* (shareware, appendix K). Their definitions are as follows:

```
TYPE UNSIGNED IS ARRAY (NATURAL RANGE <>) OF STD_LOGIC;
TYPE SIGNED IS ARRAY (NATURAL RANGE <>) OF STD_LOGIC;
```

Consequently, to make use of the UNSIGNED or SIGNED data type, one of the packages above (*numeric_std* or *std_logic_arith*) must be declared in the code. Remember, however, that they are only *partially* equivalent (the former defines logical, arithmetic, comparison, and shift operators, while the latter does not include *logical* operators; on the other hand, the latter has a wider set of data-conversion functions). For obvious reasons, these two packages cannot be used together, and because *numeric_std* is a standardized package (by the IEEE), it should be preferred.

An unsigned value ranges from 0 to $2^N - 1$ (limited to INTEGER'HIGH), where N is the number of bits. Then "0101" represents the decimal 5, while "1101" signifies 13. Signed types, on the other hand, range from -2^{N-1} to $2^{N-1} - 1$ (limited between INTEGER'LOW and INTEGER'HIGH), with negative values represented in two's complement form (section 1.8). Hence "0101" still represents the decimal 5, while "1101" now means -3.

Unfortunately, the names chosen for two other packages, *std_logic_unsigned* and *std_logic_signed* (appendix L), might lead to confusion. These packages only define unsigned and signed *operators* for STD_LOGIC_VECTOR, not the UNSIGNED and SIGNED types.

Example   Below, a, b, x, and y are declared as signals of type SIGNED, with 8 bits each, indexed in descending order, while v and w are single bits of types STD_LOGIC and BIT, respectively. Three of the operations that follow are legal, while one is illegal. $x = a + b$ is fine because the arithmetic operator "+" is allowed for SIGNED. On the other hand, y is fine only if *numeric_std* is declared, because *std_logic_arith* does not contain logical operators. For v, the operation is valid too because a scalar (single bit) of type STD_LOGIC is compatible with a scalar of type SIGNED (same base type). Finally, w is illegal because the base type of SIGNED is not BIT.

```
------------------------------------------------------
SIGNAL a, b: SIGNED(7 DOWNTO 0);
SIGNAL x, y: SIGNED(7 DOWNTO 0);
SIGNAL v: STD_LOGIC;
SIGNAL w: BIT;
x <= a + b;              --legal
y <= a AND b;            --legal only if numeric_std is used
v <= a(7) XOR b(0);      --legal because of same base type
w <= a(0);               --illegal because of type mismatch
------------------------------------------------------
```

Example 3.3: Unsigned/Signed Multiplier #1

Write a VHDL code to implement a circuit that computes $y = a*b$. Keep in mind that when the type (UN)SIGNED is used, the number of bits in the product must be equal to the sum of bits in the operands (section 5.7).

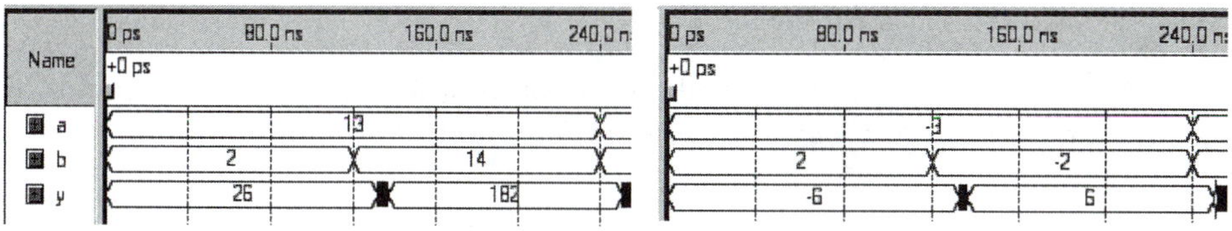

Figure 3.5
Unsigned (left) and signed (right) simulation results from the code of example 3.3.

a) Declare all signals as UNSIGNED, with 4 bits for *a* and *b* and 8 bits for *y*. Compile the code and simulate it for *a* = "1101" and *b* = "0010", followed by *b* = "1110". Observe and interpret the results.

b) Repeat the design above, now for signals of type SIGNED.

Solution

a) The code is shown below. The package *numeric_std* was employed (lines 2–3) and all signals were declared as UNSIGNED (lines 6–7). In this case, the decimal values are $a = 13$ and $b = 2$, followed by $b = 14$, from which $y = $ "00011010" $= 26$, followed by $y = $ "10110110" $= 182$, are obtained.

b) The only changes needed in the code are in lines 6–7 (SIGNED instead of UNSIGNED). Now the decimal values are $a = $ "1101" $= -3$ and $b = $ "0010" $= 2$, followed by $b = $ "1110" $= -2$, from which $y = $ "11111010" $= -6$, followed by $y = $ "00000110" $= 6$, are obtained.

Simulation results for both cases are depicted in figure 3.5. Note that the results do coincide with the expected values.

```
1   ----------------------------------------
2   LIBRARY ieee;
3   USE ieee.numeric_std.all;
4   ----------------------------------------
5   ENTITY multiplier IS
6      PORT (a, b: IN UNSIGNED(3 DOWNTO 0);
7            y: OUT UNSIGNED(7 DOWNTO 0));
8   END ENTITY;
9   ----------------------------------------
10  ARCHITECTURE multiplier OF multiplier IS
11  BEGIN
12     y <= a * b;
13  END ARCHITECTURE;
14  ----------------------------------------
```

Example 3.4: Unsigned/Signed Multiplier #2

Repeat the exercise above, this time with all ports declared as STD_LOGIC_VECTOR. In part (a), include the package *std_logic_unsigned*, and in part (b), the package *std_logic_signed*.

Solution

a) The corresponding VHDL code is shown below. The package *std_logic_1164* (line 3) is needed because it defines the type STD_LOGIC_VECTOR. The other package (*std_logic_unsigned*, line 4) is also needed because it defines arithmetic functions for that data type, which do not exist in the original (*std_logic_1164*) package.

b) The only change needed is in line 4, replacing the word *unsigned* with *signed*. Simulation results are the same as those in figure 3.5.

```
1    ------------------------------------------------
2    LIBRARY ieee;
3    USE ieee.std_logic_1164.all;
4    USE ieee.std_logic_unsigned.all;
5    ------------------------------------------------
6    ENTITY multiplier IS
7       PORT (a, b: IN STD_LOGIC_VECTOR(3 DOWNTO 0);
8             y: OUT STD_LOGIC_VECTOR(7 DOWNTO 0));
9    END ENTITY;
10   ------------------------------------------------
11   ARCHITECTURE multiplier OF multiplier IS
12   BEGIN
13      y <= a * b;
14   END ARCHITECTURE;
15   ------------------------------------------------
```

Note: In the code of example 3.4, all ports are of type STD_LOGIC_VECTOR, which is the industry standard. However, though correct, the approach in this example might lead to confusion because the nature of the operation (unsigned or signed) is not explicitly shown in the code (it depends only on the package declared in line 4). A recommended approach is to keep all ports specified as STD_LOGIC_VECTOR, but *explicitly* convert the inputs from STD_LOGIC_VECTOR to (UN)SIGNED in the code, then perform the operation(s), finally converting the result(s) back to STD_LOGIC_VECTOR. This procedure will be illustrated in example 3.9 (section 3.18) and further discussed in section 5.7.

3.8   Fixed- and Floating-Point Types

Even though the type REAL specified in the package *standard* has limited synthesis support, newer options for dealing with fixed- and floating-point numbers were introduced in VHDL 2008. Their main features are described below.

Fixed-Point Types

The main fixed-point types are the following:

```
TYPE UFIXED IS ARRAY (INTEGER RANGE <>) OF STD_LOGIC;  --unsigned
TYPE SFIXED IS ARRAY (INTEGER RANGE <>) OF STD_LOGIC;  --signed
```

The files needed to use such types in VHDL 2008 are (see IEEE 1076-2008 standard):

- *fixed_pkg.vhdl* (contains the package *fixed_pkg*)
- *fixed_generic_pkg.vhdl* (contains the package *fixed_generic_pkg*)
- *fixed_generic_pkg-body.vhdl* (contains the package body of *fixed_generic_pkg*)
- *fixed_float_types.vhdl* (contains the package *fixed_float_types*)

For previous VHDL versions (93 and 2002), compatibility files were provided (only the two files below are needed, available at www.eda.org/fphdl):

- *fixed_pkg_c.vhd* (contains a *compatible* package *fixed_pkg*)
- *fixed_float_types_c.vhd* (contains a *compatible* package *fixed_float_types*)

To implement fixed-point circuits in these previous VHDL versions, just create a folder called *ieee_proposed* in the VHDL library directory of your compiler and copy the two files above into it.

The representation of fixed-point numbers is similar to that of STD_LOGIC_VECTOR but now the location of a decimal point must also be provided. This is achieved by allowing the index to be negative, with the decimal point located after the index-zero bit.

Examples with *unsigned* fixed-point numbers:

```
x: SIGNAL UFIXED(2 DOWNTO -3);  --this is "xxx.xxx"
y: SIGNAL UFIXED(4 DOWNTO -1);  --this is "yyyyy.y"
z: SIGNAL UFIXED(-2 DOWNTO -3); --this is "0.0zz"
...
x <= "100011";  --1x2²+0x2¹+0x2⁰+0x2⁻¹+1x2⁻²+1x2⁻³=4.375
y <= "100011";  --1x2⁴+0x2³+0x2²+0x2¹+1x2⁰+1x2⁻¹=17.5
z <= "10";      --1x2⁻²+0x2⁻³=0.25
```

Examples with *signed* fixed-point numbers:

```
x: SIGNAL SFIXED(2 DOWNTO -3);  --this is "xxx.xxx"
y: SIGNAL SFIXED(4 DOWNTO -1);  --this is "yyyyy.y"
z: SIGNAL SFIXED(-2 DOWNTO -3); --this is "0.0zz"
...
x <= "100011"; --100.011 -> 2's compl=011.101 -> -3.625 (or 4.375-8=-3.625)
y <= "100011"; --10001.1 -> 2's compl=01110.1 -> -14.5 (or 17.5-32=-14.5)
z <= "10";     --0.010 = +0.25
```

The reason for using fixed- and floating-point types is that they allow operation with fractional numbers, of which numbers smaller than 1 are of particular interest. For example, observe that the signals x, y, and z specified above, for the unsigned case, have the following ranges and increment sizes:

x: range from 0 to 7.875 in steps of 0.125 (a total of $2^{bits} = 2^6 = 64$ values).

y: range from 0 to 31.5 in steps of 0.5 (a total of $2^{bits} = 2^6 = 64$ values).

z: range from 0 to 0.375 in steps of 0.125 (a total of $2^{bits} = 2^2 = 4$ values).

A large set of operators (details about operators are given in chapter 4) and type-conversion functions are defined for these new types. A simplified list is presented below.

a) Operators:
Logical: NOT, AND, NAND, OR, NOR, XOR, XNOR
Arithmetic: $+$, $-$, $*$, $/$, ABS, REM, MOD, ADD_CARRY, etc.
Comparison: $=$, $/=$, $>$, $<$, $>=$, $<=$, MAXIMUM, MINIMUM
Shift: SLL, SRL, SLA, SRA, ROR, ROL, SHIFT_LEFT, SHIFT_RIGHT
Matching: ?=, ?/=, ?>, ?<, ?>=, ?<=

b) Type-conversion functions:
TO_UFIXED, TO_SFIXED, TO_UNSIGNED, TO_SIGNED, TO_SLV (same as TO_STDLOGICVECTOR), TO_INTEGER, TO_REAL, TO_STRING, etc.

c) Text I/O functions:
WRITE (same as BWRITE or BINARY_WRITE), READ (same as BREAD or BINARY_READ), OWRITE (same as OCTAL_WRITE), OREAD (same as OCTAL_READ), HWRITE (same as HEX_WRITE), HREAD (same as HEX_READ), etc.

d) Other functions:
RESIZE, SATURATE, FIND_LEFTMOST, FIND_RIGHTMOST, UFIXED_HIGH, UFIXED_LOW, SFIXED_HIGH, etc.

Example:

```
x: SIGNAL UFIXED(4 DOWNTO -3);   --"xxxxx.xxx"
y: SIGNAL SFIXED(4 DOWNTO -3);   --"yyyyy.yyy"
z: SIGNAL SFIXED(5 DOWNTO -3);   --"zzzzzz.zzz"
...
x <= TO_UFIXED(17.5, 4, -3); --converts 17.5 to UFIXED; result="10001100"
z <= -y; --unary "-" (only for signed); result=2's compl. of y
```

The arithmetic operations listed above are constructed with vector sizes such that overflow is always prevented. Some examples are shown below, where a and b are the inputs.

$a + b$, $a - b$: Range is max(a'LEFT, b'LEFT) + 1 DOWNTO min(a'RIGHT, b'RIGHT)

$a*b$: Range is a'LEFT + b'LEFT + 1 DOWNTO a'RIGHT + b'RIGHT

a/b unsigned: Range is a'LEFT − b'RIGHT DOWNTO a'RIGHT + b'LEFT − 1

a/b signed: Range is a'LEFT − b'RIGHT + 1 DOWNTO a'RIGHT + b'LEFT

$-a$ (unary "−", for signed only): Range is a'LEFT + 1 DOWNTO a'RIGHT

A complete VHDL code involving fixed-point numbers is shown below. The inferred circuit adds (line 13) and multiplies (line 14) the signals *a* and *b*. Observe the ranges defined for *x* and *y* (lines 7–8). Note also that the only extra package declaration is for *fixed_pkg* (lines 2–3), available in the file *fixed_pkg_c.vhd*.

```
1    ------------------------------------
2    LIBRARY ieee_proposed;
3    USE ieee_proposed.fixed_pkg.all;
4    ------------------------------------
5    ENTITY fixed IS
6       PORT (a, b: IN SFIXED(3 DOWNTO -3);
7             x: OUT SFIXED(4 DOWNTO -3);
8             y: OUT SFIXED(7 DOWNTO -6));
9    END ENTITY;
10   ------------------------------------
11   ARCHITECTURE fixed OF fixed IS
12   BEGIN
13      x <= a + b;
14      y <= a * b;
15   END ARCHITECTURE;
16   ------------------------------------
```

Floating-Point Types

The main floating-point type and subtypes are the following:

```
TYPE FLOAT IS ARRAY (INTEGER RANGE <>) OF STD_LOGIC; --generic length
SUBTYPE FLOAT32 IS FLOAT(8 DOWNTO -23);     --32-bit FP of IEEE 754
SUBTYPE FLOAT64 IS FLOAT(11 DOWNTO -52);    --64-bit FP of IEEE 754
SUBTYPE FLOAT128 IS FLOAT(15 DOWNTO -112);  --128-bit FP of IEEE 754
```

The files needed to use such types in VHDL 2008 are (see IEEE 1076-2008 Standard):

- *float_pkg.vhdl* (contains the package *float_pkg*)
- *float_generic_pkg.vhdl* (contains the package *float_generic_pkg*)
- *float_generic_pkg-body.vhdl* (contains the package body of *float_generic_pkg*)
- *fixed_float_types.vhdl.* (already mentioned)

For previous VHDL versions (93 and 2002), compatibility files were provided (only the three files below are needed, available at www.eda.org/fphdl):

- *fixed_pkg_c.vhd* (already mentioned)
- *fixed_float_types_c.vhd* (already mentioned)
- *float_pkg_c.vhd* (contains a *compatible* package *float_pkg*)

To implement floating-point circuits in these previous VHDL versions, just create a folder called *ieee_proposed* in the VHDL library directory of your compiler and copy the three files above into it.

As described in Pedroni (2008), the representation of floating-point numbers in the IEEE 754 standard obeys the structure illustrated in the figure below (for the 32-bit case):

$$2^7 \; 2^6 \; 2^5 \; 2^4 \; 2^3 \; 2^2 \; 2^1 \; 2^0 \quad 2^{-1} \; 2^{-2} \; 2^{-3} \; 2^{-4} \; \ldots \; 2^{-22} \; 2^{-23}$$

S	Exponent (E)	Fraction (F)
1 bit	8 bits	23 bits

Calling x the stored number, its value is given by $x = (-1)^S (1 + F) 2^{E-N}$, where S is the sign (0 when positive, 1 when negative), F is the fraction, E is the exponent, and N is a normalization factor given by $N = (E_{\max} + 1)/2 - 1$ (for example, $N = 127$ when the exponent has 8 bits or $N = 1023$ when it has 11 bits).

Example (with 3-bit exponent and 4-bit fraction, hence a total of 8 bits):

```
x <= "10010110";   --(1)(001)(0110) = -(1+0.375)2^1-3 = -0.34375
x <= "01101000";   --(0)(110)(1000) = +(1+0.5)2^6-3 = 12.0
```

In VHDL, the minimum length of a floating-point number is 7 bits, with the following distribution: 1 bit for the sign, 3 bits for the exponent, and 3 bits for the fraction.

As with fixed-point, a large set of operators and type-conversion functions are defined for the floating-point types. A simplified list is presented below.

a) Operators:
Logical: NOT, AND, NAND, OR, NOR, XOR, XNOR
Arithmetic: +, −, *, /, ABS, REM, MOD, MAC, SQRT, etc.
Comparison: =, /=, >, <, >=, <=, MAXIMUM, MINIMUM
Matching: ?=, ?/=, ?>, ?<, ?>=, ?<=

b) Type-conversion functions:
TO_FLOAT, TO_FLOAT32, TO_FLOAT64, TO_FLOAT128, TO_UNSIGNED, TO_SIGNED, TO_SLV, etc.

c) Text I/O functions:
WRITE (BWRITE, BINARY_WRITE), READ (BREAD, BINARY_READ), OWRITE (OCTAL_WRITE), OREAD (OCTAL_READ), HWRITE (HEX_WRITE), HREAD (HEX_READ), etc.

d) Other functions:
RESIZE, FIND_LEFTMOST, FIND_RIGHTMOST, etc.

Examples

```
SIGNAL x: FLOAT(3 DOWNTO -4); --(S)(EEE)(FFFF)
SIGNAL y: FLOAT(3 DOWNTO -4);
SIGNAL z: STD_LOGIC_VECTOR(7 DOWNTO 0);
...
--Convert 12.0 to float with format (S)(EEE)(FFFF):
x <= TO_FLOAT(12.0, 3, 4); --result="01101000"
x <= TO_FLOAT(12.0, x); --same as above
--Convert std_logic_vector to float with format (S)(EEE)(FFFF):
y <= TO_FLOAT(z, 3, 4);
```

A complete VHDL code involving floating-point numbers is shown below. The inferred circuit adds (line 12, which also includes a constant) and multiplies (line 13) the signals *a* and *b*. Observe that the same range is specified for all signals (lines 6–7). Note also that the only extra package declaration is for *float_pkg* (lines 2–3), available in the file *float_pkg_c.vhd*.

```
 1  ---------------------------------------
 2  LIBRARY ieee_proposed;
 3  USE ieee_proposed.float_pkg.all;
 4  ---------------------------------------
 5  ENTITY floating IS
 6     PORT (a, b: IN FLOAT(3 DOWNTO -4);
 7           x, y: OUT FLOAT(3 DOWNTO -4));
 8  END ENTITY;
 9  ---------------------------------------
10  ARCHITECTURE floating OF floating IS
11  BEGIN
12     x <= TO_FLOAT(0.34375, 3, 4) + a + b;
13     y <= a * b;
14  END ARCHITECTURE;
15  ---------------------------------------
```

3.9   Predefined Data Types Summary

Figure 3.6 summarizes the *predefined* VHDL data types that are *synthesizable* (the types introduced in VHDL 2008, also included in the figure in the gray areas, might not be supported yet), showing the corresponding package of origin and dimensionality (based on the number of bits—figure 3.2).

Category	Package of origin	Predefined synthesizable types	Dimension
Standard	standard	BIT	Scalar
		BIT_VECTOR	1D
		BOOLEAN	Scalar
		INTEGER	1D
		NATURAL	1D
		POSITIVE	1D
		CHARACTER	1D
		STRING	1Dx1D
	standard (2008 expansion)	BOOLEAN_VECTOR	1D
		INTEGER_VECTOR	1DX1D
	numeric_bit_unsigned (2008)	(only operators for BIT and BV)	---
Standard logic	std_logic_1164	STD_(U)LOGIC	Scalar
		STD_(U)LOGIC_VECTOR	1D
	std_logic_1164 (2008 expansion)	STD_(U)LOGIC	Scalar
		STD_(U)LOGIC_VECTOR	1D
	std_logic_unsigned	(only operators for SLV)	---
	std_logic_signed	(only operators for SLV)	---
	numeric_std_unsigned (2008)	(only operators for SL and SLV)	---
Unsigned and Signed	numeric_bit	UNSIGNED (base=BIT)	1D
		SIGNED (base=BIT)	1D
	numeric_std	UNSIGNED (base=STD_LOGIC)	1D
		SIGNED (base=STD_LOGIC)	1D
	std_logic_arith	UNSIGNED (base=STD_LOGIC)	1D
		SIGNED (base=STD_LOGIC)	1D
Fixed and Floating point	fixed_pkg + associated packages (2008)	UFIXED	1D
		SFIXED	1D
	float_pkg + assoc. pack. (2008)	FLOAT	1D

Figure 3.6
Synthesizable predefined data types with corresponding package of origin and dimensionality (additions made in VHDL 2008 are inside gray areas).

3.10   User-Defined Scalar Types

In addition to the predefined data types (seen earlier), user-defined data types are also allowed in VHDL. The creation of such types is the subject of nearly all remaining sections of this chapter.

Like the predefined data types, the new types can be *scalar* (single value) or *composite* (multiple values, specified using the ARRAY or RECORD keyword—see classifications in section 3.4). The construction of the former (integer and enumerated categories) is described in this section, while ARRAY and RECORD are described in sections 3.11 and 3.14, respectively.

The most common places for TYPE declarations are the main code itself (in the declarative part of the ARCHITECTURE) or a PACKAGE (chapter 8). For a scalar type, such declarations are made using the keyword TYPE, while for composite types, the keyword ARRAY or RECORD must also be included.

Integer Types

As mentioned earlier, INTEGER is synthesizable without restrictions. All types derived from INTEGER are referred to as integer types, and can be declared using the simplified syntax below. The range bounds must fall between INTEGER'LOW and INTEGER'HIGH, whose default values are $-(2^{31} - 1)$ and $(2^{31} - 1)$, respectively.

```
TYPE type_name IS RANGE range_specifications;
```

Examples

```
----------------------------------------
TYPE negative IS RANGE INTEGER'LOW TO -1;
TYPE temperature IS RANGE 0 TO 273;
TYPE my_integer IS RANGE -32 TO 32;
----------------------------------------
```

Enumerated Types

In this case, the type values are represented by symbols, which must be explicitly listed (enumerated). This approach was used in the creation of several of the predefined data types (sections 3.5 to 3.7), like the examples below.

```
----------------------------------------------------------
TYPE BIT IS ('0', '1');
TYPE BOOLEAN IS (FALSE, TRUE);
TYPE STD_ULOGIC IS ('U','X','0','1','Z','W','L','H','X');
----------------------------------------------------------
```

Enumerated types are particularly useful in the creation of other logic systems and in the design of finite state machines (chapter 11). They can be declared using the simplified syntax below.

```
TYPE type_name IS (type_values_list);
```

Examples

```
--------------------------------------------------------
TYPE logic_01Z IS ('0', '1', 'Z');
TYPE state IS (A, B, C, D, E);
TYPE machine_state IS (idle, transmitting, receiving);
--------------------------------------------------------
```

As will be shown in chapter 11, the actual encoding for enumerated data types can be done in several ways, freely chosen by the designer. The main options are "sequential" (sequential binary encoding), "Gray" (Gray code), "Johnson" (Johnson encoding), and "one-hot" (one-hot encoding). Details will be given in section 11.4 while studying finite state machines.

3.11   User-Defined Array Types

ARRAY is a collection of same-type elements. To create an array type, the keywords TYPE and ARRAY must be employed, as shown in the simplified syntax below.

```
    TYPE type_name IS ARRAY (range_specs) OF element_type;
```

Review of Predefined Array Types

Several of the predefined synthesizable data types described earlier are array types. Examples are shown below (note the use of both TYPE and ARRAY keywords).

Standard types (section 3.5):

```
TYPE BIT_VECTOR IS ARRAY (NATURAL RANGE <>) OF BIT;
TYPE BOOLEAN_VECTOR IS ARRAY (NATURAL RANGE <>) OF BOOLEAN;
TYPE INTEGER_VECTOR IS ARRAY (NATURAL RANGE <>) OF INTEGER;
TYPE STRING IS ARRAY (POSITIVE RANGE <>) OF CHARACTER;
----------------------------------------------------------
```

Standard-logic types (section 3.6):

```
TYPE STD_ULOGIC_VECTOR IS ARRAY (NATURAL RANGE <>) OF STD_ULOGIC;
TYPE STD_LOGIC_VECTOR IS ARRAY (NATURAL RANGE <>) OF STD_LOGIC;
```

Or, in VHDL 2008:

```
TYPE STD_ULOGIC_VECTOR IS array (NATURAL RANGE <>) of STD_ULOGIC;
SUBTYPE STD_LOGIC_VECTOR IS (resolved) STD_ULOGIC_VECTOR;
-----------------------------------------------------------------
```

Unsigned/Signed types (section 3.7):

```
TYPE UNSIGNED IS ARRAY (NATURAL RANGE <>) OF STD_LOGIC;
TYPE SIGNED IS ARRAY (NATURAL RANGE <>) OF STD_LOGIC;
--------------------------------------------------------
```

Because the types above are for general use, their ranges are left unconstrained (indicated by the symbol "<>"), with the only condition that, when specified, the limits must fall within the NATURAL or POSITIVE range. Their actual range limits are established later, when they are used in a code, as shown in the examples below.

```
-------------------------------------------------
CONSTANT a: BIT_VECTOR(7 DOWNTO 0) := "10001000";
SIGNAL b: STD_LOGIC_VECTOR(1 TO 16);
VARIABLE c: SIGNED(15 DOWNTO 0);
-------------------------------------------------
```

User-Defined Integer Array Types

Here too the described data types fall in one of the following two categories: *integer* types or *enumerated* types. Arrays of integers are described in this subsection, while arrays of enumerated types are described in the next. A simplified syntax for the former is shown below. For obvious reasons, the type of the elements (*int_elements_type*) is a subtype to the new type.

```
TYPE type_name IS ARRAY (range_specs) OF int_elements_type;
```

Examples of 1D × 1D arrays   Below are two integer type arrays, called *type1* and *type2*, each followed by object (CONSTANT, in these examples, called *const1* and *const2*) assignments. The first is unconstrained (so its actual range can be anywhere from 1 to INTEGER'HIGH), while the second is constrained. Regarding the array element range, contrary to enumerated arrays, in integer arrays it can be left unconstrained (note that the last word—INTEGER or NATURAL—is not accompanied by range specifications). Note also that when the original range is unconstrained, the range bounds must be included in the object declaration. Bitwise, both are 1D × 1D arrays (figure 3.2c).

```
--------------------------------------------------------
TYPE type1 IS ARRAY (POSITIVE RANGE <>) OF INTEGER;
CONSTANT const1: type1(1 TO 4) := (5, -5, 3, 0);
--------------------------------------------------------
TYPE type2 IS ARRAY (0 TO 3) OF NATURAL;
CONSTANT const2: type2 := (2, 0, 9, 4);
--------------------------------------------------------
```

Example of 1D × 1D × 1D array   Below is another integer type array, called *type3*, followed by an object (*const3*) declaration. Bitwise, this is a 1D × 1D × 1D array (figure 3.2e).

```
---------------------------------------------------------
TYPE type3 IS ARRAY (1 TO 2) OF type2; --see type2 above
CONSTANT const3: type3 := ((5, 5, 7, 99), (33, 4, 0, 0));
---------------------------------------------------------
```

User-Defined Enumerated Array Types

The subsection above described arrays of integers. Another option is arrays of enumerated types, covered here. Some examples were already shown, including arrays of BIT, BOOLEAN, CHARACTER, and STD_(U)LOGIC symbols. Other examples are shown below, using the simplified syntax that follows. The *enum_elements_type* (which is obviously a subtype to the new type) is normally required to be constrained.

```
    TYPE type_name IS ARRAY (range_specs) OF enum_elements_type;
```

Examples of 1D arrays   Below are two enumerated type arrays, called *type1* and *type2*, each followed by object (CONSTANT, in these examples, called *const1* and *const2*) declarations. The first type is unconstrained, while the second is constrained. Note that when the original range is unconstrained, the range bounds must be included in the object declaration. Bitwise, both arrays below are 1D arrays (figure 3.2b).

```
--------------------------------------------------------
TYPE type1 IS ARRAY (NATURAL RANGE <>) OF STD_LOGIC;
CONSTANT const1: type1(4 DOWNTO 1) := "Z111";
--------------------------------------------------------
TYPE type2 IS ARRAY (7 DOWNTO 0) OF BIT;
CONSTANT const2: type2 := "00001111";
--------------------------------------------------------
```

Examples of 1D × 1D arrays   Below are two other enumerated type arrays, called *type3* and *type4*, each followed by object (*const3* and *const4*) declarations. The first array is semi-unconstrained, while the second is constrained. Again, when the original range is unconstrained, the range bounds must be included in the object declaration. Note that in both cases the element array is constrained. Two equivalent representations are shown in each case. Bitwise, both arrays are 1D × 1D arrays (figure 3.2c).

```
------------------------------------------------------------------
TYPE type3 IS ARRAY (NATURAL RANGE <>) OF BIT_VECTOR(2 DOWNTO 0);
CONSTANT const3: type3(1 DOWNTO 0) := ("000", "111");
CONSTANT const3: type3(1 DOWNTO 0) := (('0','0','0'), ('1','1','1'));
------------------------------------------------------------------
TYPE type4 IS ARRAY (1 TO 4) OF STD_LOGIC_VECTOR(2 DOWNTO 0);
CONSTANT const4: type4 := ("000", "011", "100", "100");
CONSTANT const4: type4 := (('0','0','0'), ('0','1','1'), ...);
------------------------------------------------------------------
```

Example of 2D array   Below is another enumerated type array, called *type5*, followed by an object (*const5*) declaration. Bitwise, this is a 2D array (figure 3.2d).

```
----------------------------------------------------
TYPE type5 IS ARRAY (1 TO 3, 1 TO 4) OF BIT;
CONSTANT const5: type5 := (("0000","0000","0000"));
----------------------------------------------------
```

Example of 1D × 1D × 1D array   An enumerated type array called *type6* is shown below, followed by an object (*const6*) declaration. Bitwise, this is a 1D × 1D × 1D array (figure 3.2e).

```
------------------------------------------------------------
TYPE type6 IS ARRAY (1 TO 2) OF type4; --see type4 above
CONSTANT const6: type6 := (("000","011","100","100"), ( ... ));
------------------------------------------------------------
```

Example of 3D array   Below is another enumerated type array, called *type7*, followed by an object (*const7*) declaration. Bitwise, this is a 3D array (figure 3.2f).

```
------------------------------------------------------------------------
TYPE type7 IS ARRAY (1 TO 2, 1 TO 3, 1 TO 4) OF BIT;
CONSTANT const7: type7 := (("0000","0000","0000"),("0000","0000","0000"));
------------------------------------------------------------------------
```

3.12   Integer versus Enumerated Indexing

In the examples so far, the array index was always an integer. Consider, for example, the following declaration (this is a user-defined integer type ranging from 1 to 3—section 3.10):

```
TYPE int_row IS RANGE 1 TO 3;
```

Then the declarations below are clearly equivalent:

```
TYPE test IS ARRAY (1 TO 3) OF STD_LOGIC;
TYPE test IS ARRAY (int_row) OF STD_LOGIC;
```

But in VHDL, enumerated indexing is also allowed. For example, consider now the enumerated type below (section 3.10):

```
TYPE enum_column IS ('a', 'b', 'c', 'd');
```

Then the type *matrix* below

```
TYPE matrix IS ARRAY (int_row, enum_column) OF STD_LOGIC;
```

represents a three-row by four-column array of STD_LOGIC elements, with the rows indexed by *int_row* (integers from 1 to 3) and the columns by *enum_column* (enumerated from 'a' to 'd'). Then if the following array is constructed:

```
CONSTANT my_array: matrix := ("Z101", "0011", "101Z");
```

its elements can be accessed as follows:

```
my_array(1,'a') -> 'Z',
my_array(1,'b') -> '1',
my_array(1,'c') -> '0',
...
my_array(3,'d') -> 'Z'.
```

Enumerated indexing is used extensively, for example, in the *std_logic_1164* package body, where it helps compute the functions NOT, AND, OR, XOR, and *resolution*. Indeed, a detailed example was already shown in section 3.6 while describing the STD_LOGIC resolution function from that package.

3.13 Array Slicing

Figure 3.7 shows three 3×4 (in terms of the number of bits) arrays. An 1D $\times$ 1D array with integers is shown in (a), a similar array, but with bit vectors, is presented in (b), while a 2D array of bits appears in (c). Note that all three contain the same data; that is (in decimal values), *row1* $= 3$, *row2* $= 9$, *row3* $= 13$, or, equivalently, *column1* $= 3$, *column2* $= 1$, *column3* $= 4$, *column4* $= 7$.

The figure also shows several slices. Only horizontal slices are of interest in figure 3.7a. In figure 3.7b, four slices are shown; *slice1* is a single bit (scalar), *slice2* is a section of a row (1D), *slice3* is a whole row (1D), and *slice4* is a column (1D again). Finally, similar slices are depicted in figure 3.7c.

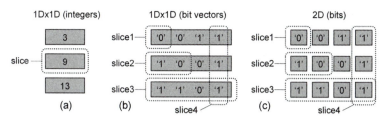

Figure 3.7
(a) 1D × 1D array of integers; (b) 1D × 1D array of bit vectors; (c) 2D array of bits. Several horizontal and vertical slices are also shown.

Slices involving only a single element (or part of it) of the array are always legal. In figure 3.7a, the individual elements are integers, so the slice shown in the figure is legal (slicing of larger arrays of integers might also be supported). In figure 3.7b, the individual elements are bit vectors, so *slice1* to *slice3* are legal (the first two are parts of a single element, while the last one is a whole single element). Finally, in figure 3.7c, the single elements are scalars, so only *slice1* is guaranteed to be supported. In conclusion, *slice4* in figure 3.7b and *slice2-slice4* in figure 3.7c might not be supported.

One way to circumvent the slicing limitation is based on the fact that scalar (single bit) slices are always supported. Therefore, if one finds a means of collecting together several of such slices, then virtually any multi-element slice of interest can be constructed (with the GENERATE statement, for example). Note in this discussion the importance of understanding and classifying the arrays according to their *bit* contents.

We show next how the slicing of figure 3.7 can be done. Each case is examined separately, using a complete VHDL code, accompanied by comments and simulation results.

Example 3.5: Slicing a 1D × 1D Array of Integers

This example shows how the slicing of figure 3.7a can be done. A VHDL code for that purpose is shown below, under the name *array_slice* (line 2). The input is *row* (line 3). Figure 3.7a has three rows, here numbered 1 to 3. The output is *slice* (line 4), which contains one of the array rows. An array type, called *oneDoneD* and containing three integers in the 0 to 15 range, was created in line 8. Next, a constant, called *table*, conforming with that data type, was declared in line 9, to which three values (3, 9, 13) were assigned. In the code proper, a slice is read from the 1D × 1D array using *row* as the index (line 11). Simulation results are depicted in figure 3.8, showing that, as expected, the values 3, 9, and 13 are indeed retrieved.

```
1   -----------------------------------------------------------
2   ENTITY array_slice IS
3      PORT (row: IN INTEGER RANGE 1 TO 3;
4              slice: OUT INTEGER RANGE 0 TO 15);
```

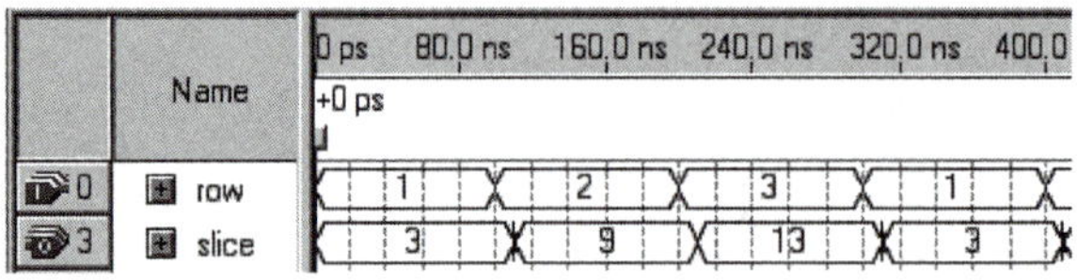

Figure 3.8
Simulation results from the code of example 3.5.

```
 5   END ENTITY;
 6   ------------------------------------------------------------
 7   ARCHITECTURE array_slice OF array_slice IS
 8      TYPE oneDoneD IS ARRAY (1 TO 3) OF INTEGER RANGE 0 TO 15;
 9      CONSTANT table: oneDoneD := (3, 9, 13);
10   BEGIN
11      slice <= table(row);
12   END ARCHITECTURE;
13   ------------------------------------------------------------
```

Example 3.6: Slicing a 1D × 1D Array of Bit Vectors

This example shows how the slicing of figure 3.7b can be done. A VHDL code for that purpose is shown below. The inputs are *row* (1 to 3) and *column* (1 to 4—declaring as 0 to 4 guarantees that three bits are assigned to that signal), while the outputs are the four slices of figure 3.7b. An array type, called *oneDoneD* and consisting of a 1D × 1D array containing bit vectors, was created in line 12. Next, a constant, called *table*, conforming with that data type, was declared in lines 13–15, to which the bit values corresponding to 1, 9, and 13 were assigned. In the code proper, the four slices were created.

Assuming that the single-element slice limitation described above applies, lines 17–19 are still synthesizable, but line 20 is not (vertical slice). Two options to circumvent that limitation are shown in lines 21 and 22–24. The case in line 21 consists simply of collecting together the scalar slices that comprise one column using the concatenation operator "&". This obviously is awkward because this list might easily get too long. The option in lines 22–24, on the other hand, uses the GENERATE statement, which builds a loop, so the code always consists of just three lines, regardless of how many elements must be collected together. (The study of GENERATE and all sorts of VHDL codes will start in chapter 5.) Simulation results are depicted in figure 3.9, showing results that coincide with the expected values.

```
 1   ------------------------------------------------------------------------
 2   ENTITY array_slices IS
 3      PORT (row: IN INTEGER RANGE 1 TO 3;
 4            column: IN INTEGER RANGE 0 TO 4; --3 bits
 5            slice1: OUT BIT;
 6            slice2: OUT BIT_VECTOR(1 TO 2);
```

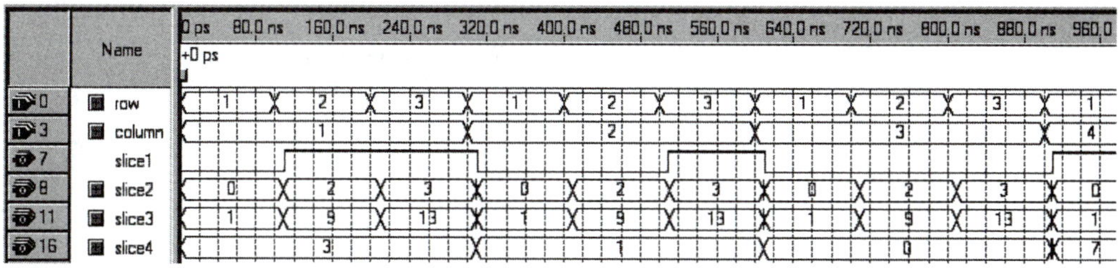

Figure 3.9
Simulation results from the code of example 3.6.

```
7              slice3: OUT BIT_VECTOR(1 TO 4);
8              slice4: OUT BIT_VECTOR(1 TO 3));
9  END ENTITY;
10 ------------------------------------------------------------------
11 ARCHITECTURE array_slices OF array_slices IS
12    TYPE oneDoneD IS ARRAY (1 TO 3) OF BIT_VECTOR(1 TO 4);
13    CONSTANT table: oneDoneD := (('0','0','0','1'),   --1
14                                 ('1','0','0','1'),   --9
15                                 ('1','1','0','1'));  --13
16 BEGIN
17    slice1 <= table(row)(column);
18    slice2 <= table(row)(1 TO 2);
19    slice3 <= table(row)(1 TO 4);
20    --slice4 <= table(1 TO 3)(column);
21    --slice4 <= table(1)(column) & table(2)(column) & table(3)(column);
22    gen: FOR i IN 1 TO 3 GENERATE
23       slice4(i) <= table(i)(column);
24    END GENERATE;
25 END ARCHITECTURE;
26 ------------------------------------------------------------------
27
```

Example 3.7: Slicing a 2D Array of Bits

This example shows how the slicing of figure 3.7c can be done. A VHDL code for that purpose is shown below. The inputs are again *row* and *column*, while the outputs are the four slices of figure 3.7c. An array type, called *twoD* and consisting of a 2D array of bits, was created in line 12. Next, a constant, called *table*, conforming with that data type, was declared in lines 13–15, to which the bit values corresponding to 1, 9, and 13 were assigned. In the code proper, the four slices were created.

Assuming that the single-element slice limitation described above applies, only line 17 is synthesizable. To circumvent that limitation, two techniques are again depicted, in lines 21 and 22–24. The former is awkward because the list can be too long, while the latter is fine

because the code size (three lines) is independent from the number of elements that must be
collected together. The same technique used for *slice4* can be extended to the others.

```
1   ----------------------------------------------------------------------
2   ENTITY array_slices IS
3      PORT (row: IN INTEGER RANGE 0 TO 3;
4            column: IN INTEGER RANGE 0 TO 4; --3 bits
5            slice1: OUT BIT;
6            slice2: OUT BIT_VECTOR(1 TO 2);
7            slice3: OUT BIT_VECTOR(1 TO 4);
8            slice4: OUT BIT_VECTOR(1 TO 3));
9   END ENTITY;
10  ----------------------------------------------------------------------
11  ARCHITECTURE array_slices OF array_slices IS
12     TYPE twoD IS ARRAY (1 TO 3, 1 TO 4) OF BIT;
13     CONSTANT table: twoD := (('0','0','0','1'),
14                              ('1','0','0','1'),
15                              ('1','1','0','1'));
16  BEGIN
17     slice1 <= table(row, column);
18     --slice2 <= table(row, 1 TO 2);
19     --slice3 <= table(row, 1 TO 4);
20     --slice4 <= table(1 TO 3, column);
21     --slice4 <= table(1, column) & table(2, column) & table(3, column);
22     gen: FOR i IN 1 TO 3 GENERATE
23        slice4(i) <= table(i, column);
24     END GENERATE;
25  END ARCHITECTURE;
26  ----------------------------------------------------------------------
```

3.14 Records

Records are collections of elements that can be of different types. Individually, the types
can be any of those discussed in this chapter, either pre- or user-defined.

Example

```
----------------------------------
TYPE memory_access IS RECORD
   address: INTEGER RANGE 0 TO 255;
   block: INTEGER RANGE 0 TO 3;
   data: BIT_VECTOR(15 DOWNTO 0);
END RECORD;
----------------------------------
```

Example A complete code is presented below. It shows a RECORD type called *pair*, with two elements, both of type NATURAL. Subsequently, a type called *stack* is constructed with four such pairs. Finally, a constant called *matrix* is declared as type *stack*, hence with four rows with two natural values each. In the code proper (after BEGIN), a 4-bit flag is produced based on the comparison between the pair values, followed by a sum between the two elements of the first pair. The expected results are therefore *flag* = "0001" and *sum* = 3. Observe, in particular, how the individual elements of the record are accessed.

```
1   ----------------------------------------------------------------
2   ENTITY record_example IS
3      PORT (flag: OUT BIT_VECTOR(1 TO 4);
4            sum: OUT NATURAL RANGE 0 TO 15);
5   END ENTITY;
6   ----------------------------------------------------------------
7   ARCHITECTURE record_example OF record_example IS
8      TYPE pair IS RECORD
9         a, b: NATURAL RANGE 0 TO 7;
10     END RECORD;
11     TYPE stack IS ARRAY (1 TO 4) OF pair;
12     CONSTANT matrix: stack := ((1, 2), (3, 4), (5, 6), (7, 0));
13  BEGIN
14     gen: FOR i IN 1 TO 4 GENERATE
15        flag(i) <= '1' WHEN matrix(i).a > matrix(i).b ELSE '0';
16     END GENERATE;
17     sum <= matrix(1).a + matrix(1).b;
18  END ARCHITECTURE;
19  ----------------------------------------------------------------
```

3.15 Subtypes

SUBTYPE is a TYPE with a constraint. The main reason for using a subtype rather than specifying a new type is that, though operations between different data types are not allowed, they are allowed between a subtype and the type from which it was derived (because they have the same *base* type).

SUBTYPE can be declared in the same places as TYPE, but it is usually done in the declarative part of the ARCHITECTURE or in a separate PACKAGE. Examples of subtype declarations are depicted below.

Example NATURAL and POSITIVE are subtypes (subsets) of INTEGER (appendix H), and so is *my_integer*.

```
-----------------------------------------------------
SUBTYPE NATURAL IS INTEGER RANGE 0 TO INTEGER'HIGH;
SUBTYPE POSITIVE IS INTEGER RANGE 1 TO INTEGER'HIGH;
SUBTYPE my_integer IS INTEGER RANGE -32 TO 32;
-----------------------------------------------------
```

Example   The new type below (*my_logic*) is a subtype of STD_LOGIC, with values '0', '1', and 'Z'.

```
-----------------------------------------------------------
TYPE STD_LOGIC IS ('X', '0', '1', 'Z', 'W', 'L', 'H', '-');
SUBTYPE my_logic IS STD_LOGIC RANGE '0' TO 'Z';
-----------------------------------------------------------
```

Example   The subtype *my_color* below, of *color*, contains only the values *green* and *blue*.

```
-----------------------------------------------
TYPE color IS (red, green, blue, white);
SUBTYPE my_color IS color RANGE green TO blue;
-----------------------------------------------
```

3.16   Specifying PORT Arrays

As we have seen, there are no predefined synthesizable data types of more than one dimension (except for STRING and INTEGER_VECTOR—see figure 3.6). However, in the specifications of the input and output ports of a circuit (made in the ENTITY), a 1D × 1D or 2D array might be needed. Since TYPE declarations are not allowed before the PORT field of the ENTITY, a typical solution consists of creating such user-defined data types in a separate PACKAGE, because this can be visible to the whole code, including the ENTITY (packages will be studied in chapter 8).

In VHDL 2008, *generic type* declarations are allowed (see section 2.6), easing the construction of port arrays.

Example 3.8: Multiplexer with 1D × 1D PORT

Say that we need to implement the multiplexer of figure 2.6, which contains 4 inputs of 8 bits each. Write a VHDL code to solve this problem with *x* (the set of all inputs) declared as a 1D × 1D signal.

Solution   *x* can be specified as a 4 × 8 array with the base type BIT or STD_LOGIC. Since this data type is needed right in the beginning of the code (in the PORT field of the ENTITY), and there is no predefined data type of that sort, it must be created in a separate PACKAGE. A corresponding VHDL code is shown below, where the package, called *my_data_types*, contains a TYPE called *oneDoneD* (line 3) that can be used later (in the main code) to specify the mux input.

The main code is also shown below. For the package to be visible, a USE clause was included in line 2. This new data type is then used in line 5 to specify *x*. As a result, the code proper (in the ARCHITECTURE) needs just one line (line 12) to implement the circuit.

```
1    -----Package:-------------------------------------------------
2    PACKAGE my_data_types IS
3       TYPE oneDoneD IS ARRAY (0 TO 3) OF BIT_VECTOR(7 DOWNTO 0);
4    END my_data_types;
5    --------------------------------------------------------------
```

```
1    -----Main code: ---------------------
2    USE work.my_data_types.all;
3    --------------------------------------------------------------
4    ENTITY mux IS
5    PORT (x: IN oneDoneD;
6          sel: INTEGER RANGE 0 TO 3;
7          y: OUT BIT_VECTOR(7 DOWNTO 0));
8    END ENTITY;
9    --------------------------------------------------------------
10   ARCHITECTURE mux OF mux IS
11   BEGIN
12      y <= x(sel);
13   END ARCHITECTURE;
14   --------------------------------------------------------------
```

3.17 Qualified Types and Overloading

Qualified expressions are used to resolve ambiguous situations. To declare a qualified type (hence resulting a qualified expression), the following syntax is employed (the use of parentheses is mandatory):

```
type_name'(expression);
```

Example Say that we want to perform the *signed* sum below, where *a*, *b*, and *sum* are all signed values:

```
sum <= a + b + "1000";
```

The problem is that there are two possible values for "1000": 8 if unsigned, -8 if signed. This conflict is resolved by the qualified expression below, which determines that "1000" is a signed value:

```
sum <= a + b + SIGNED'("1000");
```

A more common application for qualified types is to resolve ambiguous situations in *overloaded* operations. An overloaded operator is one for which more than one in-out option exists. This is indeed the case for most predefined operators (studied in the next chapter). Take, for example, the addition ("+") operator defined in the *numeric_std* package (appendix J), which has the following six versions (*L* and *R* are the left and right operands):

```
FUNCTION "+" (L, R: UNSIGNED) RETURN UNSIGNED;
FUNCTION "+" (L, R: SIGNED) RETURN SIGNED;
FUNCTION "+" (L: UNSIGNED; R: NATURAL) RETURN UNSIGNED;
FUNCTION "+" (L: NATURAL; R: UNSIGNED) RETURN UNSIGNED;
FUNCTION "+" (L: INTEGER; R: SIGNED) RETURN SIGNED;
FUNCTION "+" (L: SIGNED; R: INTEGER) RETURN SIGNED;
```

This means that the inputs to the adder can be "unsigned + unsigned", "signed + signed", and so on. Consequently, if the compiler is not able to determine the actual input types, the sum cannot be computed. As an example, say that the intended type for *a* and *b* is SIGNED, but it has not been explicitly declared. Then the qualified expression below can be used, which automatically returns a signed value (note in the list above that SIGNED is the only output option when both inputs are SIGNED):

```
sum <= SIGNED'(a) + SIGNED'(b);
```

A detailed example using type qualification will be presented in the section below. More about overloaded operators will be seen in sections 4.3 and 9.6.

3.18   Type Conversion

Direct type conversion falls in one of the three categories described next, in which the following signals are employed in the examples:

```
-------------------------------------------------
SIGNAL b: BIT;
SIGNAL bv: BIT_VECTOR(7 DOWNTO 0);
SIGNAL sl: STD_LOGIC;
SIGNAL slv1, slv2: STD_LOGIC_VECTOR(7 DOWNTO 0);
SIGNAL unsig: UNSIGNED(7 DOWNTO 0);
SIGNAL sig: SIGNED(7 DOWNTO 0);
-------------------------------------------------
```

Automatic Conversion

This is the case when dealing directly with the *base* type. For example, BIT and BIT_VECTOR have the same base type (BIT), so a single element of BIT_VECTOR is automatically compatible with an element of type BIT.

Examples

```
-------------
bv(0) <= b;
slv(7) <= sl;
-------------
```

Type Casting

UNSIGNED/SIGNED have the same base type (STD_LOGIC) and indexing (NATU-RAL) as STD_LOGIC_VECTOR, so the following direct conversions using type casting are allowed:

- (UN)SIGNED(arg), where the argument is STD_LOGIC_VECTOR
- STD_LOGIC_VECTOR(arg), where the argument is SIGNED or UNSIGNED.

Examples

```
-------------------------------
unsig <= UNSIGNED(slv);
sig <= SIGNED(slv);
slv1 <= STD_LOGIC_VECTOR(unsig);
slv2 <= STD_LOGIC_VECTOR(sig);
-------------------------------
```

Type-Conversion Functions

The last option for direct type conversion is with type-conversion functions, available in the VHDL packages. The main cases are listed in figure 3.10.

Example 3.9: Recommended Signed Multiplier Implementation (for Integers)

The code below is for a *signed* multiplier with inputs *a* and *b* and output *prod*, all of type STD_LOGIC_VECTOR (industry standard). Knowing that the type casting expression STD_LOGIC_VECTOR(arg) just described can convert an argument of type SIGNED to STD_LOGIC_VECTOR when the package *numeric_std* is used, or from SIGNED *or* STD_LOGIC_VECTOR (hence *overloaded*) to STD_LOGIC_VECTOR when the package *std_logic_arith* is employed, analyze all three architectures below (all legal) and make proper comments. Assume that *arch1* uses the package *numeric_std*, while *arch2* and *arch3* employ *std_logic_arith*. Which is the recommended approach?

From	To	Type conversion function	Package of origin
INTEGER	STD_LOGIC_VECTOR	conv_std_logic_vector(a, cs)	std_logic_arith
	UNSIGNED	to_unsigned(a, cs) conv_unsigned(a, cs)	numeric_std std_logic_arith
	SIGNED	to_signed(a, cs) conv_signed(a, cs)	numeric_std std_logic_arith
	UFIXED	to_ufixed(a, cs)	fixed_generic_pkg
	SFIXED	to_sfixed(a, cs)	fixed_generic_pkg
	FLOAT	to_float(a, cs)	float_generic_pkg
BIT_VECTOR	STD_LOGIC_VECTOR	to_stdlogicvector(a, cs)	std_logic_1164
STD_LOGIC_VECTOR	INTEGER	conv_integer(a, cs) conv_integer(a, cs) to_integer(a, cs)	std_logic_signed std_logic_unsigned numeric_std_unsigned
	BIT_VECTOR	to_bitvector(a, cs)	std_logic_1164
	UNSIGNED	unsigned(a) (*) unsigned(a) (*)	numeric_std std_logic_arith
	SIGNED	signed(a) (*) signed(a) (*)	numeric_std std_logic_arith
	UFIXED	to_ufixed(a, cs)	fixed_generic_pkg
	SFIXED	to_sfixed(a, cs)	fixed_generic_pkg
	FLOAT	to_float(a, cs)	float_generic_pkg
UNSIGNED and SIGNED	INTEGER	to_integer(a, cs) conv_integer(a, cs)	numeric_std std_logic_arith
	STD_LOGIC_VECTOR	std_logic_vector(a) (*) std_logic_vector(a) (*) conv_std_logic_vector(a, cs)	numeric_std std_logic_arith std_logic_arith
	UNSIGNED	conv_unsigned(a, cs)	std_logic_arith
	SIGNED	conv_signed(a, cs)	std_logic_arith
	UFIXED (unsigned only)	to_ufixed(a, cs)	fixed_generic_pkg
	SFIXED (signed only)	to_sfixed(a, cs)	fixed_generic_pkg
	FLOAT	to_float(a, cs)	float_generic_pkg
UFIXED and SFIXED	INTEGER	to_integer(a, cs)	fixed_generic_pkg
	STD_LOGIC_VECTOR	to_slv(a, cs)	fixed_generic_pkg
	UNSIGNED (ufixed only)	to_unsigned(a, cs)	fixed_generic_pkg
	SIGNED (sfixed only)	to_signed(a, cs)	fixed_generic_pkg
	SFIXED (ufixed only)	to_sfixed(a, cs)	fixed_generic_pkg
	FLOAT	to_float(a, cs)	float_generic_pkg
FLOAT	INTEGER	to_integer(a, cs)	float_generic_pkg
	STD_LOGIC_VECTOR	to_slv(a, cs)	float_generic_pkg
	UNSIGNED	to_unsigned(a, cs)	float_generic_pkg
	SIGNED	to_signed(a, cs)	float_generic_pkg
	UFIXED	to_ufixed(a, cs)	float_generic_pkg
	SFIXED	to_sfixed(a, cs)	float_generic_pkg

(a, cs) = (argument, conversion specifications)
cs may include vector size, left/right range constants, overflow and rounding specs, etc. (consult package)
(*) = type casting

Figure 3.10
Main type-conversion options (type casting and type-conversion functions).

```
 1  --------------------------------------------------------
 2  LIBRARY ieee;
 3  USE ieee.std_logic_1164.all;
 4  USE ieee.numeric_std.all; --for arch1
 5  --USE ieee.std_logic_arith.all; --for arch2, arch3
 6  --------------------------------------------------------
 7  ENTITY signed_multiplier IS
 8     PORT (a, b: IN STD_LOGIC_VECTOR(3 DOWNTO 0);
 9            prod: OUT STD_LOGIC_VECTOR(7 DOWNTO 0));
10  END ENTITY;
11  --------------------------------------------------------
12  ARCHITECTURE arch1 OF signed_multiplier IS
13     SIGNAL a_sig, b_sig: SIGNED(3 DOWNTO 0);
14  BEGIN
15     a_sig <= SIGNED(a);
16     b_sig <= SIGNED(b);
17     prod <= STD_LOGIC_VECTOR(a_sig * b_sig);
18  END arch1;
19  --------------------------------------------------------

12  ARCHITECTURE arch2 OF signed_multiplier IS
13     SIGNAL a_sig, b_sig: SIGNED(3 DOWNTO 0);
14     SIGNAL prod_sig: SIGNED(7 DOWNTO 0);
15  BEGIN
16     a_sig <= SIGNED(a);
17     b_sig <= SIGNED(b);
18     prod_sig <= a_sig * b_sig;
19     prod <= STD_LOGIC_VECTOR(prod_sig);
20  END arch2;
21  --------------------------------------------------------

12  ARCHITECTURE arch3 OF signed_multiplier IS
13     SIGNAL a_sig, b_sig: SIGNED(3 DOWNTO 0);
14  BEGIN
15     a_sig <= SIGNED(a);
16     b_sig <= SIGNED(b);
17     prod <= STD_LOGIC_VECTOR(SIGNED'(a_sig * b_sig));
18  END arch3;
19  --------------------------------------------------------
```

Solution   This design is called *signed_multiplier* (line 7). It has two 4-bit inputs (line 8) and one 8-bit output (line 9). In all three architectures, the inputs are *explicitly* converted from STD_LOGIC_VECTOR to SIGNED (using type casting), hence following the suggestion in the note at the end of example 3.4. In all three, type casting is also used to return the result to the STD_LOGIC_VECTOR type.

In *arch1*, the product is converted to STD_LOGIC_VECTOR in line 17. Because this architecture uses the *numeric_std* package, the argument can only be SIGNED (no overloading), so the type for the output of the expression "*a_sig * b_sig*", though not explicitly declared, is automatically assumed to be SIGNED, causing the correct transformation to occur.

In *arch2*, the *std_logic_arith* package is employed, so two types are accepted for the argument in the type casting of line 19. Even though the inputs to the expression "*a_sig * b_sig*" were declared as SIGNED, the type for the result was never explicitly declared, so the compiler does not know if SIGNED or STD_LOGIC_VECTOR should be considered. This problem is avoided by evaluating the expression in line 18, and then passing *prod_sig* to the transformation in line 19 instead of passing the expression itself. Because *prod_sig* was explicitly declared as SIGNED (line 14), the correct transformation again occurs.

Finally, note that *arch3* is similar to *arch1*, but because it uses the *std_logic_arith* package, the ambiguous situation described above again occurs, which is resolved in this case with a *qualified* expression. Note the expression "SIGNED'(*a_sig * b_sig*)" in line 17, which determines that the result of "*a_sig * b_sig*" must be SIGNED.

Because *arch1* follows the suggestion in the note at the end of example 3.4, and because it employs the standardized package *numeric_std*, it is the recommended approach. Additionally, if one is not sure whether a certain type casting/conversion expression is not overloaded, then the inclusion of an explicit computation before entering the conversion expression (as in lines 18–19 of *arch2*) is advisable. In summary, use *arch1* and do the following: delete line 17 and include lines 14, 18, and 19 or *arch2*.

3.19   Legal versus Illegal Assignments

Below is a summary of the most common mistakes made when assigning values to array type objects. In short, most mistakes fall in one (or more) of the causes listed below. A subsequent example illustrates the occurrence of such problems.

Cause 1:   Type mismatch (both sides of the assignment must be of the same type or of the same base type). Example: One side is BIT; the other is BOOLEAN.

Cause 2:   Size mismatch (both sides of the assignment must have the same number of bits or a predefined number of bits). Example: One side has 8 bits; the other has 4.

Cause 3:   Invalid value or invalid representation. Examples: BIT does not accept the value 'Z'; an integer cannot be represented with quotes; a bit vector requires double quotes.

Cause 4:   Incorrect indexing (the order and the range limits must be obeyed). Examples: The order (ascending or descending) of the index is reversed; the index values fall outside the actual range.

Cause 5:   Incorrect assignment operator ("<=" for signals, ":=" for variables, constants, and initial/default values).

Example 3.10: Legal versus Illegal Assignments

Consider the following type definitions and signal declarations:

```
------------------------------------------------------------------
TYPE row IS ARRAY (7 DOWNTO 0) OF STD_LOGIC; --1D
TYPE matrix1 IS ARRAY (0 TO 3) OF row; --1Dx1D
TYPE matrix2 IS ARRAY (0 TO 3) OF STD_LOGIC_VECTOR(3 DOWNTO 0); --1Dx1D
TYPE matrix3 IS ARRAY (0 TO 3, 7 DOWNTO 0) OF BIT; --2D
SIGNAL u: BIT;
SIGNAL v: STD_LOGIC;
SIGNAL w: row;
SIGNAL x: matrix1;
SIGNAL y: matrix2;
SIGNAL z: matrix3;
------------------------------------------------------------------
```

Explain why the assignments below are illegal (the answers are presented in the form of comments).

```
------------------------------------------------------------------
w(0) <= u; --Cause 1 (STD_LOGIC x BIT)
w <= y(0): --Cause 1 (different types) and cause 2 (8 bits x 4 bits)
x(1) <= "11ZZ"; --Cause 2 (8 bits x 4 bits)
z(0, 0) <= 'Z'; --Cause 3 (BIT can only be '0' or '1')
z(4)(0) <= '1'; --Cause 4 (wrong index and wrong use of parentheses)
z(0, 0) <= w(0); --Cause 1 (BIT x STD_LOGIC)
w(0 TO 7) <= "11110000"; --Cause 4 (incorrect index order)
y(1, 1) <= '0'; --Cause 4 (wrong use of parentheses)
y(1)(3 DOWNTO 2) <= "111"; --Cause 2 (2 bits x 3 bits)
v <= z(3, 7); --Cause 1 (STD_LOGIC x BIT)
u := '0'; --Cause 5 (should be "<=")
w(3 DOWNTO 0) <= y(0); --Cause 1 (row x STD_LOGIC_VECTOR);
--w(0) <= y(0)(0) would be fine (scalars, same base type, STD_LOGIC).
------------------------------------------------------------------
```

3.20   ACCESS Types

All data types described so far have a well-defined, known structure. However, in models for simulation with a high level of abstraction, the data structure might sometimes not be known in advance or might not be static (varying size). To deal with such situations,

VHDL provides a type class called *access types*, which acts as a *pointer* to the location in memory where the data is located rather than representing the data directly (so unknown and dynamic data structures can be used).

Access types can only be used in sequential code, and only variables can be of that type. In the example below, an access type called *int_pointer* is created, which points to objects of type INTEGER. Next, a variable of that type, called *pointer*, is declared.

```
TYPE int_pointer IS ACCESS INTEGER;
VARIABLE pointer: int_pointer := NULL;
--------------------------------------
pointer := NEW INTEGER'(16);
pointer.ALL := 0;
```

The variable *pointer* above points to integers stored in memory. However, its initial (default) value is NULL (that is, it points nowhere), which must be modified by the code. In the third line, the predefined function NEW is called to provide memory space for 16 integers, returning an access value (a pointer) to the allocated memory, which is assigned to *pointer*. Finally, in the fourth line, a value (0, in this example) is assigned to the object pointed to by *pointer*.

This description is just to give an idea on how access types work. Even though very complex models can be built with such a pointing mechanism, its regular use is for simulation, and only for very particular data structures, so a hardware designer might never need it.

3.21   FILE Types

File types were briefly described in section 3.2. Because files are particularly important for simulation, further details will be given in chapter 10, which deals specifically with that subject.

3.22   VHDL 2008

With respect to the material covered in this chapter, the main additions specified in VHDL 2008 are those seen in sections 3.3, 3.5, 3.6, and 3.8. Just as a reminder, the following occurred with the standardized VHDL packages:

1) The following packages were expanded:

Package *standard* (Appendix H)

Package *std_logic_1164* (Part II of Appendix I)

Package *numeric_bit*

Package *numeric_std* (Part II of Appendix J)

Package *textio* (Appendix M)

2) The following new packages were introduced for integer arithmetic:

Package *numeric_bit_unsigned*

Package *numeric_std_unsigned*

3) The following new packages for fixed- and floating-point arithmetic were introduced (compatibility packages for VHDL 93 and 2002 are also available):

Package *fixed_pkg*

Package *fixed_generic_pkg*

Package *float_pkg*

Package *float_generic_pkg*

Package *fixed_float_types*

3.23   Exercises

Note: For exercise solutions, please consult the book website.

Exercise 3.1: Possible Data Types #1

Say that *s1* to *s4* are four VHDL signals. Based on the assignments below, list which synthesizable predefined data types each of these signals can belong to.

```
s1 <= '0';
s2 <= 'Z';
s3 <= TRUE;
s4 <= "01000";
```

Exercise 3.2: Possible Data Types #2

Say that *s1* to *s4* below are another four VHDL signals. Based on the assignments presented, list which synthesizable predefined data types each of them can belong to.

```
s1 <= "0100Z";
s2 <= ('0','1','0','0', '0');
s3 <= (OTHERS => 'Z');
s4 <= 255;
```

Exercise 3.3: Types and Operators in the Package *standard*

a) List the *synthesizable* types defined in the package *standard* (appendix H). For each of them, list the supported operators (arithmetic, logical, comparison, shift, concatenation, matching).

b) Compare your results against those in section 3.5 and also in figure 3.6. Do they match?

Exercise 3.4: Types and Operators in the Package *std_logic_1164*

a) List the types and subtypes defined in the package *std_logic_1164* (appendix I). List also the supported operators (for example, are there arithmetic operators?).

b) Compare your results against those in section 3.6 and also in figure 3.6. Do they match?

Exercise 3.5: Types and Operators in the Package *numeric_std*

a) List the types defined in the package *numeric_std* (appendix J). List also the supported operators.

b) Compare your results against those in section 3.7 and also in figure 3.6. Do they match?

Exercise 3.6: Types and Operators in the Package *std_logic_arith*

a) List the types defined in the package *std_logic_arith* (appendix K). List also the supported operators.

b) Compare your results against those in section 3.7 and also in figure 3.6. Do they match?

c) Finally, compare the results from this exercise with those from exercise 3.5.

Exercise 3.7: Operators in the Package *std_logic_signed*

Are there any new types defined in this package (appendix L)? List all operators (arithmetic, logical, comparison, shift, concatenation) defined in it. For which type(s) are the operators intended?

Exercise 3.8: Integer versus Enumerated Types

List which among all predefined data types in sections 3.5 to 3.7 are integer-based and which are enumerated.

Exercise 3.9: Possible Packages

Consider the two sections of code below:

```
-------------------------------------------------
SIGNAL a, b, x: INTEGER RANGE 0 TO 255;
SIGNAL y: BIT;
x <= (a + b)/2;
y <= '1' WHEN a>=b ELSE '0';
-------------------------------------------------
SIGNAL a, b, x, y: STD_LOGIC_VECTOR(7 DOWNTO 0);
x <= a + b;
y <= ('1' & a(6 DOWNTO 0)) XOR b;
-------------------------------------------------
```

a) Which (if any) packages need to be explicitly included in the library/package declarations for the first code above to be valid?

b) Respond the same question, with respect to the second code above.

Exercise 3.10: Subtypes

Consider the predefined data types INTEGER and STD_LOGIC_VECTOR. Consider also the two user-defined types below. For each, write a possible SUBTYPE.

```
TYPE oneDoneD IS ARRAY (POSITIVE RANGE <>) OF INTEGER;
TYPE twoD IS ARRAY (POSITIVE RANGE <>, 3 DOWNTO 0) OF BOOLEAN;
```

Exercise 3.11: Multibit Tri-state Buffer

What needs to be changed in the code of example 3.1 to produce an 8-bit tri-state buffer instead of a single-bit buffer?

Exercise 3.12: Single Bit versus Bit Vector

Two VHDL codes are shown below. The code proper is actually the same (compare the architectures). The only difference between them is in the specifications of the entities. Draw the circuits that you expect the compiler will infer from each of these codes, then compile them and compare the results with your predictions.

```
-----------------------------------         -------------------------------------------
ENTITY and_gate IS                          ENTITY and_gate IS
   PORT (a, b: IN BIT;                          PORT (a, b: IN BIT_VECTOR(3 DOWNTO 0);
         x: OUT BIT);                                 x: OUT BIT_VECTOR(3 DOWNTO 0));
END ENTITY;                                  END ENTITY;
-----------------------------------         -------------------------------------------
ARCHITECTURE circuit OF and_gate IS         ARCHITECTURE circuit OF and_gate IS
BEGIN                                        BEGIN
   x <= a AND b;                                x <= a AND b;
END ARCHITECTURE;                            END ARCHITECTURE;
-----------------------------------         -------------------------------------------
```

Exercise 3.13: Hardware Optimization with "Don't Care" Values

Compile the code of example 3.2 and check whether the "don't care" values indeed help reduce the amount of hardware needed to build the circuit.

Exercise 3.14: 1D Array Examples

Consider the 1D array shown in figure 3.2b. Write three examples of possible array types that fall in that category (one example is already included below; complete the list with *type2* and *type3*).

```
TYPE type1 IS ARRAY (15 DOWNTO 0) OF BIT;
```

Exercise 3.15: 1D × 1D Array Examples

Consider the 1D × 1D array shown in figure 3.2c. Write three examples of possible array types that fall in that category (one example is already included below; complete the list with *type2* and *type3*).

```
TYPE type1 IS ARRAY (NATURAL RANGE <>) OF BIT_VECTOR(7 DOWNTO 0);
```

Exercise 3.16: 2D Array Examples

Consider the 2D array shown in figure 3.2d. Write three examples of possible array types that fall in that category (one example is already included below; complete the list with *type2* and *type3*).

```
TYPE type1 IS ARRAY (NATURAL RANGE <>, NATURAL RANGE <>) OF STD_LOGIC;
```

Exercise 3.17: 1D × 1D × 1D Array Examples

Consider the 1D × 1D × 1D array shown in figure 3.2e. Write three examples of possible array types that fall in that category (one example is already included below; complete the list with *type2* and *type3*).

```
TYPE matrix IS ARRAY (1 TO 4) OF BIT_VECTOR(7 DOWNTO 0);
TYPE type1 IS ARRAY (NATURAL RANGE <>) OF matrix;
```

Exercise 3.18: Resetting Arrays

Consider the data arrays of figure 3.2. Write a section of VHDL code that zeros (fills with '0's):

a) The array of figure 3.2c;

b) The array of figure 3.2d;

c) The array of figure 3.2e.

Exercise 3.19: Type Conversion by Type Casting

Write a line of VHDL code that makes a conversion using type casting for each type conversion listed below. In each case, indicate the package(s) (if any) that must be declared in the code for the conversion to be valid.

a) From STD_LOGIC_VECTOR to UNSIGNED.

b) From SIGNED to STD_LOGIC_VECTOR.

Exercise 3.20: Type Conversion by Specific Functions

Write a line of VHDL code that converts the types below using type-conversion functions. In each case, write as many options as possible, always indicating the package of origin of the function being used.

a) From INTEGER to STD_LOGIC_VECTOR;

b) From BIT_VECTOR to STD_LOGIC_VECTOR;

c) From STD_LOGIC_VECTOR to UNSIGNED;

d) From STD_LOGIC_VECTOR to INTEGER;

e) From SIGNED to STD_LOGIC_VECTOR.

Exercise 3.21: Overloaded Operator and Qualified Expression

Below is a code for a signed adder, whose ports are all of type STD_LOGIC_VECTOR (industry standard). Analyze it and answer the questions below. (Suggestion: see example 3.9.)

a) This code uses the packages *std_logic_1164* (line 3) and *std_logic_arith* (line 4). Are both necessary? Why?

b) Would the package *numeric_std* be useful here? Under which conditions could it be used?

c) There are type conversions in some lines of the code below. Which lines are they? What kinds of conversions are they (automatic, type casting, or with a type-conversion function)?

d) Do you expect any problems in the type conversion of line 16? Why?

e) Present a solution to remedy the problem in (d) by including additional computing steps in the code.

f) Present another solution for the problem in (d), this time by replacing one of the declared packages.

g) Present one more solution for that problem, now using a "qualified" expression.

h) Finally, in your opinion, what is the recommended code for such a signed adder?

```
1   -----------------------------------------------
2   LIBRARY ieee;
3   USE ieee.std_logic_1164.all;
4   USE ieee.std_logic_arith.all;
5   -----------------------------------------------
6   ENTITY signed_adder IS
```

```
 7        PORT (a, b: IN STD_LOGIC_VECTOR(3 DOWNTO 0);
 8               sum: OUT STD_LOGIC_VECTOR(3 DOWNTO 0));
 9    END ENTITY;
10    -----------------------------------------------
11    ARCHITECTURE arch OF signed_adder IS
12       SIGNAL a_sig, b_sig: SIGNED(3 DOWNTO 0);
13    BEGIN
14       a_sig <= SIGNED(a);
15       b_sig <= SIGNED(b);
16       sum <= STD_LOGIC_VECTOR(a_sig + b_sig);
17    END ARCHITECTURE;
18    -----------------------------------------------
```

Exercise 3.22: Array Slices

Say that figures 3.2b to 3.2d represent the signals *s1(3:0)*, *s2(1:3)(3:0)*, and *s3(1:3, 3:0)*, respectively. Write the values corresponding to each slice below (the first one was already answered).

a) s1(3 DOWNTO 1): "010"

b) s1(2):

c) s1:

d) s2(3)(1 DOWNTO 0):

e) s2(1):

f) s3(2, 2):

The signals and variables below are for exercises 3.23 to 3.27.

```
-----------------------------------------
SIGNAL s1: BIT;
SIGNAL s2: BIT_VECTOR(7 DOWNTO 0);
SIGNAL s3: STD_LOGIC;
SIGNAL s4: STD_LOGIC_VECTOR(7 DOWNTO 0);
SIGNAL s5: INTEGER RANGE -35 TO 35;
VARIABLE v1: BIT_VECTOR(7 DOWNTO 0);
VARIABLE v2: INTEGER RANGE -35 TO 35;
-----------------------------------------
```

Exercise 3.23: Array Dimensionality

What is the dimensionality (scalar, 1D, 1D × 1D, 2D, etc.) of each signal and variable above based on the *number of bits* (figure 3.2)?

Exercise 3.24: Legal Assignments #1

Confirm that the statements below are legal (see main causes of mistakes in section 3.19).

a) `s2(7) <= s1;`

b) `s3 <= s4(0);`

c) `s2 <= v1 XOR "10001000";`

d) `s5 <= v2/2;`

Exercise 3.25: Legal Assignments #2

Confirm that the statements below are legal (see main causes of mistakes in section 3.19).

a) `s3 <= 'Z';`

b) `s2 <= (OTHERS=>'0');`

c) `v2 := 35;`

d) `v1 := "11110000";`

Exercise 3.26: Illegal Assignments #1

Explain why the assignments below are illegal (see main causes of mistakes in section 3.19).

a) `s1(0) <= s2(0);`

b) `s3 <= s1 OR s2(2);`

c) `s2 <= (8=>'0', OTHERS=>'Z');`

d) `v2 <= -35;`

Exercise 3.27: Illegal Assignments #2

Explain why the assignments below are illegal (see main causes of mistakes in section 3.19).

a) `s3 := 'Z';`

b) `s2(7 DOWNTO 5) <= v1(3 DONWTO 0) OR "1000";`

c) `v1(7) <= s1 AND s2(0);`

d) `s4(0) <= s2(0);`

The signals and their types below are for exercises 3.28 to 3.30.

```
s1 -> TYPE type1 IS ARRAY (7 DOWNTO 0) OF BOOLEAN;
s2 -> TYPE type2 IS ARRAY (7 DOWNTO 0) OF BIT;
s3 -> TYPE type3 IS ARRAY (1 TO 4) OF INTEGER RANGE -128 TO 127;
```

```
s4 -> TYPE type4 IS ARRAY (NATURAL RANGE <>) OF BIT_VECTOR(7 DOWNTO 0);
s5 -> TYPE type5 IS ARRAY (NATURAL RANGE <>, NATURAL RANGE <>) OF STD_LOGIC;
s6 -> TYPE type6 IS ARRAY (1 TO 4) OF type4;
s7 -> TYPE type7 IS ARRAY (1 TO 4, 1 TO 4, 1 TO 4) OF STD_LOGIC;
```

Exercise 3.28: Array Dimensionality #2

a) What is the dimensionality of each type above (in term of the number of bits)?

b) To which case in figure 3.2 each type corresponds?

Exercise 3.29: Legal versus Illegal Array Slices

For each slice below, respond: to which case in figure 3.7 does it correspond? (Note that not all are represented in figure 3.7, so make a sketch for the missing ones.) Why are the first ten supported while the last five might not be?

a) s1(1)

b) s2(6 DOWNTO 1)

c) s3(4)

d) s4(0)

e) s4(1)(5 DOWNTO 3)

f) s4(2)(1)

g) s5(0,0)

h) s6(1)(0)

i) s6(2)(0)(7 DOWNTO 5)

j) s6(3)(0)(7)

k) s4(0 TO 1)(7)

l) s5(1 TO 3, 3)

m) s5(1 TO 2, 1 TO 2)

n) s5(0, 2 TO 3)

o) s5 (0 TO 1, 0 TO 1, 0 TO 1)

Exercise 3.30: Illegal Assignments #3

Explain why the assignments below are illegal (see main causes of mistakes in section 3.19).

a) s4(0)(0) <= s7(1,1,1);

b) s6(1) <= s4(1);

c) s1 <= "00000000";

d) `s7(0)(0)(0) <= 'Z';`

e) `s2(7 DOWNTO 5) <= s1(2 DOWNTO 0);`

f) `s4(1) <= (OTHERS => 'Z');`

g) `s6(1,1) <= s2;`

h) `s2 <= s3(1) AND s4(1);`

i) `s1(0 TO 1) <= (FALSE, FALSE);`

j) `s3(1) <= (3, 35, -8, 97);`

4 Operators and Attributes

4.1 Introduction

The purpose of this chapter, along with the preceding chapters, is to lay the foundations of VHDL, so in the next chapter we can start dealing with actual circuit designs. This sequence is very important because it is impossible—or unproductive, at least—to write code efficiently without knowing *data types*, *operators*, and *attributes* well.

The study of operators includes:

- All six predefined categories; namely, *logical, arithmetic, comparison, shift, concatenation,* and *matching* operators.
- User-defined and overloaded operators.

The study of attributes includes:

- All four predefined categories; namely, attributes of *scalar types*, of *array types*, of *signals*, and of *named entities*.
- User-defined and synthesis attributes.

Additionally, GROUP and ALIAS declarations are described.

4.2 Predefined Operators

VHDL provides several kinds of predefined operators:

- Assignment operators
- Logical operators
- Arithmetic operators
- Comparison (relational) operators
- Shift operators
- Concatenation operator

• Matching comparison operators

• Other operators

Each of these categories is described below.

Assignment Operators

Assignment operators are used to assign values to VHDL objects (CONSTANT, SIG-NAL, and VARIABLE). They are:

• Operator "<=": Used to assign a value to a SIGNAL.

• Operator ":=": Used to assign a value to a VARIABLE or CONSTANT. Used also for establishing default (initial) values for signals and variables. Because a GENERIC declaration (section 2.6) is also a constant, this operator is used there too.

• Operator "=>": Used to assign values to array elements, either individually or with the keyword OTHERS.

Example Three object declarations (x, y, z) are shown below, followed by several assignments. Comments follow each assignment.

```
--------------------------------------------------------
CONSTANT x: STD_LOGIC_VECTOR(7 DOWNTO 0) := "00010001";
SIGNAL y: STD_LOGIC_VECTOR(1 TO 4);
VARIABLE z: BIT_VECTOR(3 DOWNTO 0);
y(4) <= '1'; --'1' assigned to a signal using "<="
y <= "0000"; --"0000" assigned to a signal with "<="
y <= (OTHERS=>'0') --'0' assigned to all elements of y
y <= x(3 DOWNTO 0); --part of x assigned to y
z := "1000"; --"1000" assigned to a variable with ":="
z := (0=>'1', OTHERS=>'0'); --z="0001"
--------------------------------------------------------
```

Logical Operators

Logical operators are used to perform logical operations. They are:

• NOT

• AND

• NAND

• OR

• NOR

• XOR

• XNOR

The only operator with precedence over the others is NOT. The synthesizable pre-defined data types that support logical operators are BIT, BIT_VECTOR, BOOLEAN, STD_(U)LOGIC, and STD_(U)LOGIC_VECTOR. As seen in section 3.7, (UN)SIGNED can also be included in this list if the chosen definition package is *numeric_std*.

In VHDL 2008, the following new types with support for logical operations were included: BOOLEAN_VECTOR, UFIXED, SFIXED, and FLOAT.

Examples

```
-----------------------------
x <= NOT a AND b; --x=a'.b
y <= NOT (a AND b); --x=(a.b)'
z <= a NAND b; --x=(a.b)'
-----------------------------
```

Arithmetic Operators

The arithmetic operators are:

- Addition (+)
- Subtraction (−)
- Multiplication (*)
- Division (/)
- Exponentiation (**)
- Absolute value (ABS)
- Remainder (REM)
- Modulo (MOD)

As defined in the original packages (see chapter 3 or appendices), the synthesizable predefined data types that support these functions are INTEGER, NATURAL, and POSITIVE. If one of the packages for (un)signed types (*numeric_std* or *std_logic_arith*) is declared in the code, then (UN)SIGNED can also be used. If the package *std_logic_unsigned*, *std_logic_signed*, or *numeric_std_unsigned* is also declared, then STD_LOGIC_VECTOR can be employed as well.

In VHDL 2008, the following new types with support for arithmetic operations were included: UFIXED, SFIXED, and FLOAT.

There are no synthesis restrictions regarding addition, subtraction, multiplication, or division with integers. For exponentiation, expressions with static exponent are supported; if the exponent is nonstatic, then the base might be required to be static or even a power of 2 (shift operation). The other three operators (ABS, REM, MOD) are also synthesizable without restrictions for integers. The least obvious operators are explained below.

- x/y: Returns 0 when $|x| < |y|$, ± 1 when $|y| \leq |x| < 2|y|$, ± 2 when $2|y| \leq |x| < 3|y|$, etc., with the sign obviously negative when the signs of x and y are different.

Examples $3/5 = 0$, $-3/5 = 0$, $9/5 = 1$, $-9/5 = -1$, $10/5 = 2$, $-10/5 = -2$, $14/5 = 2$, $-14/5 = -2$.

- ABS x: Returns the absolute value of x.

Examples   ABS $5 = 5$, ABS $-3 = 3$.

- x REM y: Returns the remainder of x/y, with the sign of x. Its equation is x REM $y = x - (x/y)^* y$, where both operands are integers.

Examples   6 REM $3 = 0$, 7 REM $3 = 1$, 7 REM $-3 = 1$, -7 REM $3 = -1$, -7 REM $-3 = -1$.

- x MOD y: Returns the remainder of x/y, with the sign of y. Its equation is x MOD $y = x$ REM $y + a^* y$, where $a = 1$ when the signs of x and y are different or $a = 0$ otherwise. Both operands are integers.

Examples   7 MOD $3 = 1$, 7 MOD $-3 = -2$, -7 MOD $3 = 2$, -7 MOD $-3 = -1$.

Comparison Operators

Also called *relational* operators, the comparison operators are:

- Equal to ($=$)
- Not equal to ($/=$)
- Less than ($<$)
- Greater than ($>$)
- Less than or equal to ($<=$)
- Greater than or equal to ($>=$)

As defined in the original packages (see chapter 3 or appendices), the synthesizable predefined data types that support comparison operators are BIT, BIT_VECTOR, BOOLEAN, INTEGER, NATURAL, POSITIVE, CHARACTER, and STRING. If one of the packages for (un)signed types (*numeric_std* or *std_logic_arith*) is declared in the code, then (UN)SIGNED can also be used. If the package *std_logic_unsigned*, *std_logic_signed*, or *numeric_std_unsigned* is also declared, then STD_LOGIC_VECTOR can be employed as well.

In VHDL 2008, the following new types with support for comparison operations were included: BOOLEAN_VECTOR, INTEGER_VECTOR, UFIXED, SFIXED, and FLOAT.

Shift Operators

Introduced in VHDL93, shift operators are used for shifting data vectors. They are:

- Shift left logic (SLL): Positions on the right are filled with '0's.
- Shift right logic (SRL): Positions on the left are filled with '0's.
- Shift left arithmetic (SLA): Rightmost bit is replicated on the right.
- Shift right arithmetic (SRA): Leftmost bit is replicated on the left.
- Rotate left (ROL): Circular shift to the left.
- Rotate right (ROR): Circular shift to the right.

As defined in the original packages (see chapter 3 or appendices), the only synthesizable data type that supports shift operators is BIT_VECTOR. If one of the packages for (un)signed types (*numeric_std* or *std_logic_arith*) is declared in the code, then (UN)SIGNED can also be used (though the latter package contains very few of such operators—see appendix K). If the package *std_logic_unsigned*, *std_logic_signed*, or *numeric_std_unsigned* is also declared, then STD_LOGIC_VECTOR can be employed as well (also a very reduced set).

In VHDL 2008, the following new types with support for shift operations were included: BOOLEAN_VECTOR, UFIXED, and SFIXED.

The syntax for shift operators is ⟨left_operand⟩ ⟨shift_operation⟩ ⟨right_operand⟩. The left operand must be of one of the types mentioned above, while the right operand is always an INTEGER (+ or − in front of it is allowed). However, a recommended (more universal) approach for shifting data is with the *concatenation* operator (included in the example below).

Examples Say that x is a BIT_VECTOR signal with value x = "01001". Then the values produced by the assignments below are those indicated in the comments (equivalent expressions, using the *concatenation* operator, are shown between parentheses).

```
------------------------------------------------------------------
y <= x SLL 2;   --y<="00100" (y <= x(2 DOWNTO 0) & "00";)
y <= x SLA 2;   --y<="00111" (y <= x(2 DOWNTO 0) & x(0) & x(0);)
y <= x SRL 3;   --y<="00001" (y <= "000" & x(4 DOWNTO 3);)
y <= x SRA 3;   --y<="00001" (y <= x(4) & x(4) & x(4) & x(4 DOWNTO 3);)
y <= x ROL 2;   --y<="00101" (y <= x(2 DOWNTO 0) & x(4 DOWNTO 3);)
y <= x SRL -2;  --same as "x SLL 2"
------------------------------------------------------------------
```

Concatenation Operator

Used for grouping objects and values (useful also for shifting data, as shown in the example above), the concatenation operator's representation is &.

The synthesizable predefined data types for which the concatenation operator is intended are BIT_VECTOR, BOOLEAN_VECTOR (VHDL 2008), INTEGER_VECTOR (VHDL 2008), STD_(U)LOGIC_VECTOR, (UN)SIGNED, and STRING. Recall that the keyword OTHERS (seen in section 3.2) can also be helpful to make array assignments.

Example Four VHDL objects (v, x, y, z) are declared below, then several assignments are made utilizing the concatenation operator (&). The use of parentheses is optional.

```
--------------------------------------------------------------------
CONSTANT v: BIT :='1';
CONSTANT x: STD_LOGIC :='Z';
SIGNAL y: BIT_VECTOR(1 TO 4);
SIGNAL z: STD_LOGIC_VECTOR(7 DOWNTO 0);
y <= (v & "000");          --result: "1000"
y <= v & "000";            --same as above (parentheses are optional)
z <= (x & x & "11111" & x); --result: "ZZ11111Z"
z <= ('0' & "011111" & x);  --result: "0011111Z"
--------------------------------------------------------------------
```

The use of OTHERS and comma to make individual-bit assignments and concatenation is illustrated next.

Example Consider the same constants and signals above. Below is a series of individual-bit assignments using the keyword OTHERS and comma instead of the regular concatenation operator. Observe the nominal and positional mapping options. Here, parentheses are required.

```
--------------------------------------------------------------------
y <= (OTHERS=>'0');              --result: "0000"
y <= (4=>'1', OTHERS=>'0');      --result: "0001" (nominal mapping)
y <= ('1', OTHERS=>'0');         --result: "1000" (positional mapping)
y <= (4=>'1', 2=>v, OTHERS=>'0'); --result: "0101" (nominal mapping)
z <= (OTHERS=>'Z');              --result: "ZZZZZZZZ"
z <= (4=>'1', OTHERS=>'0');      --result: "00010000" (nominal mapping)
z <= (4=>x, OTHERS=>'0');        --result: "000Z0000" (nominal mapping)
z <= ('1', OTHERS=>'0');         --result: "10000000" (posit. mapping)
--------------------------------------------------------------------
```

Matching Comparison Operators

- Matching equality operator (?=)
- Matching inequality operator (?/=)
- Matching less than operator (?<)
- Matching greater than operator (?>)

- Matching less than or equal to operator (?<=)
- Matching greater than or equal to operator (?>=)

These operators were introduced in VHDL 2008. They include the types BIT (whole set), BIT_VECTOR (only equality and inequality), STD_(U)LOGIC (whole set—see part II of the package *std_logic_1164* in appendix I), STD_(U)LOGIC_VECTOR (whole set—see the new package *numeric_std_unsigned* in appendix N), and (UN)SIGNED (whole set—see part II of the package *numeric_std* in appendix J). They also include the new types UFIXED and SFIXED.

The purpose of this operator is to allow the comparison of logic values instead of enumerated symbols in STD_ULOGIC based data. For example, "IF 'H' = '1'..." returns FALSE because these symbols are different, while "IF 'H'?= '1'..." returns '1' because both 'H' and '1' are interpreted as logic value '1'. A similar reasoning is valid for 'L' and '0'. When 'X', 'Z', or 'W' are involved in the comparison, this operator (?=) returns 'X', and so on. In the case of BIT, it simply returns '1' or '0' instead of TRUE or FALSE.

Other Operators

Other operators introduced in VHDL 2008 are:

- MINIMUM and MAXIMUM operators: Return the smallest or largest value in the given set. For example, "MAXIMUM(0, 55, 23)" returns 55. These operators were defined for all VHDL types.
- Condition operator ("??"): Converts a BIT or STD_(U)LOGIC value into a BOOLEAN value. For example, "?? *a* AND *b*" returns TRUE when *a* AND *b* = '1' or FALSE otherwise.
- TO_STRING: Converts a value of type BIT, BIT_VECTOR, STD_LOGIC_VECTOR, and so on into STRING. For the types BIT_VECTOR and STD_LOGIC_VECTOR, there are also the options TO_OSTRING and TO_HSTRING, which produce an octal or hexadecimal string, respectively. This operator is useful, for example, when reporting synthesizer or simulator information, because the ASSERT statement can only report data of type STRING (this will be studied in section 9.2).

Examples

TO_STRING(58) = "58"
TO_STRING(B"1110000) = "111100000"
TO_HSTRING(B"11110000) = "F0"

Operators Summary

Figure 4.1 summarizes the predefined operators and their respective synthesizable predefined data types.

Operator type	Predefined operators	Supported synthesizable predefined data types (*)
Logical	NOT, AND, NAND, OR, NOR, XOR, XNOR	BIT, BIT_VECTOR, BOOLEAN, BOOLEAN_VECTOR[1], STD_(U)LOGIC, STD_LOGIC_(U)VECTOR, (UN)SIGNED[2], UFIXED[1], SFIXED[1], FLOAT[1]
Arithmetic	+, −, *, /, **, ABS, REM, MOD	INTEGER, NATURAL, POSITIVE, STD_(U)LOGIC_VECTOR[3], (UN)SIGNED[4], UFIXED[1], SFIXED[1], FLOAT[1]
Comparison	=, /=, >, <, >=, <=	BIT, BIT_VECTOR, BOOLEAN, BOOLEAN_VECTOR[1], INTEGER, NATURAL, POSITIVE, INTEGER_VECTOR[1], CHARACTER, STRING, STD_(U)LOGIC_VECTOR[3], (UN)SIGNED[4], UFIXED[1], SFIXED[1], FLOAT[1]
Shift	SLL, SRL, SLA, SRA, ROL, ROR	BIT_VECTOR, BOOLEAN_VECTOR[1], STD_LOGIC_(U)VECTOR[3], (UN)SIGNED[4], UFIXED[1], SFIXED[1]
Concatenation	& (",″ and OTHERS too)	BIT_VECTOR, BOOLEAN_VECTOR[1], INTEGER_VECTOR[1], STRING, STD_(U)LOGIC_VECTOR, (UN)SIGNED[4]
Matching comparison [1]	?=, ?/=, ?>, ?<, ?>=, ?<=	BIT, BIT_VECTOR[1], BOOLEAN_VECTOR[1], STD_(U)LOGIC, STD_(U)LOGIC_VECTOR, (UN)SIGNED[2], UFIXED[1], SFIXED[1], FLOAT[1]
Condition [1]	??	BIT, STD_(U)LOGIC
Min/Max and String conversion [1]	MINIMUM, MAXIMUM, TO_STRING, etc.	Nearly all VHDL types in standard packages (see appendices)

(*) Note: Some types support only a partial set of operators   (3) Requires package std_logic_(un)signed or numeric_std_unsigned
(1) Introduced or proposed in VHDL 2008   (4) Requires package numeric_std or std_logic_arith
(2) With package numeric_std

Figure 4.1
Predefined operators and corresponding synthesizable predefined data types (the gray area contains the operators introduced in VHDL 2008).

4.3 Overloaded and User-Defined Operators

Overloaded operators were already described in section 3.16. The present section is just to reinforce the idea that an operator can also be (further) overloaded *by the user*.

Let us consider again the addition operator ("+"), for which numerous options already exist in the VHDL packages. However, the addition of a single BIT to an INTEGER is not included among them. Hence we can further overload that operator by adding such an option, which can be accomplished with the FUNCTION below (operators are indeed just names of functions or procedures; specific details on how to construct them will be seen in chapter 9):

```
-----------------------------------------------
FUNCTION "+" (a: INTEGER, b: BIT) RETURN INTEGER IS
BEGIN
    IF (b='1') THEN RETURN a+1;
    ELSE RETURN a;
    END IF;
END "+";
-----------------------------------------------
```

An expression using this new version of "+" is shown below.

```
--------------------------------
SIGNAL a, sum: INTEGER RANGE ...;
SIGNAL b: BIT;
sum <= a + b;
--------------------------------
```

Recall also from section 3.16 that ambiguous situations involving overloaded operators can be resolved with *qualified expressions* (see examples in that section).

4.4 Predefined Attributes

Predefined attributes retrieve information about named entities (not to be confused with ENTITY). The *IEEE 1076-2008 Standard VHDL Language Reference Manual* defines the following four categories of predefined attributes:

- predefined attributes of scalar types (numeric, enumerated, physical)
- predefined attributes of array types
- predefined attributes of signals
- predefined attributes of named entities.

The complete lists in all categories are presented below. It is important to mention that, of all these attributes, 'EVENT (read as "tick" event) is by far the most frequently used. Another note is that some of these predefined attributes might not be supported by synthesis tools.

Predefined Attributes of Scalar Types

This attribute provides information regarding a *scalar type* (numeric, enumerated, or physical type; recall that only the first two are synthesizable; see also the classification according to the number of values in section 3.4). The scalar type is represented by T in figure 4.2.

Example Consider the following two scalar types:

```
--------------------------------
TYPE my_integer IS RANGE 0 TO 255;
TYPE state IS (a, b, c);
--------------------------------
```

The values of several predefined attributes for the *integer* type *my_integer* are shown next. Observe in the comments (and check in figure 4.2) the types required for the output signals (x, y).

Name	Result TYPE	Result
T'LEFT	Same as T	Leftmost value of T
T'RIGHT	Same as T	Rightmost value of T
T'LOW	Same as T	Lower bound of T
T'HIGH	Same as T	Upper bound of T
T'ASCENDING	BOOLEAN	TRUE if range of T is ascending, FALSE otherwise
T'IMAGE(X)	STRING	String representing the value X in T
T'VALUE(X)	Base type of T	Value of T whose string representation is X
T'POS(X)	INTEGER	Position number of the value X in T
T'VAL(X)	Base type of T	Value whose position number in T is X
T'SUCC(X)	Base type of T	Value whose position number in T is X + 1
T'PRED(X)	Base type of T	Value whose position number in T is X − 1
T'LEFTOF(X)	Base type of T	Value on the left of the position number X in T
T'RIGHTOF(X)	Base type of T	Value on the right of the position number X in T
T'BASE	Any (sub)type	Base type of T
O'SUBTYPE	Any subtype	Constrained subtype of O with constraint information

Ascending range: T'LEFT = T'LOW, T'RIGHT = T'HIGH
Descending range: T'LEFT = T'HIGH, T'RIGHT = T'LOW

Figure 4.2
Attributes of scalar types.

```
--------------------------------------------------------------------
x1 <= my_integer'LEFT;      --result=0 (type of x1 must be my_integer)
x2 <= my_integer'RIGHT;     --result=255 (type of x2 must be my_integer)
x3 <= my_integer'LOW;       --result=0 (type of x3 must be my_integer)
x4 <= my_integer'HIGH;      --result=255 (type of x4 must be my_integer)
y <= my_integer'ASCENDING;  --result=TRUE (type of y must be BOOLEAN)
--------------------------------------------------------------------
```

The values of several predefined attributes for the *enumerated* type *state* are also shown below. Observe in the comments (and check in the table above) the types required for the output signals (x, y, z). It was considered that the encoding is *sequential*; if *one-hot* had been chosen instead, the encoding would be $a =$ "001", $b =$ "010", $c =$ "100".

```
--------------------------------------------------------------------
x1 <= state'LEFT;     --result=a (="00") (type of x1 must be state)
x2 <= state'RIGHT;    --result=c (="10") (type of x2 must be state)
x3 <= state'LOW;      --result=a (="00") (type of x3 must be state)
x4 <= state'HIGH;     --result=c (="10") (type of x4 must be state)
y <= state'POS(b);    --result=1 (="01") (type of y is INTEGER)
z <= state'VAL(1);    --result=b (="01") (type of z must be state)
--------------------------------------------------------------------
```

Name	Result TYPE	Result
A'LEFT [(N)]	Type of the Nth index range of A	Left bound of the Nth index range of A
A'RIGHT [(N)]	Same as above	Right bound of the Nth index range of A
A'LOW [(N)]	Same as above	Lower bound of the Nth index range of A
A'HIGH [(N)]	Same as above	Upper bound of the Nth index range of A
A'RANGE [(N)]	Same as above	Range of the Nth index range of A
A'REVERSE_RANGE [(N)]	Same as above	Reverse range of the Nth index range of A
A'LENGTH [(N)]	INTEGER	Number of values in the Nth index range
A'ASCENDING [(N)]	BOOLEAN	TRUE if Nth index range of A is ascending, FALSE otherwise
A'ELEMENT	Element subtype of A	Element subtype of A with constraint information

Figure 4.3
Attributes of array types.

Example 4.1: Using Predefined Scalar Attributes

The code below contains an enumerated type called *color*. Several assignments to *y* are made in lines 12–16. Determine the values of *x* that produce *y* = '1'. (The solutions are included as comments.)

```
1   ------------------------------------------------------------------
2   ENTITY example IS
3     PORT (x: IN INTEGER RANGE 0 TO 3;
4           y1, y2, y3, y4, y5: OUT BIT);
5   END example;
6   ------------------------------------------------------------------
7   ARCHITECTURE example OF example IS
8     TYPE color IS (red, green, blue); --assume seq. encoding
9     SIGNAL z: color;
10  BEGIN
11    z <= red WHEN x=0 ELSE green WHEN x=1 ELSE blue;
12    y1<='1' WHEN color'VAL(x)=blue ELSE '0';      --y1='1' for x=2
13    y2<='1' WHEN color'POS(blue)=x ELSE '0';      --y2='1' for x=2
14    y3<='1' WHEN color'RIGHTOF(z)=blue ELSE '0';  --y3='1' for x=1
15    y4<='1' WHEN color'PRED(z)=green ELSE '0';    --y4='1' for x=2,3
16    y5<='1' WHEN color'PRED(green)=z ELSE '0';    --y5='1' for x=0
17  END example;
18  ------------------------------------------------------------------
```

Predefined Attributes of Array Types

In figure 4.3, *A* represents an *array type*.

Example Consider the array type *matrix* below (note that the keyword ARRAY is employed in its definition) and the signal *test* declared subsequently.

```
------------------------------------------------
TYPE matrix IS ARRAY (1 TO 4, 7 DOWNTO 0) OF BIT;
SIGNAL test: matrix;
------------------------------------------------
```

The values returned for several of the predefined attributes are shown below (the same information would be returned for *test*).

```
-------------------------------------------------------------------------
x1 <= matrix'LEFT(1);       --result=1 (type of x1 must be INTEGER or eq.)
x2 <= matrix'LEFT(2);       --result=7 (type of x2 must be INTEGER or eq.)
x3 <= matrix'RIGHT(1);      --result=4 (type of x3 must be INTEGER or eq.)
x4 <= matrix'RIGHT(2);      --result=0 (type of x4 must be INTEGER or eq.)
x5 <= matrix'LENGTH(2);     --result=8 (type of x5 must be INTEGER or eq.)
x6 <= matrix'ASCENDING(1);  --result=TRUE (type of x6 must be BOOLEAN)
matrix'RANGE(1) -> returns 1 TO 4
matrix'REVERSE_RANGE(2) -> returns 0 TO 7
matrix'ELEMENT -> returns BIT
-------------------------------------------------------------------------
```

Example Consider the signal *x* declared below. All five LOOP statements that follow are synthesizable and equivalent.

```
-----------------------------------------
SIGNAL x: STD_LOGIC_VECTOR(7 DOWNTO 0);
FOR i IN 7 DOWNTO 0 LOOP ...
FOR i IN x'RANGE LOOP ...
FOR i IN x'HIGH DOWNTO x'LOW LOOP ...
FOR i IN x'LEFT DOWNTO x'RIGHT LOOP ...
FOR i IN x'LENGTH-1 DOWNTO 0 LOOP ...
-----------------------------------------
```

Predefined Attributes of Signals

In figure 4.4, *S* represents a *signal*.

The difference between an event and a transaction is that an event is a signal edge (that is, an upward or downward transition), which might cause another signal to change. If that is the case, then a transaction is scheduled for that signal, which will occur some time later (at the end of the present process cycle).

As shown in figure 4.4, the time interval (*t*), which appears in the first three attributes, is optional. Its default value is zero. S'STABLE(0 ns) (= S'STABLE) is then equivalent

Name	Result TYPE	Result
S'DELAYED [(t)]	Base type of S	Signal equivalent to signal S delayed t units of time
S'STABLE [(t)]	BOOLEAN	TRUE when no event has occurred on signal S for t units of time, FALSE otherwise
S'QUIET [(t)]	BOOLEAN	TRUE when no transaction has been scheduled for signal S for t units of time, FALSE otherwise
S'TRANSACTION	BIT	A bit that toggles in each simulation cycle in which S becomes active (transaction scheduled)
S'EVENT	BOOLEAN	TRUE if an event has just occurred on S, FALSE otherwise
S'ACTIVE	BOOLEAN	TRUE if a transaction has just been scheduled for S, FALSE otherwise
S'LAST_EVENT	TIME	Amount of time since last event occurred on signal S
S'LAST_ACTIVE	TIME	Amount of time since last time signal S was active (transaction scheduled)
S'LAST_VALUE	Base type of S	Value of signal S previous to the last event (if no event occurred, returns current value of S)
S'DRIVING	BOOLEAN	TRUE if process is driving S, FALSE if S is disconnected from the driver.
S'DRIVING_VALUE	Base type of S	Value of the driver for S in the current process

Figure 4.4
Attributes of signals.

to S'EVENT. The use of these attributes and also of S'LAST_VALUE is illustrated in example 4.2.

Example 4.2: DFF with Several Event-Based Attributes

The code below produces a D-type flip-flop (DFF), triggered at the positive edge of the clock. To detect a positive clock edge, three equivalent synthesizable alternatives are shown in lines 13–15.

```
1    -----------------------------------------------
2    ENTITY flipflop IS
3       PORT (d, clk, rst: IN BIT;
4            q: OUT BIT);
5    END flipflop;
6    -----------------------------------------------
7    ARCHITECTURE example OF flipflop IS
8    BEGIN
9       PROCESS(clk, rst)
10      BEGIN
11         IF (rst='1') THEN
12            q <= '0';
```

Name	Result TYPE	Result
E'SIMPLE_NAME	STRING	String representing the simple name, character literal, or operator symbol of the named entity E
E'INSTANCE_NAME	STRING	String describing the hierarchical path from the top entity or architecture down to the named entity E, including the names of instantiated design entities
E'PATH_NAME	STRING	Same as above, but excluding the names of instantiated design entities

Figure 4.5
String-related attributes.

```
13        ELSIF (clk'EVENT AND clk='1') THEN
14        --ELSIF (NOT clk'STABLE AND clk='1') THEN
15        --ELSIF (clk'EVENT AND clk'LAST_VALUE='0') THEN
16           q <= d;
17        END IF;
18     END PROCESS;
19  END example;
20  ---------------------------------------------------
```

Predefined Attributes of Named Entities

The last three predefined attributes are used to produce a string related to the name of the declared entity (the entity is represented by *E* in figure 4.5).

4.5   User-Defined Attributes

Besides the predefined attributes (seen in the previous section), VHDL also allows users to create their own attributes. Such attributes are used to decorate named entities (not to be confused with ENTITY) with additional information/values.

To create a user-defined attribute, a *declaration* and a *specification* are needed, as shown in the simplified syntaxes below.

Attribute declaration:

```
ATTRIBUTE attribute_name: attribute_type;
```

Attribute specification:

```
ATTRIBUTE attr_name OF entity_tag [signature]: entity_class IS value;
```

The following appears in the syntaxes:

- *attribute_name* (same as *attribute identifier*): Name chosen for the attribute.
- *attribute_ type:* Can be any VHDL type, either predefined or user-defined.
- *entity_tag:* Name of the entity to be decorated with the attribute (can be a simple name, a character literal, or an operator symbol). Can be replaced with OTHERS or ALL. An optional signature can also be included (explained ahead).
- *entity_class:* Essentially any VHDL entity. Namely, ENTITY, ARCHITECTURE, CONFIGURATION, PACKAGE, COMPONENT, SIGNAL, VARIABLE, CONSTANT, TYPE, SUBTYPE, FUNCTION, PROCEDURE, GROUP, LABEL, LITERAL, PROPERTY, SEQUENCE, UNITS, and FILE.
- *value* (same as *expression*): Value chosen for the attribute.

Example In the first line of the code below, a SIGNAL named *number_of_pins* of type POSITIVE is declared. In the second line, an ATTRIBUTE, called *pins*, whose type is also POSITIVE, is declared. In the third line of code, this attribute, with a value of 4 (thus *pins* = 4), is associated to the entity *nand3*, which is a COMPONENT. Finally, in the fourth line, the attribute *pins* of *nand3* is passed to the signal specified in the first line, hence resulting *number_of_pins* = 4.

```
-------------------------------------------------------------------
SIGNAL number_of_pins: POSITIVE;          --signal declaration
ATTRIBUTE pins: POSITIVE;                 --attribute declaration
ATTRIBUTE pins OF nand3: COMPONENT IS 4;  --attribute specification
number_of_pins <= nand3'pins;             --attribute call (tick needed)
-------------------------------------------------------------------
```

As seen in the syntax, the specification of the entity name might include a *signature*, whose purpose is to identify among entities that have the same name (overloaded) which one is the actual target. This situation can arise when using overloaded subprograms (FUNCTION and PROCEDURE, chapter 9).

Example Say that our entity is a PROCEDURE, whose name is *sort*, which assigns the smaller and larger of *a* and *b* to *x* and *y*, respectively. However, say that two versions of *sort* were written, one for ports of type INTEGER, the other for ports of type BIT_VECTOR, that is:

```
PROCEDURE sort (a, b: IN INTEGER; x, y: OUT INTEGER) IS ...
PROCEDURE sort (a, b: IN BIT_VECTOR; x, y: OUT BIT_VECTOR) IS ...
```

In this case, to clarify matters, a signature can be used, which consists of including in the specifications the data types involved in the procedure, as shown below.

```
--------------------------------------------------------------------
ATTRIBUTE sort_attribute: STRING;
ATTRIBUTE sort_attribute OF sort [INTEGER, INTEGER, INTEGER,
   INTEGER]: PROCEDURE IS "sort_int";
ATTRIBUTE sort_attribute OF sort [BIT_VECTOR, BIT_VECTOR, BIT_VECTOR,
   BIT_VECTOR]: PROCEDURE IS "sort_bv";
--------------------------------------------------------------------
```

4.6 Synthesis Attributes

We have seen predefined and user-defined VHDL attributes, which are design-related. There are also synthesis-related attributes (normally provided by EDA vendors) whose purpose is to communicate with the compiler. The following five are described in this section:

- *enum_encoding* attribute
- *chip_pin* attribute
- *keep* attribute
- *preserve* attribute
- *noprune* attribute.

Attribute *enum_encoding*

Enumerated data types are indispensable when designing finite state machines (FSMs). (In Chapter 11, which deals exclusively with FSMs, details will be given on how to encode such data types.) For that purpose, a very helpful attribute exists, called *enum_encoding*, which allows the user to choose basically any encoding style (the only restriction is that the encoding values must employ only STD_ULOGIC symbols) for the machine states. The main options are (see details in section 11.4):

- Sequential encoding
- Gray encoding
- Johnson encoding
- One-hot encoding
- User-defined encoding.

As with any other attribute, *enum_encoding* must contain a declaration and a specification, as depicted in the example.

Example Say that the enumerated type *state* below has been created to represent an FSM's states. The results from several encoding options for such a type would then be (the last one is an user-defined encoding):

```
---------------------------------------------------------
TYPE state IS (A, B, C, D);
---------------------------------------------------------
ATTRIBUTE enum_encoding: STRING;
ATTRIBUTE enum_encoding OF state: TYPE IS "sequential";
--Result: A="00", B="01", C="10", D="11"
---------------------------------------------------------
ATTRIBUTE enum_encoding: STRING;
ATTRIBUTE enum_encoding OF state: TYPE IS "one-hot";
--Result: A="0001", B="0010", C="0100", D="1000"
---------------------------------------------------------
ATTRIBUTE enum_encoding: STRING;
ATTRIBUTE enum_encoding OF state: TYPE IS "11 00 10 01";
--Result: A="11", B="00", C="10", D="01"
---------------------------------------------------------
```

An attribute equivalent to *enum_encoding* is *fsm_state*, also used to specify the encoding scheme for the states of a FSM. Its encoding options of greatest interest are BINARY, GRAY, and ONE_HOT.

Attribute *chip_pin*

This attribute allows the user to assign device pins to the signals listed in the PORT of the design ENTITY.

Example   The section of code below causes the signal *clk* to be connected to pin N2 of the chosen device.

```
ATTRIBUTE chip_pin: STRING;
ATTRIBUTE chip_pin OF clk: SIGNAL IS "N2";
```

Example 4.3: Specifying Device Pins with the *chip_pin* Attribute

Figure 4.6 depicts a 4-bit register, with inputs *data_in* and *clk*, and output *data_out*. Design this circuit assuming that all signals must be assigned to specific, preselected pins (shown in figure 4.6—make sure to pick pins that are available as *user* pins in your target

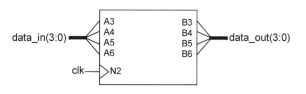

Figure 4.6
4-bit register of example 4.3.

device). This can be done in two ways: one is to select the desired pins in the compiler settings; the other is to include such assignments in the VHDL code itself. Solve this problem adopting the latter approach.

Solution A VHDL code for this circuit is shown below. The *chip_pin* attribute was placed in the declarative part of the ARCHITECTURE (lines 9–12), but could have been installed in the declarative part of the ENTITY (between lines 5 and 6) as well.

```
1   -----------------------------------------------------------------
2   ENTITY data_register IS
3      PORT (clk: IN BIT;
4            data_in: IN BIT_VECTOR(3 DOWNTO 0);
5            data_out: OUT BIT_VECTOR(3 DOWNTO 0));
6   END ENTITY;
7   -----------------------------------------------------------------
8   ARCHITECTURE data_register OF data_register IS
9      ATTRIBUTE chip_pin: STRING;
10     ATTRIBUTE chip_pin OF clk: SIGNAL IS "N2";
11     ATTRIBUTE chip_pin OF data_in: SIGNAL IS "A3, A4, A5, A6";
12     ATTRIBUTE chip_pin OF data_out: SIGNAL IS "B3, B4, B5, B6";
13  BEGIN
14     PROCESS (clk)
15     BEGIN
16        IF (clk'EVENT AND clk='1') THEN
17           data_out <= data_in;
18        END IF;
19     END PROCESS;
20  END ARCHITECTURE;
21  -----------------------------------------------------------------
```

Attribute *keep*

The purpose of the *keep* (also called *syn_keep*) attribute is to tell the compiler not to simplify (suppress) the listed nodes (in other words, it prevents certain combinational logic simplifications). Its specification must include the names of the signals (wires) that must be preserved.

Example 4.4: Construction of a Delay Line with the *keep* Attribute

Write a VHDL code from which the 4-inverter delay line of figure 4.7(a) can be inferred. Compile the code with and without the *keep* attribute and compare the results.

Solution A VHDL code for this problem is presented below. It makes use of the *keep* attribute (lines 9–10) to prevent nodes *a*, *b*, and *c* from being simplified. Figure 4.7(b) shows

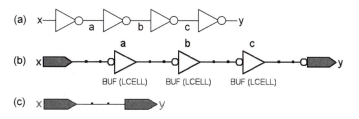

Figure 4.7
(a) Delay line of example 4.4; RTL viewer image (b) with and (c) without the *keep* attribute.

the image produced by the RTL viewer after compiling this code (note that there are four inverters, with the last one in the output buffer). The image in figure 4.7(c) is also from the RTL viewer, but for the case when lines 9–10 are commented out. If the input (x) and output (y) nodes were also included in the *keep* specification (line 10), two buffers would be inferred for them (one after the x pad, the other before the y pad).

```
1   ----------------------------------------------
2   ENTITY delay_line IS
3      PORT (x: IN BIT;
4            y: OUT BIT);
5   END ENTITY;
6   ----------------------------------------------
7   ARCHITECTURE example OF delay_line IS
8      SIGNAL a, b, c: BIT;
9      ATTRIBUTE keep: BOOLEAN;
10     ATTRIBUTE keep OF a, b, c: SIGNAL IS TRUE;
11  BEGIN
12     a <= NOT x;
13     b <= NOT a;
14     c <= NOT b;
15     y <= NOT c;
16  END ARCHITECTURE;
17  ----------------------------------------------
```

Attribute *preserve*

Two somewhat similar synthesis attributes that might be useful occasionally are *preserve* and *noprune*, described in this section and in the next.

The *preserve* attribute is the "registered" counterpart of *keep*; that is, it is used to prevent the removal of registers (flip-flops) instead of combinational logic. Example 4.5 illustrates its use.

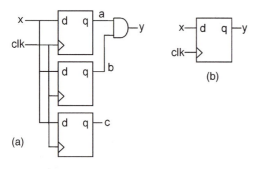

Figure 4.8
(a) Circuit with redundant registers of example 4.5; (b) Equivalent simplified circuit.

Attribute *noprune*

The difference between this and the *preserve* attribute is that *noprune* also preserves registers that do not feed the top-level ENTITY. The use of *preserve* and *noprune* is illustrated in the example below.

Example 4.5: Keeping Redundant Registers with *preserve* and *noprune* Attributes

Figure 4.8(a) shows a circuit with redundant registers. First, observe that the flip-flops for *a* and *b* can be reduced to just one flip-flop, as shown in figure 4.8(b). Second, note that the other DFF (for *c*) does not feed any output wire of the design ENTITY, so it can be simply removed. Write a VHDL code that implements the circuit of figure 4.8(a) (that is, that keeps all unnecessary registers).

Solution A VHDL code for this problem is shown below. Note the presence of the attributes described above in lines 9–12. If *preserve* is used (that is, lines 9–10 are included and lines 11–12 are commented out), flip-flops *a* and *b* will be preserved, but flip-flop *c* will not because it does not feed the design entity. On the other hand, if *noprune* is used (lines 11–12 included and lines 9–10 commented out), then even DFF *c* is preserved. The reader is invited to compile this code, with and without these attributes, and check the fitter equations to confirm these results.

```
1   -----------------------------------------------------
2   ENTITY redundant_registers IS
3      PORT (clk, x: IN BIT;
4             y: OUT BIT);
5   END ENTITY;
6   -----------------------------------------------------
7   ARCHITECTURE arch OF redundant_registers IS
8      SIGNAL a, b, c: BIT;
```

```
9       --ATTRIBUTE preserve: BOOLEAN;
10      --ATTRIBUTE preserve OF a, b, c: SIGNAL IS TRUE;
11      ATTRIBUTE noprune: BOOLEAN;
12      ATTRIBUTE noprune OF a, b, c: SIGNAL IS TRUE;
13  BEGIN
14      PROCESS (clk)
15      BEGIN
16          IF (clk'EVENT AND clk='1') THEN
17              a <= x;
18              b <= x;
19              c <= x;
20          END IF;
21      END PROCESS;
22      y <= a AND b;
23  END ARCHITECTURE;
24  ------------------------------------------------------
```

4.7   GROUP

The user-defined attributes seen in section 4.5 apply to individual named entities. To apply
an attribute to multiple entities, VHDL provides the GROUP construct. To use it, two
declarations are needed: a *template* declaration and a *group* declaration, shown in the simplified syntaxes below.

Group template declaration:

```
GROUP template_name IS (entity_class [<>], ...);
```

Group declaration:

```
GROUP group_name: template_name (constituent_name, ...);
```

The following appears in the syntaxes:

- *template_name* (same as *template identifier*): Name chosen for the template.

- *entity_class:* Essentially any VHDL entity (see list in the ATTRIBUTE syntax, section
4.5). The use of a "box" ($<>$) indicates an arbitrary number (zero or more) of constituents
from the same entity class. For example, "SIGNAL $<>$" means any number of signals in
that position in the list.

▪ *group_name* (same as *group identifier*)*:* Name chosen for the group.

▪ *constituent_name:* Name chosen for each constituent in the constituents list (can be a simple name or a character literal).

Example   Below, a template called *test_signals* is created, which contains three signals. Next, a group called *register* is declared, which employs three signals called *clock*, *reset*, and *enable*. Finally, the third line illustrates the use of such a group in an ATTRIBUTE declaration.

```
-----------------------------------------------------
GROUP test_signals IS (SIGNAL, SIGNAL, SIGNAL);
GROUP register: test_signals (clock, reset, enable);
ATTRIBUTE ... OF register: GROUP IS ...
-----------------------------------------------------
```

4.8   ALIAS

An ALIAS declaration defines an alternate name for an existing named entity (not to be confused with ENTITY). A simplified syntax is shown below.

```
ALIAS new_name [: specifications] IS original_name [signature];
```

ALIAS can be applied to nearly all named entities, except for labels, loop parameters, and generate parameters. It is divided into two groups: *object alias* (for VHDL objects; that is, CONSTANT, SIGNAL, VARIABLE, and FILE) and *non-object alias* (for all remaining entities; that is, TYPE, COMPONENT, operators, user subprograms, etc.). One difference between these two groups is that a signature cannot be used in the former, while in the latter it is required if the entity is a subprogram.

The most common places for ALIAS declarations are the declarative parts of architectures and of subprograms, and the most commonly aliased entities are subprograms, SIGNAL, and TYPE.

The use of ALIAS can be seen, for example, in the package body of the package *std_logic_1164* that accompanies the VHDL compiler.

Example   This example illustrates how an ALIAS declaration can be applied to a VHDL object (a SIGNAL, in this example) to do any of the following: (i) change its name; (ii) create a name for a section of it; (iii) change its range; (iv) create a new range for a section of it.

```
----------------------------------------------------------------------
--This is the object to which ALIAS will be applied:
SIGNAL data_bus: STD_LOGIC_VECTOR(31 DOWNTO 0);
----------------------------------------------------------------------
--bus1 is a new name for data_bus:
ALIAS bus1 IS data_bus;
----------------------------------------------------------------------
--bus2 is a new name for data_bus, but with a modified range:
ALIAS bus2: STD_LOGIC_VECTOR(32 DOWNTO 1) IS data_bus;
----------------------------------------------------------------------
--bus3 is another name for data_bus, with an ascending range:
ALIAS bus3: STD_LOGIC_VECTOR(1 TO 32) IS data_bus;
----------------------------------------------------------------------
--upper_bus1 is a new name for the upper half of data_bus
ALIAS upper_bus1 IS data_bus(31 DOWNTO 16);
----------------------------------------------------------------------
--upper_bus2 is a new name for the upper half of data_bus, but
--with a modified range:
ALIAS upper_bus2: STD_LOGIC_VECTOR(17 TO 32) IS data_bus(31 DOWNTO 16);
----------------------------------------------------------------------
--lower_bus1 is a new name for the lower half of data_bus
ALIAS lower_bus1 IS data_bus(15 DOWNTO 0);
----------------------------------------------------------------------
--lower_bus2 is a new name for the lower half of data_bus, but
--with a modified range:
ALIAS lower_bus2: STD_LOGIC_VECTOR(1 TO 16) IS data_bus(15 DOWNTO 0);
----------------------------------------------------------------------
```

As indicated in the syntax of ALIAS, the specification of the original name allows the inclusion of a *signature*. Its purpose is the same as that seen in the study of attributes (section 4.5); that is, to make possible the identification among overloaded items (with the same name). Signatures can be used in ALIAS declarations for subprograms (that is, FUNCTION and PROCEDURE).

Example This example illustrates how an ALIAS declaration can be applied to a VHDL subprogram (a PROCEDURE, in this case), which also includes the use of a signature. Let us consider again the *sort* procedure seen at the end of section 4.5, which assigns the smaller and larger of *a* and *b* to *x* and *y*, respectively. However, assume that two versions of *sort* were written, one for ports of type INTEGER, the other for ports of type BIT_VECTOR. That is:

```
PROCEDURE sort (a, b: IN INTEGER; x, y: OUT INTEGER) IS ...
PROCEDURE sort (a, b: IN BIT_VECTOR; x, y: OUT BIT_VECTOR) IS ...
```

The ALIAS declarations below define alternative names for *sort* (*sort_int* for the former, *sort_bv* for the latter), thus allowing the procedures to be easily identified.

```
ALIAS sort_int IS sort [INTEGER, INTEGER, INTEGER, INTEGER];
ALIAS sort_bv IS sort [BIT_VECTOR, BIT_VECTOR, BIT_VECTOR, BIT_VECTOR];
```

4.9 VHDL 2008

With respect to the material covered in this chapter, the main additions specified in VHDL 2008 are those listed below.

1) Regarding the logical operators:
For the types STD_(U)LOGIC and STD_(U)LOGIC_VECTOR, additional options were included in the package *std_logic_1164* (part II of appendix I). For the types (UN)SIGNED, additional options were included in the package *numeric_std* (part II of appendix J). Unary operations were also included. Logical operators were also defined for the new types UFIXED, SFIXED, and FLOAT.

2) Regarding the arithmetic operators:
For the types STD_(U)LOGIC and STD_(U)LOGIC_VECTOR, arithmetic operators were defined in the new package *numeric_std_unsigned* (appendix N). For the types (UN)SIGNED, additional options were included in the package *numeric_std* (part II of appendix J). Arithmetic operators were also specified for the new types UFIXED, SFIXED, and FLOAT.

3) Regarding the comparison operators:
Comparison operators were defined also for the new types BOOLEAN_VECTOR, INTEGER_VECTOR, UFIXED, SFIXED, and FLOAT.

4) Regarding the shift operators:
For the types STD_(U)LOGIC_VECTOR, some shift operators were included in the expansion of the package *std_logic_1164* (part II of appendix I). Other shift operators for STD_(U)LOGIC_VECTOR were introduced in the new package *numeric_std_unsinged* (appendix N). For the types (UN)SIGNED, additional shift operators were included in the expansion of the package *numeric_std* (part II of appendix J). Shift operators were also defined for the new types BOOLEAN_VECTOR, UFIXED, and SFIXED.

5) Regarding the matching comparison operators:
These operators (?=, ?/=, ?<, ?>, ?<=, ?>=) were all introduced in VHDL 2008. They include the types BIT, BIT_VECTOR (partial set), BOOLEAN_VECTOR, STD_(U)LOGIC, STD_(U)LOGIC_VECTOR (whole set if proper package used), (UN)SIGNED, UFIXED, and SFIXED.

6) Others:

The functions MINIMUM, MAXIMUM, TO_STRING, TO_OSTRING, and TO_HSTRING were also introduced in VHDL 2008, with support for nearly all VHDL types. Several new attributes for scalars, signals, etc. were included as well.

4.10   Exercises

Note: For exercise solutions, please consult the book website.

Exercise 4.1: Logical Operators and Corresponding Types

This exercise concerns the line relative to *logical* operators in figure 4.1. Check in the synthesis packages (see list in section 3.3) if the supported predefined data types match the list in the last column of figure 4.1.

Exercise 4.2: Arithmetic Operators and Corresponding Types

This exercise concerns the line relative to *arithmetic* operators in figure 4.1. Check in the synthesis packages (see list in section 3.3) if the supported predefined data types match the list in the last column of figure 4.1.

Exercise 4.3: Comparison Operators and Corresponding Types

This exercise concerns the line relative to *comparison* operators in figure 4.1. Check in the synthesis packages (see list in section 3.3) if the supported predefined data types match the list in the last column of figure 4.1.

Exercise 4.4: Logical Operators

Say that $a(7:0) = $ "00110011" and $b(3:0) = $ "1111". Determine the values produced by the assignments below.

```
a) a(7 DOWNTO 4) NAND "0111"
b) a(7 DOWNTO 4) XOR NOT b
c) "1111" NOR b
d) b(2 DOWNTO 0) XNOR "101"
```

Exercise 4.5: Arithmetic Operators #1

For the integers $x = 65$ and $y = 7$, calculate:

a) `y**2`

b) `x/y`

c) `-x/y`

d) `(x/y)*y`

e) `(x*y)/y`

f) `(x+y)y`

g) `(x-y)/y`

h) `3*((x-y)/3)`

Exercise 4.6: Arithmetic Operators #2

For the integers $x = 65$ and $y = 7$, calculate:

a) `x REM y`

b) `x REM -y`

c) `(x+2*y) REM y`

d) `(x+y) REM -x`

e) `x MOD y`

f) `x MOD -y`

g) `-x MOD -y`

h) `ABS(-y)`

Exercise 4.7: Comparison Operators

Assuming that it is a *signed* system, and given the values $v = $ "0011", $x = $ "1100", $y = $ "01000000", and $z = $ "11111111", determine if each assignment below is TRUE or FALSE.

a) `v < x`

b) `v + x <= z`

c) `y = (ABS(x))**3`

d) `(7*v) REM x = (7*v) MOD x`

Exercise 4.8: Shift and Concatenation Operators

1) For $x = $ "110010", of type BIT_VECTOR(5 DOWNTO 0), determine the values of the *shift* operations listed in the column on the left below.

2) In the column on the right, write an equivalent expression using the *concatenation* operator (the first one was already done).

a) `x SLL 3`   a) `x(2 DOWNTO 0) & "000"`

b) `x SLA -2`   b)

c) `x SRA 2`   c)

d) `x ROL 1`   d)

e) `x ROR -3`   e)

Exercise 4.9: Arithmetic Operators for Signed Types

Say that *a* is of type INTEGER, while *b*, *c*, and *x* are of type SIGNED. Recall that when SIGNED is employed, either the *numeric_std* or the *std_logic_arith* package must be included in the library/package declarations. With which (if any) of these packages are the operations below valid? (Suggestion: Check the "+" and "*" operators in appendices J–K.)

a) `x <= a + b;`

b) `x <= b + c;`

c) `x <= 3*b;`

d) `x <= 3*a + b;`

e) `x <= a + c + "1111";`

f) `x <= a + c + SIGNED'("1111");`

Exercise 4.10: Attributes of an Array Type

Consider the following data type:

```
TYPE data IS STD_LOGIC_VECTOR(31 DOWNTO 0);
```

Determine the values returned by the attributes below.

a) `data'LEFT`

b) `data'RIGHT`

c) `data'LOW`

d) `data'HIGH`

e) `data'RANGE`

f) `data'REVERSE_RANGE`

g) `data'LENGTH`

h) `data'ASCENDING`

Exercise 4.11: Attributes of an Enumerated Type #1

Consider the following enumerated data type:

```
TYPE state IS (stA, stB, stC, stD, stE);
```

Determine the values returned by the attributes below.

a) `state'LEFT`

b) `state'RIGHT`

c) `state'POS(stD)`

d) `state'VAL(3)`

e) `state'LEFTOF(stD)`

f) `state'RIGHTOF(stA)`

g) `state'PRED(stB)`

h) `state'SUCC(stC)`

Exercise 4.12: Attributes of an Enumerated Type #2

Inspect the code below, in which a series of position-related attributes were employed.

a) Determine for which values of the input (x) the outputs ($y1$, $y2$, etc.) should be '1'.

b) Compile and simulate the code to check whether the results match your predictions.

```
1   ----------------------------------------------------
2   ENTITY example IS
3     PORT (x: IN INTEGER RANGE 0 TO 7;
4           y1, y2, y3, y4, y5, y6, y7: OUT BIT);
5   END ENTITY;
6   ----------------------------------------------------
7   ARCHITECTURE example OF example IS
8     TYPE tag IS (a, b, c, d, e, f); --coded 0,1,...,5
9     SIGNAL test: tag;
10  BEGIN
11    test <=  a WHEN x=0 ELSE
12             b WHEN x=1 ELSE
13             c WHEN x=2 ELSE
14             d WHEN x=3 ELSE
15             e WHEN x=4 ELSE
16             f;
17    y1 <= '1' WHEN tag'VAL(x)=c ELSE '0';
18    y2 <= '1' WHEN tag'POS(c)=x ELSE '0';
19    y3 <= '1' WHEN tag'RIGHTOF(temp)=c ELSE '0';
20    y4 <= '1' WHEN tag'LEFTOF(temp)=e ELSE '0';
21    y5 <= '1' WHEN tag'PRED(temp)=e ELSE '0';
22    y6 <= '1' WHEN tag'PRED(b)=temp ELSE '0';
23    y7 <= '1' WHEN tag'SUCC(a)=temp ELSE '0';
24  END ARCHITECTURE;
25  ----------------------------------------------------
```

Exercise 4.13: The *enum_encoding* Attribute

Consider the following enumerated data type:

```
TYPE fsm_state IS (a, b, c, d, e, f);
```

Determine the values assigned by the compiler to represent *fsm_state* in each of the following encoding options (the first one has already been answered—see details in section 11.4):

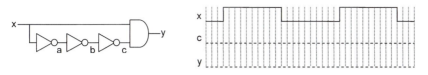

Figure 4.9

a) Sequential: a = "000", b = "001", c = "010", d = "011", e = "100", f = "101"

b) Gray

c) Johnson

d) One-hot

Exercise 4.14: The *chip_pin* Attribute

Using the *chip_pin* attribute, make the changes needed in the code of example 4.4 in order to have its input and output ports automatically assigned to user I/O pins (check your device's pin list to make a proper selection).

Exercise 4.15: The *keep* Attribute

Figure 4.9 shows a typical circuit used for pulse-shortening in pulse-based flip-flop implementations (Pedroni 2008).

a) Explain how this circuit works, then sketch the waveforms for c and y on the right of figure 4.9. Assume that the propagation delay in each inverter and in the AND gate is 1 ns, which is the distance between the vertical lines in the waveform plots.

b) Write a VHDL code from which this circuit can be inferred. Do not include the *keep* attribute yet. Compile the code and examine the fitter equations to verify what was indeed inferred.

c) Repeat part (b) above, now with *keep* included (for a, b, c). After compiling your code and examining the fitter equations, simulate it to observe the pulse-shortening effect.

Exercise 4.16: *preserve* versus *keep* Attributes

a) If in example 4.5 we remove the *preserve* and *noprune* attributes and include the *keep* attribute, what circuit do you expect will be inferred?

b) Make such a modification in the code and compile it. Then check the fitter equations to verify whether the result matches your prediction.

Exercise 4.17: Capturing Digits from a Decimal Number

Say that *abc* is a three-digit decimal number between 000 and 999. Using one of the predefined operators, show how each individual digit (that is, a, b, and c) can be separated from the others. (Hint: Think of REM or MOD.)

5 Concurrent Code

5.1 Introduction

Having finished laying out the foundations of VHDL (chapters 1 to 4), we can now concentrate on the design (code) proper.

A *combinational* logic circuit is one in which the outputs depend solely on the current inputs, therefore exhibiting no memory, as in the feed-forward model of figure 5.1(a). In contrast, a *sequential* logic circuit is one in which the outputs do depend on previous system state(s), so storage elements are needed, along with a clock signal to control the system evolution and possibly a reset too, as in the model of figure 5.1(b) (the storage elements are usually D-type flip-flops (DFFs)).

VHDL code can be concurrent (parallel) or sequential. Only statements placed inside a PROCESS, FUNCTION, or PROCEDURE (the last two are called *subprograms*) are executed sequentially. However, because VHDL is inherently concurrent, a PROCESS, as a whole, is also concurrent with respect to any other (external) statements. In other words, a process body or a subprogram call is also a concurrent statement.

Two other concurrent statements are the BLOCK statement and COMPONENT instantiations. These, however, can be viewed as just different ways of organizing the code, without any new internal statements (hence they will be studied in chapter 8, which is in the system-level part of the book).

Apart from the pieces of code mentioned above, there are three purely *concurrent* statements (they can only be used *outside* sequential code—that is, outside PROCESS or subprograms), which are WHEN, SELECT, and GENERATE. (*Note*: See in section 5.10 the new options for WHEN and SELECT specified in VHDL 2008.)

Following the same reasoning, there are four purely *sequential* statements (they can only be used *inside* sequential code), which are IF, WAIT, LOOP, and CASE.

While *concurrent* code is intended only for the design of *combinational* circuits, *sequential* code can be used indistinctly to design both sequential *and* combinational circuits.

Statements for concurrent code (WHEN, SELECT, GENERATE) are studied in this chapter, while statements for sequential code (IF, WAIT, LOOP, CASE) are described in the next.

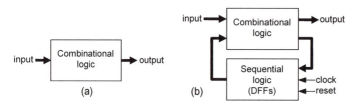

Figure 5.1
Models for (a) combinational and (b) sequential logic circuits.

Figure 5.2
4×1 multiplexer of example 5.1.

Remember that in a concurrent code the order of the statements does not matter. For example, if a code uses three concurrent statements, called *stat1*, *stat2*, and *stat3*, then any of the following sequences will render the same physical circuit: $\{stat1, stat2, stat3\} = \{stat3, stat2, stat1\} = \{stat1, stat3, stat2\}$.

5.2   Using Operators

Operators were discussed in section 4.2 and summarized in figure 4.1. Basically any kind of circuit can be designed using only operators. This approach, however, is viable only for arithmetic circuits or simple logic circuits. In example 5.1, a multiplexer is designed using only logical operators.

Example 5.1: Multiplexer Implemented with Operators

Implement the 4×1 (four inputs of one bit each) multiplexer of figure 5.2 using only logical operators.

Solution   This multiplexer's logical equation (Pedroni 2008) is $y = sel_1' \cdot sel_0' \cdot x_0 + sel_1' \cdot sel_0 \cdot x_1 + sel_1 \cdot sel_0' \cdot x_2 + sel_1 \cdot sel_0 \cdot x_3$, which employs only AND, OR, and NOT operators. Its implementation is in lines 13–16 of the code below.

```
1   ---------------------------------------------
2   LIBRARY ieee;
3   USE ieee.std_logic_1164.all;
```

```
 4   -------------------------------------------
 5   ENTITY mux IS
 6      PORT (x0, x1, x2, x3: IN STD_LOGIC;
 7              sel: IN STD_LOGIC_VECTOR(1 DOWNTO 0);
 8              y: OUT STD_LOGIC);
 9   END mux;
10   -------------------------------------------
11   ARCHITECTURE operators_only OF mux IS
12   BEGIN
13      y <= (NOT sel(1) AND NOT sel(0) AND x0) OR
14           (NOT sel(1) AND sel(0) AND x1) OR
15           (sel(1) AND NOT sel(0) AND x2) OR
16           (sel(1) AND sel(0) AND x3);
17   END operators_only;
19   -------------------------------------------
```

5.3 The WHEN Statement

WHEN is the simplest conditional statement. It is approximately equivalent to the sequential statement IF. A simplified syntax for WHEN is presented below.

```
assignment_expression WHEN conditions ELSE
    assignment_value WHEN conditions ELSE
    ...;
```

Examples

```
x <= '0' WHEN rst='0' ELSE
     '1' WHEN a='0' OR b='1' ELSE
     '-';   --don't care
y <= "00" WHEN (a AND b)="01" ELSE
     "11" WHEN (a AND b)="10" ELSE
     "ZZ";   --high impedance
```

Note that multiple conditions (boolean expressions) are accepted in the WHEN statement, which are grouped using AND, OR, and NOT.

The WHEN statement does not require that all input values be specified. However, when implementing combinational circuits (truth tables), it is a good practice to always cover all input options in order to prevent the inference of latches. For such, the keyword OTHERS is usually helpful.

Another sometimes useful keyword for concurrent code is UNAFFECTED, which should be used when no action is to take place. Note, however, that this usually causes the inference of latches, so this keyword should only be used when memorization of the previous system state is indeed wanted.

Example   Both codes below implement a positive-level D-type latch (Pedroni 2008). The code on the right, however, covers all possible input values, explicitly documenting the fact that a memory is indeed wanted and the inference of a latch was not by accident.

```
q <= '0' WHEN rst='1' ELSE        q <= '0' WHEN rst='1' ELSE
       d WHEN clk='1';                   d WHEN clk='1' ELSE
                                         UNAFFECTED;
```

In VHDL 2008, WHEN can also be used in sequential code and allows boolean tests. (See details in section 5.10.)

5.4   The SELECT Statement

SELECT is another concurrent statement. It is approximately equivalent to the sequential statement CASE. A simplified syntax for SELECT is presented below.

```
WITH identifier SELECT
   assignment_expression WHEN values,
       assignment_value WHEN values,
       ...;
```

Examples

```
WITH control SELECT              WITH (a AND b) SELECT
  y <= "000" WHEN 0 | 1,           y <= "00" WHEN "001",
       "100" WHEN 2 TO 5,               "11" WHEN "100",
       "Z--" WHEN OTHERS;               UNAFFECTED WHEN OTHERS;
```

As shown in the first example, SELECT allows the use of multiple values (instead of multiple conditions), which can only be grouped with "|" (means "or") or "TO" (for range), as follows:

```
WHEN value1 | value2 |...   --value1 or value2 or ...
WHEN value1 TO value2       --range
```

The SELECT statement requires that all input values be covered (complete truth table), for which the keyword OTHERS is often helpful.

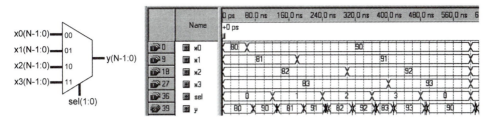

Figure 5.3
$4 \times N$ multiplexer of example 5.2 and respective simulation results.

UNAFFECTED is another sometimes useful keyword, already described in section 5.3. See, however, the observation about the inference of latches made in that section.

As a last remark, any signal assignment (like those in this chapter, for example) can be preceded by a label, which was omitted in the simplified syntaxes above because that is a rarely used practice.

In VHDL 2008, SELECT can also be used in sequential code and the matching SELECT? statement was introduced, which allows the use of don't care inputs. (See details in section 5.10.)

Example 5.2: Multiplexer Implemented with WHEN and SELECT

Implement the same multiplexer of example 5.1, but now with N-bit inputs instead of single bit, as shown in figure 5.3. Specify N using GENERIC (section 2.6). Present two solutions: with WHEN and with SELECT. Show also simulation results.

Solution A VHDL code (with two architectures) for this circuit is presented below, under the title *mux* (line 5). N is entered as a generic parameter (line 6), which is used in lines 7 and 9 to establish the size of the input-output buses. Only STD_LOGIC_VECTOR ports (industry standard) are employed in the code. In the first architecture (called *with_WHEN*), the WHEN statement is employed, while in the second architecture (called *with_SELECT*), the SELECT statement is used instead. Note that in both cases all possible input values are covered.

```
1   ------------------------------------------------------------
2   LIBRARY ieee;
3   USE ieee.std_logic_1164.all;
4   ------------------------------------------------------------
5   ENTITY mux IS
6      GENERIC (N: INTEGER := 8);
7      PORT (x0, x1, x2, x3: IN STD_LOGIC_VECTOR(N-1 DOWNTO 0);
8            sel: IN STD_LOGIC_VECTOR(1 DOWNTO 0);
9            y: OUT STD_LOGIC_VECTOR(N-1 DOWNTO 0));
```

```
10   END ENTITY;
11   ------------------------------------------------------------
12   ARCHITECTURE with_WHEN OF mux IS
13   BEGIN
14      y <= x0 WHEN sel="00" ELSE
15           x1 WHEN sel="01" ELSE
16           x2 WHEN sel="10" ELSE
17           x3;
19   END ARCHITECTURE;
20   ------------------------------------------------------------

12   ARCHITECTURE with_SELECT OF mux IS
13   BEGIN
14      WITH sel SELECT
15         y <= x0 WHEN "00",
16              x1 WHEN "01",
17              x2 WHEN "10",
19              x3 WHEN OTHERS;
20   END ARCHITECTURE;
21   ------------------------------------------------------------
```

Simulation results (for $N = 8$), confirming the correct circuit operation, are included in figure 5.3. Recall that the apparent glitches in the waveform for y are expected because multiple signals are considered at once, whose actual values neither change instantaneously nor change all exactly at the same time.

Only one entity-architecture pair can be synthesized at a time. Therefore, if we want to write more than one architecture for the same entity (as in the example above), we need either to comment all but one out (with "--") or include a CONFIGURATION declaration in the code in order to direct the compiler to the desired units. In the example above, both architectures can be included in the same code if followed by a declaration like that below (placed outside any entities or architectures).

```
----------------------------------
CONFIGURATION which_mux OF mux IS
   FOR with_WHEN
   END FOR;
END CONFIGURATION;
----------------------------------
```

This declaration causes the compiler to choose the pair *mux-with_WHEN*. After compiling and simulating this code, just change the name of the architecture in the configuration declaration above to test the other architecture. One (minor) disadvantage of this approach is that it does not automatically prevent the compiler from checking the syntax in the unselected unit. (Details about CONFIGURATION will be seen in chapter 8.)

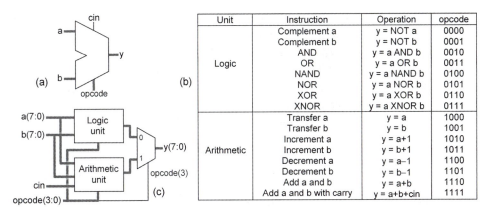

Figure 5.4
ALU of example 5.3. (a) ALU symbol; (b) truth table; (c) a possible implementation.

Example 5.3 also uses the concurrent statement SELECT.

Example 5.3: ALU

An ALU (Arithmetic Logic Unit) is shown in figure 5.4(a), having *a*, *b*, *cin* (carry in), and *opcode* (operation code) as inputs, and *y* as output. The desired functionality is expressed in the truth table of figure 5.4(b), where each function is selected by a different value of *opcode*. Note that the upper eight instructions are logical, while the lower eight are arithmetic. Design this circuit using the concurrent statement SELECT, satisfying the following conditions:

1) The arithmetic operations must be *signed*.
2) The number of bits for inputs *a* and *b* must be *generic*.
3) All ports must be of type STD_LOGIC(_VECTOR) (industry standard).
4) Simulation results must also be included in the solution.

Solution Figure 5.4(c) shows a possible ALU implementation (among several other options). The circuit contains two main sections, called *logic* and *arithmetic* units, each controlled by the same three LSBs of *opcode*. The MSB of *opcode* is employed to control a multiplexer, letting the *logic* result out when low or the arithmetic result out if high.

A VHDL code for this circuit is presented below, under the title *alu* (line 6). The number of bits in *a* and *b* is a generic parameter (line 7), and all ports (lines 8–11) are of type STD_LOGIC(_VECTOR). Because the arithmetic operations were asked to be *signed*, the package *numeric_std* (line 4) was included in the library/package declarations. The code proper is divided according to figure 5.4(c)—that is, a logic unit (lines 22–30), an arithmetic unit (lines 32–43), and a multiplexer (lines 45–47).

The implementation of the logic unit is straightforward. However, because the arithmetic unit must be *signed*, the same procedure used in the recommended solution of example 3.9 (see also recommendations in section 5.7) is adopted here; that is, the inputs are explicitly converted from STD_LOGIC_VECTOR to SIGNED (by type casting, in lines 32–33), they are then processed, and finally the result is converted back to STD_LOGIC_VECTOR (at the mux input, line 47, again by type casting).

Note also that because *cin* is STD_LOGIC, not SIGNED, NATURAL, or INTEGER, it could not participate directly in the sum of line 43 (observe in the package *numeric_std*, in appendix J, that the overloaded operator "+" does not contain the SIGNED+ STD_LOGIC option), so a small integer (lines 19 and 34) was created to allow the sum.

```
1    -------------------------------------------------
2    LIBRARY ieee;
3    USE ieee.std_logic_1164.all;
4    USE ieee.numeric_std.all;
5    -------------------------------------------------
6    ENTITY alu IS
7       GENERIC (N: INTEGER := 8); --word bits
8       PORT (a, b: IN STD_LOGIC_VECTOR(N-1 DOWNTO 0);
9             cin: IN STD_LOGIC;
10            opcode: IN STD_LOGIC_VECTOR(3 DOWNTO 0);
11            y: OUT STD_LOGIC_VECTOR(N-1 DOWNTO 0));
12   END ENTITY;
13   -------------------------------------------------
14   ARCHITECTURE alu OF alu IS
15      SIGNAL a_sig, b_sig: SIGNED(N-1 DOWNTO 0);
16      SIGNAL y_sig: SIGNED(N-1 DOWNTO 0);
17      SIGNAL y_unsig: STD_LOGIC_VECTOR(N-1 DOWNTO 0);
19      SIGNAL small_int: INTEGER RANGE 0 TO 1;
20   BEGIN
21      ------Logic unit:--------------
22      WITH opcode(2 DOWNTO 0) SELECT
23         y_unsig <= NOT a WHEN "000",
24                    NOT b WHEN "001",
25                    a AND b WHEN "010",
26                    a OR b WHEN "011",
27                    a NAND b WHEN "100",
28                    a NOR b WHEN "101",
29                    a XOR b WHEN "110",
30                    a XNOR b WHEN OTHERS;
31      ------Arithmetic unit:---------
32      a_sig <= SIGNED(a);
33      b_sig <= SIGNED(b);
34      small_int <= 1 WHEN cin='1' ELSE 0;
```

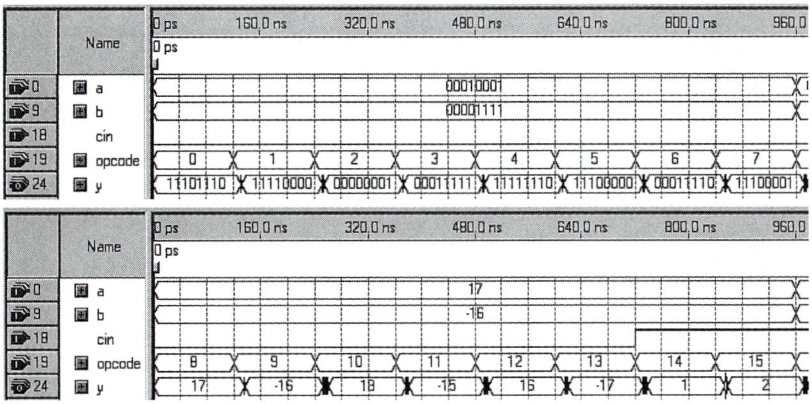

Figure 5.5
Simulation results from the ALU of example 5.3 (logic instruction in the upper graph, arithmetic instructions in the lower graph).

```
35      WITH opcode(2 DOWNTO 0) SELECT
36         y_sig <= a_sig WHEN "000",
37                  b_sig WHEN "001",
38                  a_sig + 1 WHEN "010",
39                  b_sig + 1 WHEN "011",
40                  a_sig - 1 WHEN "100",
41                  b_sig - 1 WHEN "101",
42                  a_sig + b_sig WHEN "110",
43                  a_sig + b_sig + small_int WHEN OTHERS;
44      ------Mux:--------------------
45      WITH opcode(3) SELECT
46         y <= y_unsig WHEN '0',
47              STD_LOGIC_VECTOR(y_sig) WHEN OTHERS;
48   END ARCHITECTURE;
49   --------------------------------------------------------
```

Simulation results are depicted in figure 5.5. The upper graph is for logic instructions, while the lower graph exhibits results from arithmetic operations. The reader is invited to examine both to check the correct circuit operation.

5.5   The GENERATE Statement

GENERATE is another *concurrent* statement. In its most popular form (*unconditional* GENERATE), it is equivalent to the *sequential* statement LOOP (chapter 6) in the sense that it too is employed to have a section of code repeated a number of times. GENERATE also allows the inclusion of an IF condition (*conditional* GENERATE), hence with some similarity to the combination of the *sequential* statements LOOP and IF. Both forms can be nested inside one another.

Unconditional GENERATE (also called FOR-GENERATE) is used to create multiple instances of a section of code. A simplified syntax for it is shown below. Notice that a label is required, and that the word BEGIN is only needed when declarations are made.

```
label: FOR identifier IN range GENERATE
  [declarative_part
BEGIN]
  concurrent_statements_part
END GENERATE [label];
```

Example   Below, three signals are declared, then three equivalent sections of code utilizing the GENERATE statement are presented. In all three the label is *gen*, the identifier is *i*, and the range is 0-to-7 or 7-downto-0.

```
----------------------------------------
SIGNAL a, b, x: BIT_VECTOR(7 DOWNTO 0);
----------------------------------------
gen: FOR i IN 0 TO 7 GENERATE
   x(i) <= a(i) XOR b(7-i);
END GENERATE;
----------------------------------------
gen: FOR i IN a'RANGE GENERATE
   x(i) <= a(i) XOR b(7-i);
END GENERATE;
----------------------------------------
gen: FOR i IN a'REVERSE_RANGE GENERATE
   x(i) <= a(i) XOR b(7-i);
END GENERATE;
----------------------------------------
```

Conditional GENERATE (also called IF-GENERATE) includes an IF statement in the GENERATE loop. A simplified syntax is shown below.

```
label: IF condition GENERATE
  [declarative_part
BEGIN]
  concurrent_statements_part
END GENERATE [label];
```

This version of GENERATE is of limited interest. To make it more interesting, additional features were specified in VHDL 2008, allowing the use of ELSIF/ELSE, plus the use of alternative labels and the END keyword before END GENERATE. Another option, called CASE-GENERATE, was also introduced in VHDL 2008 (see details in section 5.10).

Another important remark about GENERATE (and the same is true for LOOP, which will be studied in the next chapter) is that both range limits are normally required to be static. For example, say that in the section of code below x is an input (therefore, a non-static parameter). Then this code might not be synthesizable.

```
---------------------------------
NotOK: FOR i IN 0 TO x GENERATE
   ...
END GENERATE;
---------------------------------
```

It is also important to be aware of multiply-driven signals. As will be seen in chapters 6 and 7, multiple assignments to the same VARIABLE are fine because its value is updated immediately, but that is not allowed for SIGNAL. This problem is illustrated in the example below. In section 7.7, a way of circumventing such a limitation will be introduced.

Example The first of the three sections of code below is fine because values are assigned to each bit of x (a signal) only once. However, the second is not correct, because a value is assigned to y several (up to four) times. The same applies to the third section of code, where z might also receive up to four assignments.

```
-----------------------------------------------
SIGNAL a, b, x, y: BIT_VECTOR(3 DOWNTO 0);
SIGNAL z: INTEGER RANGE 0 TO 7;
-----------------------------------------------
OK: FOR i IN x'RANGE GENERATE
   x(i)<='1' WHEN (a(i) AND b(i))='1' ELSE '0';
END GENERATE;
-----------------------------------------------
NotOK: FOR i IN y'LOW TO y'HIGH GENERATE
   y <="1111" WHEN (a(i) AND b(i))='1' ELSE
       "0000";
END GENERATE;
-----------------------------------------------
NotOK: For i IN 0 TO 3 GENERATE
   z <= z + 1 WHEN a(i)='1';
END GENERATE;
-----------------------------------------------
```

We conclude this section by presenting two complete design examples using GENERATE.

Example 5.4: Generic Address Decoder with GENERATE

Redesign the generic address decoder of example 2.4, this time using only STD_LOGIC-based ports (industry standard).

Solution A VHDL code for this problem is presented below. All ports are STD_LOGIC-based (lines 8–10). The GENERATE statement is employed in lines 17–19, containing just one assignment (using WHEN, line 18). Note that *address* was converted into an integer in line 16 using a functions available in the package *std_logic_unsigned* (see figure 3.10). Lines 14 and 16 can obviously be suppressed if we choose to write "... WHEN i=conv_integer(address) ..." in line 18. The size of the code is fixed, regardless of the number of input bits. Simulation results are similar to those in figure 2.7.

```
1   ------------------------------------------------------------
2   LIBRARY ieee;
3   USE ieee.std_logic_1164.all;
4   USE ieee.std_logic_unsigned.all;
5   ------------------------------------------------------------
6   ENTITY address_decoder IS
7      GENERIC (N: NATURAL := 3); --number of address bits
8      PORT (address: IN STD_LOGIC_VECTOR (N-1 DOWNTO 0);
9            ena: IN STD_LOGIC;
10           word_line: OUT STD_LOGIC_VECTOR(2**N-1 DOWNTO 0));
11  END ENTITY;
12  ------------------------------------------------------------
13  ARCHITECTURE decoder OF address_decoder IS
14     SIGNAL addr: NATURAL RANGE 0 TO 2**N-1;
15  BEGIN
16     addr <= conv_integer(address);
17     gen: FOR i IN word_line'RANGE GENERATE
18        word_line(i)<='0' WHEN i=addr AND ena='1' ELSE '1';
19     END GENERATE;
20  END ARCHITECTURE;
21  ------------------------------------------------------------
```

A very useful application for GENERATE is in the instantiation of *components* to build larger, *structural* circuits. Even though COMPONENT will be studied in chapter 8, a preliminary example is presented below in order to illustrate the usage of GENERATE in this kind of design.

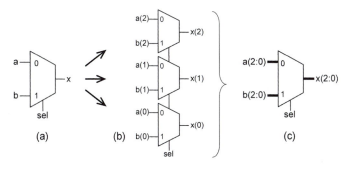

Figure 5.6
(a) A 2 × 1 mux that (b) instantiated three times (c) creates a 2 × 3 mux.

Example 5.5: COMPONENT Instantiation with GENERATE

Figure 5.6 illustrates the construction of a larger multiplexer using multiple instances of a basic unit. In (a), a 2 × 1 (two inputs of one bit each) mux is shown, which is instantiated three times in (b), resulting the 2 × 3 mux of (c). Design this circuit using a *structural* approach (that is, with COMPONENT instantiations), with GENERATE employed to make the instantiations.

Solution A VHDL code for this circuit is shown below, consisting of two parts. The first part builds the basic unit (*mux2x1*), which is then instantiated in the second part (main code) using the GENERATE statement (lines 18–20). (Details on COMPONENT construction and usage will be seen in chapter 8.)

```
1   -----The component (mux2x1):-----
2   LIBRARY ieee;
3   USE ieee.std_logic_1164.all;
4   --------------------------------
5   ENTITY mux2x1 IS
6      PORT (a, b, sel: IN STD_LOGIC;
7             x: OUT STD_LOGIC);
8   END ENTITY;
9   --------------------------------
10  ARCHITECTURE mux2x1 OF mux2x1 IS
11  BEGIN
12     x<=a WHEN sel='0' ELSE b;
13  END ARCHITECTURE;
14  --------------------------------
```

```
1   -------Main code:-------------------------------
2   LIBRARY ieee;
```

```
 3   USE ieee.std_logic_1164.all;
 4   --------------------------------------------------------
 5   ENTITY mux2x3 IS
 6      PORT (a, b: IN STD_LOGIC_VECTOR(2 DOWNTO 0);
 7             sel: IN STD_LOGIC;
 8             x: OUT STD_LOGIC_VECTOR(2 DOWNTO 0));
 9   END ENTITY;
10   --------------------------------------------------------
11   ARCHITECTURE mux2x3 OF mux2x3 IS
12      ---Component declaration:-----
13      COMPONENT mux2x1 IS
14         PORT (a, b, sel: IN STD_LOGIC; x: OUT STD_LOGIC);
15      END COMPONENT;
16   BEGIN
17      ---Component instantiation:---
18      generate_mux2x3: FOR i IN 0 TO 2 GENERATE
19         comp: mux2x1 PORT MAP (a(i), b(i), sel, x(i));
20      END GENERATE generate_mux2x3;
21   END ARCHITECTURE;
22   --------------------------------------------------------
```

5.6   Implementing Sequential Circuits with Concurrent Code

In principle, only combinational circuits should be implemented with concurrent code. We know, however, that using only NAND or NOR gates *any* digital circuit can be constructed. Since sequential logic circuits are digital circuits, and since NAND or NOR gates can be easily constructed with concurrent code, then sequential circuits can obviously also be constructed with pure concurrent code. However, this approach is only viable for simple circuits, because in general the code would be much longer, more complex to write and debug, and unnatural to follow. In conclusion, the use of concurrent code to design sequential circuits is in general *not* recommended.

Example 5.6: DFF Implemented with Concurrent Code

Figure 5.7 shows, on the left, a pair of multiplexers (combinational circuits), whose connections emulate a DFF (a sequential circuit), shown on the right. Write a VHDL code for this circuit and examine the expressions inferred by the compiler.

Solution   A corresponding VHDL code is presented below. The intermediate signal *p* is specified in line 8, and the multiplexers are implemented in lines 10–11 using the WHEN statement. Looking at the compilation report (fitter equations), the expected result is the inference of two serially connected latches, because a multiplexer with a feedback loop (as in figure 5.7) emulates a latch. In conclusion, the resulting circuit should present the cor-

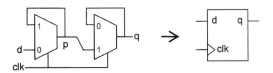

Figure 5.7
DFF implemented with multiplexers (example 5.7).

rect functionality, but because of its construction, it is slower than an actual, prefabricated DFF (the latter is optimized to operate specifically as a DFF).

```
1   --------------------------------------------
2   ENTITY concurrent_dff IS
3      PORT (d, clk: IN BIT;
4            q: BUFFER BIT);
5   END ENTITY;
6   --------------------------------------------
7   ARCHITECTURE concurrent OF concurrent_dff IS
8      SIGNAL p: BIT;
9   BEGIN
10     p <= d WHEN clk='0' ELSE p; --1st mux
11     q <= p WHEN clk='1' ELSE q; --2nd mux
12  END ARCHITECTURE;
13  --------------------------------------------
```

5.7   Implementing Arithmetic Circuits with Operators

Arithmetic circuits are normally constructed using only arithmetic operators (+, −, *, /, **, ABS, REM, and MOD, described in section 4.2). The first four in this list are by far the most frequently used, so special attention is dedicated to them in this section.

The main *interface* (PORT) types for arithmetic circuits are INTEGER (should be avoided) and STD_LOGIC_VECTOR (industry standard). On the other hand, *internally*, the preferred types are UNSIGNED and SIGNED (as seen in chapter 3, these types are defined in the packages *numeric_std* [preferred] and *std_logic_arith*).

To use such operators, it is necessary to know the *size* (number of bits) required for the result as a function of the operands' sizes. For example, below are definitions for four functions (+, −, *, /), copied from the package *numeric_std*, for the case of signed inputs and signed output (*L* and *R* represent the left and right operands).

```
------------------------------------------------------------
FUNCTION "+" (L, R: SIGNED) RETURN SIGNED;
--Result SUBTYPE: SIGNED(MAX(L'LENGTH, R'LENGTH)-1 DOWNTO 0)
```

```
--------------------------------------------------------------
FUNCTION "-" (L, R: SIGNED) RETURN SIGNED;
--Result SUBTYPE: SIGNED(MAX(L'LENGTH, R'LENGTH)-1 DOWNTO 0)
--------------------------------------------------------------
FUNCTION "*" (L, R: SIGNED) RETURN SIGNED;
--Result SUBTYPE: SIGNED((L'LENGTH+R'LENGTH-1) DOWNTO 0)
--------------------------------------------------------------
FUNCTION "/" (L, R: SIGNED) RETURN SIGNED;
--Result SUBTYPE: SIGNED(L'LENGTH-1 DOWNTO 0)
--------------------------------------------------------------
```

The conclusions from the definitions above are:

1) For "+" and "−": The size of the result must be equal to the size of the largest operand.

2) For "*": The size of the result must be equal to the sum of the operands' sizes.

3) For "/": The size of the result must be equal to the size of the numerator.

As already mentioned, a fundamental aspect of arithmetic circuits is their *nature*, which can be *unsigned* or *signed*. As seen in section 1.8, the range covered with N bits in the former is from 0 to $2^N - 1$, while in the latter it goes from -2^{N-1} to $2^{N-1} - 1$. For example, with 4 bits, the range is from 0 to 15 when unsigned or from -8 to $+7$ if signed.

Negative numbers are represented in two's complement form. Consequently, addition and subtraction are essentially the same function because the latter is just the former preceded by a two's complement operation. Some examples are shown below for a 4-bit *signed* system.

$5 + 2 = $ "0101" $+ $ "0010" $= $ "0111" $= 7$

$5 - 2 = 5 + (-2) = $ "0101" $+ $ "1110" $= $ "0011" $= 3$

$5 - (-2) = 5 + 2 = $ "0111" $= 7$

$-5 + 2 = $ "1011" $+ $ "0010" $= $ "1101" $= -3$

Multiplication and division with signed numbers involve again two's complement operations. A negative number must be two's complemented to attain its absolute value. Then the multiplication or division is performed, with the result two's complemented again if the result is to be negative (that is, if the signs of the operands are different). Some examples are shown below, again for a *signed* system with 4-bit inputs.

$5*3 = $ "0101" $ * $ "0011" $= $ "00001111" $= 15$

$-5*3 = -(5*3) = -($ "0101" $ * $ "0011" $) = -($ "00001111" $) = $ "11110001" $= -15$

$5/3 = $ "0101" $ / $ "0011" $= $ "0001" $= 1$

$-5/3 = -(5/3) = -($ "0101" $ / $ "0011" $) = -($ "0001" $) = $ "1111" $= -1$

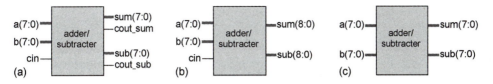

Figure 5.8
Adder/subtracter with the carry-out bits (a) separated from *sum* and *sub*, (b) grouped with *sum* and *sub*, and (c) inexistent. The first two are equivalent and not subject to overflow, whereas in the last one overflow can occur.

Because multiplication and division are not subject to overflow (due to the number of bits adopted in the respective functions), the corresponding VHDL code is straightforward. However, that is not the case for addition and subtraction, in which there are also carry-in/out bits to be taken care of.

Figure 5.8 shows three diagrams for an adder/subtracter. In (a), the operands are *a* and *b* (eight bits each), the carry-in bit is *cin*, the sum and its carry-out bit are *sum* and *cout_sum*, and the subtraction and its carry-out are *sub* and *cout_sub*. An equivalent representation appears in (b), in which the carry-out bits are combined with *sum* and *sub* (note that now they have 9 bits; the MSB is *cout*). Finally, in (c), there are no carry bits, so this circuit is subject to overflow.

Assume that it is an *unsigned* system. Then the sum in figure 5.8(a) can be computed as follows:

```
------------------------------------------------------
SIGNAL a_uns, b_uns, sum: UNSIGNED(7 DOWNTO 0);
SIGNAL sum_uns: UNSIGNED(9 DOWNTO 0);
SIGNAL cin, cout_sum: STD_LOGIC;
sum_uns <= ('0' & a_uns & cin) + ('0' & b_uns & '1');
sum <= sum_uns(8 DOWNTO 1);
cout_sum <= sum_uns(9);
------------------------------------------------------
```

Note in the code above that the operands for *sum_uns* have 10 bits, attained by appending *cin* or '1' on the right and '0' on the left. The LSB of *sum_uns* is discarded and its MSB is *cout_sum*.

Assuming now that the system is *signed*, the following can be done:

```
------------------------------------------------------------
SIGNAL a_sig, b_sig, sum: SIGNED(7 DOWNTO 0);
SIGNAL sum_sig: SIGNED(9 DOWNTO 0);
SIGNAL cin, cout_sum: STD_LOGIC;
sum_sig <= (a_sig(7) & a_sig & cin) + (b_sig(7) & b_sig & '1');
sum <= sum_sig(8 DOWNTO 1);
cout_sum <= sum_sig(9);
------------------------------------------------------------
```

Note that the only difference in this code with respect to the previous code is in the *sign-extension* bit, which must now be a copy of the leftmost bit (see binary arithmetic algorithms in Pedroni (2008)).

The problem with the above approaches is that the expressions do not work for subtraction, which is often necessary, especially in signed systems. Consequently, a more general approach is needed. A solution is shown below. Note in the expression for *sub_sig* that all three operands (including *cin*) were sign-extended (a '0' is used for *cin* because it is a non-negative number), so both additions and subtractions can now be performed. A complete design example will be shown shortly.

```
------------------------------------------------------------------
SIGNAL a_sig, b_sig, sum, sub: SIGNED(7 DOWNTO 0);
SIGNAL sum_sig, sub_sig: SIGNED(8 DOWNTO 0);
SIGNAL cin, cout_sum, cout_sub: STD_LOGIC;
sum_sig <= (a_sig(7) & a_sig) + (b_sig(7) & b_sig) + ('0' & cin);
sum <= sum_sig(7 DOWNTO 0);
cout_sum <= sum_sig(8);
sub_sig <= (a_sig(7) & a_sig) - (b_sig(7) & b_sig) + ('0' & cin);
sub <= sub_sig(7 DOWNTO 0);
cout_sub <= sub_sig(8);
------------------------------------------------------------------
```

Contrary to the $+$ and $-$ operators for integer arithmetic, fixed-point and floating-point arithmetic are guaranteed to be overflow free because the size of the output vector in both $+$ and $-$ operations is defined to be one unit larger than the largest input (see comments on arithmetic operators in section 3.8). The case of integer arithmetic is further covered in the recommendations and design example below.

Based on the above discussion, plus the note at the end of example 3.4 and the contents of example 3.9, the following is recommended:

1) For the interfaces (PORT specifications), use only STD_LOGIC(_VECTOR) (industry standard).

2) Internally, use only (UN)SIGNED.

3) For the type above, use the package *numeric_std* (standardized by IEEE).

4) Before performing any computation, explicitly convert the data from STD_LOGIC(_VECTOR) to (UN)SIGNED. This can be done with type casting (section 3.18).

5) Make the computations.

6) Finally, return the result to STD_LOGIC(_VECTOR). Type casting can again be used.

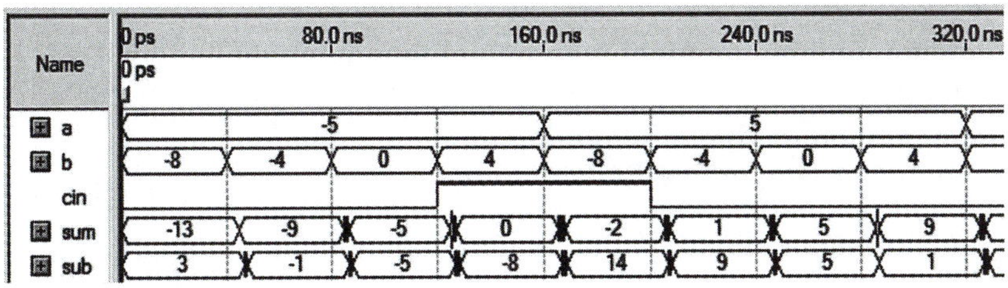

Figure 5.9
Simulation results from the adder (or subtracter) of example 5.7.

Example 5.7: Recommended Adder/Subtracter Implementation

Write a VHDL code that implements the adder/subtracter of figure 5.8(a) or 5.8(b) (physically, they are equal). In the design, follow the recommendations just presented. Assume that it is part of a *signed* system.

Solution A VHDL code for this circuit is presented below, under the title *signed_add_sub* (line 6). Because the circuit is signed, the package *numeric_std* was included in line 4. The number of bits in the operands is a *generic* parameter (line 7). The type of all ports is STD_LOGIC(_VECTOR) (lines 8–12). The inputs are *a*, *b*, and *cin*, while the outputs are *sum* and *sub*.

The code proper (lines 20–33) is organized in four parts. In the first part (lines 21–22), the operands are explicitly converted to SIGNED. In the second part (lines 24–25), they are added and subtracted, with carry-in included. In the third part (lines 27–28), the signals are converted back to STD_LOGIC_VECTOR, with the carry-out bits grouped with *sum* and *sub*, as in figure 5.8(b). The forth part (lines 30–33) is equivalent to the third, just with the carry-out bits separated from *sum* and *sub*, as in figure 5.8(a). To have the code resemble figure 5.8(a) instead of 5.8(b), just comment out lines 10, 27, and 28 and uncomment lines 11–12 and 30–33. Simulation results are shown in figure 5.9.

```
1    -------------------------------------------------------------------
2    LIBRARY ieee;
3    USE ieee.std_logic_1164.all;
4    USE ieee.numeric_std.all;
5    -------------------------------------------------------------------
6    ENTITY signed_add_sub IS
7       GENERIC (N: INTEGER := 4); --number of input bits
8       PORT (a, b: IN STD_LOGIC_VECTOR(N-1 DOWNTO 0);
9             cin: IN STD_LOGIC;
10            sum, sub: OUT STD_LOGIC_VECTOR(N DOWNTO 0));
11            --sum, sub: OUT STD_LOGIC_VECTOR(N-1 DOWNTO 0);
12            --cout_sum, cout_sub: OUT STD_LOGIC);
```

```
13   END ENTITY;
14   ------------------------------------------------------------------------
15   ARCHITECTURE signed_add_sub OF signed_add_sub IS
16      SIGNAL a_sig, b_sig: SIGNED(N-1 DOWNTO 0);
17      SIGNAL sum_sig, sub_sig: SIGNED(N DOWNTO 0);
19   BEGIN
20      -----convert to signed:--------------
21      a_sig <= signed(a);
22      b_sig <= signed(b);
23      -----add and subtract:--------------
24      sum_sig <= (a_sig(N-1) & a_sig) + (b_sig(N-1) & b_sig) + ('0' & cin);
25      sub_sig <= (a_sig(N-1) & a_sig) - (b_sig(N-1) & b_sig) + ('0' & cin);
26      -----output option #1:--------------
27      sum <= std_logic_vector(sum_sig);
28      sub <= std_logic_vector(sub_sig);
29      -----output option #2:--------------
30      --sum <= std_logic_vector(sum_sig(N-1 DOWNTO 0));
31      --cout_sum <= std_logic(sum_sig(N));
32      --sub <= std_logic_vector(sub_sig(N-1 DOWNTO 0));
33      --cout_sub <= std_logic(sub_sig(N));
34   END ARCHITECTURE;
35   ------------------------------------------------------------------------
```

5.8   Preventing Combinational-Logic Simplification

A series of synthesis attributes were described in section 4.6; namely:

- *enum_encoding* attribute
- *chip_pin* attribute
- *keep* attribute
- *preserve* attribute
- *noprune* attribute.

As shown there, the *keep* attribute can be used to tell the compiler not to simplify (suppress) specific nodes that would otherwise be removed during the optimization process. Its application was illustrated in example 4.4, in which a delay line was constructed. Another example is included (example 5.8), followed by another (less objective) approach that does not make use of the *keep* attribute.

Example 5.8: Short-Pulse Generator with the *keep* Attribute

Figure 5.10(a) shows a typical circuit used for pulse-shortening in pulse-based flip-flop implementations (Pedroni 2008). Note that the delay line contains three inverters, which

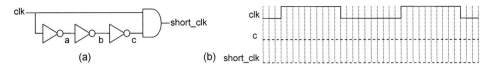

Figure 5.10
Short-pulse generator of example 5.8.

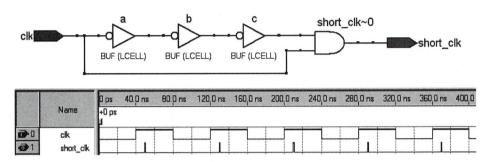

Figure 5.11
RTL view and simulation results from the short-pulse generator of example 5.8.

would be reduced to just one by the compiler if no measure were taken to prevent it. Design this circuit using the *keep* attribute to preserve nodes *a*, *b*, and *c*. Include simulation results in your solution. Before starting, draw in figure 5.10(b) the expected waveforms at nodes *c* and *short_clk*. Assume that the propagation delay through each inverter and through the AND gate is 1 ns, which is the distance between the vertical lines in the figure.

Solution   It is left to the reader to fill figure 5.10(b). A VHDL code for this circuit is presented below. Note the use of *keep* in lines 9–10, telling the compiler to preserve nodes *a*, *b*, and *c*. Observe also that the code proper is not altered by the presence of this attribute. The RTL view produced by the compiler is shown in figure 5.11, along with simulation results. Does the overall shape of *short_clk* in figure 5.11 coincide with your sketch in figure 5.10(b)?

```
1   --------------------------------------------
2   ENTITY short_pulse_gen IS
3      PORT (clk: IN BIT;
4             short_clk: OUT BIT);
5   END ENTITY;
6   --------------------------------------------
7   ARCHITECTURE short_pulse OF short_pulse_gen IS
8      SIGNAL a, b, c: BIT;
9      ATTRIBUTE keep: BOOLEAN;
10     ATTRIBUTE keep OF a, b, c: SIGNAL IS TRUE;
```

```
11    BEGIN
12      a <= NOT clk;
13      b <= NOT a;
14      c <= NOT b;
15      short_clk <= clk AND c;
16    END ARCHITECTURE;
17    ----------------------------------------------
```

Another (but less generic, shown here for Quartus II) solution for the problem above can be devised with the LCELL primitive, which consists of a buffer that can be inserted into the signal path. This primitive is instantiated as a COMPONENT (chapter 8), with the following specifications:

```
COMPONENT LCELL
   PORT (a_in: IN STD_LOGIC;
         a_out: OUT STD_LOGIC);
END COMPONENT;
```

Example 5.9: Short-Pulse Generator with the LCELL Primitive

Redesign the circuit of example 5.8, this time employing the LCELL primitive instead of the *keep* attribute.

Solution   A VHDL code for this problem is shown below. The COMPONENT declaration for LCELL (lines 12–15) is in the declarative part of the architecture (this megafunction is available in the *altera_mf_components.vhd* file). Six auxiliary signals are declared in line 11 instead of three because the wires must be broken in order to insert the buffers. In the code proper, three COMPONENT instances of LCELL are created (lines 18, 20, 22). The inferred circuit is similar to that in the previous example.

```
1    ----------------------------------------------
2    LIBRARY ieee;
3    USE ieee.std_logic_1164.all;
4    ----------------------------------------------
5    ENTITY short_pulse_gen IS
6      PORT (clk: IN STD_LOGIC;
7            short_clk: OUT STD_LOGIC);
8    END ENTITY;
9    ----------------------------------------------
10   ARCHITECTURE short_pulse OF short_pulse_gen IS
11     SIGNAL a1, a2, b1, b2, c1, c2: STD_LOGIC;
12     COMPONENT LCELL IS
13       PORT (a_in: IN STD_LOGIC;
14             a_out: OUT STD_LOGIC);
15     END COMPONENT;
```

```
16  BEGIN
17     a1 <= NOT clk;
18     buffer_a: COMPONENT LCELL PORT MAP (a1, a2);
19     b1 <= NOT a2;
20     buffer_b: COMPONENT LCELL PORT MAP (b1, b2);
21     c1 <= NOT b2;
22     buffer_c: COMPONENT LCELL PORT MAP (c1, c2);
23     short_clk <= clk AND c2;
24  END ARCHITECTURE;
25  -----------------------------------------------
```

5.9   Allowing Multiple Signal Assignments

As will be seen in section 7.4, only one assignment can be made to a SIGNAL, while a VARIABLE allows multiple assignments. One way of circumventing such a limitation will be introduced in section 7.7. To solve some of the exercises in section 5.11, a preview of section 7.7 might be helpful.

5.10   VHDL 2008

With respect to the material covered in this chapter, the main additions specified in VHDL 2008 are those listed below.

1) The concurrent WHEN and SELECT statements can be used also in sequential code. For example, WHEN can replace IF or can be used inside IF, while SELECT can replace CASE.

2) WHEN allows boolean tests (the only consequence of this is sometimes a slightly shorter code, at the cost of reduced code clarity). For example, the two codes below are equivalent (the traditional format is on the left).

```
x <= '0' WHEN rst='0' ELSE          x <= '0' WHEN NOT rst ELSE
     '1' WHEN a='0' OR b='1' ELSE        '1' WHEN NOT a OR b ELSE
     '-';                                '-';
```

3) The matching "SELECT?" statement was introduced, which allows the use of don't care inputs. An example is shown below.

```
WITH interrupt SELECT?
   priority <= 4 WHEN "1---",
               3 WHEN "01--",
               2 WHEN "001-",
               1 WHEN "0001",
               0 WHEN OTHERS;
```

4) In the conditional IF-GENERATE statement, the use of ELSIF/ELSE is allowed. A simplified syntax is shown below (on the left).

```
label: IF condition GENERATE
  [declarative_part
  BEGIN]
  concurrent_statements_part
[ELSIF condition GENERATE
  [declarative_part
  BEGIN]
  concurrent_statements_part]
[ELSE GENERATE
  [declarative_part
  BEGIN]
  concurrent_statements_part]
END GENERATE [label];
```

```
label: CASE expression GENERATE
WHEN condition_1 =>
  [declarative_part
  BEGIN]
  concurrent_statements_part
WHEN condition1_2 =>
  [declarative_part
  BEGIN]
  concurrent_statements_part
...
END GENERATE [label];
```

5) A new form of conditional GENERATE, called CASE-GENERATE, was also introduced. A simplified syntax is shown above (on the right).

6) The use of alternative labels for IF-GENERATE and the use of END before END GENERATE are both allowed.

5.11   Exercises

The exercises proposed in this section are to be solved using only truly *concurrent* code (that is, with WHEN, SELECT, and GENERATE, plus operators, of course). See also the comment in section 5.9.

Note: For exercise solutions, please consult the book website.

Exercise 5.1: Circuit with 'Don't Care' Outputs

Figure 5.12 shows a diagram for a combinational circuit that must compute the function described in the accompanying truth table. Note that some outputs are marked as 'don't care'.

a) Using Karnaugh maps, derive the optimal boolean expressions for both output bits (y_1 and y_0).

b) Which data type should be used in the VHDL code in order to take advantage of the 'don't care' states?

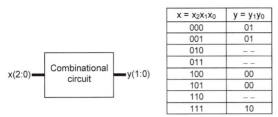

$x = x_2x_1x_0$	$y = y_1y_0$
000	01
001	01
010	– –
011	– –
100	00
101	00
110	– –
111	10

Figure 5.12

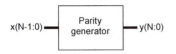

Figure 5.13

c) Design this circuit using WHEN or SELECT. After compiling and testing the code, compare the equations implemented by the fitter against those obtained above.

Exercise 5.2: Dual-Edge Flip-Flop

Using concurrent code, design the dual-edge flip-flop of example 7.6 (figure 7.4).

Exercise 5.3: Generic AND and NAND Gates

Using concurrent code, solve exercises 7.12 and 7.13.

Exercise 5.4: Generic Parity Generator

The circuit in figure 5.13 has an N-bit input x, from which an $(N+1)$-bit output y must be produced. The circuit must detect the parity of x, then add an extra bit to it (on the left) such that the final parity (number of '1's) is odd. Design this circuit using concurrent code. Enter N as a *generic* parameter, so the code can be easily adjusted to any input size. (Suggestion: see section 7.7.)

Exercise 5.5: Parity Generator with Automated Pin Allocation

Assume the case of $N = 4$ in the exercise above (so the circuit has a total of nine ports). Using the *chip_pin* attribute (section 4.6), make proper pin assignments in the code such that all nine ports are automatically routed to the desired pins (check the target device's pin list to select pins that are available to the user).

Exercise 5.6: Generic Binary-to-Gray Converter

The *regular binary code*, which consists of code words ordered according to their increasing unsigned decimal values, constitutes the most commonly used digital code. In some

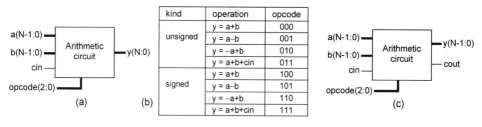

kind	operation	opcode
unsigned	y = a+b	000
	y = a–b	001
	y = –a+b	010
	y = a+b+cin	011
signed	y = a+b	100
	y = a–b	101
	y = –a+b	110
	y = a+b+cin	111

Figure 5.14

applications, however, *gray code* (Pedroni 2008), which is an UDC (Unit Distance Code) because any two adjacent code words differ by just one bit, might be preferred (in electromechanical applications, for example). This exercise deals with the design of a circuit capable of converting regular binary code into gray code.

a) Just to practice with gray code, make a table with two columns, placing the 16 four-bit binary entries in the first column, and the corresponding gray values in the second column.

b) In order to implement a *generic* converter, find a closed-form expression for binary-to-gray conversion. Use it to check your entries in the second column for part (a).

c) Write a VDHL code from which this converter can be inferred. Enter *N* (number of bits) as a GENERIC parameter, and use the closed-form expression obtained above to do the computations. Include simulation results in your solution.

Exercise 5.7: Hamming Weight with GENERATE

The *Hamming weight* of a vector is the number of '1's in it. Design a circuit that computes that number for a *generic-length* vector using only concurrent code. (Suggestion: see section 7.7.)

Exercise 5.8: Binary Sorter with GENERATE

Using concurrent code, design a circuit capable of ordering the bits of a bit vector. The ordering should be from left to right, with all '1's coming first (for example, "00011001" would become "11100000"). (Suggestions: solve exercise 5.7 first; see section 7.7)

Exercise 5.9: ALU with WHEN

Redesign the ALU of example 5.3 using the WHEN statement instead of SELECT.

Exercise 5.10: Arithmetic Circuit with INTEGER

Figure 5.14(a) shows an arithmetic circuit that must be designed to produce the computations specified in the truth table of figure 5.14(b) (it is a "mini-ALU", with just the arithmetic unit). Write a VHDL code for this circuit under the following constraints:

Figure 5.15

1) The code must be truly concurrent;

2) All ports must be specified as INTEGER. Note that in this case the output can have more bits than the inputs, so no separate carry-out computation is needed. Note also that some operations are signed.

Exercise 5.11: Arithmetic Circuit with STD_LOGIC

Figure 5.14(c) shows an arithmetic circuit similar to that in the previous exercise, the only difference being that now the output has the same number of bits as the inputs, so a separate wire is needed for the carry-out bit. Design this circuit such that it performs the same operations listed in figure 5.14(b), under the following constraints:

1) Again, the code must be truly concurrent;

2) All ports must be specified as STD_LOGIC(_VECTOR) (industry standard).

Exercise 5.12: Barrel Shifter with INTEGER and BIT_VECTOR

A *barrel shifter* (Pedroni 2008) is a circuit capable of shifting an input word to the right or to the left by a certain number of bit positions. A top level diagram is shown in figure 5.15, with inputs x (word to be shifted) and *shift* (number of positions to be shifted), and output y (shifted word). Assuming that our shifter is unregistered (that is, does not contain memory, so it is not dependent on a clock signal) and that the empty positions must be filled with zeros, write a VHDL code from which this circuit can be inferred. Represent the number of bits by N in x and y and M in *shift* (where $2^M = N$). These parameters must be *generic*, and the total amount of shift should be allowed to be as large as $N - 1$. Solve this exercise with *shift* declared as INTEGER and x and y as BIT_VECTOR. Are the shift operators SLL and SRL helpful here?

Exercise 5.13: Barrel Shifter with STD_LOGIC_VECTOR

Repeat the exercise above, this time with all ports specified as STD_LOGIC_VECTOR.

Exercise 5.14: Recommended Unsigned Adder/Subtracter Implementation

Write a VHDL code that implements the adder/subtracter of figure 5.8(a) or 5.8(b) for an *unsigned* system. In the design, follow the recommendations presented in section 5.7.

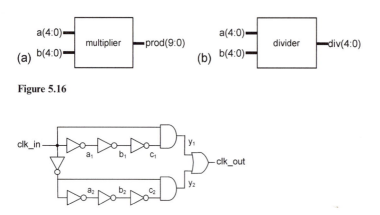

Figure 5.16

Figure 5.17

Exercise 5.15: Recommended Unsigned Multiplier Implementation

Figure 5.16(a) shows a multiplier. Write a VHDL code that implements this circuit for an *unsigned* system. In the design, follow the recommendations presented in section 5.7.

Exercise 5.16: Recommended Signed Multiplier Implementation

Figure 5.16(a) shows a multiplier. Write a VHDL code that implements this circuit for an *signed* system. In the design, follow the recommendations presented in section 5.7.

Exercise 5.17: Recommended Unsigned Divider Implementation

Figure 5.16(b) shows a divider. Write a VHDL code that implements this circuit for an *unsigned* system. In the design, follow the recommendations presented in section 5.7.

Exercise 5.18: Recommended Signed Divider Implementation

Figure 5.16(b) shows a divider. Write a VHDL code that implements this circuit for an *signed* system. In the design, follow the recommendations presented in section 5.7.

Exercise 5.19: Frequency Multiplier with the *keep* Attribute

Figure 5.17 shows two short-pulse generators similar to that in figure 5.10, connected in parallel and operating with complementary versions of the clock.

a) Examine this circuit then draw the expected waveforms for c_1, y_1, c_2, y_2, and *clk_out*.

b) Design this circuit with concurrent code, using *keep* to preserve internal nodes. Does the resulting waveform for *clk_out* match yours?

Exercise 5.20: Generic Multiplexer

Consider the multiplexer implemented in example 5.2 (figure 5.3), which has four inputs with an arbitrary number of bits (N) per input. To make the design truly generic, consider the case where the number of inputs is also arbitrary (M). Redesign that circuit with M and N declared as GENERIC constants, in the following two situations:

a) Using a 1D × 1D data array for the input (recall that the only predefined data types with dimension 1D × 1D are STRING and INTEGER_VECTOR, which are of no interest in the present example).

b) Using a predefined data type.

Exercise 5.21: INOUT bus

Solve exercise 13.1.

Exercise 5.22: INOUT versus BUFFER

Solve exercise 13.2.

Exercise 5.23: Floating-Point Adder

This exercise concerns the "backward" construction of a floating-point (FP) adder. (Suggestion: see first the example in section 3.8.)

a) Figure 5.18 shows simulation results from a FP adder that computes $x = a + b$, where all signals are expressed with 4 bits for the exponent and also 4 bits for the fraction, that is, (S)(EEEE)(FFFF). Examine each result in the figure and check its correctness. The non-exact values obtained in the simulation are wrong or is this something inherent of FP numbers?

b) Design such an adder, then simulate it and check whether similar results are obtained.

Name	0 ps	80,0 ns	160,0 ns	240,0 ns	320,0 ns	400,0 ns	480,0 ns
a				010010110			
b	001101000	010010110	011100001	011101110	111100001	111101110	
x	010011001	010100110	011100010	011101111	111100000	111101101	

Expected:	Expected:	Expected:	Expected:	Expected:	Expected:
5.5+0.75 = 6.25	5.5+5.5 = 11	5.5+136 = 141.5	5.5+240 = 245.5	5.5−136 = −130.5	5.5−240 = −234.5
Obtained:	Obtained:	Obtained:	Obtained:	Obtained:	Obtained:
6.25 (exact)	11 (exact)	144 (not exact)	248 (not exact)	-128 (not exact)	-232 (not exact)

Figure 5.18

Exercise 5.24: Floating-Point Adder, Subtracter, Multiplier, Divider

This exercise concerns the design of a single-precision (32-bit option of IEEE 754) adder, subtracter, multipler, and divider circuit. (Suggestion: see first the example in section 3.8.)

a) Among the four implementations, that is, $a + b$, $a - b$, $a*b$, and a/b, which ones do you expect to require the smallest and the largest amount of hardware?

b) Design a single-precision FP adder that computes $x = a + b$. Compile the circuit in a low-cost (say, Cyclone II or Spartan 3A) and also in a high-end (say, Stratix III or Virtex 5) device. In each case, write down the number of logic cells or LUTs or slices needed by the circuit.

c) Simply modify the output equation to $x = a - b$ and repeat the compilations, writing down the new amounts of hardware.

d) Repeat the procedure for $x = a*b$.

e) Finally, do it for $x = a/b$.

f) Make a table with all the values obtained above. Compare the results against your predictions in part (a).

6 Sequential Code

6.1 Introduction

As described in section 5.1, *concurrent* code is intended only for the design of *combinational* circuits, while *sequential* code can be used indistinctly to design both sequential *and* combinational circuits.

The statements intended only for completely concurrent code, referred to as *concurrent statements*, are WHEN, SELECT, and GENERATE (seen in the previous chapter), while those for sequential code, referred to as *sequential statements*, are IF, WAIT, LOOP, and CASE (described in this chapter).

In VHDL, there are three kinds of sequential code: PROCESS, FUNCTION and PROCEDURE (the last two are called *subprograms*). Because PROCESS is intended for the architecture body (main code, for example), it will be described in this chapter, while subprograms, being intended mainly for libraries, will be seen in chapter 9, which deals with system-level code. Recall that, as a whole, a PROCESS or a subprogram call is a concurrent statement.

A crucial point when dealing with sequential code is to fully understand the differences between SIGNAL and VARIABLE (these are the two VHDL objects for dealing with nonstatic values). For that reason, an entire chapter (chapter 7) will be dedicated to the matter. Moreover, an introduction has already been made in section 3.2, needed for the discussion on data types in chapter 3. However, for the discussions on sequential code in this chapter, it is necessary to preview some of their fundamental properties.

Main properties of SIGNAL:

- A signal can only be declared *outside* sequential code (though it can be *used* there).

- A signal is not updated immediately (when a value is assigned to a signal inside sequential code, the new value will only be ready after the conclusion of that run).

- A signal assignment, when made at the transition of another signal, will cause the inference of registers (given that the signal affects the design entity).

• Only a single assignment is allowed to a signal in the whole code (even though the compiler might accept multiple assignments to the same signal in PROCESS or subprograms, only the last one will be effective, so again it is just one assignment).

Main properties of VARIABLE:

• A variable can only be declared and used *inside* a PROCESS or subprogram (if it is a *shared* variable, then the declaration is made elsewhere, but it still should only be modified inside a sequential unit).

• A variable is updated immediately (hence the new value can be used/tested in the next line of code).

• A variable assignment, when made at the transition of another signal, will cause the inference of registers (assuming that the variable's value affects a signal, which in turn affects the design entity).

• Multiple assignments are fine.

6.2 Latches and Flip-Flops

Because flip-flops are indispensable building blocks for sequential circuits, a brief review is made in this section. Additionally, because latches are also needed occasionally, they too are included.

As shown in Pedroni (2008), there are two fundamental types of latches, called *SR latch* (SRL) and *D latch* (DL), and there are four types of flip-flops, called *SR flip-flop* (SRFF), *D flip-flop* (DFF), *T flip-flop* (TFF), and *JK flip-flop* (JKFF). DL and DFF are the most commonly used in their categories, but overall DFF is by far the most common. For instance, there are thousands of such units in FPGA devices.

The fundamental difference between latches and flip-flops is that the former are *level*-sensitive, while the latter are *edge*-sensitive. This means that a latch is transparent (input copied to the output) during the whole time in which the clock is '1' (or '0') and opaque (thus retaining the last value of the transparent cycle) when the clock is '0' (or '1'), while a flip-flop is transparent only during one of the clock transitions, either from '0' to '1' (called *positive-edge DFF*) or from '1' to '0' (*negative-edge DFF*).

Two DL symbols are shown in figures 6.1a–b. The first is transparent while *clk* is high; the second DL is transparent while *clk* is low. Both have a *reset* input, which immediately zeros the output when asserted.

Four DFF symbols are depicted in figures 6.1c–f. The first circuit operates at the positive clock edge, while the second copies the input at the negative clock edge. These two circuits have a *reset* input (in our context, reset is always *asynchronous*—that is, whenever *rst* = '1' occurs, the output is immediately zeroed, regardless of the clock). The third and fourth circuits are similar to the first and second, respectively, but they have a *clear* input instead of reset (in our context, clear is always *synchronous*, meaning that after *clr* is

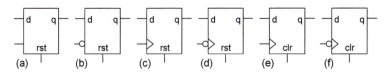

Figure 6.1
Latches and flip-flops. (a) Positive-level DL with reset; (b) Negative-level DL with reset; (c) Positive-edge DFF with reset; (d) Negative-edge DFF with reset; (e) Positive-edge DFF with clear; (f) Negative-edge DFF with clear.

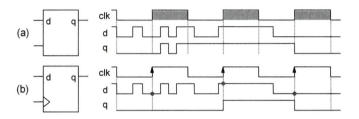

Figure 6.2
(a) DL and (b) DFF operation examples.

asserted we still need to wait until the proper clock transition occurs for the output to be zeroed). DFFs will be present in almost all designs described in the book from this point on.

Figure 6.2 illustrates the operation of DL and DFF circuits (this is a *functional* analysis because internal propagation delays were neglected). The same clock (*clk*) and the same data (*d*) are applied to both circuits, which respond with *q*. In both cases the initial state was considered to be $q = '0'$. To ease the analysis of figure 6.2, the portions of the *clk* waveform during which the DL is transparent were marked with gray shades, while the moments at which the DFF is transparent were marked with arrows. Note that from the same *d* waveform very distinct waveforms are obtained for *q*.

6.3 PROCESS

PROCESS is a *sequential* section of VHDL code, located in the statements part of an architecture. Inside it, only sequential statements (IF, WAIT, LOOP, CASE) are allowed. A simplified syntax is shown below.

```
[label:] PROCESS [(sensitivity_list)] [IS]
   [declarative_part]
BEGIN
   sequential_statements_part
END PROCESS [label];
```

As shown in the syntax, the label, whose purpose is to improve readability in long codes, is optional. The sensitivity list is mandatory (but is forbidden when WAIT is used), and causes the process to be run every time a signal in the list changes (or the condition associated with WAIT is fulfilled).

The declarative part of PROCESS can contain the following: subprogram declaration, subprogram body, type declaration, subtype declaration, constant declaration, variable declaration, file declaration, alias declaration, attribute declaration, attribute specification, use clause, group template declaration, and group declaration. Signal declaration is not allowed, while variable is by far the most common declaration (see its syntax in section 3.2).

In the statements part of PROCESS, only sequential statements are allowed (besides operators, of course, seen in chapter 4, because these can go in any kind of code).

Example   The (partial) process below is executed whenever *clk* or *rst* changes. It contains three variable declarations (a, b, c), the first two specified as INTEGER, the last one as BIT_VECTOR. Only for c a default value (optional) was entered.

```
PROCESS (clk, rst)
    VARIABLE a, b: INTEGER RANGE 0 TO 255;
    VARIABLE c: BIT_VECTOR(7 DOWNTO 0) := "00001111";
BEGIN
...
END PROCESS;
```

In VHDL 2008, the following is allowed in the declarative part of PROCESS besides the items already listed above: subprogram instantiation declaration, package declaration, package body, and package instantiation declaration. Additionally, the keyword ALL was introduced for the sensitivity list (to reduce errors when implementing combinational circuits with sequential code). See other details in section 6.10.

6.4   The IF Statement

As mentioned earlier, IF, WAIT, LOOP, and CASE are the statements intended for sequential code (they can only be used inside a PROCESS or subprogram), of which IF is by far the most common. Though this could, in principle, have a negative impact (because the IF-ELSE statement might infer an unnecessary priority decoder), the synthesizer will be able to simplify the structure, so essentially the same hardware will result as with basically any other statements. A simplified syntax for IF is shown below ("conditions" can be optionally surrounded by parentheses).

```
[label:] IF conditions THEN
   assignments;
ELSIF conditions THEN
   assignments;
...
ELSE
   assignments;
END IF [label];
```

Example

```
IF (x<y) THEN
  temp:= "00001111";
ELSIF (x=y AND w='0') THEN
  temp:= "11110000";
ELSE
  temp:=(OTHERS => '0');
END IF;
```

In VHDL 2008, the concurrent statements WHEN and SELECT are allowed inside the IF statement and IF allows boolean tests.

Example 6.1: DFFs with Reset and Clear

Employing the IF statement, write a code that implements the DFFs of figures 6.1c and 6.1e.

Solution A code for this circuit is shown below. The inputs are *d1*, *clk*, and *rst* for the first flip-flop, and *d2*, *clk*, and *clr* for the second, while the outputs are *q1* for the first and *q2* for the second DFF. Even though both DFFs could be designed with just one process, two processes were employed to make the code easier to inspect (this does not affect the inferred circuit).

The first DFF is in the process of lines 13–20, under the (optional) label *with_reset*. Note that *clk* and *rst* are in the sensitivity list (line 13), so if any of them changes the process is run. Note also that *rst* has precedence over *clk* in the IF statement (lines 15–19). To detect a clock edge, the 'EVENT attribute (line 17), seen in section 4.4, is used, which returns TRUE when an event occurs on *clk* (clk'EVENT) *and* this event is an upward transition (AND *clk* = '1').

The second process is in lines 22–31, under the label *with_clear*. Only *clk* is in the sensitivity list, so the process is run only when *clk* changes (in this particular example, the presence of *clr* in the sensitivity list would not affect the result). The synchronism of *clr* is established in the IF statements of lines 24–30, because *clr* is only tested (line 25) when a positive clock edge occurs (line 24).

```
1   ------------------------------------------------
2   LIBRARY ieee;
3   USE ieee.std_logic_1164.all;
4   ------------------------------------------------
5   ENTITY flipflops IS
6      PORT (d1, d2, clk, rst, clr: IN STD_LOGIC;
7            q1, q2: OUT STD_LOGIC);
8   END ENTITY;
9   ------------------------------------------------
10  ARCHITECTURE flipflops OF flipflops IS
11  BEGIN
12     ---DFF of Figure 6.1(c):---
13     with_reset: PROCESS (clk, rst)
14     BEGIN
15        IF (rst='1') THEN
16           q1 <= '0';
17        ELSIF (clk'EVENT AND clk='1') THEN
18           q1 <= d1;
19        END IF;
20     END PROCESS with_reset;
21     ---DFF of Figure 6.1(e):---
22     with_clear: PROCESS (clk)
23     BEGIN
24        IF (clk'EVENT AND clk='1') THEN
25           IF (clr='1') THEN
26              q2 <= '0';
27           ELSE
28              q2 <= d2;
29           END IF;
30        END IF;
31     END PROCESS with_clear;
32  END ARCHITECTURE;
33  ------------------------------------------------
```

Simulation results are displayed in figure 6.3. The reader is invited to examine the plots to check the (correct) operation of both DFFs (observe particularly the effects of *rst* and *clr*).

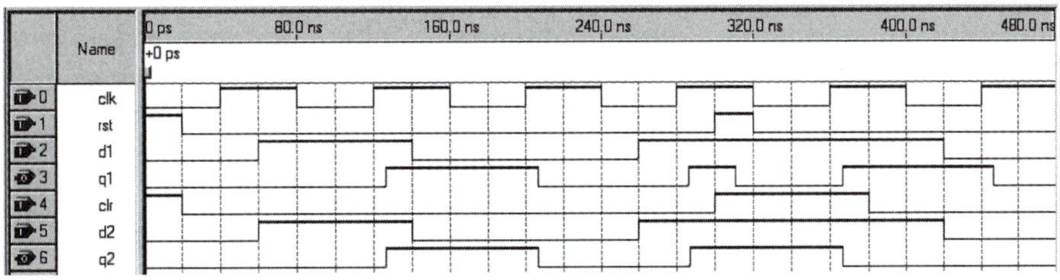

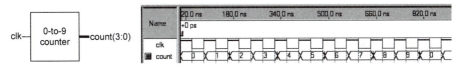

Figure 6.3
Simulation results from the DFFs of example 6.1.

Figure 6.4
Counter (with simulation results) of example 6.2.

Example 6.2: Basic Counter

Figure 6.4 shows, on the left, a diagram for a regular binary 0-to-9 counter. Write a VHDL code that implements this circuit.

Solution A VHDL code for this counter is presented below, with input *clk* (line 3) and output *count* (line 4). A PROCESS (lines 9–19), with the IF statement playing the central role, is used to construct the circuit. In it, a VARIABLE, called *temp* (line 10), is employed, whose value is eventually passed to the actual output, *count* (line 18). Because a variable is updated *immediately*, the comparison in line 14 must be against 10 instead of 9 (state 10 actually never occurs because the variable never reaches line 18 with that value), so the actual range of the counter is from 0 to 9. Simulation results are included in figure 6.4.

```
1   -------------------------------------------
2   ENTITY counter IS
3      PORT (clk: IN BIT;
4            count: OUT INTEGER RANGE 0 TO 9);
5   END ENTITY;
6   -------------------------------------------
7   ARCHITECTURE counter OF counter IS
8   BEGIN
9      PROCESS(clk)
```

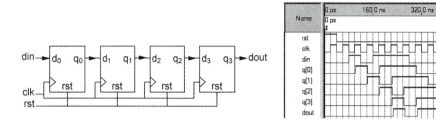

Figure 6.5
Shift register (with simulation results) of example 6.3.

```
10          VARIABLE temp: INTEGER RANGE 0 TO 10;
11       BEGIN
12          IF (clk'EVENT AND clk='1') THEN
13             temp := temp + 1;
14             IF (temp=10) THEN
15                temp := 0;
16             END IF;
17          END IF;
18          count <= temp;
19       END PROCESS;
20    END ARCHITECTURE;
21    --------------------------------------------
```

Example 6.3: Shift Register

Figure 6.5 shows a shift register, which consists of a string of serially connected DFFs. The input is *din* and the output is either q_0 to q_3 or just *dout*, depending on the application. For example, the former can be used to convert data from serial to parallel form, while the latter can be used to implement a delay line. Design this circuit using VHDL. The number of stages should be *generic*.

Solution A VHDL code for this circuit is presented below. The number of stages (N) is generic (line 6). The data input and output ports are *din* (line 7) and *dout* (line 8), respectively. A process is used to implement the circuit (lines 13–22), with *clk* and *rst* in the sensitivity list (line 13). Note that the shift register is obtained by simply shifting the whole vector q one position to the right at every positive clock transition, with the rightmost value discarded and the leftmost position taken by *din*. Simulation results (for $N = 4$) are included in figure 6.5. As can be seen, the whole data vector does move one position to the right at every rising edge of the clock.

```
1   -----------------------------------------------
2   LIBRARY ieee;
3   USE ieee.std_logic_1164.all;
4   -----------------------------------------------
5   ENTITY shift_register IS
6      GENERIC (N: INTEGER := 4); --number of stages
7      PORT (din, clk, rst: IN STD_LOGIC;
8            dout: OUT STD_LOGIC);
9   END ENTITY;
10  -----------------------------------------------
11  ARCHITECTURE shift_register OF shift_register IS
12  BEGIN
13     PROCESS (clk, rst)
14        VARIABLE q: STD_LOGIC_VECTOR(0 TO N-1);
15     BEGIN
16        IF (rst='1') THEN
17           q := (OTHERS => '0');
18        ELSIF (clk'EVENT AND clk='1') THEN
19           q := din & q(0 TO N-2);
20        END IF;
21        dout <= q(N-1);
22     END PROCESS;
23  END ARCHITECTURE;
24  -----------------------------------------------
```

6.5   The WAIT Statement

WAIT is another sequential statement. It is available in three forms, of which two are for synthesis and one is for simulation. When WAIT is employed, the PROCESS cannot have a sensitivity list. Simplified syntaxes for all three forms follow.

```
[label:] WAIT UNTIL condition;
```

```
[label:] WAIT ON sensitivity_list;
```

```
[label:] WAIT FOR time_expression;
```

WAIT UNTIL: This statement causes the process or subprogram to hold until the expressed condition is fulfilled. In the example below, two equivalent processes for a DFF with synchronous clear are shown, one with IF, the other with WAIT UNTIL. Note that the process has no sensitivity list when WAIT is used.

```
---DFF process with IF:-----------        ---DFF process with WAIT UNTIL:-------
PROCESS (clk)                             PROCESS
BEGIN                                     BEGIN
   IF (clk'EVENT AND clk='1') THEN           WAIT UNTIL (clk'EVENT AND clk='1');
      IF (clr='1') THEN                         IF (clr='1') THEN
         q <= '0';                                 q <= '0';
      ELSE                                      ELSE
         q <= d;                                   q <= d;
      END IF;                                   END IF;
   END IF;                                   END PROCESS;
END PROCESS;                              -------------------------------------
----------------------------------
```

WAIT ON: This statement causes the process or subprogram to hold until any listed signal changes. In the example below, WAIT ON monitors the clock. Since a single WAIT ON statement at the beginning or at the end of a process is equivalent to using a process with the same signals listed in the sensitivity list, the two processes below for a DFF with synchronous clear are equivalent. Note again that no sensitivity list is allowed when WAIT is used and that WAIT ON and IF are quite similar.

```
---DFF process with IF:-----------        ---DFF process with WAIT ON:------
PROCESS (clk)                             PROCESS
BEGIN                                     BEGIN
   IF (clk'EVENT AND clk='1') THEN           IF (clk'EVENT AND clk='1') THEN
      IF (clr='1') THEN                         IF (clr='1') THEN
         q <= '0';                                 q <= '0';
      ELSE                                      ELSE
         q <= d;                                   q <= d;
      END IF;                                   END IF;
   END IF;                                   END IF;
END PROCESS;                                 WAIT ON clk;
----------------------------------        END PROCESS;
                                          ----------------------------------
```

WAIT FOR: This statement is for simulations, so it will be studied in chapter 10. The declaration below creates a clock waveform with period 80 ns.

```
WAIT FOR 40ns;
clk <= NOT clk;
```

6.6   The LOOP Statement

As the name says, LOOP is used when a piece of code must be instantiated several times. It is the counterpart of the concurrent statement GENERATE. Like IF, WAIT, and CASE, LOOP also can only be used in sequential code (PROCESS and subprograms).

There are five cases involving the LOOP statements: unconditional, with FOR, with WHILE, with EXIT, and with NEXT. To simplify matters, all five are grouped under the LOOP designation. Simplified syntaxes for all five cases are presented below. LOOP with FOR (also called FOR-LOOP) is by far the most frequently used.

Unconditional LOOP:

```
[label:] LOOP
   sequential_statements
END LOOP [label];
```

LOOP with FOR:

```
[label:] FOR identifier IN range LOOP
   sequential_statements
END LOOP [label];
```

LOOP with WHILE:

```
[label:] WHILE condition LOOP
   sequential_statements
END LOOP [label];
```

LOOP with EXIT:

```
[loop_label:] [FOR identifier IN range] LOOP
   ...
   [exit_label:] EXIT [loop_label] [WHEN condition];
   ...
END LOOP [loop_label];
```

LOOP with NEXT:

```
[loop_label:] [FOR identifier IN range] LOOP
   ...
   [next_label:] NEXT [loop_label] [WHEN condition];
   ...
END LOOP [loop_label];
```

Example of unconditional LOOP:

```
LOOP
   WAIT UNTIL clk='1';
   count := count + 1;
END LOOP;
```

Example of LOOP with FOR: In the code below, which employs the FOR-LOOP version of LOOP, the loop is repeated unconditionally until i reaches 5 (that is, 6 times).

```
FOR i IN 0 TO 5 LOOP
   x(i) <= a(i) AND b(5-i);
   y(0, i) <= c(i);
END LOOP;
```

An important remark regarding FOR-LOOP (similar to that made for GENERATE in chapter 5) is that both range bounds are normally required to be static. Thus a declaration of the type "FOR i IN 0 TO x LOOP", where x is an input (therefore a nonstatic parameter), is generally not synthesizable.

Example of LOOP with WHILE: The loop below will keep repeating while $i < 10$.

```
WHILE (i<10) LOOP
   WAIT UNTIL clk'EVENT AND clk='1';
   ...
END LOOP;
```

Example of LOOP with EXIT: The loop will be terminated if a value different from '0' is found in *data*.

```
FOR i IN data'RANGE LOOP
   CASE data(i) IS
     WHEN '0' => count:=count+1;
     WHEN OTHERS => EXIT;
   END CASE;
END LOOP;
```

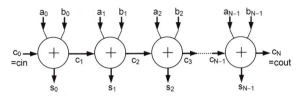

Figure 6.6
Carry-ripple adder.

Example of LOOP with NEXT: NEXT will cause LOOP to skip one iteration if $i = skip$ occurs.

```
FOR i IN 0 TO 15 LOOP
   NEXT WHEN i=skip;
   ...
END LOOP;
```

In VHDL 2008, some tests associated with LOOP statements can be boolean. For example, "`WHILE ena LOOP`" can be used instead of "`WHILE ena='1' LOOP`".

Example 6.4 illustrates the use of the FOR-LOOP version of LOOP, which is the most frequently used. Additionally, it illustrates the construction of a fully combinational circuit with sequential code.

Example 6.4: Carry-Ripple Adder

Adders are combinational circuits. The simplest multibit architecture, shown in figure 6.6, is called *carry-ripple adder* (Pedroni 2008). Each individual cell, called *full-adder* (FA), was seen in section 1.5 (figure 1.2). Write a VHDL code from which this adder can be inferred. Employ sequential code and enter the number of bits (stages) as a *generic* parameter, so the code can be easily adjusted to any adder size. Assume that the adder is unsigned.

Solution   Each FA cell computes the sum and carry-out bits according with:

Sum: $s_k = a_k \oplus b_k \oplus c_k$

Carry: $c_{k+1} = a_k \cdot b_k + a_k \cdot c_k + b_k \cdot c_k$

These two expressions appear in lines 20–22 of the code below. The FOR-LOOP statement (lines 19–23) is employed to instantiate the expressions N times. The number of bits is entered using a GENERIC declaration (line 6). Simulation results, for $N = 8$, are displayed in figure 6.7. As expected, whenever the sum is higher than 255, the carry-out bit is asserted and the sum is subtracted of 256.

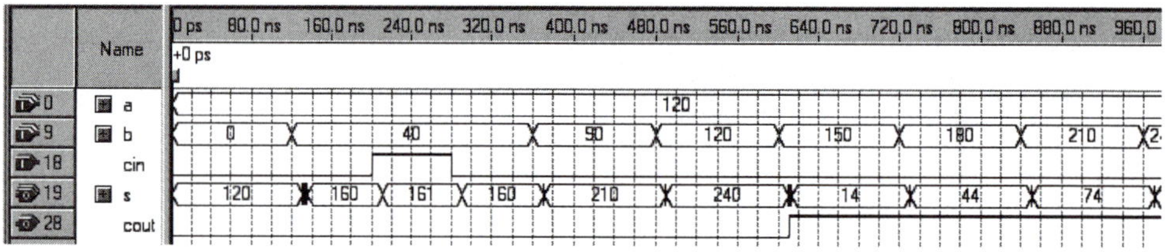

Figure 6.7
Simulation results from the carry-ripple adder of example 6.4.

```
1   -----------------------------------------------------------
2   LIBRARY ieee;
3   USE ieee.std_logic_1164.all;
4   -----------------------------------------------------------
5   ENTITY carry_ripple_adder IS
6      GENERIC (N : INTEGER := 8); --number of bits
7      PORT (a, b: IN STD_LOGIC_VECTOR(N-1 DOWNTO 0);
8            cin: IN STD_LOGIC;
9            s: OUT STD_LOGIC_VECTOR(N-1 DOWNTO 0);
10           cout: OUT STD_LOGIC);
11  END ENTITY;
12  -----------------------------------------------------------
13  ARCHITECTURE structure OF carry_ripple_adder IS
14  BEGIN
15     PROCESS(a, b, cin)
16        VARIABLE c: STD_LOGIC_VECTOR(N DOWNTO 0);
17     BEGIN
18        c(0) := cin;
19        FOR i IN 0 TO N-1 LOOP
20           s(i) <= a(i) XOR b(i) XOR c(i);
21           c(i+1) := (a(i) AND b(i)) OR (a(i) AND c(i)) OR
22                      (b(i) AND c(i));
23        END LOOP;
24        cout <= c(N);
25     END PROCESS;
26  END ARCHITECTURE;
27  -----------------------------------------------------------
```

Example 6.5 illustrates the use of the LOOP with EXIT version of LOOP.

Example 6.5: Leading Zeros

Design a circuit that counts the number of leading zeros in a binary vector, starting from its left end (MSB).

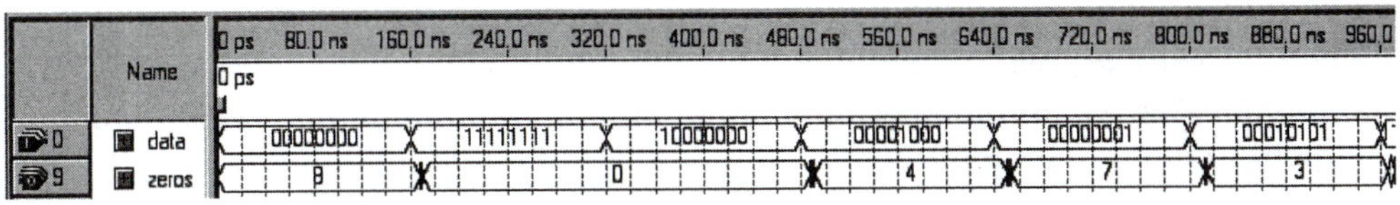

Figure 6.8
Simulation results from the leading-zeros counter of example 6.5.

Solution A VHDL code for this problem is shown below. LOOP is in lines 16–21, and can be repeated up to $N = 8$ times. If a '1' is found in the data vector, then EXIT (line 19) will terminate the loop. Simulation results are displayed in figure 6.8, illustrating the correct operation of the inferred circuit.

```
1   ------------------------------------------------
2   LIBRARY ieee;
3   USE ieee.std_logic_1164.all;
4   ------------------------------------------------
5   ENTITY leading_zeros IS
6      PORT (data: IN STD_LOGIC_VECTOR (7 DOWNTO 0);
7             zeros: OUT INTEGER RANGE 0 TO 8);
8   END ENTITY;
9   ------------------------------------------------
10  ARCHITECTURE behavior OF leading_zeros IS
11  BEGIN
12     PROCESS (data)
13        VARIABLE count: INTEGER RANGE 0 TO 8;
14     BEGIN
15        count := 0;
16        FOR i IN data'RANGE LOOP
17           CASE data(i) IS
18              WHEN '0' => count := count + 1;
19              WHEN OTHERS => EXIT;
20           END CASE;
21        END LOOP;
22        zeros <= count;
23     END PROCESS;
24  END ARCHITECTURE;
25  ------------------------------------------------
```

6.7 The CASE Statement

Along with IF, WAIT, and LOOP, CASE too is only allowed inside sequential code (PROCESS or subprogram). A simplified syntax is shown below.

```
[label:] CASE expression IS
   WHEN value => assignments;
   WHEN value => assignments;
   ...
END CASE;
```

Example

```
CASE control IS
 WHEN "000" => x<=a; y<=b;
 WHEN "000" | "111" => x<=b; y<= '0';
 WHEN OTHERS => x<='0'; y<='1';
END CASE;
```

Like SELECT, CASE too allows the use of multiple values, which can be grouped with "|" (means "or") or "TO" (for range), as shown.

```
WHEN value1 | value2 |...      --value1 or value2 or ...
WHEN value1 TO value2          --range (for enumerated types only)
```

Like SELECT, CASE too requires that all input values be covered (complete truth table), for which the keyword OTHERS is often helpful. Another important keyword is NULL (the counterpart of UNAFFECTED, used with SELECT), which should be used when no action is to take place (recall, however, the note on latch inference at the end of section 5.3).

Even though CASE can only be used in *sequential* code, its fundamental role is to allow the construction of *combinational* circuits (truth tables) without having to leave the PROCESS or subprogram. In other words, the main application for CASE is very similar to that of its concurrent counterpart SELECT. The use of CASE is illustrated in example 6.6.

In VHDL 2008, the concurrent statement WHEN is allowed inside the CASE statement. Also, the matching CASE? statement was introduced to allow the use of don't care inputs, and the use of UNAFFECTED was extended to sequential code.

Example 6.6: Slow 0-to-9 Counter with SSD

Add an SSD (Seven-Segment Display—described in section 12.1, see figure 12.2) to the output of the 0-to-9 counter designed in example 6.2, such that the state of the counter can be visually inspected. This arrangement is depicted in figure 6.9. Besides the counter and the SSD, an SSD driver is also shown, which is a *combinational* circuit that converts the 4-bit output from the counter (called *count*) into a 7-bit signal (called *ssd*) to feed the seven segments of the display. Assume that the clock frequency is 50 MHz, which

Figure 6.9
One-second-per-state 0-to-9 counter with SSD of example 6.6.

should be entered as a *generic* parameter, and that the counter must remain one second in each state. Write a VHDL code that implements such a circuit.

Solution A VHDL code for this circuit is presented below. The inputs are *clk* and *rst* (line 4), while *ssd* is the output (line 5). The clock frequency, *fclk*, was entered using a GENERIC declaration (line 3).

A process (lines 10–42) is employed to build the circuit. In its first part (lines 15–27), the counter is built using the EVENT attribute (line 18) combined with the IF statement. Note that in fact two counters are constructed, for which two variables (lines 11–12) are used. The first counter sends and enable signal (lines 20–22) to the second counter after every *fclk* clock pulses, which lasts just one clock period, hence allowing the second counter to be incremented every one second. In the second part of the process (lines 29–41), the CASE statement is employed to build the SSD driver. Note that the signal *counter2* in the code corresponds to the signal *count* in figure 6.9.

```
 1   ------------------------------------------------------
 2   ENTITY slow_counter IS
 3      GENERIC (fclk: INTEGER := 50_000_000); --50MHz
 4      PORT (clk, rst: IN BIT;
 5            ssd: OUT BIT_VECTOR(6 DOWNTO 0));
 6   END ENTITY;
 7   ------------------------------------------------------
 8   ARCHITECTURE counter OF slow_counter IS
 9   BEGIN
10      PROCESS (clk, rst)
11         VARIABLE counter1: NATURAL RANGE 0 TO fclk := 0;
12         VARIABLE counter2: NATURAL RANGE 0 TO 10 := 0;
13      BEGIN
14         ------counter:---------
15         IF (rst='1') THEN
16            counter1 := 0;
17            counter2 := 0;
18         ELSIF (clk'EVENT AND clk='1') THEN
19            counter1 := counter1 + 1;
20            IF (counter1=fclk) THEN
21               counter1 := 0;
22               counter2 := counter2 + 1;
```

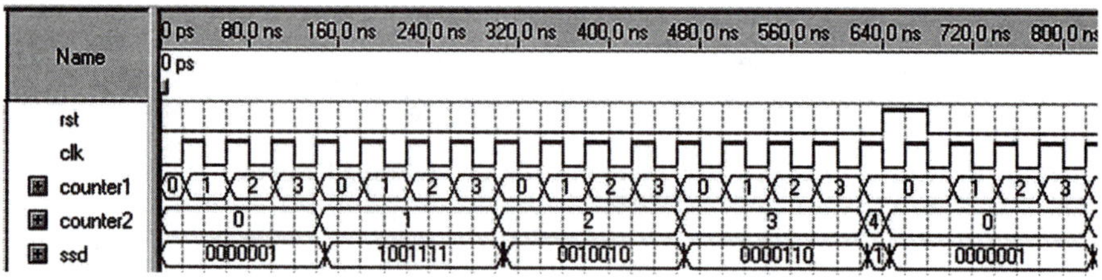

Figure 6.10
Simulation results from the counter of example 6.6.

```
23                  IF (counter2=10) THEN
24                      counter2 := 0;
25                  END IF;
26              END IF;
27          END IF;
28          ------SSD driver:------
29          CASE counter2 IS
30              WHEN 0 => ssd<="0000001";   --"0" on SSD
31              WHEN 1 => ssd<="1001111";   --"1" on SSD
32              WHEN 2 => ssd<="0010010";   --"2" on SSD
33              WHEN 3 => ssd<="0000110";   --"3" on SSD
34              WHEN 4 => ssd<="1001100";   --"4" on SSD
35              WHEN 5 => ssd<="0100100";   --"5" on SSD
36              WHEN 6 => ssd<="0100000";   --"6" on SSD
37              WHEN 7 => ssd<="0001111";   --"7" on SSD
38              WHEN 8 => ssd<="0000000";   --"8" on SSD
39              WHEN 9 => ssd<="0000100";   --"9" on SSD
40              WHEN OTHERS => ssd<="0110000";   --"E"rror
41          END CASE;
42      END PROCESS;
43  END ARCHITECTURE;
44  -------------------------------------------------------
```

Some simulation results are presented in figure 6.10, where the digit changes after every four clock pulses (to ease the inspection of the results). Note that this design (like many others that will come) is interesting to be tested in an actual device (FPGA board).

6.8   CASE versus SELECT

CASE and SELECT are very similar. However, while the latter is for concurrent code, the former is for sequential code. Their main similarities and differences are summarized in figure 6.11.

	SELECT	CASE
Statement type	Concurrent (*)	Sequential
Location	Outside sequential code (*)	Inside sequential code
Input values that must be tested	All	All
Number of tests per entry	One test, multiple values	One test, multiple values
Number of assignments per test	One	Any
No-action keyword	UNAFFECTED (*)	NULL (*)
Matching statement	SELECT? (introd. in VHDL 2008)	CASE? (introd. in VHDL 2008)
(*) See Section 6.10 for extended options in VHDL 2008.		

Figure 6.11
Comparison between SELECT and CASE.

Example   The codes below are equivalent.

```
----With SELECT:----------          ----With CASE:--------------------
WITH sel & ena SELECT               CASE sel & ena IS
   x <= a WHEN "00" | "11",             WHEN "00" => x<=a; y<="1--1";
         b WHEN "01",                   WHEN "01" => x<=b; y<="1--1";
         c WHEN OTHERS;                 WHEN "11" => x<=a; y<="0000";
WITH sel & ena SELECT                   WHEN OTHERS => x<=c; y<="1--1";
   y <= "0000" WHEN "11",           END CASE;
        "1--1" WHEN OTHERS;         ----------------------------------
--------------------------
```

6.9   Implementing Combinational Circuits with Sequential Code

We have already seen that sequential code can implement sequential as well as combinational circuits. In the former, the inference of registers is necessary, but in the latter it should be avoided, so the following should be observed:

1) The circuit's truth table should always be completely specified.

2) All input signals that are used (read) in the PROCESS should appear in its sensitivity list.

   Failing to comply with (2) will cause the compiler to issue a warning saying that a certain signal is read in the process but is not in the sensitivity list. Failing to comply with (1), however, results in a more serious consequence, because the compiler will infer (unnecessary) latches in order to hold previous circuit states. This fact is illustrated in example 6.7.

Example 6.7: Incomplete Combinational Design

Consider a circuit whose top-level diagram is that in figure 6.12a, for which the specifications in figure 6.12b were given, saying that x should behave as a multiplexer—that is,

Figure 6.12
(a) Top-level diagram for the circuit of example 6.7; (b) Specifications provided; (c) Implemented truth table; (d) Correct approach.

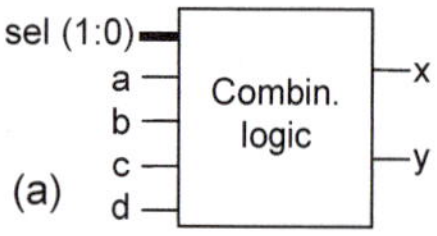

should be equal to the input selected by *sel*, while *y* should be equal to '0' when *sel* = "00" or '1' if sel = "01". Design such a circuit using VHDL.

Solution This is a combinational circuit for which only partial specifications were provided for *y* (figure 6.12b). Using just those specifications, the code could be as follows.

```
1    -------Poor design:-------------------
2    LIBRARY ieee;
3    USE ieee.std_logic_1164.all;
4    ---------------------------------------
5    ENTITY poor_design IS
6       PORT (a, b, c, d: IN STD_LOGIC;
7              sel: IN INTEGER RANGE 0 TO 3;
8              x, y: OUT STD_LOGIC);
9    END ENTITY;
10   ---------------------------------------
11   ARCHITECTURE example OF poor_design IS
12   BEGIN
13      PROCESS (a, b, c, d, sel)
14      BEGIN
15         IF (sel=0) THEN x<=a; y<='0';
16         ELSIF (sel=1) THEN x<=b; y<='1';
17         ELSIF (sel=2) THEN x<=c;
18         ELSE x<=d;
19         END IF;
20      END PROCESS;
21   END ARCHITECTURE;
22   ---------------------------------------
```

After compiling this code, the compiler will report that (as expected) no registers (flip-flops) were inferred. However, when we look at the simulation results (figure 6.13), we notice something peculiar about *y*. Observe that, for the same value of the input (*sel* = 3), two different results are obtained for *y* (when *sel* = 3 is preceded by *sel* = 0, *y* = '0' results, while *y* = '1' occurs when *sel* = 3 is preceded by *sel* = 1). This means that some sort of

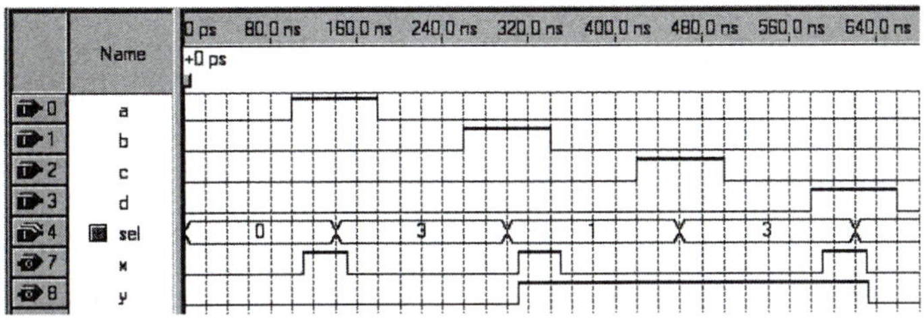

Figure 6.13
Simulation results from the circuit of example 6.7.

memory was implemented. If we now inspect the actual *equations* implemented by the fitter, a latch will be found. In summary, extra (unnecessary) logic was inferred, resulting the
truth table of figure 6.12c. In this kind of situation, the 'don't care' value ("-") should be
used to complete the specifications for *y* (see figure 6.12d).

6.10 VHDL 2008

With respect to the material covered in this chapter, the main additions specified in VHDL
2008 are those listed below.

1) In the declarative part of PROCESS, the following additional declarations are allowed:
subprogram instantiation declaration, package declaration, package body, and package
instantiation declaration.

2) The keyword ALL is allowed in the PROCESS's sensitivity list (to reduce errors when
implementing combinational circuits with sequential code). For example:

```
PROCESS (ALL)
BEGIN
...
END PROCESS;
```

3) The concurrent statements WHEN and SELECT are allowed in sequential code (the
only consequence of this is a slightly shorter code). An example with IF and WHEN is
shown below (the traditional format is on the left).

```
IF (clk'EVENT AND clk='1') THEN        IF (clk'EVENT AND clk='1') THEN
  IF (clr='1') THEN q <= '0';            q <= '0' WHEN clr='1' ELSE d;
  ELSE q <= d;                         END IF;
  END IF;
END IF;
```

4) IF allows boolean tests (the only consequence of this is sometimes a slightly shorter code, at the cost of reduced code clarity). For example, the two lines below are equivalent (the traditional format is in the first).

```
IF (a='0' AND b='0') OR c='1' THEN ...
IF (NOT a AND NOT b) OR c THEN ...
```

5) The matching CASE? statement was introduced, which has the same purpose of SELECT?—that is, to allow the use of don't care inputs. Two equivalent codes are presented.

```
WITH interrupt SELECT?              CASE? interrupt IS
   priority <= 4 WHEN "1---",          WHEN "1---" => priority <= 4;
                 3 WHEN "01--",         WHEN "01--" => priority <= 3;
                 2 WHEN "001-",         WHEN "001-" => priority <= 2;
                 1 WHEN "0001",         WHEN "0001" => priority <= 1;
                 0 WHEN OTHERS;         WHEN OTHERS => priority <= 0;
```

6) The UNAFFECTED keyword was extended to sequential code (IF and CASE statements).

7) Some tests associated with LOOP statements can be boolean. For example, "WHILE ena LOOP" can be used instead of "WHILE ena = '1' LOOP".

6.11   Exercises

The problems in this section are to be solved with *sequential* VHDL code (with IF, WAIT, LOOP, and/or CASE, always located inside a PROCESS). Recall that sequential code can implement both sequential and combinational circuits. Simulation results should always be included (unless they do not apply).

Note: For exercise solutions, please consult the book website.

Exercise 6.1: Latch and Flip-Flop

Figure 6.14 shows a DL and a DFF.

a) Given the waveforms for *clk* and *d*, draw the waveform for *q* in each case. Assume $q =$ '0' as the initial state.

b) Write a VHDL code that implements these two units, then simulate it with the same waveforms. Compare the actual results against your sketches.

Exercise 6.2: Gray Counter

Design a 0-to-*max* counter with gray-encoded outputs. Enter *max* as a generic parameter (Suggestion: see exercise 5.6.)

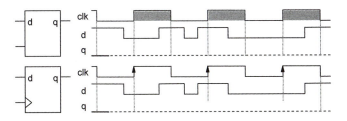

Figure 6.14

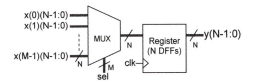

Figure 6.15

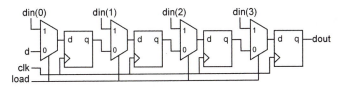

Figure 6.16

Exercise 6.3: Registered Multiplexer

Figure 6.15 shows an $M \times N$ (M inputs with N bits each) multiplexer, followed by an N-flip-flop register. Write a VHDL code from which this circuit, with $N = 8$ and M generic, can be inferred (see example 5.2).

Exercise 6.4: Generic Registered Multiplexer

In the exercise above, M is generic, but N is not. Redesign the circuit with both parameters declared as GENERIC (hence a truly generic mux—see exercise 5.20).

Exercise 6.5: Shift Register with Load

Figure 6.16 depicts a 4×1 (four stages of one bit each) shift register with data-load capability (Pedroni 2008). When $load = $ '0', it operates as a regular shift register. However, if $load = $ '1', din is loaded into the DFFs at the next positive clock transition (thus the initial state of the flip-flops can be programmed). Design this circuit using VHDL.

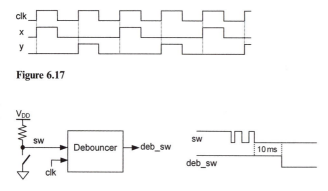

Figure 6.17

Figure 6.18

Exercise 6.6: Hamming Weight with LOOP

The *Hamming weight* of a vector is the number of '1's in it. Similar to exercise 5.7, design a circuit that computes that number for a *generic-length* vector using *sequential* code.

Exercise 6.7: Binary Sorter with LOOP

Similar to exercise 5.8, design a circuit capable of ordering the bits of a bit vector, using *sequential* code. The ordering should be from left to right, with all '1's coming first (for example, "00011001" would become "11100000"). Enter the number of bits as a *generic* parameter.

Exercise 6.8: Signal Generator

Given the clock waveform of figure 6.17, design a circuit capable of generating from it the signals *x* and *y* included in the figure. Recall that in a signal generator glitches are not acceptable.

Exercise 6.9: Switch Debouncer

Figure 6.18 shows a mechanical switch that produces a '0' (when closed) or '1' (when open) to some circuit. Because this type of switch is subject to bounces, a debouncer is inserted into the signal path in order to "clean" the signal *sw* produced by the switch, resulting a bounce-free signal *deb_sw* (debounced switch). This procedure is illustrated in the waveforms included in the figure. Design this digital debouncer using VHDL. Consider that a new value should only be assigned to *deb_sw* if *sw* stays in the new position for at least 10 ms. Enter the clock frequency (*fclk*) as a generic parameter.

Exercise 6.10: Two-Digit Timer

Figure 6.19 shows a two-digit timer, which is an extension of that seen in Example 6.6. The counter is the *sequential* part of the system. It must count seconds from 00 to 60, start-

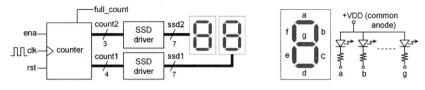

Figure 6.19

ing whenever the enable (*ena*) input is asserted, and stopping whenever 60 is reached or the enable switch is turned off. It must also have an asynchronous reset switch that zeros the system. If 60 is reached, besides stopping, the *full_count* output must be asserted. The SSD drivers comprise the *combinational* part of the system. They must convert the 4-bit and 3-bit outputs from the counters (*count1*, *count2*) into 7-bit signals (*ssd1*, *ssd2*) to feed the display (assume common-anode SSDs, as shown in the figure).

a) Write (and explain) an equation, as a function of *fclk* (clock frequency), for the number of flip-flops that will be needed to implement this circuit. Note that here a decision must be made: will be signal *full_count* be registered (stored) in your design? What are the pros and cons of storing it? (Suggestion: think of glitches.)

b) Design this circuit with VHDL. Enter *fclk* (say, 50 MHz) using GENERIC. After compilation, check whether the number of DFFs inferred by the compiler matches your prediction. Test it also for other values of *fclk*.

c) Physically implement the circuit in your FPGA development board. Connect *ena* and *rst* to two toggle switches (it is not necessary to debounce them), *clk* to the board clock, *full_count* to an LED, and *ssd1* and *ssd2* to two SSDs (check whether they are common-anode in your board).

Exercise 6.11: Frequency Meter (with SSDs)

Solve exercise 12.4. Note that a similar design is presented in Section 12.4, but using an LCD instead of SSDs. Observe in lines 84–86 of that design that a function is used to implement the display driver, which should not be employed here yet (functions will be studied in chapter 9).

Exercise 6.12: Programmable Signal Generator

Figure 6.20 shows a signal generator that must produce, from the clock, a square wave called *sig_out* with 50% duty cycle and frequency 1,000 Hz or 2,000 Hz ... or 10,000 Hz (every time the pushbutton switch is pressed the next frequency must be selected). Assume that the clock frequency is *fclk* = 50 MHz and that an error of up to ±1 Hz is acceptable in the generated frequencies. Assume also that the pushbotton has already been debounced. Note that the frequency of the generated signal must be measured by a 5-digit frequency meter, but that will be treated in exercis 6.13.

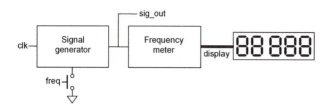

freq

Figure 6.20

a) Determine the values of the dividing coefficients that must be employed in the circuit and calculate the error in each case. Should they be odd or even (or any)? Note that it was not informed that the system clock is symmetric (50% duty cycle); if that is the case, how can that information affect your design?

b) Design the signal generator. To implement the table of dividing coefficients, there are three typical approaches: (i) with the WHEN or SELECT statement, (ii) with the CASE statement, and (iii) with a ROM memory. In the present design, use the last option (see ROM implementations in section 13.4).

c) Simulate the design. To ease the inspection of the results, employ just the following three dividers: 2, 4, and 6 (with $fclk = 50$ MHz). Present just one (well chosen) set of simulation plots.

Exercise 6.13: Programmable Signal Generator with Frequency Meter

a) Using the same technique of exercise 6.11, design the frequency meter (with SSDs) of figure 6.20.

b) Add to this design the signal generator development in exercise 6.12 (with its original ten dividing coefficients) to attain the complete system.

c) Physically implement it in your FPGA development board and test its operation.

Note: See more on this in exercise 8.9.

Exercise 6.14: Digital Wall Clock (with SSDs)

Design the digital wall clock described in section 12.5. Observe that in that design a function is used to implement the display driver, which should not be employed here yet (functions will be studied in chapter 9). Only after solving or trying to solve this exercise see the solution in section 12.5.

Exercise 6.15: Data Serializer

Data serializers are described in section 14.2, with two implementations suggested in figure 14.3. The material seen so far suffices to implement that kind of circuit, except for the PLL instantiation, which is explained in appendix G, and also shown in the example on page 379. After examining that material, design the data serializer of exercise 14.4.

7 SIGNAL and VARIABLE

7.1 Introduction

VHDL provides three objects for dealing with (numeric or enumerated) values, called CONSTANT, SIGNAL, and VARIABLE (FILE is mainly for simulation). A popular example in the CONSTANT category is the GENERIC declaration, which allows global constants to be specified in the ENTITY header. PORT is an example in the SIGNAL category because all items listed in it are signals by definition.

CONSTANT and GENERIC were seen in sections 3.2 and 2.6, respectively, while SIGNAL and VARIABLE, also introduced in section 3.2 (because they were needed in the discussions on data types that followed in that chapter) are described in detail here.

7.2 SIGNAL

SIGNAL serves to pass values in and out of the circuit, as well as between its internal units. In other words, a signal represents circuit interconnects (wires). For example, all ports of an entity are signals.

SIGNAL declarations can be made in the declarative part of ENTITY, ARCHITECTURE, PACKAGE, BLOCK, and GENERATE. A simplified syntax for signal declarations is repeated below.

```
SIGNAL signal_name: signal_type [range] [:= default_value];
```

Example

```
-------------------------------------
SIGNAL flag: STD_LOGIC := '0';
SIGNAL address: NATURAL RANGE 0 TO 2**N-1;
SIGNAL data: BIT_VECTOR(15 DOWNTO 0);
-------------------------------------
```

SIGNAL has important properties, like those introduced in section 6.1, described with more detail in section 7.4.

Resolved Signals

A special feature of a signal is that a "resolution" function can be associated to it. Its purpose is to define the resulting logic value in case multiple drivers feed it (analogous to what was done with the STD_LOGIC type, seen in section 3.6, which is a *resolved* subtype of STD_ULOGIC). A simplified syntax for signal declarations with a resolution function included is shown below.

```
SIGNAL signal_name: resolution_function signal_type [:= def_value];
```

Example

```
SIGNAL x: my_resolution_function my_data_type;
```

Guarded Signals

When the high-impedance state ('Z') is needed, the STD_LOGIC(_VECTOR) type is employed. However, if other types are used, VHDL still provides an indirect means for attaining the high-impedance state, which consists of *disconnecting* selected drivers (using the NULL keyword). A signal with such a feature is called a *guarded* signal, attained by including the keyword REGISTER or BUS in its declaration, as follows.

```
SIGNAL signal_name: ... signal_type [REGISTER | BUS] [:= def_value];
```

Only resolved signals can be guarded. The difference between the REGISTER and BUS options is observed when all drivers are disconnected. A REGISTER signal keeps the value that it had before the last disconnection, while a BUS signal has its value determined by a resolution function. In large designs, the STD_LOGIC(_VECTOR) type (industry standard) is generally employed, so guarded signals are not needed.

7.3 VARIABLE

VARIABLE is a very valuable object for use in sequential code. Because it is visible only inside the sequential unit in which it was created, it represents just local information (the only exception is *shared* variables, explained shortly). It can be declared in the declarative part of PROCESS, FUNCTION, PROCEDURE, PACKAGE, and PACKAGE BODY, using the simplified syntax below.

```
VARIABLE variable_name: variable_type [range] [:= default_value];
```

Example

```
------------------------------------------------
VARIABLE flag: STD_LOGIC := '0';
VARIABLE address: NATURAL RANGE 0 TO 2**N-1;
VARIABLE data: BIT_VECTOR(15 DOWNTO 0);
------------------------------------------------
```

VARIABLE too has important properties, such as those introduced in section 6.1, and described in more detail in section 7.4.

Shared Variable

When declared as *shared*, a variable can be accessed by more than one sequential code and also by concurrent code, though only one sequential unit should modify its value. Additionally, the value of a shared variable can be passed to a signal in an assignment made *outside* the sequential code. A shared variable can be declared in ENTITY, ARCHITECTURE, BLOCK, GENERATE, and PACKAGE (the package must be not in a process or subprogram). Its use is illustrated in example 7.1.

Example 7.1: Counter with SHARED VARIABLE

Design a 00-to-99 counter employing shared variables for both digits.

Solution The code below implements a circuit that counts from 00 to 99 and then automatically restarts from 00. Even though a single process would do, two were employed in order to illustrate the use of shared variables (*temp1* and *temp2*, declared in line 8). Note that each variable is modified by only one process (*proc1* for *temp1*, *proc2* for *temp2*) and that the passing of their values to signals can be done outside the processes (lines 35–36).

```
 1    ----------------------------------------------------------
 2    ENTITY counter_with_sharedvar IS
 3       PORT (clk: IN BIT;
 4             digit1, digit2: OUT INTEGER RANGE 0 TO 9);
 5    END ENTITY;
 6    ----------------------------------------------------------
 7    ARCHITECTURE counter OF counter_with_sharedvar IS
 8       SHARED VARIABLE temp1, temp2: INTEGER RANGE 0 TO 9;
 9    BEGIN
10       ----------------------------------
11       proc1: PROCESS (clk)
```

```
12      BEGIN
13         IF (clk'EVENT AND clk='1') THEN
14            IF (temp1=9) THEN
15               temp1 := 0;
16            ELSE
17               temp1 := temp1 + 1;
18            END IF;
19         END IF;
20      END PROCESS proc1;
21      ---------------------------------
22      proc2: PROCESS (clk)
23      BEGIN
24         IF (clk'EVENT AND clk='1') THEN
25            IF (temp1=9) THEN
26               IF (temp2=9) THEN
27                  temp2 := 0;
28               ELSE
29                  temp2 := temp2 + 1;
30               END IF;
31            END IF;
32         END IF;
33      END PROCESS proc2;
34      ---------------------------------
35      digit1 <= temp1;
36      digit2 <= temp2;
37   END ARCHITECTURE;
38   -------------------------------------------------------
```

7.4 SIGNAL versus VARIABLE

Choosing between SIGNAL and VARIABLE is not always straightforward. Their main differences and usage are summarized in the six rules that follow.

Rule 1: Local of Declaration

SIGNAL: Can be declared in the declarative part of ENTITY, ARCHITECTURE, PACKAGE, BLOCK, or GENERATE.

VARIABLE: Can only be declared in sequential units (PROCESS and subprograms). The only exception is for *shared* variables, which can be declared in the same places as SIGNAL, but should only be modified by one sequential unit.

Rule 2: Scope (Local of Use)

SIGNAL: Can be global (seen and modified in the whole code, including in sequential units).

VARIABLE: Always local (seen and modified only inside the sequential unit where it was created). To leave that unit, its value must be passed directly or indirectly to a signal. The only exception is for a *shared* variable, which can be global (seen by more than one sequential unit and also by concurrent statements, though it should be modified only by one sequential unit).

Rule 3: Update

SIGNAL: A new value is only available after the conclusion of the present run of the process or subprogram.

VARIABLE: Updated immediately, so its new value is ready to be used in the next line of code.

Rule 4: Assignment Operator

SIGNAL: Values are assigned using "<=" (example: sig <= 5;).

VARIABLE: Values are assigned using ":=" (example: var := 5;).

Rule 5: Multiple Assignments

SIGNAL: Only one effective assignment is allowed in the whole code.

VARIABLE: Because its update is immediate, multiple assignments are fine.

Rule 6: Inference of Registers

SIGNAL: Flip-flops are inferred when an assignment to a signal occurs at the transition of another signal.

Rule	SIGNAL	VARIABLE
1. Local of declaration	ENTITY, ARCHITECTURE, BLOCK, GENERATE, and PACKAGE (declaration in sequential code is forbidden)	Only in sequential units (PROCESS and subprograms), except shared variables (declared in ENTITY, ARCHITECTURE, BLOCK, GENERATE, PACKAGE)
2. Scope (local of use and of modification)	Can be global (used and modified anywhere in the code)	Local (used and modified only inside its own sequential unit), except shared variables (can be global, but modified by only one sequential unit)
3. Update	New value available only at the end of the current cycle	Updated immediately (new value ready to be used in the next line of code)
4. Assignment operator	Values are assigned using "<=" Example: sig<=5;	Values are assigned using ":=" Example: var:=5;
5. Multiple assignments	Only one assignment is allowed	Multiple assignments are fine (because update is immediate)
6. Inference of registers	Flip-flops are inferred when an assignment to a signal occurs at the transition of another signal	Flip-flops are inferred when an assignment to a variable occurs at the transition of a signal and this variable's value eventually affects a signal's value

Figure 7.1
Summary of SIGNAL versus VARIABLE comparison.

VARIABLE: Flip-flops are inferred when an assignment to a variable occurs at the transition of another signal and this variable's value is eventually passed directly or indirectly to a signal.

These rules are summarized in figure 7.1. Application examples follow.

Example 7.2: SIGNAL versus VARIABLE Usage

Consider the section of code shown below, which contains a SIGNAL (*sig*, declared in the declarative part of the architecture, line 4) and a VARIABLE (*var*, declared in the declarative part of the process, line 7). Check whether any of the six rules above was violated.

```
 1   ...
 2   -----------------------------------------
 3   ARCHITECTURE example OF example IS
 4      SIGNAL sig: INTEGER RANGE -8 TO 7;
 5   BEGIN
 6      PROCESS (clock)
 7         VARIABLE var INTEGER RANGE -8 TO 7;
 8      BEGIN
 9         sig <= 0;
10         var := 0;
11         IF (clock'EVENT AND clock='1') THEN
12            sig <= sig + 1;
13            var := var + 1;
14            IF (sig=a) THEN ...
15            ELSIF (var=b) THEN ...
16            END IF;
17         END IF;
18         ...
19      END PROCESS;
20   END example:
21   -----------------------------------------
```

Solution

Rule 1: Was not violated because both *sig* and *var* were declared in right places (lines 4 and 7).

Rule 2: Is fine as well because *var* was used only inside the process (*sig* can be used anywhere in the architecture code, including in the process).

Rule 3: The assignment in line 12 increments *sig*, but the new value will only be ready at the conclusion of the present process cycle, so the test in line 14 indeed compares *a* to an outdated value of *sig*. If that was not done on purpose (although not a recommended practice, one might want to actually compare *sig* to *a* − 1), then an incorrect circuit will be inferred. Regarding *var*, being a variable, its update is immediate, so the value assigned in line 13 is already available to be used in line 15.

Rule 4: Was not violated because all assignments to *sig* and *var* (lines 9–10 and 12–13) employed the proper operators.

Rule 5: Was violated because lines 9 and 12 together make multiple assignments to *sig*. However, because this is a sequential code, such repetitions might be accepted by the compiler, but only the last assignment will survive, thus producing an incorrect circuit. Regarding *var*, no violation occurred because for variables multiple assignments (lines 10, 13) are fine.

Rule 6: This is not exactly a rule, but rather a consequence. In this example, *sig* will cause DFFs to be inferred because a value is assigned to it (line 12) at the transition of another signal (*clk*, line 11). The assignment to *var* in line 13 will infer registers if *var* affects a signal that leaves the process (which is the case in regular codes).

Example 7.3: Counters with SIGNAL and VARIABLE

Write a VHDL code that implements a regular 0-to-9 binary counter. Develop two solutions: with a signal and with a variable. If the same tests and assignments are made in both codes, will the resulting counters have the same counting range?

Solution   A VHDL code for these counters is presented below. Counter 1 employs a signal, while counter 2 uses a variable. Even though they could be designed with just one process, two processes were used to make the code easier to inspect (this does not affect the inferred circuit).

The only input is *clk* (line 3), while the outputs are *count1* and *count2* (line 4), with a range from 0 to 9. The purpose of the range is just for the compiler to know the number of bits that it should use to represent each object, so 0-to-9 and 0-to-15, for example, are equivalent.

The process for counter 1 is in lines 11–20, under the (optional) label *with_sig*. It employs a signal named *temp1*, specified in the declarative part of the architecture (line 8). Every time a rising clock edge occurs (line 13), *temp1* is incremented (line 14). When *temp1* reaches 10 (line 15), it is zeroed (line 16). The value of *temp1* is eventually passed to the *count1* output (line 19).

The process for counter 2 is in lines 22–32, under the (optional) label *with_var*. Instead of a signal, a variable, called *temp2*, is now used. Its specification is made in the declarative part of the process (line 23). Every time a rising clock edge occurs (line 25), *temp2* is incremented (line 26). When *temp2* reaches 10 (line 27), it is zeroed (line 28). The value of *temp2* is eventually passed to the *count2* output (line 31).

As requested, the same assignments and tests were made in both processes, so now we need to determine the expected results. Based on rule 3, we know that a variable is updated immediately, while the transaction scheduled for a signal will only be concluded at the end

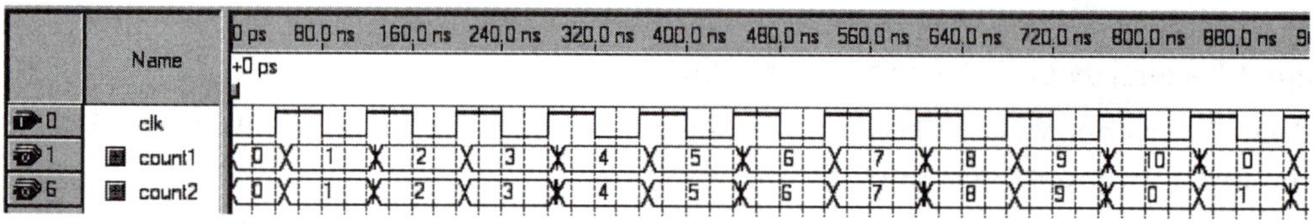

Figure 7.2
Simulation results from the counters of example 7.3.

of the process cycle. Consequently, *temp1* = 10 will only occur in the eleventh process run, meaning that counter 1 will count from 0 to 10 instead of from 0 to 9. Counter 2, on the other hand, will have the correct range. This can be observed in the simulation results of figure 7.2.

```
1   -----------------------------------------------
2   ENTITY counter IS
3      PORT (clk: IN BIT;
4             count1, count2: OUT INTEGER RANGE 0 TO 9;
5   END ENTITY;
6   -----------------------------------------------
7   ARCHITECTURE dual_counter OF counter IS
8      SIGNAL temp1: INTEGER RANGE 0 TO 10;
9   BEGIN
10     -----counter 1: with signal:-----
11     with_sig: PROCESS(clk)
12     BEGIN
13        IF (clk'EVENT AND clk='1') THEN
14           temp1 <= temp1 + 1;
15           IF (temp1=10) THEN
16              temp1 <= 0;
17           END IF;
18        END IF;
19        count1 <= temp1;
20     END PROCESS with_sig;
21     -----counter 2: with variable:-----
22     with_var: PROCESS(clk)
23        VARIABLE temp2: INTEGER RANGE 0 TO 10;
24     BEGIN
25        IF (clk'EVENT AND clk='1') THEN
26           temp2 := temp2 + 1;
27           IF (temp2=10) THEN
28              temp2 := 0;
```

```
29              END IF;
30            END IF;
31            count2 <= temp2;
32      END PROCESS with_var;
33  END ARCHITECTURE;
34  -------------------------------------------------
```

7.5   The Inference of Registers

This section discusses the number of flip-flops inferred by the compiler for a given code. The purpose is not only to understand which approaches require fewer registers, but also to make sure that the code does implement the intended circuit.

As described in rule 6 of figure 7.1, a signal generates flip-flops whenever a value is assigned to it at the *transition* of another signal. Such assignment, being synchronous, should be made inside a section of *sequential* code (PROCESS or subprogram), usually following a declaration of the type "IF clock'EVENT..." or "WAIT UNTIL clock'EVENT...".

A variable also generates flip-flops when a value is assigned to it at the transition of another signal, and this value affects a signal that affects the design (which is always the case in regular designs).

Example   Assume the following object specifications:

```
--------------------------------------
SIGNAL clk: BIT;
SIGNAL sig1: BIT_VECTOR(7 DOWNTO 0);
SIGNAL sig2: INTEGER RANGE 0 TO 7;
VARIABLE var: BIT_VECTOR(3 DOWNTO 0);
--------------------------------------
```

Then in the code below all three objects would be registered (stored) because assignments are made to all three at the transition of another signal (*clk*). A total of $8 + 3 + 4 = 15$ DFFs would then be inferred.

```
------------------------------
IF (clk'EVENT AND clk='1') THEN
   sig1 <= x;
   sig2 <= y;
   var := z;
END IF;
------------------------------
```

In the next code, only *sig2* and *var* would be stored ($3 + 4 = 7$ DFFs).

```
----------------------------------
PROCESS (clk)
BEGIN
    IF (clk'EVENT AND clk='1') THEN
        sig2 <= y;
        var := z;
    END IF; ...
    sig1 <= x;
END PROCESS;
----------------------------------
```

Additional (complete) examples are presented next to further illustrate when and why registers are inferred from SIGNAL and VARIABLE assignments.

Example 7.4: DFF with *q* and *qbar*

Figure 7.3 shows four implementations involving DFFs with *q* and *qbar* outputs. The circuits in (a) and (b) are equivalent, with the only difference that the latter does not have a built-in *qbar* output, so an inverter is needed to construct it. On the other hand, the circuit in (c) is only *functionally* equivalent to (a)–(b) and that in (d) is not equivalent at all (*qbar* is one clock period behind *q*). Examine the codes below and determine which of these circuits each code implements. Then compile the codes and check in the RTL viewer and/or fitter equations if the inferred circuits match you answers.

```
 1  ------------------------------          ------------------------------
 2  ENTITY flipflop IS                  10  proc2: PROCESS (clk)
 3  PORT (d, clk: IN BIT;               11  BEGIN
 4        q: BUFFER BIT;                12  IF clk'EVENT AND clk='1' THEN
 5        qbar: OUT BIT);               13      q <= d;
 6  END ENTITY;                         14      qbar <= NOT q;
 7  ------------------------------      15  END IF;
 8  ARCHITECTURE arch OF flipflop IS    16  END PROCESS proc2;
 9  BEGIN                                   ------------------------------
10  proc1: PROCESS (clk)                10  proc3: PROCESS (clk)
11  BEGIN                               11  BEGIN
12     IF clk'EVENT AND clk='1' THEN    12  IF clk'EVENT AND clk='1' THEN
13         q <= d;                      13      q <= d;
14         qbar <= NOT d;               14  END IF;
15     END IF;                          15  END PROCESS proc3;
16  END PROCESS proc1;                  16  qbar <= NOT q;
17  END ARCHITECTURE;                       ------------------------------
18  ------------------------------
```

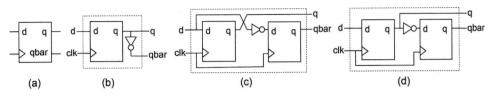

Figure 7.3
Flip-flops of example 7.4.

Solution   From the code of *proc1*, DFFs are inferred for *q* and *qbar* because values are assigned to both (lines 13–14) at the transition of another signal (*clk*, line 12). Since both assignments are related to *d*, the circuit of figure 7.3c is expected to result. Applying the same reasoning to *proc2*, again two DFFs are inferred; however, the assignment to *qbar* (line 14) is related to *q*, thus causing the circuit of figure 7.3d to be inferred. Finally, in the third code, the assignment to *qbar* (line 16) is no longer under the IF clk'EVENT ... test, so *qbar* is not registered, resulting the circuit of figure 7.3b. It is important to mention that, depending on the compiler being used and the target CPLD/FPGA device, during the optimization phase the compiler might be able to simplify *proc1* to produce the same circuit as *proc3* (but the approach of *proc3* is always recommended).

Example 7.5: Over-registered Counter

Consider the counter designed in example 6.6.

a) How many flip-flops are needed to implement it?

b) What happens if line 27 (END IF;) of that code is moved to the position between lines 41 and 42?

Solution

a) To obtain the 1 Hz signal (*counter1*), $\lceil \log_2 50M \rceil = 26$ DFFs are needed. For the 0-to-9 counter (*counter2*), four DFFs are required. Hence the total is 30 DFFs.

b) This causes the CASE statement to be under the IF clk'EVENT ... test, so *ssd* will be registered. In summary, in figure 6.9 an extra block, containing seven DFFs, would be included between the SSD driver block and the display (then totaling 37 flip-flops). Note that these DFFs are not necessary because the input to the SSD driver is already registered (this is a very popular mistake). The reader is invited to compile the code to check these answers (exercise 7.6).

7.6   Dual-Edge Circuits

Say that in a certain application dual-edge flip-flops (that is, DFFs that store data at both clock transitions) are needed. If the target CPLD/FPGA is equipped with only single-edge

DFFs (which is the case for nearly all current devices), the compiler will not be able to synthesize a dual-edge code directly into a flip-flop.

Two attempts that one might think of are shown below. The process on the left contains a counter that needs to be incremented at both clock edges, so clk'EVENT is invoked twice. The process on the right tries the same thing but in a slightly different way; because *clk* is in the sensitivity list, the process will execute every time *clk* changes, but because no AND condition is associated with clk'EVENT, one might expect again both clock edges to increment the counter. For single-edge technologies, both codes will cause the compiler to halt (one exception for the second process below is the case of compilers that place a *default* AND condition to go with clk'EVENT, so the code will be synthesized, but still in a single-edge circuit).

```
----------Not OK:----------------------        --------Not OK:----------
PROCESS (clk)                                   PROCESS (clk)
   VARIABLE count: INTEGER RANGE 0 TO 15;       BEGIN
BEGIN                                              IF (clk'EVENT) THEN
   IF (clk'EVENT AND clk='1') THEN                    count <= count + 1;
      count := count + 1;                          END IF;
   ELSIF (clk'EVENT AND clk='0') THEN              ...
      count := count + 1;                        END PROCESS;
   END IF;                                       ------------------------
   ...
END PROCESS;
------------------------------------------
```

This does not mean, however, that a process or subprogram cannot operate at both clock edges. The code below, for example, is fine, because now *distinct* variables (could be signals) are employed, one operating only at the positive transitions of *clk* and the other only at the negative transitions.

```
----------OK:----------------------------------
PROCESS (clk)
    VARIABLE count1, count2: INTEGER RANGE ...;
BEGIN
   IF (clk'EVENT AND clk='1') THEN
      count1 := count1 + 1;
   ELSIF (clk'EVENT AND clk='0') THEN
      count2 := count2 + 1;
   END IF;
   ...
END PROCESS;
------------------------------------------------
```

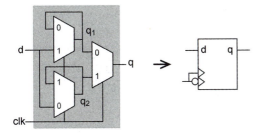

Figure 7.4
Dual-edge DFF implemented with multiplexers (the first two operate as DLs).

Example 7.6: Dual-Edge Flip-Flop

Write a VHDL code from which a circuit that resembles a dual-edge flip-flop can be inferred (assuming that no dual-edge DFFs are available in the target device).

Solution   One alternative to construct a dual-edge DFF is shown in figure 7.4 (Pedroni 2008), which employs two D-type latches (DLs) connected in parallel, followed by a multiplexer (in figure 7.4, the DLs too are implemented with multiplexers, with a feedback loop).

A code for this circuit is presented below, employing sequential code with the IF statement (one for each mux). Note that although a flip-flop will result from this code, a fully concurrent code could have been employed because the individual units (muxes) are combinational circuits, in which case the WHEN statement would render a shorter code.

```
1   ------------------------------------------
2   ENTITY dual_edge_dff IS
3   PORT (d, clk: IN BIT;
4         q: OUT BIT);
5   END ENTITY;
6   ------------------------------------------
7   ARCHITECTURE structure OF dual_edge_dff IS
8      SIGNAL q1, q2: BIT;
9   BEGIN
10  PROCESS(clk, d)
11     BEGIN
12         ---mux for q1:------
13         IF (clk='0') THEN q1 <= q1;
14         ELSE q1 <= d;
15         END IF;
16         ---mux for q2:------
```

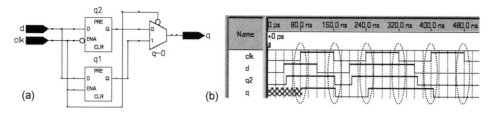

Figure 7.5
Dual-edge DFF of example 7.6. (a) RTL view; (b) Simulation results showing that both clock edges are active.

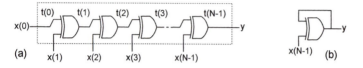

Figure 7.6
Parity detector.

```
17              IF (clk='0') THEN q2 <= d;
18              ELSE q2 <= q2;
19              END IF;
20              ---mux for q:-------
21              IF (clk='0') THEN q <= q1;
22              ELSE q <= q2;
23              END IF;
24          END PROCESS;
25      END ARCHITECTURE;
26      -------------------------------------------
```

The RTL view produced by the compiler is shown in figure 7.5a, which matches our circuit perfectly. Simulation results are depicted in figure 7.5b, where it can be observed that both clock transitions (highlighted) are indeed active.

7.7 Making Multiple Signal Assignments

According to rule 5 of figure 7.1, only a single assignment can be made to a signal in the entire code. Let us consider the design of the parity detector shown in figure 7.6a, which computes the function $y = x_0 \oplus x_1 \oplus x_2 \oplus \cdots \oplus x_{N-1}$, hence producing $y = $ '1' when the number of '1's in x in odd. Two codes for this circuit are presented below.

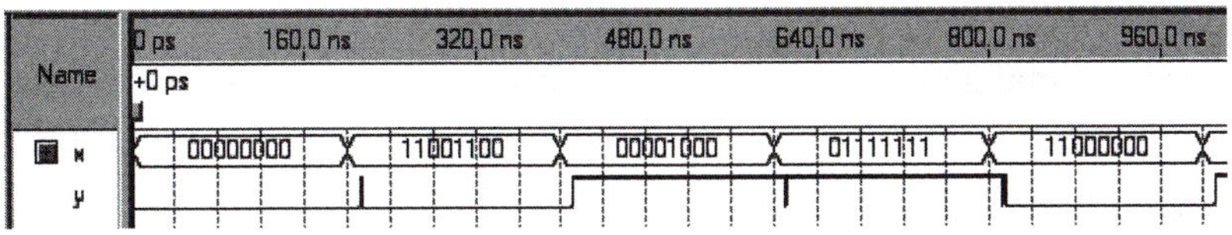

Figure 7.7
Simulation results from the parity detector of figure 7.6.

```
1    ---------------------------------        7    --------------------------
2    ENTITY parity_det IS                      8    ARCHITECTURE not_ok OF ...
3      GENERIC (N: POSITIVE := 8);             9       SIGNAL temp: BIT;
4      PORT (x: IN BIT_VECTOR(N-1 DOWNTO 0);  10    BEGIN
5            y: OUT BIT);                      11      PROCESS (x)
6    END ENTITY;                               12      BEGIN
7    ---------------------------------        13        temp <= x(0);
8    ARCHITECTURE not_ok OF parity_det IS     14        FOR i IN 1 TO N-1 LOOP
9       SIGNAL temp: BIT;                     15          temp <= temp XOR x(i);
10   BEGIN                                     16        END LOOP;
11     temp <= x(0);                           17      y <= temp;
12     gen: FOR i IN 1 TO N-1 GENERATE         18     END PROCESS;
13       temp <= temp XOR x(i);               19   END ARCHITECTURE;
14     END GENERATE;                           20   --------------------------
15     y <= temp;
16   END ARCHITECTURE;
17   ---------------------------------
```

The first code above is concurrent (with GENERATE), while the second is sequential (with PROCESS and LOOP). In both, *temp* is a SIGNAL (line 9), and in both multiple assignments are made to it (lines 11 and 13 in the first code, lines 13 and 15 in the second), so neither is expected to render the correct circuit.

With the concurrent code, the compiler will normally issue an error message saying that *temp* is multiply driven and then quit compilation. The sequential code, on the other hand, is synthesizable because the assignments are made sequentially; however, because a new value of a signal is only ready at the conclusion of the process run, only the last assignment will actually occur—that is, $y = y \oplus x_{N-1}$, thus resulting the circuit of figure 7.6b, which is not what we wanted. Note also that when $x_{N-1} = \,'1'$ this circuit becomes a ring oscillator.

A solution for this problem is presented below, which consists of using a signal whose dimension is one unit higher than the dimension of the signal being computed. In this example, *y* is a single bit (hence a scalar, according to the definition in figure 3.2); consequently, a 1D signal must be employed, shown in line 9 (so *temp* was replaced with $temp(N-1:0)$, which is the internal vector represented as *t* in figure 7.6a). Note that now each assignment involves a *different part* of *temp*, so rule 5 is no longer violated and the correct circuit results. Simulation results from this code are shown in figure 7.7.

```
 7  ----------------------------------------
 8  ARCHITECTURE ok OF parity_det IS
 9     SIGNAL temp: BIT_VECTOR(N-1 DOWNTO 0);
10  BEGIN
11     temp(0) <= x(0);
12     gen: FOR i IN 1 TO N-1 GENERATE
13        temp(i) <= temp(i-1) XOR x(i);
14     END GENERATE;
15     y <= temp(N-1);
16  END ARCHITECTURE;
17  ----------------------------------------
```

Note: It is important to remember that this kind of problem does not occur when VARIABLE is used because its value is updated immediately, so multiple assignments are fine.

Example 7.7: Generic Hamming Weight with Concurrent Code

Exercise 5.7 asked to design a circuit that computes the number of '1's in a binary vector using the GENERATE statement. Exercise 6.6 asked the same thing, but using the LOOP statement. While the latter is simple, the former requires some additional effort. That exercise is an excellent opportunity to illustrate the extra dimension required in the auxiliary signal in order to allow multiple assignments to be made to it, because now a *different part* of the signal is involved in each iteration. Design such a circuit using only concurrent code (operators, WHEN, SELECT, GENERATE).

Solution Two solutions are presented below, both with N (size of the input vector) declared as a generic parameter (line 3). In the first solution, *temp* is declared as an integer in the 0 to N range, because that is the range of y (the value of *temp* must be passed to y eventually—line 15). For the reasons explained above, this solution is not fine (it contains multiple assignments to *temp*—lines 11 and 13). The second solution adopts the described approach, which increases the dimension of *temp* by one unit (from 1D to 1D $\times$ 1D). This is done in lines 9–10. Now the assignments in lines 12 and 14 are no longer to the same portions of *temp*. Note that y is still 1D, so only the last value of *temp* is passed to y (line 16).

```
 1  ------------------------------------------------------------
 2  ENTITY hamm_weight IS
 3     GENERIC (N: POSITIVE := 8);
 4     PORT (x: IN BIT_VECTOR(N-1 DOWNTO 0);
 5           y: OUT INTEGER RANGE 0 TO N);
 6  END ENTITY;
 7  ------------------------------------------------------------
```

```
 8  ARCHITECTURE not_ok OF hamm_weight IS
 9     SIGNAL temp: INTEGER RANGE 0 TO N;
10  BEGIN
11     temp <= 0;
12     gen: FOR i IN 1 TO N-1 GENERATE
13        temp <= temp + 1 WHEN x(i)='1' ELSE temp;
14     END GENERATE;
15     y <= temp;
16  END ARCHITECTURE;
17  --------------------------------------------------------------

 8  ARCHITECTURE ok OF hamm_weight IS
 9     TYPE oneDoneD is ARRAY (0 TO N-1) OF INTEGER RANGE 0 TO N;
10     SIGNAL temp: oneDoneD;
11  BEGIN
12     temp(0) <= 0;
13     gen: FOR i IN 1 TO N-1 GENERATE
14        temp(i) <= temp(i-1) + 1 WHEN x(i)='1' ELSE temp(i-1);
15     END GENERATE;
16     y <= temp(N-1);
17  END ARCHITECTURE;
18  --------------------------------------------------------------
```

7.8 Exercises

Note: For exercise solutions, please consult the book website.

Exercise 7.1: SIGNAL versus VARIABLE #1

Consider the two sections of code below.

a) Is *count* a signal or a variable? Is *count* the name of an actual circuit output?

b) What is the range of the counter in each code below?

c) For the code on the left, what are the consequences of declaring *count* as a signal versus declaring it as a variable?

d) Make the same analysis for the code on the right.

```
IF (clk'EVENT AND clk='1') THEN      IF (clk'EVENT AND clk='1') THEN
   count := count + 1;                  IF (count=25) THEN
   IF (count=25) THEN                      count := 1;
      count := 1;                       ELSE
   END IF;                                 count := count + 1;
END IF;                                 END IF;
pointer <= count;                    END IF;
                                     pointer <= count;
```

Exercise 7.2: SIGNAL versus VARIABLE #2

Consider the 0-to-9 counter of example 7.3.

a) Check in the complete code the application of the 6 rules described in figure 7.1 (as in example 7.2).

b) In each process, which lines of code are responsible for the inference of registers?

c) How many flip-flops are inferred in each process? Why?

Exercise 7.3: Latches and Flip-Flops

Examine the three codes given below.

a) Draw the circuit that you expect will be inferred in each case.

b) Compile the codes in order to verify your answers.

```
1  ----------------------------          11      IF (clk'EVENT AND clk='1') THEN
2  ENTITY test IS                        12          IF (rst='1') THEN
3     PORT (d, clk, rst: IN BIT;         13             q <= '0';
4          q: OUT BIT);                  14          ELSE
5  END ENTITY;                           15             q <= d;
6  -----code 1:-----------------         16          END IF;
7  ARCHITECTURE circuit OF test IS       17      END IF;
8  BEGIN                                 18  END PROCESS;
9     PROCESS (d, clk, rst)                  ----------------------------------
10    BEGIN
11       IF (rst='1') THEN                    -----code 3:---------------------
12          q <= '0';                      9  PROCESS (clk)
13       ELSIF (clk='1') THEN            10  BEGIN
14          q <= d;                      11     IF (clk='1') THEN
15       END IF;                         12         IF (rst='1') THEN
16    END PROCESS;                       13            q <= '0';
17 END ARCHITECTURE;                     14         ELSE
18 ----------------------------          15            q <= d;
                                         16         END IF;
   -----code 2:-----------------         17     END IF;
9  PROCESS (clk)                         18  END PROCESS;
10 BEGIN                                     ----------------------------------
```

Exercise 7.4: Combinational versus Sequential Circuits #1

Consider the carry-ripple adder seen in example 6.4.

a) Is it a combinational or sequential circuit? Why?

b) Looking at the code, explain why it does not infer registers.

Consider now the leading-zeros counter of example 6.5.

c) Is it a combinational or sequential circuit? Why?

d) Explain why this code does not infer registers.

Exercise 7.5: Combinational versus Sequential Circuits #2

Consider the counter designed in example 6.6.

a) What portions of the code generate a sequential circuit and which generate a combinational circuit? Why?

b) Could the portion implemented by CASE be moved outside the process? If so, which statement should be used? Would CASE still be OK?

Exercise 7.6: Over-registered Counter

In example 7.5, an analysis of the counter designed in example 6.6 was made. Compile the code presented in the latter to confirm the answers presented in the former.

Exercise 7.7: Registered Circuits

a) Examine the three codes below, all involving registers, and draw the circuit that you expect will be inferred in each case. How many DFFs are needed?

b) Compile the codes in order to verify your answers.

c) Comment: Are any of these circuits functionally equivalent to each other? Which one would you choose? Why?

```
1   ----------------------------------          -----code 2:---------------------------
2   ENTITY test IS                               11  PROCESS (clk)
3     PORT (clk: IN BIT;                         12    VARIABLE temp: BIT_VECTOR(7 DOWNTO 0);
4           x: IN BIT_VECTOR(7 DOWNTO 0);        13  BEGIN
5           sel: IN INTEGER RANGE 0 TO 7;        14    IF clk'EVENT AND clk='1' THEN
6           y: OUT BIT);                         15      temp := x;
7   END ENTITY;                                  16    END IF;
8   -----code 1:------------------------         17    y <= temp(sel);
9   ARCHITECTURE circuit OF test IS              18  END PROCESS;
10  BEGIN                                            ---------------------------------------
11    PROCESS (clk)
12      VARIABLE temp: BIT;                          -----code 3:---------------------------
13    BEGIN                                      11  PROCESS (clk)
14      temp := x(sel);                          12  BEGIN
15      IF clk'EVENT AND clk='1' THEN            13    IF clk'EVENT AND clk='1' THEN
16        y <= temp;                             14      y <= x(sel);
17      END IF;                                  15    END IF;
18    END PROCESS;                               16  END PROCESS;
19  END ARCHITECTURE;                                ---------------------------------------
    ----------------------------------
```

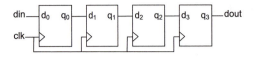

Figure 7.8

Exercise 7.8: Shift Register

Figure 7.8 shows a 4×1 shift register. Examine the three codes below and answer:

a) Which of these codes implement the circuit of figure 7.8?

b) Compile the codes in order to verify your answer.

Note: The recommended approach to design shift registers is that seen in example 6.3.

```
1  --------------------------------
2  ENTITY test IS
3    PORT (din, clk: IN BIT;
4          dout: OUT BIT);
5  END ENTITY;
6  -----code 1:--------------------
7  ARCHITECTURE circ1 OF test IS
8    SIGNAL q0, q1, q2: BIT;
9  BEGIN
10   PROCESS (clk)
11   BEGIN
12     IF clk'EVENT AND clk='1' THEN
13       q0 <= din;
14       q1 <= q0;
15       q2 <= q1;
16       dout <= q2;
17     END IF;
18   END PROCESS;
19 END ARCHITECTURE;
20 --------------------------------

6  -----code 2:--------------------
7  ARCHITECTURE circ2 OF test IS
8  BEGIN
9    PROCESS (clk)
10     VARIABLE q0, q1, q2, q3: BIT;
11     BEGIN
```

```
12     IF (clk'EVENT AND clk='1') THEN
13       q0 := din;
14       q1 := q0;
15       q2 := q1;
16       q3 := q2;
17     END IF;
18     dout <= q3;
19   END PROCESS;
20 END ARCHITECTURE;
21 ------------------------------------

6  -----code 3:------------------------
7  ARCHITECTURE circ3 OF test IS
8  BEGIN
9    PROCESS (clk)
10     VARIABLE q0, q1, q2, q3: BIT;
11     BEGIN
12     IF clk'EVENT AND clk='1' THEN
13       q3 := q2;
14       q2 := q1;
15       q1 := q0;
16       q0 := din;
17     END IF;
18     dout <= q3;
19   END PROCESS;
20 END ARCHITECTURE;
21 ------------------------------------
```

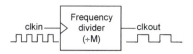

Figure 7.9

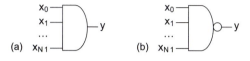

Figure 7.10

Exercise 7.9: Frequency Divider with VARIABLE

Figure 7.9 depicts a circuit that must divide the clock frequency by a generic integer M. In this exercise, there are no restrictions regarding the phase of *clkout* (that is, it is not required to be symmetric or with any other particular duty cycle).

a) Estimate the number of DFFs that will be needed for $M = 4$ and $M = 5$.

b) Write a VHDL code for this circuit. Enter M as a *generic* parameter and use a VARIABLE to implement the counter. Compile and simulate it for $M = 4$ and $M = 5$, checking also whether the numbers of registers match your predictions.

c) Assume that *clkout* is required to be glitch free. Check if your solution is subject to glitches or not. If it is, provide a way of cleaning it. (Hint: Check whether *clkout* comes directly from a flip-flop.)

Exercise 7.10: Frequency Divider with SIGNAL

If in the code for exercise 7.9 above the variable that implements the counter is replaced with a signal, without changing any numeric value in the code, by which integer value will the clock frequency be divided?

Exercise 7.11: Frequency Divider with Symmetric Phase

Redesign the frequency divider of exercise 7.9 such that it now produces an output with symmetric phase (50% duty cycle), even when M is odd (in this case, the circuit needs to operate at *both* clock transitions). Will the number of DFFs change?

Exercise 7.12: Generic AND with Concurrent Code

Figure 7.10a shows an AND gate with a *generic* number of inputs (N). Design this circuit using *concurrent* code (chapter 5). If necessary, use the technique introduced in section 7.7.

Exercise 7.13: Generic NAND with Concurrent Code

Figure 7.10b shows a NAND gate with a *generic* number of inputs (N). Design this circuit using *concurrent* code (chapter 5). If necessary, use the technique introduced in section 7.7.

II SYSTEM-LEVEL VHDL

8 PACKAGE and COMPONENT

8.1 Introduction

This chapter and the next deal with four VHDL units whose main purpose is to allow code partitioning, code sharing, and code reuse: PACKAGE, COMPONENT, FUNCTION, and PROCEDURE. Since these units can be (and often are) located outside the main code (in libraries), we refer to them as *system-level* units. The first two are studied in this chapter, while the other two will be covered in the next.

The relationship between the system-level units and the main code is illustrated in figure 8.1. If a certain library is declared in the main code, then its parts can be used in the design. A library can contain *packages* (with *functions*, *procedures*, and several kinds of *declarations*) and other designs (these are instantiated in the main code using the keyword COMPONENT). The construction and use of libraries are essential to the designs presented in this chapter and in the next.

8.2 PACKAGE

To construct a package, two sections of code might be needed, called PACKAGE and PACKAGE BODY, shown in the simplified syntax below.

```
PACKAGE package_name IS
  declarative_part
END [PACKAGE] [package_name];
------------------------------------
[PACKAGE BODY package_name IS
  [subprogram_body]
  [deferred_constant_specifications]
END [PACKAGE BODY] [package_name]];
```

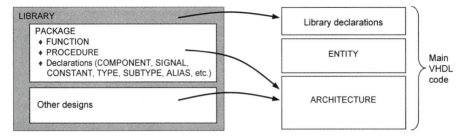

Figure 8.1
Relationship between the main code and the units intended mainly for system-level design (libraries).

The first part (PACKAGE) must contain only *declarations*, which can include subprogram declaration, type declaration, subtype declaration, constant declaration, signal declaration, shared variable declaration, file declaration, alias declaration, component declaration, attribute declaration, attribute specification, disconnection specification, use clause, group template declaration, and group declaration.

The second part (PACKAGE BODY) is needed only when a subprogram (FUNCTION or PROCEDURE) or a deferred constant (with undefined value—see section 3.2) is declared in the upper part. In this case, the full subprogram body or the full constant specification must be exhibited. The full subprogram header, exactly as it appears in the PACKAGE, must be repeated in the PACKAGE BODY. The latter can contain the same kinds of declarations as the former, with the exception of signal declaration, component declaration, attribute declaration, attribute specification, and disconnection specification. The former, on the other hand, can obviously not contain subprogram bodies.

Example   The PACKAGE below is called *my_package* and contains only TYPE, SIGNAL, and (complete) CONSTANT declarations (thus PACKAGE BODY is not necessary).

```
-------------------------------------------------
PACKAGE my_package IS
   TYPE matrix IS ARRAY (1 TO 3, 1 TO 3) OF BIT;
   SIGNAL x: matrix;
   CONSTANT max1, max2: INTEGER := 255;
END PACKAGE;
-------------------------------------------------
```

In VHDL 2008, a header was added to PACKAGE, allowing the declaration of GENERIC constants. In the declarative part, the following additional declarations are allowed: subprogram instantiation declaration, package declaration, package instantiation declaration, and PSL declarations. PACKAGE BODY is still for constructing subprogram bodies and for specifying deferred constants.

Example 8.1: PACKAGE with FUNCTION and Deferred CONSTANT

The code below shows, in lines 1–9, a PACKAGE called *my_package*, which contains a deferred constant (called *flag*) and also a function (called *down_edge*), so in this case a PACKAGE BODY is needed. Knowing that the constant value must be '1' and that this function should be equivalent to "clk'EVENT AND clk = '0'", write a code for the PACKAGE BODY.

```
1   ---------------------------------------------------------------
2   LIBRARY ieee;
3   USE ieee.std_logic_1164.all;
4   ---------------------------------------------------------------
5   PACKAGE my_package IS
6      CONSTANT flag: STD_LOGIC;
7      FUNCTION down_edge(SIGNAL s: STD_LOGIC) RETURN BOOLEAN;
8   END my_package;
9   ---------------------------------------------------------------
10  PACKAGE BODY my_package IS
11     CONSTANT flag: STD_LOGIC := '1';
12     FUNCTION down_edge(SIGNAL s: STD_LOGIC) RETURN BOOLEAN IS
13     BEGIN
14        RETURN (s'EVENT AND s='0');
15     END down_edge;
16  END my_package;
17  ---------------------------------------------------------------
```

Solution   The code was included in lines 10–16, with the deferred constant specified in line 11 and the function body in lines 12–15 (this function returns TRUE when a negative clock edge occurs). Note that the function header (line 12) is an exact copy of the function declaration (line 7). Observe also that the package *std_logic_1164* was included in lines 2–3 because the data type STD_LOGIC was used in the code.

8.3   COMPONENT

COMPONENT is simply a *conventional* code (that is, library/package declarations + ENTITY + ARCHITECTURE). However, its declaration as COMPONENT allows reusability and also the construction of *hierarchical* designs. Commonly used digital subsystems, like adders, multipliers, multiplexers, and the like are often compiled using this technique. Designs based on components are referred to as *structural* designs.

COMPONENT can be declared in ARCHITECTURE, PACKAGE, GENERATE, and BLOCK. To use it, two sections of code are needed: one for *declaring* it, the other for *instantiating* it. Both are shown in the simplified syntaxes below.

COMPONENT declaration:

```
COMPONENT component_name [IS]
    [GENERIC (
        const_name: const_type := const_value;
        const_name: const_type := const_value;
        ...);]
    PORT (
        port_name: port_mode signal_type;
        port_name: port_mode signal_type;
        ...);
END COMPONENT [component_name];
```

COMPONENT instantiation:

```
label: [COMPONENT] component_name
        [GENERIC MAP (generic_list)]
        PORT MAP (port_list);
```

The *component declaration*, shown in the first syntax above, must be an exact copy of the ENTITY of the design being instantiated. The GENERIC specification only needs to be included if the component contains a GENERIC list and one or more values in that list must be changed by the instantiating design (done with GENERIC MAP).

The *component instantiation*, shown in the second syntax above, starts with a mandatory label, followed by the optional word COMPONENT, then the component's name (name of the ENTITY in the design being instantiated), an optional GENERIC MAP declaration (explained in section 8.4), and finally the respective PORT MAP declaration, which associates the port names in the new design to the port names in the design being instantiated. GENERIC MAP is necessary when the original component has a GENERIC specification in the ENTITY header and one or more of its values must be overwritten by the new design.

Example   Say that a 3-input NAND gate, called *nand3*, has been previously designed and we now want to use it as part of a new design. This is illustrated in the code below. The first part contains the *component declaration*, which must be exactly as in the original entity. The second part shows the *component instantiation*, labeled *nand_gate*, with the current *x1*, *x2*, *x3*, and *y* ports assigned to the original *a1*, *a2*, *a3*, and *b* ports, respectively.

This is called *positional* mapping, because the first signal in one list corresponds to the first signal in the other, the second in one to the second in the other, and so on.

```
-----COMPONENT declaration:-------------------------
COMPONENT nand3 IS
   PORT (a1, a2, a3: IN STD_LOGIC; b: OUT STD_LOGIC);
END COMPONENT;

-----COMPONENT instantiation:-------------
nand_gate: nand3 PORT MAP (x1, x2, x3, y);
```

Mapping Options

PORT MAP is simply a list relating the ports of the actual circuit to the ports of the predesigned circuit (component being instantiated). Such a mapping can be *positional* or *nominal*, as illustrated in the example below, which employs the *nand3* circuit again. (An expansion specified in VHDL 2008 allows the use of expressions in PORT MAP assignments.)

```
-------- Component instantiation: ----------------------
nand3_1: nand3 PORT MAP (x1, x2, x3, y); --positional mapping
nand3_2: nand3 PORT MAP (a1=>x1, a2=>x2, a3=>x3, b=>y); --nominal mapping
nand3_3: nand3 PORT MAP (x1, x2, x3, OPEN); --positional mapping
nand3_4: nand3 PORT MAP (a1=>x1, a2=>x2, a3=>x3, b=>OPEN); --nominal mapping
```

The first two instantiations are equivalent, just using distinct mapping options. The last two instantiations are also equivalent, now with the keyword OPEN employed to indicate that the circuit's output must be left unconnected.

COMPONENT Declaration Options

Figure 8.2 shows two common ways of using components. In (a), the components are in a library, with the main code including their declarations and also their instantiations. In (b), the declarations too are in a library (in a specific package), so only the instantiations are needed in the main code.

File/Project Assembling Options

A code employing COMPONENT can be entered into the compiling environment in several ways.

• *Method 1*:   The complete code is entered in a single file, saved with the same name as the main entity's name. In this case, the component declarations are also normally included in the main code (as in figure 8.2a). In summary, everything goes in just one (big) file.

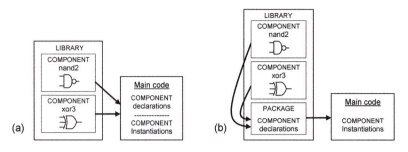

Figure 8.2
Typical COMPONENT usage: (a) With the declarations and instantiations in the main code; (b) With the declarations in a separate package, thus with only the instantiations in the main code.

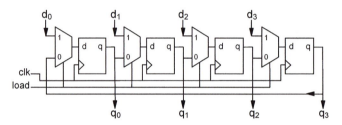

Figure 8.3
Programmable circular shift register of example 8.2.

▪ *Method 2*:   Each component is previously designed in a separate project, which is compiled either into its own work library or into the current work library. If the former is adopted, then USE clauses pointing to those libraries are needed in the main code. The component declarations are placed in the main code (as in figure 8.2a).

▪ *Method 3*:   This is the same as method 2, but with the component declarations in a separate package (as in figure 8.2b). In this case, a USE clause pointing to that package is needed in the main code.

The example below illustrates the use of method 1.

Example 8.2: Circular Shift Register with COMPONENT

Figure 8.3 shows a circular shift register with a programmable rotating sequence. The inputs are *clk*, *load*, and $d = d_0d_1d_2d_3$, with the latter employed to load the desired sequence into the flip-flops when *load* = '1'. The output is $q = q_0q_1q_2q_3$, which displays the instantaneous value of the rotating sequence. Note that all cells in this circuit are alike and composed of a multiplexer plus a DFF. Design this circuit using COMPONENT to instantiate the multiplexers and DFFs. Adopt the following approach: use method 1 of section 8.3 to enter the code, with *positional* mapping in it.

Solution A corresponding VHDL code is presented below. It consists of just one VHDL file (called *circular_shift.vhd* in this example), having all component declarations included in the main code (lines 11–17), hence complying with method 1. The labels chosen for the component instances are *mux1*, *mux2*, and so on for the multiplexers (lines 20–23), and *dff1*, *dff2*, and so on for the DFFs (lines 24–27). An internal signal called *i*(0:3) was declared in line 9 to provide interface between the muxes and the flip-flops.

```
 1  -----Multiplexer:-----------------------------
 2  ENTITY mux IS
 3     PORT (a, b, sel: IN BIT;
 4            x: OUT BIT);
 5  END ENTITY;
 6  ---------------------------------------------
 7  ARCHITECTURE mux OF mux IS
 8  BEGIN
 9    x <= a WHEN sel='0' ELSE b;
10  END ARCHITECTURE;
11  ---------------------------------------------
```

```
 1  -----Flip-flop:------------------------------
 2  ENTITY flipflop IS
 3     PORT (d, clk: IN BIT;
 4            q: OUT BIT);
 5  END ENTITY;
 6  ---------------------------------------------
 7  ARCHITECTURE flipflop OF flipflop IS
 8  BEGIN
 9     PROCESS (clk)
10     BEGIN
11       IF (clk'EVENT AND clk='1') THEN
12          q <= d;
13       END IF;
14     END PROCESS;
15  END ARCHITECTURE;
16  ---------------------------------------------
```

```
 1  -----Main code:------------------------------
 2  ENTITY circular_shift IS
 3     PORT (clk, load: IN BIT;
 4            d: IN BIT_VECTOR(0 TO 3);
 5            q: BUFFER BIT_VECTOR(0 TO 3));
 6  END ENTITY;
 7  ---------------------------------------------
 8  ARCHITECTURE structural OF circular_shift IS
```

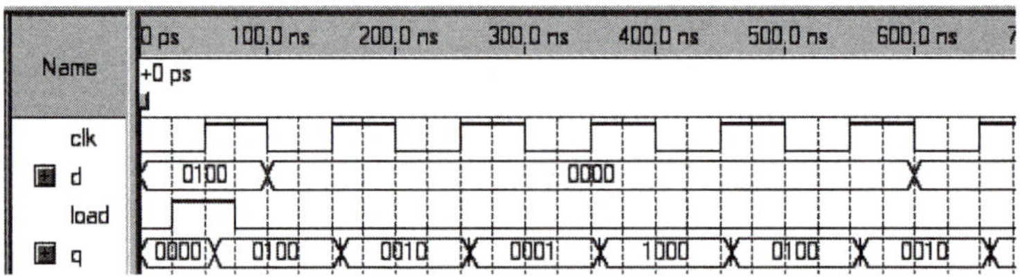

Figure 8.4
Simulation results from the circular shift register of example 8.2.

```
9      SIGNAL i: BIT_VECTOR(0 TO 3);
10     ----------------------
11     COMPONENT mux IS
12        PORT (a, b, sel: IN BIT; x: OUT BIT);
13     END COMPONENT;
14     ----------------------
15     COMPONENT flipflop IS
16        PORT (d, clk: IN BIT; q: OUT BIT);
17     END COMPONENT;
18     ----------------------
19   BEGIN
20      mux1: mux PORT MAP (q(3), d(0), load, i(0));
21      mux2: mux PORT MAP (q(0), d(1), load, i(1));
22      mux3: mux PORT MAP (q(1), d(2), load, i(2));
23      mux4: mux PORT MAP (q(2), d(3), load, i(3));
24      dff1: flipflop PORT MAP (i(0), clk, q(0));
25      dff2: flipflop PORT MAP (i(1), clk, q(1));
26      dff3: flipflop PORT MAP (i(2), clk, q(2));
27      dff4: flipflop PORT MAP (i(3), clk, q(3));
28   END ARCHITECTURE;
29   -----------------------------------------------
```

Simulation results are shown in figure 8.4. Note that the input sequence is $d =$ "0100", which is loaded into the shift register when *clk* goes up with *load* = '1', and then rotates, moving one position to the right at every positive clock transition.

In VHDL 2008, expressions are allowed in PORT MAP assignments. See other details in section 8.8.

8.4   GENERIC MAP

Completely generic code (for libraries, for example) can be attained using GENERIC declarations (seen in section 2.6 and used extensively in the examples). When a COMPO-

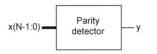

Figure 8.5
Generic parity detector of example 8.3.

NENT containing such declarations is instantiated, the values originally given to the generic parameters can be overwritten by including a GENERIC MAP declaration in the component instantiation. The corresponding syntax was already presented in section 8.3.

Example A component called *and_gate* is declared below, which has a generic parameter called *inputs*, whose (optional) default value is 8. The two instantiations shown next (*a1*, *a2*) are equivalent. In the first, the mapping is *positional*, while in the second, it is *nominal*. In either case, the default value of *inputs* (8) is replaced with 16.

```
-----Component declaration:-------------------------------
COMPONENT and_gate IS
   GENERIC (inputs: POSITIVE := 8); --see Note below
   PORT (a: IN BIT_VECTOR(1 TO inputs);
         b: OUT BIT);
END COMPONENT;
----------------------------------------------------------

-----Component instantiation:-----------------------------
a1: and_gate GENERIC MAP (16) PORT MAP (x, y);
a2: and_gate GENERIC MAP (inputs=>16) PORT MAP (a=>x, b=>y);
----------------------------------------------------------
```

Note: GENERIC does not need to be included in the component declaration above if GENERIC MAP is not used in the component instantiation and the GENERIC values specified in the main code coincide with those in the component code (this is fine even if they do not have the same name). Also, in the component code and in the component declaration, the GENERIC values can be left unspecified, but then GENERIC MAP is obviously required in order to specify them.

The use of GENERIC MAP is illustrated in example 8.3.

Example 8.3: Parity Detector with COMPONENT and GENERIC MAP

The generic *N*-bit parity detector of figure 8.5 must produce $y = $ '1' when the number of '1's in *x* is odd, or $y = $ '0' otherwise (an implementation for this kind of circuit was seen in section 7.7). Write a VHDL code to solve this problem, where the parity detector is entered as a COMPONENT that employs a GENERIC declaration to define *N*. Include a GENERIC MAP declaration in your code to overwrite the original value of *N*. Adopt the following approach: use method 1 of section 8.3 to enter the code, with *nominal* mapping in it.

Solution A VHDL code for this circuit is presented below. First, the component (*par_detector*) is shown, which has a generic parameter called *bits* (line 3) that defines the size of the input vector (left unspecified in this example). In the main code (*parity_detector*), this component is declared in lines 10–14 and then used in lines 17–18, with GENERIC MAP (line 17) defining the actual value of *bits*. Note in lines 17–18 that nominal mappings were utilized.

```
1    ----------The component:-------------------------
2    ENTITY par_detector IS
3       GENERIC (bits: POSITIVE);
4       PORT (input: IN BIT_VECTOR(bits-1 DOWNTO 0);
5             output: OUT BIT);
6    END par_detector;
7    --------------------------------------------------
8    ARCHITECTURE behavior OF par_detector IS
9    BEGIN
10      PROCESS(input)
11         VARIABLE temp: BIT;
12      BEGIN
13         temp := '0';
14         FOR i IN input'RANGE LOOP
15            temp := temp XOR input(i);
16         END LOOP;
17         output <= temp;
18      END PROCESS;
19   END behavior;
20   --------------------------------------------------
```

```
1    ----------Main code:-------------------------
2    ENTITY parity_detector IS
3       GENERIC (N: POSITIVE := 8);
4       PORT (x: IN BIT_VECTOR(N-1 DOWNTO 0);
5             y: OUT BIT);
6    END parity_detector;
7    --------------------------------------------------
8    ARCHITECTURE structural OF parity_detector IS
9       ----------------------------
10      COMPONENT par_detector IS
11         GENERIC (bits: POSITIVE);
12         PORT (input: IN BIT_VECTOR(bits-1 DOWNTO 0);
```

```
13                 output: OUT BIT);
14      END COMPONENT;
15      ---------------------------
16   BEGIN
17      det: par_detector GENERIC MAP (bits=>N)
18         PORT MAP(input=>x, output=>y);
19   END structural;
20   ----------------------------------------------------
```

8.5 COMPONENT Instantiation with GENERATE

As mentioned in section 8.3, one of the places where COMPONENT can be used is in
GENERATE loops (studied in section 5.5). This method is very helpful when a large or
an arbitrary number of instantiations of the same component must be made. The typical
use is shown below, where *gen* and *comp* are the labels chosen for the GENERATE statement and for the COMPONENT instantiation, respectively.

```
----------------------------------------------
gen: FOR i IN 0 TO max GENERATE
   comp: my_component PORT MAP (x(i), y(i), z(i));
END GENERATE gen;
----------------------------------------------
```

An example of component instantiation with GENERATE was already presented in
chapter 5 (see example 5.5). Below is another example.

Example 8.4: Shift Register with COMPONENT and GENERATE

Figure 8.6 shows a truly generic $M \times N$ shift register with data-load capability (Pedroni
2008) (a related circuit was described in example 8.2). The inputs are *clk*, *load*, *x*, and *d*
(the flip-flops are synchronously loaded with the values of *d* when *load* = '1'), while the
(only) output is *y*. As in example 8.2, design this circuit using COMPONENT to instantiate the multiplexers and flip-flops. Adopt the following approach: use method 3 of section
8.3 to enter the code, with *positional* mapping in it.

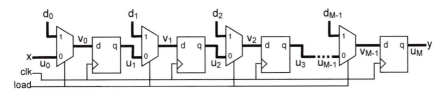

Figure 8.6
Generic shift register of example 8.4.

Solution   Note in figure 8.6 that the circuit has *M* stages with *N* bits each, hence totaling $M \times N$ mux-DFF pairs. A VHDL code for this circuit is presented below. The multiplexer and flip-flop circuits are those seen in example 8.2, here compiled in separate projects (because method 3 is to be used to enter the design). As can be seen in the code below, the component declarations are in a separate package (called *my_declarations*), so a USE clause pointing to it was included in line 2 of the main code. Note that the package contains also a TYPE declaration (*twoD*, line 3) to create a 2D array of BIT elements, which is used in the main code to specify the *d* input (line 9) and also the internal signals *u* and *v* (lines 14–15). The code proper contains two sections, both with the GENERATE statement. The first section of code (lines 18–21) transfers *x* to the first slice of *u* and the last slice of *u* to *y* (if one prefers, the transfer of *y* can be done in a separate section at the end of the code). The second section of code (lines 23–28) computes the values for the internal 2D arrays *u* and *v*.

```
1    -----Package:------------------------------------------------------
2    PACKAGE my_declarations IS
3       TYPE twoD IS ARRAY (NATURAL RANGE <>, NATURAL RANGE <>) OF BIT;
4       ----------------
5       COMPONENT mux IS
6          PORT (a, b, sel: IN BIT; x: OUT BIT);
7       END COMPONENT;
8       ----------------
9       COMPONENT flipflop IS
10         PORT (d, clk: IN BIT; q: OUT BIT);
11      END COMPONENT;
12      ----------------------
13   END PACKAGE;
14   ------------------------------------------------------------------
```

```
1    -----Main code:---------------------------------------------------
2    USE work.my_declarations.all;
3    ------------------------------------------------------------------
4    ENTITY shift_register IS
5       GENERIC (M: POSITIVE := 4;
6                N: POSITIVE := 8);
7       PORT (clk, load: IN BIT;
8             x: IN BIT_VECTOR(N-1 DOWNTO 0);
9             d: IN twoD(0 TO M-1, N-1 DOWNTO 0);
10            y: OUT BIT_VECTOR(N-1 DOWNTO 0));
11   END ENTITY;
12   ------------------------------------------------------------------
13   ARCHITECTURE structural OF shift_register IS
```

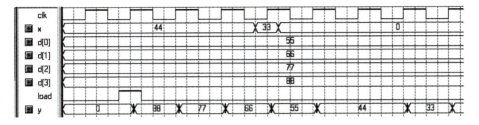

Figure 8.7
Simulation results from the shift register of example 8.4.

```
14      SIGNAL u: twoD(0 TO M, N-1 DOWNTO 0);
15      SIGNAL v: twoD(0 TO M-1, N-1 DOWNTO 0);
16   BEGIN
17      ----Transfer x->u and u->y:---------
18      gen1: FOR i IN N-1 DOWNTO 0 GENERATE
19         u(0,i) <= x(i);
20         y(i) <= u(M,i);
21      END GENERATE gen1;
22      ----Update internal array:----------
23      gen2: FOR i IN 0 TO M-1 GENERATE
24         gen3: FOR j IN N-1 DOWNTO 0 GENERATE
25            mux1: mux PORT MAP (u(i,j), d(i,j), load, v(i,j));
26            dff1: flipflop PORT MAP (v(i,j), clk, u(i+1,j));
27         END GENERATE gen3;
28      END GENERATE gen2;
29   END ARCHITECTURE;
30   -------------------------------------------------------------------
```

Simulation results are displayed in figure 8.7 (for $N = 8$ and $M = 4$). Note that d is loaded into the shift register at the first positive clock transition after *load* is asserted. Observe also that each individual value of d lies in the 0-to-255 range (because $N = 8$) and that a new input value (33, for example, in this simulation) reaches the output at the fourth positive clock transition (because $M = 4$).

8.6 CONFIGURATION

CONFIGURATION provides architecture-entity bindings, which can be helpful in projects with multiple architectures or in complex hierarchical designs. Simplified syntaxes for these two cases are presented below.

Direct binding:

```
CONFIGURATION config_name OF entity_name IS
  FOR arch_name
  END FOR;
END [CONFIGURATION] [config_name];
```

Binding in component instantiations:

```
CONFIGURATION config_name OF entity_name IS
  FOR arch_name
    FOR label: component_name
    --or FOR OTHERS/ALL: component_name
      USE ENTITY entity_name [(arch_name)];
    END FOR;
  END FOR;
END [CONFIGURATION] [config_name];
```

The direct ENTITY-ARCHITECTURE binding starts with the word CONFIGURA-TION, followed by a chosen name for this configuration setup, then the name of the EN-TITY (project name) to which the architecture will be bound. In the second line, FOR determines the name of the ARCHITECTURE to be bound. The configuration code must be located outside any entities or architectures.

Example   The code below contains an entity (called *test*) and two architectures (*arch1* and *arch2*). The configuration declaration (called *config1*) defines the binding as *test-arch1* (that is, *arch1* is to be used with *test*).

```
-------------------------------
ENTITY test ...
END test;
-------------------------------
ARCHITECTURE arch1 ...
END arch1;
-------------------------------
ARCHITECTURE arch2 ...
END arch2;
-------------------------------
```

```
CONFIGURATION config1 OF test IS
   FOR arch1
   END FOR;
END CONFIGURATION;
--------------------------------
```

The second syntax above allows more complex bindings. For example, a component can be declared and instantiated in the code using an arbitrary name, with the actual binding between the used names and the actual entity-architecture names made later with a CONFIGURATION declaration.

Example   The code below is from example 8.3. Note, however, that the name of the component in the component declaration (lines 10–14) and component instantiation (line 17) was changed to *detector*, which does not correspond to any entity available in our design. However, a configuration declaration was also included in the code (lines 21–27), saying the following: its name is *my_config* and it relates to the entity *parity_detector* (line 21); the architecture named *structural* (line 22) of that entity must employ, for the component instantiation labeled *det* of component *detector* (line 23), the entity *par_detector* with architecture *behavior* (line 24), both available in the current *work* library (project's directory).

```
 1   ---------The component:------------------------
 2   ENTITY par_detector IS
..   ... (see Example 8.3)
19   END behavior;
20   -----------------------------------------------

 1   ---------Main code:----------------------------
 2   ENTITY parity_detector IS
 3      GENERIC (N: POSITIVE := 8);
 4      PORT (x: IN BIT_VECTOR(N-1 DOWNTO 0);
 5            y: OUT BIT);
 6   END parity_detector;
 7   -----------------------------------------------
 8   ARCHITECTURE structural OF parity_detector IS
 9      ---------------------------
10      COMPONENT detector IS
11         GENERIC (bits: POSITIVE);
12         PORT (input: IN BIT_VECTOR(bits-1 DOWNTO 0);
13            output: OUT BIT);
14      END COMPONENT;
15      ---------------------------
16   BEGIN
17      det: detector GENERIC MAP (N) PORT MAP(x, y);
```

```
18   END structural;
19   ----------------------------------------------------

20   ----------------------------------------------------
21   CONFIGURATION my_config OF parity_detector IS
22      FOR structural
23        FOR det: detector
24          USE ENTITY work.par_detector(behavior);
25        END FOR;
26      END FOR;
27   END my_config;
28   ----------------------------------------------------
```

It is important to mention that in regular codes (with just one architecture for each entity, and with component instantiations fully specified using the syntax of section 8.3) there is no need to use CONFIGURATION.

8.7   BLOCK

BLOCK is a concurrent statement whose main purpose is to provide a means for code *partitioning* (hence *system-level* design). BLOCK can be used in an ARCHITECTURE body to cluster related portions of code, making the overall code more readable and more manageable (which can be helpful in long designs).

Because it is a concurrent statement, BLOCK cannot be used in sequential code, but PROCESS can be located inside BLOCK, because, as a whole, PROCESS is a concurrent statement. A simplified syntax for BLOCK is shown below.

```
label: BLOCK [(guard_expression)] [IS]
  [declarative_part]
BEGIN
  concurrent_statements_part
END BLOCK [label];
```

As seen in the syntax, the label is mandatory, while the guard expression (explained later) is optional. The declarative part can contain the following kinds of declarations: GENERIC, GENERIC MAP, PORT, PORT MAP, plus all kinds of declarations allowed in the architecture's declarative part.

Complex block arrangements can be constructed because block statements can be nested inside one another. The example below shows the simplest possible application for BLOCK (just a code partitioner).

```
--------------------------
ARCHITECTURE example OF ...
BEGIN
   ...
   controller: BLOCK
   BEGIN
   ...
   END BLOCK controller;
   ...
END example;
--------------------------
```

A *guarded* BLOCK is one in which the optional *guard expression* mentioned in the syntax above is included. Because the statements contained in a block are only evaluated when the guard expression is TRUE, a disconnecting mechanism is automatically established.

Example 8.5 illustrates how a guard expression can affect a design. Note, however, that this is just an illustrative example, not a recommended design approach for latches or any other circuits, because the actual purpose of the guard expression is to allow the connection/disconnection of drivers.

Example 8.5: Latch Implemented with a Guarded BLOCK

The code below implements a D-type latch. The whole architecture body is included in a block called *blk*, which contains the guard expression $clk = $ '1' (line 12). When this condition is fulfilled, the block is evaluated. Because the output is defined as $q = $ GUARDED d (line 14), q receives the value of d when the guard expression is TRUE. In other words, $q = d$ results while $clk = $ '1', which is the logic equation for a D-type latch.

```
1    -----------------------------------
2    LIBRARY ieee;
3    USE ieee.std_logic_1164.all;
4    -----------------------------------
5    ENTITY latch IS
6       PORT (d, clk: IN STD_LOGIC;
7             q: OUT STD_LOGIC);
8    END ENTITY;
9    -----------------------------------
10   ARCHITECTURE block_latch OF latch IS
11   BEGIN
12      blk: BLOCK (clk='1')
13      BEGIN
14         q <= GUARDED d;
15      END BLOCK blk;
16   END ARCHITECTURE;
17   -----------------------------------
```

2Chapter 8

8.8 VHDL 2008

With respect to the material covered in this chapter, the main additions specified in VHDL 2008 are those listed below.

1) The declarative part of a PACKAGE can contain these additional kinds of declarations: subprogram instantiation declaration, package declaration, package instantiation declaration, and PSL declaration.

2) The use of GENERIC (section 2.6) in the PACKAGE header is allowed, as indicated in the simplified syntax below. An example is shown subsequently.

```
PACKAGE package_name IS
  [GENERIC (generic_list);]
  declarative_part
END [PACKAGE] [package_name];
```

```
PACKAGE generic_type IS
  GENERIC (CONSTANT words: NATURAL;
          TYPE: word_type);
  TYPE gen_type IS ARRAY 1 TO words OF word_type;
END PACKAGE;
```

3) A package with a generic list is called an *uninstantiated package*, which must be instantiated with a *package instantiation* declaration, shown in the simplified syntax below. An example of instantiation for the package *generic_type* above is also presented below.

```
PACKAGE package_name IS NEW uninstantiated_package_name
    GENERIC MAP (instantiation_list);
```

```
LIBRARY ieee;
USE ieee.std_logic_1164;
PACKAGE memory_array IS NEW work.generic_type
  GENERIC MAP (words => 256, word_type => STD_LOGIC_VECTOR(15 DOWNTO 0));
```

4) Expressions are allowed in PORT MAP assignments, as in the example below.

```
cir: my_circuit PORT MAP(inp => a AND b, outp => c);
```

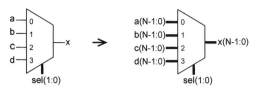

Figure 8.8

8.9 Exercises

Note: For exercise solutions, please consult the book website.

Exercise 8.1: Mux with COMPONENT and GENERATE

Say that the 4×1 (four inputs of one bit each) multiplexer on the left of figure 8.8 is available, with which we want to implement the $4 \times N$ multiplexer on the right of figure 8.8. Write a VHDL code to solve this exercise, using GENERATE to make the COMPONENT (mux) instantiations.

Exercise 8.2: Circular Shift with COMPONENT #1

Modify the code of example 8.2 such that the circuit operates with *generic* length (M) and *fixed* width ($N = 8$).

Exercise 8.3: Circular Shift with COMPONENT #2

Modify the code of example 8.2 such that the circuit operates with *fixed* length ($M = 4$) and *generic* width (N).

Exercise 8.4: Parity Detector with COMPONENT and GENERIC MAP

Simulate the design of example 8.3 in order to verify its (correct) operation.

Exercise 8.5: Adder with COMPONENT and GENERATE

Redesign the carry-ripple adder of example 6.4 using a *structural* code (that is, a code based on *components*). The FA unit must be the component, which should be instantiated N times with GENERATE.

Exercise 8.6: Synchronous Counter with COMPONENT

Figure 8.9 shows a modulo-8 synchronous counter with serial enable (Pedroni 2008). Since it employs a standard cell, the use of COMPONENT to implement it is appropriate. Write a VHDL code to solve this problem. Each gray cell must be entered as a component.

a) Enter the code using method 1 of section 8.3 and *positional* mapping.

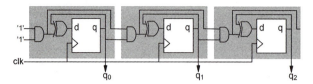

Figure 8.9

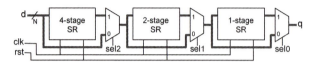

Figure 8.10

b) Repeat the exercise with the code entered using method 3 of section 8.3 and *nominal* mapping.

Exercise 8.7: Synchronous Counter with COMPONENT and GENERATE

Suppose that the counter in the exercise above must now be a 32-bit counter instead of 3 bits. In this case, the easiest way to instantiate the standard cell (32 times) is by means of GENERATE. Write a code that solves this problem. Enter the number of bits (N) using GENERIC, so the code can be easily adjusted to any counter size.

Exercise 8.8: Tapped Delay Line with COMPONENT and GENERIC MAP

A *tapped delay line* is shown in figure 8.10 (Pedroni 2008). Note that all cells are of the same type ($M \times N$ shift register followed by a $2 \times N$ multiplexer). There is, however, an interesting particularity: the value of M varies from one cell to another. Design this circuit using COMPONENT to construct the SR and mux cells. Adopt $N = 1$ and use GENERIC MAP to define the values of M. Compile your code and check whether the number of flip-flops inferred is seven. Also, simulate it to make sure that the correct functionality was attained.

Exercise 8.9: Programmable Signal Generator with Frequency Meter

a) Design the programmable signal generator of exercise 6.12.

b) Design the accompanying frequency meter of exercise 6.13.

c) Using the COMPONENT construct, write a main code that instantiates these two circuits to produce the complete system.

d) Physically implement the circuit in your FPGA development board and test its operation.

9 FUNCTION and PROCEDURE

9.1 Introduction

FUNCTION and PROCEDURE (called *subprograms*) are very similar to PROCESS (studied in section 6.3), in the sense that these three are the only sections of VHDL code that are interpreted *sequentially* (like regular computer programs). Consequently, only sequential statements (IF, WAIT, LOOP, CASE) are allowed (plus operators, of course, because these can be used in any kind of code).

On the other hand, contrary to PROCESS, which is intended for the ARCHITECTURE body (regular codes), subprograms can be constructed in a PACKAGE, ENTITY, ARCHITECTURE, or PROCESS. Because PACKAGE is the most common location, with which the VHDL libraries are built, in our context subprograms are considered *system-level* units (along with PACKAGE and COMPONENT, studied in chapter 8).

9.2 The ASSERT Statement

Before we start discussing subprograms, let us see the ASSERT statement, which is very useful for checking subprogram inputs (it is also very helpful in simulations, as will be shown in the next chapter).

ASSERT is a statement that can be used in both concurrent and sequential code. Its purpose is not to create circuits, but simply to assert that certain basic requirements are met during synthesis or simulation. Its syntax is shown below.

```
[label:] ASSERT boolean_expression
  [REPORT string_expression]
  [SEVERITY severity_level];
```

The string expression can be a constant or a signal of type STRING. The use of the concatenation operator (&) is allowed.

Example   Say that *s* is a string whose value is *idle*. Then the statement below

```
REPORT "Attention: s=" & s & "!"
```

will cause the message *Attention: s = idle!* to be printed on the screen. Other cases will be presented ahead.

The SEVERITY level can be NOTE (to pass information from the compiler/simulator), WARNING (to inform that something unusual has occurred), ERROR (to inform that a serious unusual condition has been found), or FAILURE (a completely unacceptable condition has occurred). The message is issued when the condition is FALSE. Generally, the compiler/simulator is set up to halt when one of the last two (ERROR or FAILURE) occurs, having NOTE as the default value.

Example   Say that a certain function receives two vectors, called *a* and *b*, which must have the same size. Then the following test could be done (the use of parentheses is optional):

```
ASSERT (a'LENGTH=b'LENGTH)
   REPORT "Signals a and b do not have the same length!"
   SEVERITY FAILURE;
```

The option above, which is the option generally used, is called *conditional* ASSERT. An *unconditional* version also exists, shown in the syntax below.

```
[label:] ASSERT FALSE
   [REPORT string_expression]
   [SEVERITY severity_level];
```

Because the message is issued when the condition is FALSE, this syntax forces the message to be issued. Note, however, that only when this version of ASSERT is used by itself is it truly unconditional, because if it is combined with an IF statement, for example (as in example 9.3), a conditional test still results. The truly unconditional case is useful when the user wants to know when the compiler/simulator has reached particular points in the design (NOTE is usually chosen as the severity level in such a case).

A helpful attribute when constructing more complex ASSERT statements is T'IMAGE(X), seen in section 4.4. This attribute determines the STRING type representation for the value X of type T. It is useful because the REPORT section of the ASSERT statement accepts only strings. Recall, however, that T is restricted to the numeric, enumerated, and physical types (TIME is the main physical type, used in simulations). An example is presented below, which is a code for simulation (chapter 10).

```
-----------------------------------------------------
USE ieee.std_logic_unsigned.all;
...
SIGNAL x, y: STD_LOGIC_VECTOR(3 DOWNTO 0);
SIGNAL n: INTEGER RANGE 0 TO 255;
SIGNAL t: TIME RANGE 0ns TO 200ns;
...
ASSERT (x=y AND n=ref)
  REPORT "Mismatch at t=" & TIME'IMAGE(t) &
    " (for n=" & INTEGER'IMAGE(n) &
    ", x=" & INTEGER'IMAGE(conv_integer(x)) &
    ", y=" & INTEGER'IMAGE(conv_integer(y)) & ")."
  SEVERITY FAILURE;
...
-----------------------------------------------------
```

In the code above, x and y are STD_LOGIC_VECTOR signals, n is INTEGER, and t is TIME. For them to be used in the REPORT section of ASSERT, they must first be converted to the type STRING. Such conversion can be made within the REPORT statement itself, using the T'IMAGE(X) attribute. In the first line, a direct conversion for t, from TIME to STRING, is made. In the second line, another direct conversion is made, for n, now from INTEGER (a numeric type) to STRING. However, in the cases of x and y, a direct conversion is not possible because STD_LOGIC_VECTOR is not allowed in the T'IMAGE(X) attribute. Consequently, a conversion to INTEGER is made first, using the function *conv_integer( )*, available in the *std_logic_unsigned* package (see the package declaration at the beginning of the code), so such integers can then be converted to STRING. If anyone of the two boolean conditions is not satisfied, the following message will be issued (assuming $x = $ "0011", $y = $ "1111", $n = 8$, and $t = 5$ns): *Mismatch at t = 5000 ps (for n = 8, x = 3, y = 15)*.

The use of ASSERT will be illustrated in the examples with FUNCTION in section 9.3. It will also be used in the next chapter, which deals with simulation.

In VHDL 2008, the TO_STRING type-conversion function was introduced, which can be used instead of the T'IMAGE(X) attribute. The advantage of TO_STRING is that it supports a wider set of data types: BOOLEAN, BIT, BIT_VECTOR, INTEGER, NATURAL, POSITIVE, CHARACTER, STD_(U)LOGIC_VECTOR, (UN)SIGNED, REAL, TIME, SFIXED, UFIXED, and FLOAT.

9.3 FUNCTION

FUNCTION is a section of *sequential* VHDL code whose main purpose is to allow the creation and storage in libraries of solutions for commonly encountered problems, like data-type conversions, logical and arithmetic operations, etc.

FUNCTION is similar to PROCESS (section 6.3) in the sense that it too is sequential and therefore can only use the same statements (IF, WAIT, LOOP, and CASE—see expansion related to WHEN and SELECT in VHDL 2008 in section 9.7). Moreover, the items that can be declared in its declarative part are the same as those for PROCESS (see list in section 6.3; again, signal declarations are not allowed). A simplified syntax for the construction of functions is shown below.

```
[PURE | IMPURE] FUNCTION function_name [(input_list)]
RETURN return_value_type IS
   [declarative_part]
BEGIN
   statement_part
  [label:] RETURN expression;
END [FUNCTION] [function_name];
```

The syntax above starts with an optional PURE or IMPURE declaration (described later). If left unspecified, the default value (PURE) is assumed.

The input list can contain any number of parameters (including zero), which are all of mode IN (that is, all parameters are *inputs* to the function). The list can only contain the objects CONSTANT (default), SIGNAL, and FILE (VARIABLE is not allowed), declared as:

```
[CONSTANT] constant_name: constant_type;
SIGNAL signal_name: signal_type;
```

Regardless of the number of input parameters, a function always returns *one* parameter value, whose type must be specified after the keyword RETURN in the function header.

Example   The function below, named *positive_edge*, receives a signal called *s*, returning TRUE when a positive transition occurs on *s*.

```
---------------------------------------------------------------
FUNCTION positive_edge (SIGNAL s: STD_LOGIC) RETURN BOOLEAN IS
BEGIN
   RETURN (s'EVENT AND s='1');
END FUNCTION positive_edge;
---------------------------------------------------------------
```

A FUNCTION can be constructed (using the syntax above) in a PACKAGE, ENTITY, ARCHITECTURE, PROCESS, BLOCK, or another subprogram, with PACKAGE as the most common location (for libraries).

Recall from chapter 8 that when a subprogram is constructed in a PACKAGE, a PACKAGE BODY is required (only the subprogram declaration goes in the PACKAGE; the subprogram body goes in the PACKAGE BODY). This situation is illustrated below.

```
------Package:---------------------------------------------
PACKAGE my_subprograms IS
   FUNCTION positive_edge (SIGNAL s: STD_LOGIC) RETURN BOOLEAN;
END PACKAGE;
-----Package body:-----------------------------------------
PACKAGE BODY my_subprograms IS
   FUNCTION positive_edge (SIGNAL s: STD_LOGIC) RETURN BOOLEAN IS
   BEGIN
       RETURN (s'EVENT AND s='1');
   END FUNCTION positive_edge;
END PACKAGE BODY;
-----------------------------------------------------------
```

Function Call

A function can be called basically anywhere (in combinational or sequential code, inside subprograms, inside GENERATE, etc.). As will be shown in the examples, a function call is always part of an expression.

Example   Say that the function *positive_edge* seen above is part of our design. Then the first line below, which contains a call to that function, is equivalent to the second line.

```
IF positive_edge(clk) THEN ...
IF clk'EVENT AND clk='1' THEN...
```

Positional versus Nominal Mapping

Similarly to PORT MAP in COMPONENT instantiations (section 8.3), the mapping between a function call and the corresponding function declaration can be *positional* or *nominal*.

Example   Three equivalent function calls are shown below.

```
-----Function declaration:---------------------------
FUNCTION my_function (SIGNAL a, b: BIT) RETURN BIT;
-----Equivalent function calls:----------------------
y <= my_function (x1, x2);        --positional mapping
y <= my_function (a=>x1, b=>x2);  --nominal mapping
y <= my_function (b=>x2, a=>x1);  --nominal mapping
-----------------------------------------------------
```

Pure versus Impure Functions

A function is said to be *pure* when it can only modify its own variables. Consequently, any call to that function passing the same parameters will receive exactly the same result. An *impure* function, on the other hand, may also modify signals or variables from the architecture, process, or subprogram where it is declared, so in this case different results might occur among calls made with the same parameters at different times. The latter option can be helpful in some cases, but one must be very careful when declaring a function as impure.

In VHDL 2008, a GENERIC list is allowed in a FUNCTION. See other details in section 9.7.

Complete examples illustrating the construction and usage of functions are presented next.

Example 9.1: FUNCTION *max* in an ARCHITECTURE

Write a function that returns the largest of three integers. Assume that all in/out signals are required to have the same range (use the ASSERT statement to check this). Construct the function directly in the ARCHITECTURE of the main code (in its declarative part).

Solution A code for this circuit is shown below, under the title *comparator* (line 2). The FUNCTION, called *max*, is in lines 8–23, located in the declarative part of the ARCHITECTURE. Note that it is divided into two parts. The first part (lines 11–14) contains an ASSERT statement that checks whether the sizes of a, b, c, and y are all equal, printing the message "Signal sizes are not all equal!" on the screen if the result of the assertion test is FALSE, which also causes the compilation to stop (due to the chosen severity level, FAILURE). The second part (lines 16–22) determines the largest of the three inputs. A function call is made in the main code (line 25), which passes the largest input to the output (y). Positional versus nominal mapping is illustrated in lines 25–26. Simulation results are depicted in figure 9.1.

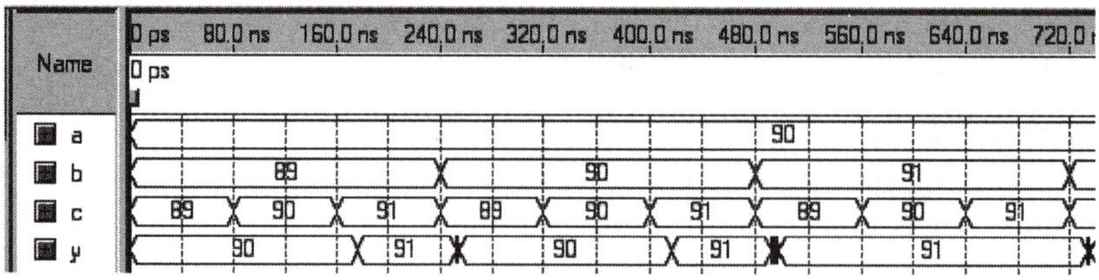

Figure 9.1
Simulation results from the code of example 9.1.

```
 1   ----------------------------------------------------------------------
 2   ENTITY comparator IS
 3      PORT (a, b, c: IN INTEGER RANGE 0 TO 255;
 4          y: OUT INTEGER RANGE 0 TO 255);
 5   END ENTITY;
 6   ----------------------------------------------------------------------
 7   ARCHITECTURE comparator OF comparator IS
 8      FUNCTION max (in1, in2, in3: INTEGER) RETURN INTEGER IS
 9      BEGIN
10          -----Check in-out signals:-----
11          ASSERT (y'LEFT=a'LEFT AND y'LEFT=b'LEFT AND y'LEFT=c'LEFT
12             AND y'RIGHT=a'RIGHT AND y'RIGHT=b'RIGHT AND y'RIGHT=c'RIGHT)
13             REPORT "Signal sizes are not all equal!"
14             SEVERITY FAILURE;
15          -----Find maximum:-------------
16          IF (in1>=in2 AND in1>=in3) THEN
17             RETURN in1;
18          ELSIF (in2>=in1 AND in2>=in3) THEN
19             RETURN in2;
20          ELSE
21             RETURN in3;
22          END IF;
23      END FUNCTION;
24   BEGIN
25      y <= max(a, b, c); --positional mapping
26      --y <= max(in1=>a, in2=>b, in3=>c); --nominal mapping
27   END ARCHITECTURE;
28   ----------------------------------------------------------------------
```

Example 9.2: FUNCTION *order_and_fill* in a PACKAGE

Write a function that reorganizes a binary word, such that the indexing is always descending and ending in zero, regardless of the original specification (for example: $a(5:2) \rightarrow b(3:0)$, $a(1:4) \rightarrow b(3:0)$, $a(3:0) \rightarrow b(3:0)$, etc.). Moreover, the vector must be filled with zeros (on the left) until a predefined size is attained. Include the ASSERT statement in your solution to assure that the size of the input word is not bigger than the size wanted for the final vector (after filling). Construct your function in a PACKAGE (most common option).

Solution The FUNCTION, called *order_and_fill*, is shown below, constructed in a PACKAGE called *my_package*. Recall that when a subprogram is constructed in a package, only the declaration goes in the package, because the subprogram body must go in the PACKAGE BODY. That can be seen below, with the function declaration in lines 6–7 and the function body in lines 11–36. The function receives two parameters, called *input* (of type UNSIGNED, which is the vector to be reorganized and filled) and *bits* (of type NATURAL, which is the size wanted for *input* after reorganization and filling). Note that the function body (in the package body) is broken into three parts. The first part (lines 17–19) contains an ASSERT statement that checks whether the size of *input* is not larger than *bits*, printing the message "Improper input size!" on the screen if the result of the assertion test is FALSE, which also causes the compilation to stop (due to the chosen severity level, FAILURE). The second part (lines 21–27) reorders the vector, such that its indexing always becomes "*input_length* – 1 DOWNTO 0". Finally, the third part (lines 29–34) fills the vector with zeros (on the left) to attain a final size equal to *bits*. The result is returned to the calling expression in line 35.

An example of application is also included below (main code). The function is used (called) in line 14, causing the input vector *x* to be reorganized and filled with zeros, producing an output *y* with a total of *size* bits.

```
1  ------------------------------------------------------------
2  LIBRARY ieee;
3  USE ieee.numeric_std.all;
4  ------------------------------------------------------------
5  PACKAGE my_package IS
6     FUNCTION order_and_fill (input: UNSIGNED; bits: NATURAL)
7     RETURN UNSIGNED;
8  END PACKAGE;
9  ------------------------------------------------------------
10 PACKAGE BODY my_package IS
11    FUNCTION order_and_fill (input: UNSIGNED; bits: NATURAL)
12    RETURN UNSIGNED IS
13       VARIABLE a: UNSIGNED(input'LENGTH-1 DOWNTO 0);
14       VARIABLE result: UNSIGNED(bits-1 DOWNTO 0);
15    BEGIN
16       -----Check input size:--------
17       ASSERT (input'LENGTH <= bits)
18          REPORT "Improper input size!"
19          SEVERITY FAILURE;
20       -----Organize input:----------
21       IF (input'LEFT>input'RIGHT) THEN
22          a := input;
23       ELSE
24          FOR i IN a'RANGE LOOP
25             a(i) := input(input'LEFT + i);
26          END LOOP;
27       END IF;
```

```
28        -----Fill with zeros:----------
29        IF (a'LENGTH < bits) THEN
30            result(bits-1 DOWNTO a'LENGTH) := (OTHERS => '0');
31            result(a'LENGTH-1 DOWNTO 0) := a;
32        ELSE
33            result:=a;
34        END IF;
35        RETURN result;
36     END FUNCTION;
37 END PACKAGE BODY;
38 ------------------------------------------------------------

1  -----Main code:---------------------------
2  LIBRARY ieee;
3  USE ieee.numeric_std.all;
4  USE work.my_package.all;
5  ------------------------------------------
6  ENTITY organizer IS
7     GENERIC (size: NATURAL := 5);
8     PORT (x: IN UNSIGNED(2 TO 5);
9           y: OUT UNSIGNED(size-1 DOWNTO 0));
10 END ENTITY;
11 ------------------------------------------
12 ARCHITECTURE organizer OF organizer IS
13 BEGIN
14    y <= order_and_fill(x, size);
15 END ARCHITECTURE;
16 ------------------------------------------
```

Example 9.3: FUNCTION *slv_to_integer* in an ENTITY

Write a FUNCTION that converts a signal of type STD_LOGIC_VECTOR to type INTEGER. Include an ASSERT statement in your code to ensure that no symbols other than '0', 'L' (both synthesized as '0'), '1' or 'H' (both synthesized as '1') are present at the input. This time, construct the function directly in the code's ENTITY.

Solution A VHDL code that solves this problem is shown below. The FUNCTION, called *slv_to_integer*, was constructed in the code's ENTITY (after PORT, lines 7–24). An ALIAS (section 4.8), called *ss*, was used in line 9 to normalize the range of *s* to "1 TO *s*'LENGTH", which then simplifies the writing of the function. The ASSERT statement in lines 17–20 checks if only the proper values occur at the input (if any of the other five STD_LOGIC values occurs, an error message is issued and the compilation is interrupted). Note that an unconditional ASSERT is employed in lines 18–20, but because it is associated to the IF statement in line 17, the overall test is still conditional.

```
1  -----------------------------------------------------
2  LIBRARY ieee;
3  USE ieee.std_logic_1164.all;
4  -----------------------------------------------------
5  ENTITY ...
6     PORT (...)
7     FUNCTION slv_to_integer (SIGNAL s: STD_LOGIC_VECTOR)
8        RETURN INTEGER IS
9        ALIAS ss: STD_LOGIC_VECTOR(1 TO s'LENGTH) IS s;
10       VARIABLE result: INTEGER RANGE 0 TO 2**s'LENGTH-1;
11    BEGIN
12       result := 0;
13       FOR i IN 1 TO s'LENGTH LOOP
14          result := result * 2;
15          IF (ss(i)='1' OR ss(i)='H') THEN
16             result := result + 1;
17          ELSIF (ss(i)/='0' AND ss(i)/='L') THEN
18             ASSERT FALSE
19             REPORT "There is an invalid input!"
20             SEVERITY FAILURE;
21          END IF;
22       END LOOP;
23       RETURN result;
24    END FUNCTION slv_to_integer;
25 -----------------------------------------------------
26 ARCHITECTURE ...
27 -----------------------------------------------------
```

9.4 PROCEDURE

The purpose, construction, and usage of PROCEDURE are similar to those of FUNC-
TION. Their main difference is that a PROCEDURE can return *more* than one value. A
syntax for the construction of procedures is shown below.

```
PROCEDURE procedure_name (input_output_list) IS
   [declarative_part]
BEGIN
   statement_part
END [PROCEDURE] [procedure_name]
```

The input-output list can contain CONSTANT, SIGNAL, and VARIABLE. Their mode can be IN, OUT, or INOUT; if it is IN, then CONSTANT is the default object, while for OUT and INOUT the default is VARIABLE. Their declarations are as follows:

```
CONSTANT constant_name: mode constant_type;
SIGNAL signal_name: mode signal_type;
VARIABLE variable_name: mode variable_type;
```

Both FUNCTION and PROCEDURE are sequential codes, so only sequential statements are allowed (but see the note about the extension of WHEN and SELECT in VHDL 2008 in section 9.7). They can be constructed and used in the same way, with PACKAGE (plus the corresponding PACKAGE BODY) as the most common location (for libraries).

Like function calls, procedure calls can be made basically anywhere (in sequential as well concurrent code, in subprograms, etc.). However, the former is called as part of an expression, while the latter is a statement on its own. Examples of procedure calls are shown below.

```
-------------------------------------------------
sort (a1, a2, a3, b1, b2, b3);
-------------------------------------------------
divide (dividend, divisor, quotient, remainder);
-------------------------------------------------
IF (x>y) THEN get_max (x1, x2, x3, x4, y1, y2);
-------------------------------------------------
```

Finally, regarding the mapping between a procedure call and the corresponding procedure declaration, it is similar to the mapping for functions—that is, *positional* or *nominal* (see section 9.3).

In VHDL 2008, a GENERIC list is allowed in a PROCEDURE and the OUT mode can be read by the containing PROCEDURE.

Example 9.4: PROCEDURE *min_max* in a PACKAGE

A diagram for a 3-input, 2-output circuit (called *min_max*) is depicted in the top-left corner of figure 9.2. The circuit must detect the smallest and largest values among *a*, *b*, and *c*, and assign them to *min* and *max*, respectively. Write a PROCEDURE capable of implementing such a functionality. Construct it in a PACKAGE, then write a main code with a call to this procedure to test it (with negative numbers allowed).

Solution A VHDL code for this problem is shown below. The procedure (called *min_max*) is declared in lines 3–4 of a PACKAGE (called *my_package*) and its body (lines 8–30) is constructed in the corresponding PACKAGE BODY. The code is based on the flowchart included in figure 9.2. A call to the PROCEDURE is made in line 11 of the main code (note that it is a statement on its own).

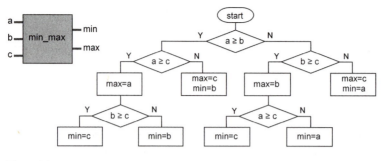

Figure 9.2
Top-level diagram and respective flowchart for the procedure *min_max* of example 9.4.

```
1    ---------Package:-----------------------------------------------
2    PACKAGE my_package IS
3       PROCEDURE min_max (SIGNAL a, b, c: IN INTEGER;
4          SIGNAL min, max: OUT INTEGER);
5    END PACKAGE;
6    ----------------------------------------------------------------
7    PACKAGE BODY my_package IS
8       PROCEDURE min_max (SIGNAL a, b, c: IN INTEGER RANGE 0 TO 255;
9          SIGNAL min, max: OUT INTEGER RANGE 0 TO 255) IS
10      BEGIN
11         IF (a>=b) THEN
12            IF (a>=c) THEN max <= a;
13               IF (b>=c) THEN min <= c;
14               ELSE min <= b;
15               END IF;
16            ELSE
17               max <= c;
18               min <= b;
19            END IF;
20         ELSE
21            IF (b>=c) THEN max <= b;
22               IF (a>=c) THEN min <= c;
23               ELSE min <= a;
24               END IF;
25            ELSE
26               max <= c;
27               min <= a;
28            END IF;
29         END IF;
30      END PACKAGE BODY;
```

```
31   END my_package;
32   ----------------------------------------------------------------

1    ---------Main code:--------------------------
2    USE work.my_package.all;
3    ---------------------------------------------
4    ENTITY comparator IS
5       PORT (a, b, c: IN INTEGER RANGE -256 TO 255;
6              min, max: OUT INTEGER RANGE -256 TO 255);
7    END ENTITY;
8    ---------------------------------------------
9    ARCHITECTURE comparator OF comparator IS
10   BEGIN
11      min_max(a, b, c, min, max);
12   END ARCHITECTURE;
13   ---------------------------------------------
```

9.5   FUNCTION versus PROCEDURE Summary

• Location: Both can be located in the declarative part of an ENTITY, ARCHITEC-TURE, or of another subprogram. The most usual location, however, is in a PACKAGE (in which case a PACKAGE BODY is also needed).

• Call: Both can be called basically anywhere (in sequential as well concurrent code, in subprograms, etc.). However, the first is called as part of an expression, while the second is a statement on its own.

• Statements: Only sequential statements are allowed (IF, WAIT, LOOP, CASE).

• In/Out parameters for functions: Any number of input parameters are allowed, but only (exactly) one value is returned. The in/out objects can only be CONSTANT (default) or SIGNAL.

• In/Out parameters for procedures: Any number of in/out parameters are allowed, which can be CONSTANT (default for input), SIGNAL, or VARIABLE (default for output).

9.6   Overloading

As seen in section 3.17, an *overloaded* operator is one for which more than one in-out option exists. As an example, the *numeric_std* package (appendix J) has six versions of "+", listed below (*L* and *R* are the left and right operands, respectively).

```
FUNCTION "+" (L, R: UNSIGNED) RETURN UNSIGNED;
FUNCTION "+" (L, R: SIGNED) RETURN SIGNED;
```

```
FUNCTION "+" (L: UNSIGNED; R: NATURAL) RETURN UNSIGNED;
FUNCTION "+" (L: NATURAL; R: UNSIGNED) RETURN UNSIGNED;
FUNCTION "+" (L: INTEGER; R: SIGNED) RETURN SIGNED;
FUNCTION "+" (L: SIGNED; R: INTEGER) RETURN SIGNED;
```

Based on the involved data types, the compiler determines which of these functions to use (note that an operator is just a subprogram's name).

In section 4.3, an example of overloading (done by the user) was already presented. Example 9.5 is more detailed.

Example 9.5: Overloaded "+" Operator

According to section 4.2 and figure 4.1, arithmetic operators were not originally defined for the type STD_LOGIC_VECTOR (see package *std_logic_1164* in appendix I). Write a FUNCTION that further overloads the "+" (addition) operator, such that STD_LOGIC_ VECTOR inputs are also supported, returning a value of the same type.

Solution The requested "+" function is shown below. Its declaration is in line 6 of a PACKAGE, and its body is in lines 10–22 of the corresponding PACKAGE BODY. ALIAS declarations (section 4.8) were employed in lines 11–12 to normalize the range of *a* and *b* to "1 TO *a*'LENGTH" and "1 TO *b*'LENGTH", which helps write the function. Notice that, contrary to example 9.3, invalid STD_LOGIC inputs are not tested in this example. Note also that the logic operations in lines 17–19 require that the vector sizes be equal, which is not tested either (see exercise 9.1). An example of call to this FUNCTION is illustrated in line 14 of the main code succeeding the package, which adds several objects of type STD_LOGIC_VECTOR.

```
1   ---------Package:----------------------------------------------------
2   LIBRARY ieee;
3   USE ieee.std_logic_1164.all;
4   ----------------------------------------------------------------------
5   PACKAGE my_package IS
6      FUNCTION "+" (a, b: STD_LOGIC_VECTOR) RETURN STD_LOGIC_VECTOR;
7   END PACKAGE;
8   ----------------------------------------------------------------------
9   PACKAGE BODY my_package IS
10     FUNCTION "+" (a, b: STD_LOGIC_VECTOR) RETURN STD_LOGIC_VECTOR IS
11        ALIAS aa: STD_LOGIC_VECTOR(1 TO a'LENGTH) IS a;
12        ALIAS bb: STD_LOGIC_VECTOR(1 TO b'LENGTH) IS b;
13        VARIABLE result: STD_LOGIC_VECTOR(1 TO a'LENGTH);
14        VARIABLE carry: STD_LOGIC := '0';
15     BEGIN
16        FOR i IN result'REVERSE_RANGE LOOP
17           result(i) := aa(i) XOR bb(i) XOR carry;
```

```
18              carry := (aa(i) AND bb(i)) OR (aa(i) AND carry) OR (bb(i) AND carry);
19                OR (bb(i) AND carry);
20            END LOOP;
21            RETURN result;
22          END FUNCTION "+";
23  END PACKAGE BODY;
24  -------------------------------------------------------------------------------
1   ---------Main code:-----------------------------------------------------------
2   LIBRARY ieee;
3   USE ieee.std_logic_1164.all;
4   USE work.my_package.all;
5   -------------------------------------------------------------------------------
6   ENTITY add_stdlogic IS
7     PORT (x: IN STD_LOGIC_VECTOR(7 DOWNTO 0);
8           y: OUT STD_LOGIC_VECTOR(7 DOWNTO 0));
9   END ENTITY;
10  -------------------------------------------------------------------------------
11  ARCHITECTURE adder OF add_stdlogic IS
12    CONSTANT const: STD_LOGIC_VECTOR(7 DOWNTO 0) := "00001111";
13  BEGIN
14    y <= x + const + "01111111"; --overloaded "+" operator
15  END ARCHITECTURE;
16  -------------------------------------------------------------------------------
```

We close this chapter with a final example on the construction of functions. This example is somehow the opposite of that above: while the above example overloads an operator by including a new function, the next example uses a function equivalent to one that already exists, but which does not overload the corresponding operator.

Example 9.6: Non-overloaded "AND" Operator

According to section 4.2 and figure 4.1, logical operators were already defined for the type STD_LOGIC_VECTOR in its package of origin (*std_logic_1164*, appendix I). Write a FUNCTION that computes the AND function and, contrary to the previous example, does not overload the AND operator.

Solution   Not overloading an operator is trivial: just give the function a different name (like *my_and*, used in the code below). The FUNCTION was again declared in a PACKAGE (line 6) and then constructed in the corresponding PACKAGE BODY (lines 25–34). Note the specification of a new data type, called *stdlogic_table* (line 10), followed by a CONSTANT, called *and_table* (lines 11–24—see the PACKAGE BODY of the *std_logic_1164* package in the VHDL libraries that accompany your VHDL compiler), which conforms with that data type. Observe that this constant is a table that works as a resolution function for the AND function, which is accessed using *enumerated* indexing (section

3.12). Two ALIAS declarations (section 4.8) appear in lines 26–27, with the purpose of normalizing the range of *a* and *b* to "1 TO *a*'LENGTH" and "1 TO *b*'LENGTH", respectively, which helps write the function. The function proper is in the LOOP statement of lines 30–32, which accesses the *and_table* for each bit of *aa* and *bb*. Note that *a* and *b* must have the same length, a condition that was not tested in this example (see exercise 9.2). A main code, with a call to this function, is also included in the code below.

```
1    ---------Package:-------------------------------------------------------
2    LIBRARY ieee;
3    USE ieee.std_logic_1164.all;
4    ------------------------------------------------------------------------
5    PACKAGE my_package IS
6       FUNCTION my_and (a, b: STD_LOGIC_VECTOR) RETURN STD_LOGIC_VECTOR;
7    END PACKAGE;
8    ------------------------------------------------------------------------
9    PACKAGE BODY my_package IS
10      TYPE stdlogic_table IS ARRAY(STD_ULOGIC, STD_ULOGIC) OF STD_ULOGIC;
11      CONSTANT and_table: stdlogic_table := (
12      -----------------------------------------------
13      -- U    X    0    1    Z    W    L    H    -
14      -----------------------------------------------
15      ( 'U', 'U', '0', 'U', 'U', 'U', '0', 'U', 'U' ),  --| U |
16      ( 'U', 'X', '0', 'X', 'X', 'X', '0', 'X', 'X' ),  --| X |
17      ( '0', '0', '0', '0', '0', '0', '0', '0', '0' ),  --| 0 |
18      ( 'U', 'X', '0', '1', 'X', 'X', '0', '1', 'X' ),  --| 1 |
19      ( 'U', 'X', '0', 'X', 'X', 'X', '0', 'X', 'X' ),  --| Z |
20      ( 'U', 'X', '0', 'X', 'X', 'X', '0', 'X', 'X' ),  --| W |
21      ( '0', '0', '0', '0', '0', '0', '0', '0', '0' ),  --| L |
22      ( 'U', 'X', '0', '1', 'X', 'X', '0', '1', 'X' ),  --| H |
23      ( 'U', 'X', '0', 'X', 'X', 'X', '0', 'X', 'X' )); --| - |
24      -----------------------------------------------
25      FUNCTION my_and (a, b: STD_LOGIC_VECTOR) RETURN STD_LOGIC_VECTOR IS
26         ALIAS aa: STD_LOGIC_VECTOR(1 TO a'LENGTH) IS a;
27         ALIAS bb: STD_LOGIC_VECTOR(1 TO b'LENGTH) IS b;
28         VARIABLE result: STD_LOGIC_VECTOR(1 TO a'LENGTH);
29      BEGIN
30         FOR i IN result'RANGE LOOP
31            result(i) := and_table (aa(i), bb(i));
32         END LOOP;
33         RETURN result;
34      END FUNCTION;
35   END PACKAGE BODY;
36   ------------------------------------------------------------------------
```

```
 1   ---------Main code:----------------------------
 2   LIBRARY ieee;
 3   USE ieee.std_logic_1164.all;
 4   USE work.my_package.all;
 5   ------------------------------------------------
 6   ENTITY myand IS
 7      PORT (x1, x2: IN STD_LOGIC_VECTOR(7 DOWNTO 0);
 8            y: OUT STD_LOGIC_VECTOR(7 DOWNTO 0));
 9   END ENTITY;
10   ------------------------------------------------
11   ARCHITECTURE myand OF myand IS
12   BEGIN
13      y <= my_and(x1, x2);
14   END ARCHITECTURE;
15   ------------------------------------------------
```

9.7   VHDL 2008

With respect to the material covered in this chapter, the main additions specified in VHDL 2008 are listed below.

1) The TO_STRING type-conversion function was introduced, which eases the construction of ASSERT statements because it supports a much wider set of types than T'IMAGE(X). The types supported by TO_STRING are BOOLEAN, BIT, BIT_ VECTOR, INTEGER, NATURAL, POSITIVE, CHARACTER, STD_(U)LOGIC_ VECTOR, (UN)SIGNED, REAL, TIME, SFIXED, UFIXED, and FLOAT.

2) GENERIC lists are allowed in FUNCTION and PROCEDURE. The corresponding syntaxes are depicted below.

```
[PURE | IMPURE] FUNCTION
function_name
  [GENERIC (generic_list)]
  [(input_list)]
  RETURN return_value_type IS
  [declarative_part]
BEGIN
  statements_part
  [label:] RETURN expression;
END [FUNCTION] [function_name];
```

```
PROCEDURE procedure_name
  [GENERIC (generic_list)]
  (input_output_list) IS
  [declarative_part]
BEGIN
  statements_part
END [PROCEDURE] [procedure_name]
```

3) In PROCEDURE, an OUT mode object can be read internally.

4) Recall that in VHDL 2008 the WHEN and SELECT statements can also be used inside sequential code.

9.8   Exercises

Exercise 9.1: ASSERT Statement #1

Include an ASSERT statement in example 9.5 to check whether the inputs have the same size.

Exercise 9.2: ASSERT Statement #2

Include an ASSERT statement in example 9.6 to check whether the inputs have the same size.

Exercise 9.3: Function *integer_to_slv*

Example 9.3 shows a function called *slv_to_integer*, which converts an object of type STD_LOGIC_VECTOR to type INTEGER. Do now the opposite—write a function that makes the conversion from INTEGER to STD_LOGIC_VECTOR, following the same overall approach of example 9.3. The function should be constructed in an ENTITY.

Exercise 9.4: Function *shift_logical_left*

According to section 4.2 and figure 4.1, shift operators were not originally defined for the type STD_LOGIC_VECTOR (see package *std_logic_1164* in appendix I). Write a function that further overloads the "SLL" (shift left logical) operator such that STD_LOGIC_VECTOR inputs are also supported, returning a value of the same type. Install your function in a PACKAGE, then make a call to it in the main code in order to test (simulate) its operation.

Exercise 9.5: Function *my_not*

In example 9.6 we developed a function called *my_and* for the type STD_ULOGIC_VECTOR, which was equivalent to the original AND operator defined in the package *std_logic_1164*. Following the same reasoning, write a function called *my_not* that is equivalent to the NOT operator defined in the same package. Also show an application for that function in the main code. The conversion table (called *not_table*) needed in this case is presented below.

```
------------------------------------------------
CONSTANT not_table: stdlogic_1d :=
------------------------------------------------
-- U    X    0    1    Z    W    L    H    -
------------------------------------------------
( 'U', 'X', '1', '0', 'X', 'X', '1', '0', 'X');
------------------------------------------------
```

Exercise 9.6: Function *bcd_to_ssd*

The conversion from BCD (binary coded decimal) to SSD (seven-segment display) was needed in example 6.6 (see lines 29–41 of the code in that example) and also in other designs in chapter 12. Write a function to perform such a conversion, which should be located in a PACKAGE. Include an ASSERT statement to check whether the input's length is exactly four bits. Also include an application (main code) to test (simulate) your converter.

Exercise 9.7: Function *binary_to_gray*

A binary-to-gray code converter was proposed in exercise 5.6. Write a function that implements such a converter, which should be constructed in the declarative part of the ARCHITECTURE (main code). In the code proper (architecture body), include a call to that function in order to test (simulate) it.

Exercise 9.8: Procedure *mean_and_median*

Write a procedure that receives three 8-bit STD_LOGIC_VECTOR signals and returns their arithmetic mean and their median, of the same type. The system should be considered to be *signed*. Install your procedure in the declarative part of the ARCHITECTURE (main code). In the code proper (architecture body), include a call to that procedure in order to test (simulate) it.

Exercise 9.9: Procedure *equal_length*

Write a procedure that receives two vectors, called *a* and *b*, of type STD_LOGIC_VECTOR, whose sizes can be different, and returns both vectors with the same size. Only the shortest of the received vectors should be modified (filled with zeros on the left) until its size becomes equal to the size of the other vector. Construct this procedure in a PACKAGE.

10 Simulation with VHDL Testbenches

10.1 Introduction

Everything in this book concerns synthesis, with the exception of chapter 10. Separating the simulation-only code from the synthesis-only code is one of the most crucial steps for a fast, correct, and broad understanding of the VHDL language.

Figure 10.1 (borrowed from chapter 1) shows a simplified view of the design flow with VHDL. The synthesis steps are shown on the left, while the simulation options are depicted on the right. Even though a design can contain more than three simulation points, the main ones are those included in the figure.

The RTL (register transfer level) simulation is based on the VHDL code, containing no timing or other device information; its purpose is to check the design functionalities, so it is said to be a *functional* simulation. The next simulation is also *functional* (no timing information yet) and is executed after synthesis; its purpose is to verify that the functionalities hold after the synthesis process. The final simulation is after fitting and includes internal cell and routing delays, thus representing the actual physical device; because time information is now included, it is a *timing* simulation. For instance, in all designs shown in the other chapters, the *timing* simulation was performed (and presented).

Time information is annotated in an SDF (standard delay format) file, which complies with the VITAL standards. Such information allows simulations under best- and worst-case operating conditions (temperature, voltage, etc.). For example, Xilinx calls such scenarios MIN, TYP, and MAX, which can be selected by the user, while Altera employs worst-case in the regular timing simulation and best-case in what it calls *fast* timing simulation.

Below is a list of well-known VHDL simulators (ModelSim, described in appendix D, will be employed in the examples in this chapter):

- From Altera: Quartus II graphical simulator
- From Xilinx: ISE simulator
- From Mentor Graphics: ModelSim
- From Synopsys: VCS

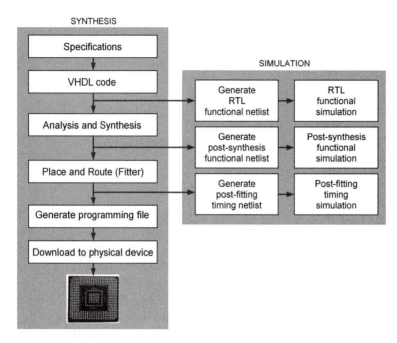

Figure 10.1
Simplified VHDL design flow.

• From Cadence: NC-Sim
• From Aldec: Active-HDL

We close this introduction with some simple (but very useful) simulation recommendations.

1) In-system testing might not be enough to catch all design problems, so also include timing simulation, at the least.

2) Keep the projects that are for simulation separate from those that are for synthesis (in the same way that the material in this chapter was kept apart from that in all the other chapters).

3) Remember that the fundamental time-related statements and function are AFTER, WAIT FOR, and NOW. The attribute 'LAST_EVENT is also helpful sometimes.

4) Never use these statements and function in code that is for synthesis.

5) Use GENERIC to enter arbitrary design/simulation parameters whenever appropriate.

6) Do not hesitate to use type-conversion functions whenever convenient (see figure 3.10).

7) To convert an INTEGER type into a TIME type, just multiply the former by one unit of the desired time scale. For example, *time_value* = *int_value**1ns.

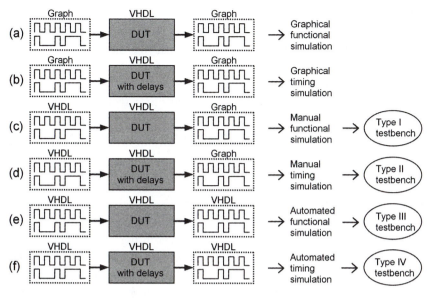

Figure 10.2
Simulation interface options.

8) Make sure to feed the same clock to all units that belong to the same clock domain.

9) In the simulation of large, complex systems, use files to enter/store data (see sections 10.3, 10.4, and 10.13).

10) Do not display unimportant data.

11) Before starting a simulation, make sure that you have the correct design code and that you have understood its functionalities.

12) Finally, always provide a means for the simulation to end. For example, close the code with a WAIT FOR statement with an additional time interval such that the total simulation time (in the code) is larger than the total time set up in the simulation software, so the latter will be able to close the simulation unconditionally when its time limit is reached (this can be done, for example, with a simple "WAIT;" at the end of the code).

The material in this chapter is complemented by a tutorial on ModelSim in appendix D.

10.2   Simulation Types

Figure 10.2 shows six simulation interface options. On the left, the input stimuli are depicted, entered using either a graphics interface or a VHDL code. In the center, the DUT (design under test) is shown, always designed with VHDL (without or with propagation delays included). Finally, the circuit responses are depicted on the right; these can be graphical (visually inspected) or checked by a VHDL code (automated verification).

In figures 10.2a–b, the input and output are both graphical (the user draws the input waveforms using a graphics interface, for which the simulator calculates and plots the corresponding output waveforms, which are then visually inspected by the user), so they are referred to as *graphical* simulations. Because in (a) the circuit's propagation delays are not considered but in (b) they are, the former is a *functional* simulation while the latter is a *timing* simulation. For example, the last of these two was the option employed to check the designs in the previous chapters.

In figures 10.2c–d, the stimuli are produced by a VHDL code, while the output is still graphical. Because only the stimuli are automated, they are referred to as *stimulus-only* (or *manual*) simulations. Again, the *functional* and *timing* options are depicted. These two simulation options are referred to as *Type I* and *Type II* testbenches, respectively.

Finally, in figures 10.2e–f, both input and output are treated using VHDL (at the output, the code provides means for comparing the obtained results against expected values). Because now the input stimuli and the result analysis are both automated, these simulations are referred to as *automated* simulations. Again, the *functional* and *timing* options are depicted. They are referred to as *Type III* and *Type IV* testbenches, respectively.

The four testbench types (Pedroni 2008) are summarized below.

• Type I testbench (*manual functional simulation*): The DUT's internal delays are not considered and the output is manually verified (normally by visual inspection). This is the simplest kind of VHDL code for simulation.

• Type II testbench (*manual timing simulation*): The DUT's internal delays are taken into account, but the output is still manually verified.

• Type III testbench (*automated functional simulation*): The DUT's internal delays are not considered, but the output is automatically verified by the simulator (the data for comparison can be included, for example, in the test file itself or in a separate file).

• Type IV testbench (*automated timing simulation*, also called *full bench*): The DUT's internal delays are taken into account and the output is automatically verified by the simulator. This is obviously the most complete and also the most complex type of simulation with testbenches.

It is also indispensable to know which files must be provided to run a simulation. Say that the DUT is called *mydesign*. To graphically simulate it (as in figures 10.2a–b), two files are needed from the user. The first is the *design* file (*mydesign.vhd*), containing the VHDL code for the circuit to be tested. The second is a *test* file (*mydesign.vwf* if using Quartus II, for example), prepared using the waveform editor (graphics interface).

Different files are needed when using VHDL testbenches to test the circuit. However, only two need to be prepared by the user, the others being produced automatically by the synthesizer (when under the proper setup), as listed below.

1) For functional simulation (testbenches I and III, figures 10.2c and 10.2e):

- Design file (prepared by the user).

- Test file (this is the testbench, also prepared by the user).

2) For timing simulation (testbenches II and IV, figures 10.2d and 10.2f):

- Design file (prepared by the user—same as above).

- Test file (testbench, also prepared by the user—same as above).

- Postsynthesis file (generated by the synthesizer).

- SDF file (also generated by the synthesizer).

The preparation of these files will be described later.

10.3 Writing Data to Files

Because files are very helpful for storing data used in simulations, knowing how to manage them with VHDL is indispensable for the development of complex tests. Even though VHDL does not allow data from files to be directly loaded into the synthesis environment, such action is possible for simulation. Typical procedures for writing to files are described below, accompanied by a complete working example. Procedures for reading from files will be seen in the next section.

The main VHDL procedures (extracted from the packages *textio*, appendix M, and *standard*, appendix H) for writing data to files are summarized below, where the following data types are employed:

```
TYPE LINE IS ACCESS STRING;
TYPE TEXT IS FILE OF STRING;
TYPE SIDE IS (left, right);
SUBTYPE WIDTH IS NATURAL;
```

1) To open a file in "write" mode, where *f* is a file identifier (essentially, any name can be used):

```
FILE f: TEXT OPEN WRITE_MODE IS "file_name";
```

2) To write a value *val* to a variable *l* of type LINE (two options), where *data_type* can be BOOLEAN, BIT, BIT_VECTOR, INTEGER, REAL, TIME, CHARACTER, or STRING:

```
PROCEDURE WRITE(l: INOUT LINE; val: IN data_type);
PROCEDURE WRITE(l: INOUT LINE; val: IN data_type;
   justified: IN SIDE := right; field: IN WIDTH := 0);
```

3) To write the line *l* to the file identified by *f*:

```
PROCEDURE WRITELINE(FILE f: TEXT; l: INOUT LINE);
```

4) The file modes and error flags are:

```
TYPE file_open_kind IS (read_mode, write_mode, append_mode);
TYPE file_open_status IS (open_ok, status_error, name_error, mode_error);
```

The use of these procedures is illustrated in example 10.1.

Example 10.1: Writing Values to a File

Write a VHDL code that generates a clock with a 100ns period and writes, at every positive clock transition, a value to a table in a text file containing in each line the time value at the clock transition followed by an integer (pointer) *i* from 0 to 7, as shown below.

```
t=50ns  i=0
t=150ns i=1
t=250ns i=2
t=350ns i=3
t=450ns i=4
t=550ns i=5
t=650ns i=6
t=750ns i=7
```

Solution   A VHDL code for this exercise is presented below. Note the presence of the *textio* package in line 2. As with all simulation codes, the entity (lines 4–5) is empty (optionally, GENERIC can be used).

The declarative part of the architecture (lines 8–10) contains a constant representing the desired clock period (100ns), a signal declaration for the clock (with initial value '0'), and finally a file declaration for the output file. Such a file, named *test_file.txt* and identified by *f*, is declared to be opened in *write* mode.

In the code proper (lines 11–28), a process is employed to create the clock and the file. The declarative part of the process (lines 13–17) contains two constants of type STRING to provide the proper fixed characters for the table. It also contains a variable *l*, of type LINE, to represent the file's lines, and a variable *t*, or type TIME, to represent the time (must include a time unit; ns in this case). Finally, it contains a variable *i*, of type NATURAL, to represent the requested 0-to-7 pointer. The process body (lines 18–27) produces the desired clock (lines 19–20 plus 25–26), the pointer (line 24), and the file (line 22 creates a file line, which is written to the actual file in line 23). Before starting the writing procedure, the file is automatically cleared.

If simulated with ModelSim (appendix D), for example, the waveform and file contents depicted in figure 10.3 will be obtained (in the file, the time unit, ns, and the space between

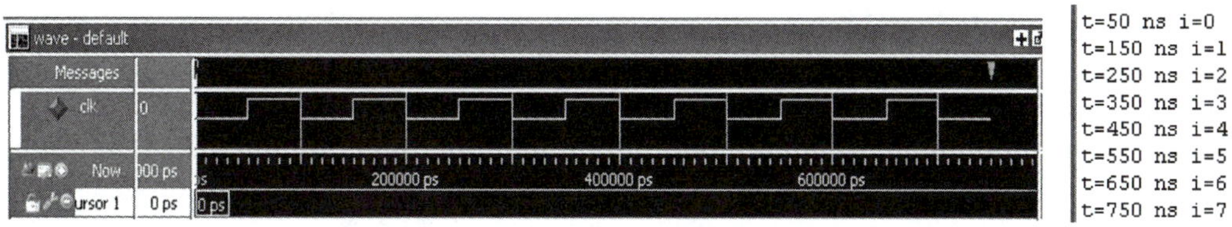

```
t=50 ns i=0
t=150 ns i=1
t=250 ns i=2
t=350 ns i=3
t=450 ns i=4
t=550 ns i=5
t=650 ns i=6
t=750 ns i=7
```

Figure 10.3
Simulation results from example 10.1.

it and the time value are automatically inserted by the simulator). To perform such a simulation, follow the procedure in sections D.1 to D.3 of appendix D, but using just one file.

```
1    ----------------------------------------------------------------
2    USE std.textio.all;
3    ----------------------------------------------------------------
4    ENTITY write_to_file IS
5    END ENTITY;
6    ----------------------------------------------------------------
7    ARCHITECTURE write_to_file OF write_to_file IS
8       CONSTANT period: TIME := 100ns;
9       SIGNAL clk: BIT := '0';
10      FILE f: TEXT OPEN WRITE_MODE IS "test_file.txt";
11   BEGIN
12      PROCESS
13         CONSTANT str1: string(1 TO 2) := "t=";
14         CONSTANT str2: string(1 TO 3) := " i=";
15         VARIABLE l: LINE;
16         VARIABLE t: TIME RANGE 0ns TO 800ns;
17         VARIABLE i: NATURAL RANGE 0 TO 7 := 0;
18      BEGIN
19         WAIT FOR period/2;
20         clk <= '1';
21         t := period/2 + i*period;
22         WRITE(l, str1); WRITE(l, t); WRITE(l, str2); WRITE(l, i);
23         WRITELINE(f, l);
24         i := i + 1;
25         WAIT FOR period/2;
26         clk <='0';
27      END PROCESS;
28   END ARCHITECTURE;
29   ----------------------------------------------------------------
```

10.4   Reading Data from Files

The main VHDL procedures (extracted from the packages *textio*, appendix M, and *standard*, appendix H) for reading data from files are summarized below, where the following data types are employed:

```
TYPE LINE IS ACCESS STRING;
TYPE TEXT IS FILE OF STRING;
TYPE SIDE IS (left, right);
SUBTYPE WIDTH IS NATURAL;
```

1) To open a file in "read" mode, where *f* is a file identifier (essentially any name can be used):

```
FILE f: TEXT OPEN READ_MODE IS "file_name";
```

2) To read a line from the file identified by *f* and assign it to the variable *l*:

```
PROCEDURE READLINE(FILE identifier: TEXT; l: OUT LINE);
```

3) To read a value from the line *l* and assign it to the variable *val* (two options), where *data_type* can be BOOLEAN, BIT, BIT_VECTOR, INTEGER, REAL, TIME, CHARACTER, or STRING:

```
PROCEDURE READ(l: INOUT LINE; val: OUT data_type);
PROCEDURE READ(l: INOUT LINE; val: OUT data_type; good: OUT BOOLEAN]);
```

The *good* test returns FALSE when the type of the value read from the file does not match the type of the object to which such a value is being assigned.

4) Typical check for end-of-file:

```
WHILE NOT ENDFILE (f) LOOP ...
```

Even though ENDFILE is not a true VHDL command, it is supported by any VHDL simulator.

5) Use the ASSERT statement (section 9.2) to compare design responses against expected values and also to detect improper values when reading data from a file.

   The use of these procedures is illustrated in example 10.2.

Example 10.2: Reading Values from a File

Write a VHDL code that generates a clock with a 100ns period and reads, at every positive clock transition, one line from the file *test_file.txt* written by the code of example 10.1. Display in the waveforms the signals *t* (time) and *i* (pointer) read from that file.

Solution A VHDL code for this example is presented below. Note again the presence of
the *textio* package (line 2) and that the entity (lines 4–5) is empty.

The declarative part of the architecture (lines 8–11) contains a file declaration for the
input file, followed by three signal declarations concerning the signals that must be displayed in the waveforms. The input file is named *test_file.txt*, identified by *f*, now opened
in *read* mode.

In the code proper (lines 12–31), a process is employed to read the file and produce the
three output signals. The declarative part of the process (lines 14–18) contains essentially
the same objects seen in the previous example. The process body (lines 19–30) produces the desired clock (lines 20–21 plus 28–29) and reads the file. While it is not end-
of-file (line 22), a line is read (line 23) after every positive clock transition, with the four
parameters assigned to proper variables (line 24), of which *t* and *i* are assigned to *t_out*
(line 25) and *i_out* (line 26), respectively.

```
1   -----------------------------------------------------------------
2   USE std.textio.all;
3   -----------------------------------------------------------------
4   ENTITY read_from_file IS
5   END ENTITY;
6   -----------------------------------------------------------------
7   ARCHITECTURE read_from_file OF read_from_file IS
8      FILE f: TEXT OPEN READ_MODE IS "test_file.txt";
9      SIGNAL clk: BIT := '0';
10     SIGNAL t_out: TIME RANGE 0ns TO 800ns;
11     SIGNAL i_out: NATURAL RANGE 0 TO 7;
12  BEGIN
13     PROCESS
14        VARIABLE l: LINE;
15        VARIABLE str1: string(1 TO 2);
16        VARIABLE str2: string(1 TO 3);
17        VARIABLE t: TIME RANGE 0ns TO 800ns;
18        VARIABLE i: NATURAL RANGE 0 TO 7;
19     BEGIN
20        WAIT FOR 50ns;
21        clk <= '1';
22        IF NOT ENDFILE(f) THEN
23            READLINE(f, l);
24            READ(l, str1); READ(l, t); READ(l, str2); READ(l, i);
25            t_out <= t;
26            i_out <= i;
27        END IF;
```

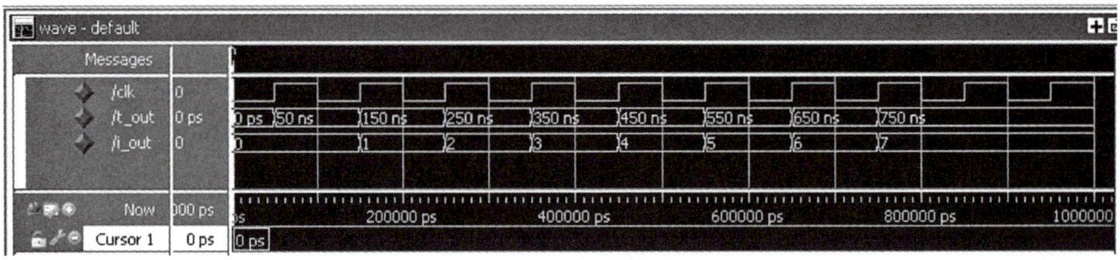

Figure 10.4
Simulation results from example 10.2.

```
28          WAIT FOR 50ns;
29          clk <= '0';
30      END PROCESS;
31  END ARCHITECTURE;
32  ----------------------------------------------------------------
```

A more sophisticated code is shown below, which includes a verification for the second parameter (*t*) read from the file. If its type does not match the type specified for *t_out*, the message "Bad value at i = 5!" (assuming that the previous value of *i* is 4) is issued and the software quits reading the file (see details about ASSERT and IMAGE in section 9.2). The same check could obviously also be included for the other values read from the file.

```
22          IF NOT ENDFILE(f) THEN
23              READ(l, str1);
24              READ(l, t, good_value);
25              ASSERT good_value
26                  REPORT "Bad value at i=" & INTEGER'IMAGE(i+1) & "!"
27                  SEVERITY FAILURE;
28              READ(l, str2); READ(l, i);
29              t_out <= t;
30              i_out <= i;
31          END IF;
32          WAIT FOR 50ns;
33          clk <= '0';
34      END PROCESS;
35  END ARCHITECTURE;
36  ----------------------------------------------------------------
```

If simulated with ModelSim (appendix D), for example, the waveforms of figure 10.4 will be obtained, which display *clk*, *t_out*, and *i_out*. To perform such a simulation, follow the procedure in sections D.1 to D.3 of appendix D, but using just one file.

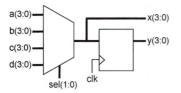

Figure 10.5
Registered multiplexer.

10.5   Graphical Simulation (Preparing the Design)

Figure 10.5 shows a registered 4×4 multiplexer (note that the registered output is y, while x is unregistered). This circuit will be used in the examples with VHDL testbenches in later sections, so before proceeding it will be designed in this section. It will also be simulated here using *graphical* simulation (no VHDL simulation yet) in both *functional* and *timing* modes.

A VHDL code for this circuit is presented below, under the name *reg_mux* (line 5). Note that all ports (lines 6–9) are of type STD_LOGIC(_VECTOR) (industry standard). The 4×4 multiplexer (lines 15–20) constitutes the combinational part of the circuit, while the 4-bit register (lines 21–26) is the sequential part.

```
1    ------------------------------------------------------
2    LIBRARY ieee;
3    USE ieee.std_logic_1164.all;
4    ------------------------------------------------------
5    ENTITY reg_mux IS
6       PORT (a, b, c, d: IN STD_LOGIC_VECTOR(3 DOWNTO 0);
7              sel: IN STD_LOGIC_VECTOR(1 DOWNTO 0);
8              clk: IN STD_LOGIC;
9              x, y: OUT STD_LOGIC_VECTOR(3 DOWNTO 0));
10   END ENTITY;
11   ------------------------------------------------------
12   ARCHITECTURE reg_mux OF reg_mux IS
13      SIGNAL mux: STD_LOGIC_VECTOR(3 DOWNTO 0);
14   BEGIN
15      mux <= a WHEN sel="00" ELSE
16             b WHEN sel="01" ELSE
17             c WHEN sel="10" ELSE
19             d;
20      x <= mux;
21      PROCESS (clk)
22      BEGIN
```

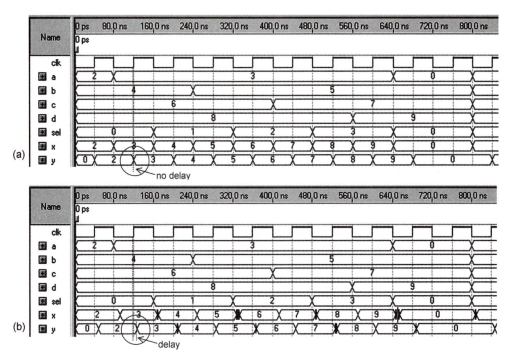

Figure 10.6
(a) Functional and (b) timing simulation results from the registered multiplexer of figure 10.5.

```
23          IF (clk'EVENT AND clk='1') THEN
24             y <= mux;
25          END IF;
26       END PROCESS;
27    END ARCHITECTURE;
28    -------------------------------------------------------
```

Say that the waveforms plotted for *clk*, *a*, *b*, *c*, *d* and *sel* in figure 10.6 are the desired test inputs, from which the simulator must determine the outputs *x* and *y*. Say also that we are using Quartus II (the conclusions would be exactly the same with any other simulator); then the setups for functional and timing simulations are those described below.

Functional Simulation
Select Processing > Generate Functional Simulation Netlist. After the program finishes producing the netlist, select Assignments > Settings > Simulator Settings. In the Simulation Mode list, choose Functional. Finally, in the Simulation Input box, enter the name of the waveform file (*reg_mux.vwf*) and click OK. Recompile the code and run the simulation. The results for *reg_mux* are those displayed in figure 10.6a. Notice that there are no propagation delays (the output changes immediately when the clock changes).

Timing Simulation

Select Assignments > Settings > Simulator Settings. In the Simulation Mode list, choose Timing. In the Simulation Input box, enter the name of the waveform file (*reg_mux.vwf*) and click OK. Recompile the code and run the simulation. The results are now those displayed in figure 10.6b. Observe the propagation delay between the clock edge and the settling of *y*.

Note: If the timing simulation is performed using a third-party software (ModelSim, for example) instead of the Quartus II simulator, postsynthesis and SDF files specific for that simulator must be generated by the synthesizer, hence requiring appropriate setups (described in the ModelSim tutorial, appendix D).

In the example above, the simulation results were *visually* inspected. This approach is fine for individual system units or for small systems, but not for large systems. As described ahead, the use of VHDL testbenches allows automated verification to be performed. It also allows the testing code to be reused and the testing procedure to be more effectively documented.

10.6   Stimulus Generation

Stimulus generation is a fundamental part of any testbench. As mentioned earlier, AFTER, WAIT FOR, and NOW are usually used to construct testbenches. AFTER is a concurrent statement, while WAIT FOR is sequential. Therefore, the former can only be used outside sequential code, while the latter can only be used within it (that is, inside PROCESS, FUNCTION, or PROCEDURE). Equivalent codes can be written with these two statements.

NOW represents the present simulation time, therefore being useful for checking particular simulation points. For example, the IF statement below

```
IF (NOW<50ns) THEN
    WAIT FOR 50ns-NOW;
    y <= x;
END IF;
```

causes the process to wait until the time 50ns is reached before assigning the value of *x* to *y*.

The 'LAST_EVENT attribute is also helpful sometimes. It returns the time interval since an event (change) has occurred on a signal or variable. For example,

```
IF (s'LAST_EVENT>20ns) THEN ...
```

The use of the VHDL constructs above will be illustrated in this and in the remaining sections of this chapter.

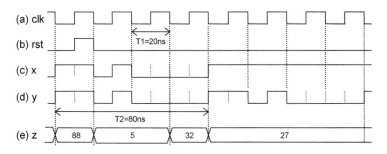

Figure 10.7
Typical stimuli: (a) Periodic signal (clock); (b) Single-pulse signal (reset); (c) Irregular finite signal; (d) Irregular periodic signal; (e) Multibit irregular signal.

The stimulus generation procedure will be illustrated with the help of figure 10.7, which exhibits five waveforms, representing common stimulus types. The waveform in (a) is regular and periodic, hence typical of clock. That in (b) has a single pulse, so is typical of reset. The waveform in (c) is irregular and finite, while that in (d) is irregular but periodic. Finally, (e) shows a multibit waveform. The generation of these signals, with AFTER and with WAIT FOR, is shown below. A complete code is presented subsequently in example 10.3.

Generation of *clk* (figure 10.7a):

```
------------------------------------------
SIGNAL clk: STD_LOGIC := '0';
---option 1:-------------------------------
clk <= NOT clk AFTER 10ns;
---option 2:-------------------------------
WAIT FOR 10ns;
clk <= NOT clk;
---option 3:-------------------------------
WAIT FOR 10ns;
clk <= '1';
WAIT FOR 10ns;
clk <= '0';
------------------------------------------
```

Generation of *rst* (figure 10.7b):

```
------------------------------------------
SIGNAL rst: STD_LOGIC := '0';
---option 1:-------------------------------
rst <= '0', '1' AFTER 10ns, '0' AFTER 20ns;
```

```
---option 2:-----------------------------
rst <= '1' AFTER 10ns, '0' AFTER 20ns;
---option 3:-----------------------------
rst <= '0'; --optional
WAIT FOR 10ns;
rst <= '1';
WAIT FOR 10ns;
rst <= '0';
WAIT;
-----------------------------------------
```

Generation of *x* (figure 10.7c):

```
-----------------------------------------
SIGNAL x: STD_LOGIC := '1';
---option 1:-----------------------------
x <= '1', --optional
     '0' AFTER 20ns,
     '1' AFTER 30ns,
     '0' AFTER 40ns,
     '1' AFTER 80ns;
---option 2:-----------------------------
x <= '1'; --optional
WAIT FOR 20ns;
x <= '0';
WAIT FOR 10ns;
x <= '1';
WAIT FOR 10ns;
x <= '0';
WAIT FOR 40ns;
x <= '1';
WAIT;
---option 3:-----------------------------
CONSTANT template: STD_LOGIC_VECTOR(1 TO 9)
   := "110100001";
FOR i IN template'RANGE LOOP
   x <= template(i);
   WAIT FOR 10ns;
END LOOP;
WAIT;
-----------------------------------------
```

Generation of *y* (figure 10.7d): See exercise 10.1.

Generation of *z* (figure 10.7e):

```
------------------------------------------
SIGNAL z: NATURAL RANGE 0 TO 255 := 88;
---option 1:-------------------------------
z <= 88, --optional
      5 AFTER 20ns,
      32 AFTER 60ns,
      27 AFTER 80ns;
---option 2:-------------------------------
z <= 88; --optional
WAIT FOR 20ns;
x <= 5;
WAIT FOR 40ns;
x <= 32;
WAIT FOR 20ns;
x <= 27;
WAIT;
------------------------------------------
```

Example 10.3: Stimuli Generation

Write a complete VHDL code that generates the signals *clk* and *rst* of figure 10.7. Employ AFTER for the former and WAIT FOR for the latter.

Solution A VHDL code for this exercise is shown below. As with all codes for simulation, the entity (lines 5–6) has no PORT declarations. The declarative part of the architecture (lines 9–10) contains declarations relative to the two signals to be generated, with respective initial values. In the architecture body (lines 11–24), *clk* is generated using AFTER in line 13 (outside the process, because AFTER is concurrent), while *rst* is created in lines 15–22. Note that because WAIT FOR is sequential a process is needed, causing the code to be much longer than that for *clk* (the whole process could be replaced with just line 23). Simulation results from this code, compiled with ModelSim (appendix D), are depicted in figure 10.8. To check them, follow the procedure in sections D.1 to D.3 of appendix D, but using just one file.

```
1    ------------------------------------------
2    LIBRARY ieee;
3    USE ieee.std_logic_1164.all;
4    ------------------------------------------
5    ENTITY testbench IS
6    END ENTITY;
7    ------------------------------------------
8    ARCHITECTURE testbench OF testbench IS
```

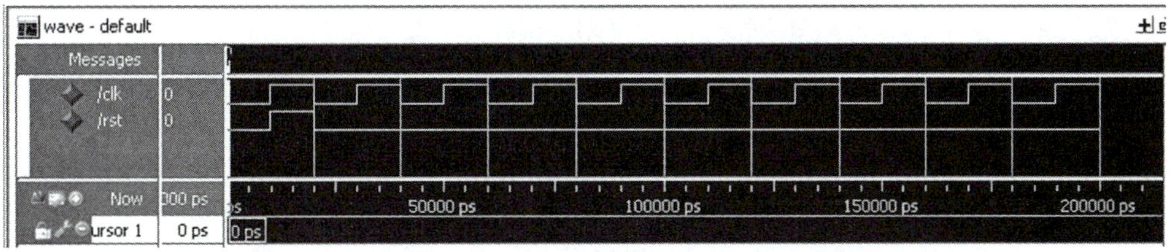

Figure 10.8
Simulation results from example 10.3.

```
 9      SIGNAL clk: STD_LOGIC := '0';
10      SIGNAL rst: STD_LOGIC := '0';
11   BEGIN
12      --Generation of clk with AFTER:
13      clk <= NOT clk AFTER 10ns;
14      --Generation of rst with WAIT FOR:
15      PROCESS
16      BEGIN
17         WAIT FOR 10ns;
18         rst <= '1';
19         WAIT FOR 10ns;
20         rst <= '0';
21         WAIT;
22      END PROCESS;
23      --rst <= '1' AFTER 10ns, '0' AFTER 20ns;
24   END ARCHITECTURE;
25   ------------------------------------------
```

10.7 General VHDL Template for Testbenches

A VHDL code for testbench generation was already seen in example 10.3. A generalization is presented in figure 10.9. As indicated on the left of the figure, it is similar to a regular VHDL code—that is, library declarations, entity, and architecture.

The particularity in the entity is that it is empty (except for GENERIC, which can be used optionally). The particularity in the architecture is that it is not for hardware inference (synthesis), but for simulation, so testbenches are generated and can also be, optionally, compared against expected values.

In the declarative part of the architecture (lines 10–18), the DUT is declared, along with the signals needed to test it, here called *a_tb*, *b_tb* (inputs) and *y_tb* (output). In the architecture body (lines 19–32), first the DUT is instantiated (line 21–22), then the stimuli are generated (lines 24–25), and finally an optional code section is shown (lines 27–31), which

Library/pack. declarations

Entity (no PORT)

Architecture

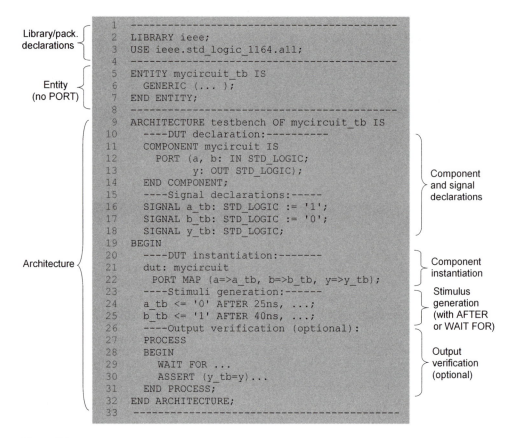

Component and signal declarations

Component instantiation

Stimulus generation (with AFTER or WAIT FOR)

Output verification (optional)

```
1  -------------------------------------------------
2  LIBRARY ieee;
3  USE ieee.std_logic_1164.all;
4  -------------------------------------------------
5  ENTITY mycircuit_tb IS
6    GENERIC (... );
7  END ENTITY;
8  -------------------------------------------------
9  ARCHITECTURE testbench OF mycircuit_tb IS
10    ----DUT declaration:----------
11    COMPONENT mycircuit IS
12      PORT (a, b: IN STD_LOGIC;
13             y: OUT STD_LOGIC);
14    END COMPONENT;
15    ----Signal declarations:-----
16    SIGNAL a_tb: STD_LOGIC := '1';
17    SIGNAL b_tb: STD_LOGIC := '0';
18    SIGNAL y_tb: STD_LOGIC;
19  BEGIN
20    ----DUT instantiation:-------
21    dut: mycircuit
22      PORT MAP (a=>a_tb, b=>b_tb, y=>y_tb);
23    ----Stimuli generation:------
24    a_tb <= '0' AFTER 25ns, ...;
25    b_tb <= '1' AFTER 40ns, ...;
26    ----Output verification (optional):
27    PROCESS
28    BEGIN
29      WAIT FOR ...
30      ASSERT (y_tb=y)...
31    END PROCESS;
32  END ARCHITECTURE;
33  -------------------------------------------------
```

Figure 10.9
VHDL template for testbenches.

is used when one wishes the simulator to automatically compare the values obtained for
y_tb against expected values for *y* (automated verification). Note that the inputs are generated using the AFTER statement (lines 24–25), but WAIT FOR could also be employed.
In some cases, the stimulus generation and the output verification codes are mixed (as in
examples 10.7 and 10.8 ahead).

The general template of figure 10.9 will be used in all remaining sections of this chapter,
in which complete designs of types I–IV testbenches are presented.

10.8   Type I Testbench (Manual Functional Simulation)

This section shows the construction of a complete Type I testbench. Recall from figure
10.2c that in this case the analysis is *functional* (circuit delays not included) and *manual*
(response checked manually, usually by visual inspection).

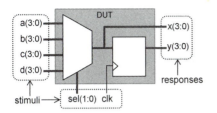

Figure 10.10
Registered multiplexer of example 10.4.

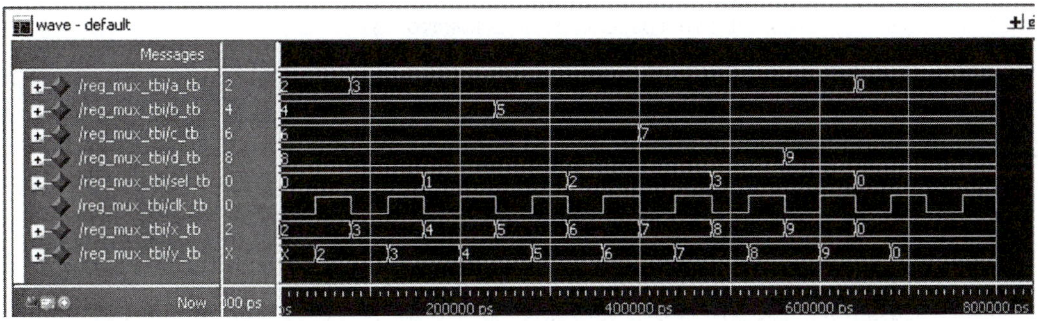

Figure 10.11
Simulation results from example 10.4.

Example 10.4: Type I Testbench for a Registered Mux

This example concerns the development of a *manual functional simulation*. The design to be tested is that seen in section 10.5 (repeated in figure 10.10, with the stimuli and responses highlighted). Develop a Type I testbench to test this circuit using the same waveforms of figure 10.6a.

Solution The *design* file for this circuit (*reg_mux.vhd*) was already prepared in section 10.5. We need now to write the *test* file (to be saved as *reg_mux_tb.vhd*), after which the simulation can be performed.

A VHDL code for the test file is presented below. As usual, the entity (lines 5–6) is empty (GENERIC was not needed in this example). The declarative part of the architecture (lines 9–24) contains a declaration for the design to be tested (DUT = *reg_mux.vhd* of section 10.5) and for the stimuli. Finally, the architecture body (lines 25–46) contains the DUT instantiation and the stimuli generation.

Simulation results, using ModelSim, are depicted in figure 10.11 (just follow the procedure in sections D.1 to D.3 of appendix D). Note that the waveforms are the same as those in figure 10.6a. Just to ease the inspection, the waveforms were displayed using integers instead of binary values.

```
1   -----------------------------------------------------------
2   LIBRARY ieee;
3   USE ieee.std_logic_1164.all;
4   -----------------------------------------------------------
5   ENTITY reg_mux_tb IS
6   END ENTITY;
7   -----------------------------------------------------------
8   ARCHITECTURE testbench OF reg_mux_tb IS
9       ----DUT declaration:--------
10      COMPONENT reg_mux IS
11        PORT (a, b, c, d: IN STD_LOGIC_VECTOR(3 DOWNTO 0);
12               sel: IN STD_LOGIC_VECTOR(1 DOWNTO 0);
13               clk: IN STD_LOGIC;
14               x, y: OUT STD_LOGIC_VECTOR(3 DOWNTO 0));
15      END COMPONENT;
16      ----Signal declarations:----
17      SIGNAL a_tb: STD_LOGIC_VECTOR(3 DOWNTO 0) := "0010";
18      SIGNAL b_tb: STD_LOGIC_VECTOR(3 DOWNTO 0) := "0100";
19      SIGNAL c_tb: STD_LOGIC_VECTOR(3 DOWNTO 0) := "0110";
20      SIGNAL d_tb: STD_LOGIC_VECTOR(3 DOWNTO 0) := "1000";
21      SIGNAL sel_tb: STD_LOGIC_VECTOR(1 DOWNTO 0) := "00";
22      SIGNAL clk_tb: STD_LOGIC := '0';
23      SIGNAL x_tb: STD_LOGIC_VECTOR(3 DOWNTO 0);
24      SIGNAL y_tb: STD_LOGIC_VECTOR(3 DOWNTO 0);
25  BEGIN
26      ---DUT instantiation:-------
27      dut: reg_mux PORT MAP (
28         a => a_tb,
29         b => b_tb,
30         c => c_tb,
31         d => d_tb,
32         clk => clk_tb,
33         sel => sel_tb,
34         x => x_tb,
35         y => y_tb);
36      ---Stimuli generation:------
37      clk_tb <= NOT clk_tb AFTER 40ns;
38      a_tb <= "0011" AFTER 80ns, "0000" AFTER 640ns;
39      b_tb <= "0101" AFTER 240ns;
40      c_tb <= "0111" AFTER 400ns;
41      d_tb <= "1001" AFTER 560ns;
42      sel_tb <= "01" AFTER 160ns,
43                "10" AFTER 320ns,
44                "11" AFTER 480ns,
```

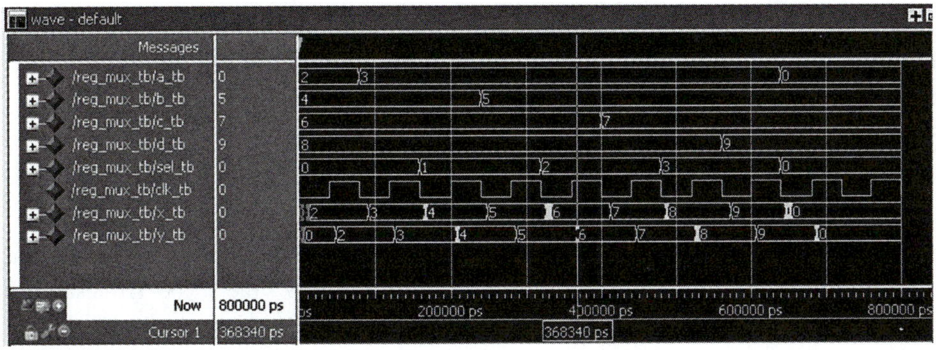

Figure 10.12
Simulation results from example 10.5.

```
45                    "00" AFTER 640ns;
46   END ARCHITECTURE;
47   -------------------------------------------------------------
```

10.9   Type II Testbench (Manual Timing Simulation)

Type II is similar to Type I, with the only exception being that now the DUT's internal propagation delays are taken into account. Because it involves time but the output is still manually verified, it is a *manual timing simulation*.

Example 10.5: Type II Testbench for a Registered Mux

Modify the Type I simulation of example 10.4 in order to produce a Type II simulation.

Solution   As explained in section 10.2, besides the two files prepared by the user (*reg_mux.vhd* and *reg_mux_tb.vhd*), two other files (postsynthesis and SDF), produced by the synthesizer, are also needed now.

For example, if using Quartus II to synthesize the design and ModelSim to simulate it, follow the procedure in sections D.1, D.2, and D.4 of appendix D. After the simulation is finished, the waveforms of figure 10.12 will be displayed. Note that the only difference with respect to the simulation results in figure 10.11 is the inclusion of propagation delays. For example, the cursor shows a transition at 368.34ns, in response to a clock transition at 360ns, hence with a propagation delay of 8.34ns.

10.10   Type III Testbench (Automated Functional Simulation)

Because the output values are automatically compared against expected values, but internal circuit delays are not taken into account, this is an *automated functional simulation*.

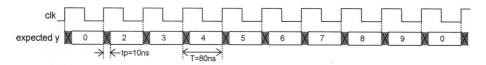

Figure 10.13
Expected output values in example 10.6.

Hence in the same way that Type I is a particular case of Type II, Type III is a particular case of Type IV. Consequently, after understanding Type IV, the Type III testbench becomes relatively straightforward.

10.11   Type IV Testbench (Automated Timing Simulation)

Type IV is the most complete and also the most complex type of testbench. Because the output is automatically checked against expected values and the circuit's internal delays are taken into account, it is an *automated timing simulation*, also known as *full bench*.

Example 10.6: Type IV Testbench for a Registered Mux

Develop a Type IV testbench to test the registered output (y) of the multiplexer seen in section 10.5. Assume that the expected maximum propagation delay between the clock and y (t_{pCQ}) is 10ns. Use the same stimuli employed in the previous examples (from figure 10.6), so the expected output values are those depicted in figure 10.13, where the propagation delays are highlighted by gray shades.

Solution   A VHDL code for a full bench to test this circuit is presented below. The only major difference with respect to the test file in example 10.4 is the inclusion of the optional code section that automatically checks the output values against expected results (lines 42–68). The expected output signal (called *expected*) was declared in line 27 and then constructed in lines 44–52. The comparison is made in the process of lines 54–68 by means of the ASSERT statement (see details in section 9.2), and occurs every 10ns (line 56). Note in line 7 that the acceptable propagation delay was specified as a generic parameter.

The IF statement in lines 57–67 causes the process to be run until the time reaches 800ns. If a mismatch is detected, the first ASSERT (lines 58–62) terminates the simulation and issues the message "Mismatch at t = xxx y_tb = xxx y_exp = xxx", where *xxx* represents the corresponding actual value. If no errors are found, the simulation is terminated when the time reaches 800ns, with the second ASSERT (lines 64–66) forcing the message "No error found (t = 800000 ps)" to be issued.

Looking at the simulation results obtained in example 10.5 (figure 10.12), delays under 10ns are observed, so we do not expect this simulation to find any data mismatches. However, if the value of the parameter in line 7 is reduced, a problem will eventually occur, which is one way of estimating the limiting speed of this circuit.

To simulate this design (with ModelSim), just follow the procedure in sections D.1, D.2, and D.4 of appendix D. Try to play with t_p and *expected* in order to better understand the automated comparison process.

```
1   ------------------------------------------------------------------
2   LIBRARY ieee;
3   USE ieee.std_logic_1164.all;
4   USE ieee.std_logic_unsigned.all;
5   ------------------------------------------------------------------
6   ENTITY reg_mux_tb IS
7      GENERIC (tp: TIME := 10ns);
8   END ENTITY;
9   ------------------------------------------------------------------
10  ARCHITECTURE testbench OF reg_mux_tb IS
11     ----DUT declaration:----------
12     COMPONENT reg_mux IS
13        PORT (a, b, c, d: IN STD_LOGIC_VECTOR(3 DOWNTO 0);
14              sel: IN STD_LOGIC_VECTOR(1 DOWNTO 0);
15              clk: IN STD_LOGIC;
16              x, y: OUT STD_LOGIC_VECTOR(3 DOWNTO 0));
17     END COMPONENT;
18     ----Signal declarations:------
19     SIGNAL a_tb: STD_LOGIC_VECTOR(3 DOWNTO 0) := "0010";
20     SIGNAL b_tb: STD_LOGIC_VECTOR(3 DOWNTO 0) := "0100";
21     SIGNAL c_tb: STD_LOGIC_VECTOR(3 DOWNTO 0) := "0110";
22     SIGNAL d_tb: STD_LOGIC_VECTOR(3 DOWNTO 0) := "1000";
23     SIGNAL sel_tb: STD_LOGIC_VECTOR(1 DOWNTO 0) := "00";
24     SIGNAL clk_tb: STD_LOGIC := '0';
25     SIGNAL x_tb: STD_LOGIC_VECTOR(3 DOWNTO 0);
26     SIGNAL y_tb: STD_LOGIC_VECTOR(3 DOWNTO 0);
27     SIGNAL expected: STD_LOGIC_VECTOR(3 DOWNTO 0) := "0000";
28  BEGIN
29     ---DUT instantiation:------------
30     dut: reg_mux PORT MAP (a_tb, b_tb, c_tb, d_tb, sel_tb,
31        clk_tb, x_tb, y_tb);
32     ---Stimuli generation:-----------
33     clk_tb <= NOT clk_tb AFTER 40ns;
34     a_tb <= "0011" AFTER 80ns, "0000" AFTER 640ns;
35     b_tb <= "0101" AFTER 240ns;
36     c_tb <= "0111" AFTER 400ns;
37     d_tb <= "1001" AFTER 560ns;
38     sel_tb <= "01" AFTER 160ns,
39               "10" AFTER 320ns,
40               "11" AFTER 480ns,
```

```
41                    "00" AFTER 640ns;
42        ---Output verification:----------
43        ---(i)Generate template:
44        expected <= "0010" AFTER 40ns+tp,
45                    "0011" AFTER 120ns+tp,
46                    "0100" AFTER 200ns+tp,
47                    "0101" AFTER 280ns+tp,
48                    "0110" AFTER 360ns+tp,
49                    "0111" AFTER 440ns+tp,
50                    "1000" AFTER 520ns+tp,
51                    "1001" AFTER 600ns+tp,
52                    "0000" AFTER 680ns+tp;
53        ---(ii)Make comparison:
54        PROCESS
55        BEGIN
56           WAIT FOR tp;
57           IF (NOW<800ns) THEN
58              ASSERT (y_tb=expected)
59                 REPORT "Mismatch at t=" & TIME'IMAGE(NOW) &
60                    " y_tb=" & INTEGER'IMAGE(conv_integer(y_tb)) &
61                    " y_exp=" & INTEGER'IMAGE(conv_integer(expected))
62                 SEVERITY FAILURE;
63           ELSE
64              ASSERT FALSE
65                 REPORT "No error found (t=" & TIME'IMAGE(NOW) & ")"
66                 SEVERITY NOTE;
67           END IF;
68        END PROCESS;
69     END ARCHITECTURE;
70     ----------------------------------------------------------------
```

10.12   Testbenches with Record Types

The type RECORD is useful for testing codes whose input-output values are expressed by means of arbitrary (but not too long) lookup tables located in the test code itself. For large data sets, the use of files is recommended. The use of RECORD is illustrated in this section, while the use of files is illustrated in the next.

Example 10.7: Type IV Testbench with a Record Type

This example concerns the development of a full bench for a binary-to-gray converter (figure 10.14a) with the input stimuli and expected results stored in a table in the test file itself, constructed with the help of RECORD. Recall that a gray code is one in which neigh-

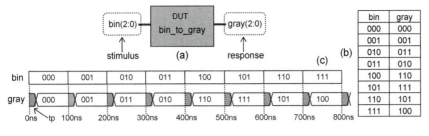

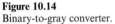

Figure 10.14
Binary-to-gray converter.

boring words differ by just one bit. To ease the analysis, just three bits will be used. Their values are listed in figure 10.14b. This exercise is divided into three parts:

1) Write a VHDL code for the *design* file. In it, use a closed-form expression to make the binary-to-gray conversion.

2) Write a VHDL code for the *test* file. In it, to illustrate the use of tables and RECORD, enter the data (stimuli plus expected results) using a lookup table instead of an expression.

3) Simulate the design, for *timing* analysis. Change the input data (*bin*) every 100ns, then give some time (t_p) for the output (*gray*) to settle. Next, make the comparison between the actual and the expected results. This approach is illustrated in figure 10.14c, where the gray areas highlight the output settling period (maximum propagation delay allowed).

Solution   Part (1): A VHDL code for the design file, under the title *bin_to_gray* (line 5), is shown below. Only concurrent code is employed (lines 13–16), with closed-form expressions used to make the conversion.

```
 1   -----------------------------------------------------
 2   LIBRARY ieee;
 3   USE ieee. std_logic_1164.all;
 4   -----------------------------------------------------
 5   ENTITY bin_to_gray IS
 6      GENERIC (N: NATURAL := 3);
 7      PORT (bin: IN STD_LOGIC_VECTOR(N-1 DOWNTO 0);
 8            gray: OUT STD_LOGIC_VECTOR(N-1 DOWNTO 0));
 9   END ENTITY;
10   -----------------------------------------------------
11   ARCHITECTURE bin_to_gray OF bin_to_gray IS
12   BEGIN
13      gray(N-1) <= bin(N-1);
14      gen: FOR i IN 1 TO N-1 GENERATE
15          gray(N-1-i) <= bin(N-i) XOR bin(N-1-i);
```

```
16     END GENERATE;
17  END ARCHITECTURE;
18  ----------------------------------------------------
```

Part (2): A full testbench for the binary-to-gray converter is presented next, under the title
bin_to_gray_tb (line 5). As before, the entity (lines 5–8) has no PORT declarations; it contains only generic parameters representing the distance (*period* = 100ns) between the data
samplings and the waiting time interval (t_p = 15ns) before the output is read.

The declarative part of the architecture (lines 11–26) contains a component (DUT) declaration followed by signal, type, and constant declarations. A RECORD (section 3.14),
called *data_pair* and with two STD_LOGIC_VECTOR values named *col1* and *col2*, is
created in lines 19–22. Next, in line 23, a type called *table* is declared as having eight such
pairs. Finally, a constant called *templates*, of type *table*, is specified in lines 24–26, containing the values listed in figure 10.14b.

The code proper (lines 27–45) contains two parts. The first part (line 29) consists of the
DUT instantiation, while the second part (lines 31–44) is a process responsible for generating the stimuli and also for checking the results. Observe that, contrary to the previous
example, here the stimulus generation and the output verification are done together (a
stimulus is applied in line 34, then some time is given in line 35 for the output to settle,
and finally a comparison is made by the ASSERT statement in lines 36–38). If a mismatch
is detected, the simulation is terminated and the following message is issued (assuming
i = 5): "Mismatch at iteration i = 5". On the other hand, if no mismatch occurs, the simulator eventually reaches line 41, whose ASSERT statement forces the message "No error
found!" to be displayed. (For details on ASSERT and the usage of IMAGE, see section
9.2.)

```
1   -----------------------------------------------------------------------
2   LIBRARY ieee;
3   USE ieee. std_logic_1164.all;
4   -----------------------------------------------------------------------
5   ENTITY bin_to_gray_tb IS
6      GENERIC (period: TIME := 100ns;
7               tp: TIME := 15ns);
8   END ENTITY;
9   -----------------------------------------------------------------------
10  ARCHITECTURE testbench OF bin_to_gray_tb IS
11     ----DUT declaration:--------------------
12     COMPONENT bin_to_gray IS
13        PORT (bin: IN STD_LOGIC_VECTOR(2 DOWNTO 0);
14              gray: OUT STD_LOGIC_VECTOR(2 DOWNTO 0));
15     END COMPONENT;
16     ----Signal declarations:----------------
```

```
17      SIGNAL b: STD_LOGIC_VECTOR(2 DOWNTO 0); --binary in
18      SIGNAL g: STD_LOGIC_VECTOR(2 DOWNTO 0); --gray out
19      TYPE data_pair IS RECORD
20         col1: STD_LOGIC_VECTOR(2 DOWNTO 0);
21         col2: STD_LOGIC_VECTOR(2 DOWNTO 0);
22      END RECORD;
23      TYPE table IS ARRAY (1 TO 8) OF data_pair;
24      CONSTANT templates: table := (
25         ("000", "000"), ("001", "001"), ("010", "011"), ("011", "010"),
26         ("100", "110"), ("101", "111"), ("110", "101"), ("111", "100"));
27   BEGIN
28      ---DUT instantiation:--------------------
29      dut: bin_to_gray PORT MAP (bin => b, gray => g);
30      ---Stimuli generation and comparison:----
31      PROCESS
32      BEGIN
33         FOR i IN table'RANGE LOOP
34            b <= templates(i).col1;
35            WAIT FOR tp;
36            ASSERT g=templates(i).col2
37               REPORT "Mismatch at iteration=" & INTEGER'IMAGE(i)
38               SEVERITY FAILURE;
39            WAIT FOR period-tp;
40         END LOOP;
41         ASSERT FALSE
42            REPORT "No error found!"
43         SEVERITY NOTE;
44      END PROCESS;
45   END ARCHITECTURE;
46   --------------------------------------------------------------------
```

Part (3): Since it is a timing simulation, we must include in the design the postsynthesis and SDF files generated by the synthesizer. This procedure was already described in the previous example. If using ModelSim, just follow the procedure in sections D.1, D.2, and D.4 of appendix D. Simulation results (from ModelSim) are depicted in figure 10.15. Observe at the cursors' feet that the input-output propagation delay is 7.92ns (the device used in this design was a Cyclone II FPGA).

10.13   Testbenches with Data Files

Automated data comparison often involves the examination of large data sets, in which case the use of data files is recommended. This section illustrates how a VHDL code can

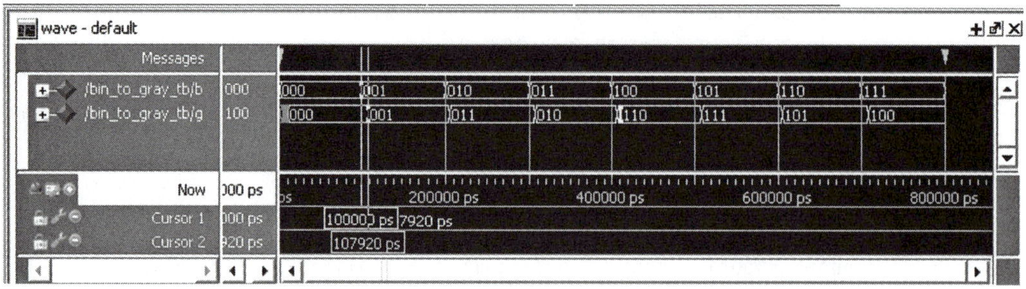

Figure 10.15
Simulation results from example 10.7.

deal with a file during a Type IV simulation. Variants of the code below can be easily attained using the procedures described in sections 10.3 and 10.4.

Example 10.8: Type IV Testbench with a Data File

This example concerns the development of a full bench with the in-out data stored in a file. Redo example 10.7, this time with the stimuli (*bin*) and expected values (*gray*) stored in a file (call it *template.txt*). The design file is obviously still the same.

Solution A test file for a Type IV testbench for this exercise is presented below. Because a data file will now be used, the *textio* package (line 4) was included in the package declarations. The entity (lines 6–9) is similar to that in the previous example.

As usual, the declarative part of the architecture (lines 12–20) contains the DUT declaration followed by signal declarations. In the latter, not only three signals (*b*, *g*, and *gtest*, needed to represent *bin* and *gray*) are declared (lines 18–19), but also a file (line 20), called *template.txt*, identified by *f* and opened in *read* mode.

Again, as in the previous design, the architecture body (lines 21–55) consists of the DUT instantiation (line 23) followed by a process (lines 25–54) for stimulus generation plus output verification. Note that the overall file-reading process is relatively similar to that seen earlier in section 10.4.

In the declarative part of the process (lines 26–29), five variables are specified. The first (*l*) is used to store a file line during the file-read procedure. The second (*good_value*) is used to check whether the type of the read value matches the type of the object to which the value is being assigned. The third (*space*) is just to suppress the line space between *bin* and *gray*. Finally, the last two (*bfile*, *gfile*) receive the values of *bin* and *gray* read from the file.

In the process body (lines 30–54), a file line is read and assigned to *l* (line 32), then the first value is separated from *l* and assigned to *bfile* (line 33), and is subsequently tested using ASSERT (lines 34–36). Because the procedure READ does not accept the type

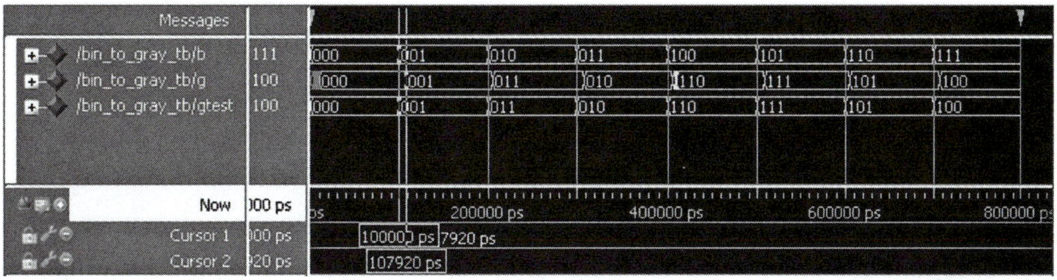

Figure 10.16
Simulation results from example 10.8.

STD_LOGIC_VECTOR (see section 10.4), a type-conversion function was employed to convert *bfile* (BIT_VECTOR) into *b* (STD_LOGIC_VECTOR) (see type-conversion functions in the table of figure 3.10). If an improper value is found in the file, the message "Improper value for 'bin' in file!" is issued (line 35) and the simulation is terminated (line 36). A similar construction is used for the value assigned to *gtest*. Finally, some time is given (line 44) for the output to settle, after which the expected value, *gtest*, is compared against the value attained from the circuit, *g*. If they do not match, the message "Data mismatch!" is issued (line 46) and the simulation is terminated (line 47). Only when no errors are found does the simulator reach line 50, where the last ASSERT forces the message "No errors found!" to be issued, concluding the simulation.

A final remark regards the *good_value* tests. If, for example, the check in lines 34–36 is deleted, and an improper value is found in the file, the simulator will still report the problem. But because that will cause just the loop to be terminated, lines 50–52 will still be executed, giving the user the false information that no error was found (indeed, the circuit might be correct; the problem is that the tests were incomplete).

Simulation results, using ModelSim, are depicted in figure 10.16. To check them, just follow the procedure in sections D.1, D.2, and D.4 of appendix D.

```
1    -----------------------------------------------------
2    LIBRARY ieee;
3    USE ieee.std_logic_1164.all;
4    USE std.textio.all;
5    -----------------------------------------------------
6    ENTITY bin_to_gray_tb IS
7       GENERIC (period: TIME := 100ns;
8                tp: TIME := 15ns);
9    END ENTITY;
10   -----------------------------------------------------
11   ARCHITECTURE testbench OF bin_to_gray_tb IS
12      ----DUT declaration:----------
```

```
13     COMPONENT bin_to_gray IS
14       PORT (bin: IN STD_LOGIC_VECTOR(2 DOWNTO 0);
15             gray: OUT STD_LOGIC_VECTOR(2 DOWNTO 0));
16     END COMPONENT;
17     ----Signal declarations:------
18     SIGNAL b: STD_LOGIC_VECTOR(2 DOWNTO 0);
19     SIGNAL g, gtest: STD_LOGIC_VECTOR(2 DOWNTO 0);
20     FILE f: TEXT OPEN READ_MODE IS "template.txt";
21   BEGIN
22     ---DUT instantiation:------------------
23     dut: bin_to_gray PORT MAP (bin => b, gray => g);
24     ---Output verification:----------------
25     PROCESS
26       VARIABLE l: LINE;
27       VARIABLE good_value: BOOLEAN;
28       VARIABLE space: CHARACTER;
29       VARIABLE bfile, gfile: BIT_VECTOR(2 DOWNTO 0);
30     BEGIN
31       WHILE NOT ENDFILE (f) LOOP
32         READLINE(f, l);
33         READ(l, bfile, good_value);
34         ASSERT (good_value)
35           REPORT "Improper value for 'bin' in file!"
36           SEVERITY FAILURE;
37         b <= to_stdlogicvector(bfile);
38         READ(l, space);
39         READ(l, gfile, good_value);
40         ASSERT (good_value)
41           REPORT "Improper value for 'gray' in file!"
42           SEVERITY FAILURE;
43         gtest <= to_stdlogicvector(gfile);
44         WAIT FOR tp;
45         ASSERT (gtest=g)
46           REPORT "Data mismatch!"
47           SEVERITY FAILURE;
48         WAIT FOR period-tp;
49       END LOOP;
50       ASSERT FALSE
51         REPORT "No errors found!"
52         SEVERITY NOTE;
53       WAIT;
54     END PROCESS;
55   END ARCHITECTURE;
56   ----------------------------------------------------
```

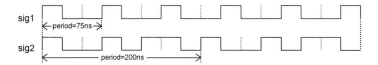

Figure 10.17

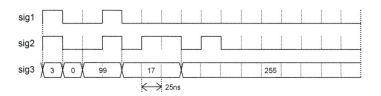

Figure 10.18

time	inp	outp
0ns	1	100
50ns	2	110
100ns	0	011
150ns	3	111
200ns	4	010
250ns	6	101
300ns	7	000
350ns	5	001

Figure 10.19

10.14   Exercises

Exercise 10.1: Generation of Periodic Stimuli

Write and simulate a VHDL code that generates the signals *sig1* and *sig2* of figure 10.17 and the signal *y* of figure 10.7d.

Exercise 10.2: Generation of Nonperiodic Stimuli

Write and simulate a VHDL code that generates the signal *sig1*, *sig2*, and *sig3* of figure 10.18.

Exercise 10.3: Writing to a File

Write and simulate a VHDL code that writes the table of figure 10.19 to a file.

Exercise 10.4: Reading from a File

Write and simulate a VHDL code that reads the table of figure 10.19 from a file. Say that *inp* is INTEGER and *outp* is STD_LOGIC_VECTOR. Include good-value tests in your code.

Exercise 10.5: Type I Testbench for a LUT-Based Design

This exercise concerns the design and test of a circuit that produces the signal *outp* specified in the table of figure 10.19 when it receives the corresponding *time* and *inp* stimuli.

a) Write a VHDL code for the *design* file (call it *lut.vhd*) and *graphically* test it (as in section 10.5).

b) Write a VHDL code for the *test* file (call it *lut_tb.vhd*), which must consist of a Type I testbench (as in example 10.4). Run the simulation from 0 to 400ns.

Exercise 10.6: Type II Testbench for a LUT-Based Design

a) Make the changes needed in the solution to exercise 10.5 in order to turn it into a Type II testbench (see example 10.5). Again, run the simulation from 0 to 400ns.

b) What is the propagation delay of your design for the chosen device?

Exercise 10.7: Type III Testbench for a LUT-Based Design

Make the changes needed in the solution to exercise 10.5 in order to turn it into a Type III testbench. Run the simulation from 0 to 400ns.

Exercise 10.8: Type IV Testbench for a LUT-Based Design with a Record

a) Make the changes needed in the solution (test file *lut_tb.vhd*) to exercise 10.7 in order to turn it into a Type IV testbench. Enter the values of figure 10.19 using a constant of type RECORD, located in the declarative part of the architecture of the test file (as in example 10.7). Adopt 15ns as the maximum acceptable propagating delay (call it t_p), which should be entered as a generic parameter. Run the simulation from 0 to 400ns.

b) What is the propagation delay of your design for the chosen device?

c) What will eventually happen if the value of t_p is gradually reduced?

Exercise 10.9: Type IV Testbench for a LUT-Based Design with a File

Redo exercise 10.8, this time with the table of figure 10.19 stored in a file (call it *data_file.txt*). This procedure was illustrated in example 10.8. Again, run the simulation from 0 to 400ns.

Exercise 10.10: Type I Testbench for a Binary-to-Gray Converter

a) Consider the solution to example 10.7, which is a Type IV testbench for a binary-to-gray converter with the values of figure 10.14b entered by means of a constant located in the test code. How many other options can you find for entering that data?

b) Simplify (and simulate) the solution to example 10.7 in order to turn it into a Type I testbench. Enter the stimuli of figure 10.14b using one of the methods that you have listed in part (a) above.

Exercise 10.11: Type I Testbench for an Address Decoder

An address decoder was designed in example 2.4. Develop a Type I testbench to test that circuit (as in example 10.4). Use the same stimuli employed in the graphical simulation of figure 2.7.

Exercise 10.12: Type IV Testbench for an Address Decoder

Enhance the solution to exercise 10.11 in order to turn it into a Type IV testbench (see examples 10.6–10.8). Use again the same stimuli employed in the graphical simulation of figure 2.7.

Exercise 10.13: Type I Testbench for a Carry-Ripple Adder

A carry-ripple adder was designed in example 6.4. Develop a Type I testbench to test that circuit (as in example 10.4). Use the same stimuli employed in the graphical simulation of figure 6.7.

Exercise 10.14: Type IV Testbench for a Carry-Ripple Adder

Enhance the solution to exercise 10.13 in order to turn it into a Type IV testbench (see examples 10.6–10.8). Use again the same stimuli employed in the graphical simulation of figure 6.7.

Exercise 10.15: Type I Testbench for a Shift Register

A 4×1 shift register was designed in example 6.3. Develop a Type I testbench to test that circuit (as in example 10.4). Use the same stimuli employed in the graphical simulation of figure 6.5.

Exercise 10.16: Type IV Testbench for a Shift Register

Enhance the solution to exercise 10.15 in order to turn it into a Type IV testbench (see examples 10.6–10.8). Use again the same stimuli employed in the graphical simulation of figure 6.5.

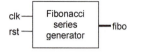

Figure 10.20

Exercise 10.17: Type I Testbench for a Fibonacci Series Generator

Figure 10.20 shows the top-level diagram of a Fibonacci series generator (the series starts with 0 and 1, with each subsequent element obtained by summing the preceding two elements, so the following results: 0, 1, 1, 2, 3, 5, 8, . . .). The circuit must produce a new value at every positive clock transition, starting from "000. . .0" (upon reset) and stopping when the largest 8-bit Fibonacci value is reached (assume that *fibo* is an 8-bit signal, so the largest value is 233). If *rst* is asserted, the output must return to zero, with the series restarting automatically after *rst* returns to '0'. Assume 100ns for the clock period.

a) Write a VHDL code for the *design* file (call it *fibonacci.vhd*) and *graphically* test it (as in section 10.5).

b) Write a VHDL code for the *test* file (call it *fibonacci_tb.vhd*), which must consist of a Type I testbench (as in example 10.4). Make sure to observe in the simulations the main circuit features (stop, reset, etc.).

(A VHDL code and other Fibonacci series details can be seen in Pedroni 2008.)

Exercise 10.18: Type II Testbench for a Fibonacci Series Generator

a) Make the changes needed in the solution to exercise 10.17 in order to turn it into a Type II testbench (see example 10.5).

b) What is the propagation delay of your design for the chosen device?

Exercise 10.19: Type IV Testbench for a Fibonacci Series Generator

a) Make the changes needed in the solution to exercise 10.17 or 10.18 in order to turn it into a Type IV testbench (see examples 10.6–10.8). Adopt 15ns as the maximum acceptable propagation delay (t_p), which should be entered as a generic parameter.

b) What is the propagation delay of your design for the chosen device?

c) What will eventually happen if the value of t_p is gradually reduced?

III EXTENDED AND ADVANCED DESIGNS

11 VHDL Design of State Machines

11.1 Introduction

Finite state machine (FSM) is a special modeling technique for sequential logic circuits. This approach can be very helpful in designing circuits whose tasks form a well-defined list, containing all possible system states and the necessary conditions for the system to move from one state to another, as well as the output values that the system must produce in each state. Digital controllers are classical examples of circuits that fall in this category.

There are two fundamental representations for FSMs, one relative to the specifications (called *state transition diagram*) and the other relative to the hardware (*combinational* versus *sequential* logic).

State Transition Diagram

The specifications of a finite state machine can be translated using a *state transition diagram*. An example is shown in figure 11.1a, which says the following: the machine has three states, called *A*, *B*, and *C*, one input (besides *clock*, of course, and possibly *reset*), called *x*, and one output, called *y*; the system starts at state *A* (upon reset), progressing to *B* if $x = 2$ at the moment when a (positive) clock edge occurs, or remaining in *A* otherwise; when in *B*, it must return to *A* if $x = 1$, move to *C* if $x = 0$, or remain in *B* otherwise; finally, when in *C*, it must move to *A* if $x = 1$ or remain in *C* otherwise; the output must be $y = \text{'0'}$ when in state *A* or *B*, or $y = \text{'1'}$ if in state *C*.

The three state transition diagrams in figure 11.1 are equivalent. The case in (a) explicitly declares all progressing conditions; the option in (b) uses "else" to indicate the remaining conditions; finally, the case in (c) suppresses the "else" conditions altogether, which then become implicit. These representations can (and will) be used interchangeably throughout the book.

Hardware-Based Representation

An FSM representation from a hardware perspective is shown in figure 11.2a, which shows the system divided into two sections. The lower section is *sequential* (contains the flip-flops,

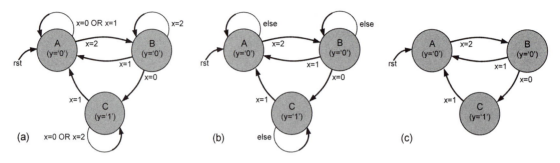

Figure 11.1
Three equivalent state transition diagrams: (a) Explicitly specified; (b) Using the "else" keyword; (c) With implicit "else" conditions.

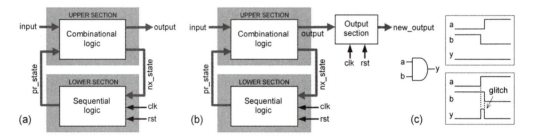

Figure 11.2
FSM representation from a hardware perspective (a) without and (b) with output register; (c) Illustration of glitch formation.

all of type D), while the upper section is *combinational* (contains the combinational circuits). The signal presently stored in the flip-flops is called *pr_state*, while that to be stored at the next (say, positive) clock edge is called *nx_state*.

One limitation of this architecture is that the output, generally produced by the combinational circuit (there are exceptions in which just wires connecting the flip-flop outputs to the actual outputs suffice), might be subject to glitches. If in that particular application glitches are not acceptable, then some kind of solution must be provided.

Glitches are brief voltage or current spikes that corrupt a signal. Their occurrence is illustrated in figure 11.2c, which shows a simple AND gate. When any input is low, the output must be low, as in the upper plot; however, depending on the propagation delays, *b* might arrive after *a*, causing a glitch to occur in *y*, as shown in the second plot.

One solution to avoid glitches is presented in the FSM model of figure 11.2b. It consists of using an extra section at the output, containing only DFFs, hence guaranteeing glitch-free signals. It is important to note that when this approach is adopted the new output will be one clock cycle delayed with respect to the original output when the same clock edge (say, positive) is used, or one-half of a clock cycle if opposite clock edges are employed instead.

A last comment concerns the classification of FSMs. If, besides the stored state, the output depends also on an external input, it is called a Mealy machine. Otherwise, if it depends solely on the stored state, it is a Moore machine. The latter is obviously a particular case of the former. Conventional counters are classical examples of Moore machines, but most implementations fall in the Mealy category.

11.2   VHDL Template for FSMs

A VHDL template for the design of FSMs is presented below, which directly resembles the diagram of figure 11.2b. The design of the lower (sequential) section is in the process of lines 18–25, while the upper (combinational) section is in the process of lines 27–48. Note that only the former requires clock and reset because all flip-flops are in that section. Recall also that the latter, being combinational, can be designed with concurrent code (WHEN or SELECT statement) as well. The optional (sequential) output section of figure 11.2b, for glitch removal, is included in the process of lines 50–57. In line 12, an enumerated data type, called *state*, is created, then the signals *pr_state* and *nx_state* are declared in line 13 as conforming with that data type. An optional attribute for enumerated data types (explained in section 11.4) appears in lines 14–15. All FSMs in the book will be designed according with this VHDL template.

Note: An extension to this template will be presented in section 11.6 to deal with complex, timed machines.

```
1    -----------------------------------------------------
2    LIBRARY ieee;
3    USE ieee.std_logic_1164.all;
4    -----------------------------------------------------
5    ENTITY <entity_name> IS
6       PORT (clk, rst: IN STD_LOGIC;
7             input: IN <data_type>;
8             output: OUT <data_type>);
9    END <entity_name>;
10   -----------------------------------------------------
11   ARCHITECTURE <architecture_name> OF <entity_name> IS
12      TYPE state IS (A, B, C, ...);
13      SIGNAL pr_state, nx_state: state;
14      ATTRIBUTE ENUM_ENCODING: STRING; --optional attribute
15      ATTRIBUTE ENUM_ENCODING OF state: TYPE IS "sequential";
16   BEGIN
```

```
17        ------Lower section of FSM:------------
18      PROCESS (clk, rst)
19      BEGIN
20         IF (rst='1') THEN
21            pr_state <= A;
22         ELSIF (clk'EVENT AND clk='1') THEN
23            pr_state <= nx_state;
24         END IF;
25      END PROCESS;
26      ------Upper section of FSM:------------
27      PROCESS (pr_state, input)
28      BEGIN
29         CASE pr_state IS
30            WHEN A =>
31               output <= <value>;
32               IF (input=<value>) THEN
33                  nx_state <= B;
34               ...
35               ELSE
36                  nx_state <= A;
37               END IF;
38            WHEN B =>
39               output <= <value>;
40               IF (input=<value>) THEN
41                  nx_state <= C;
42               ...
43               ELSE
44                  nx_state <= B;
45               END IF;
46            WHEN ...
47         END CASE;
48      END PROCESS;
49      ------Output section (optional):-------
50      PROCESS (clk, rst)
51      BEGIN
52         IF (rst='1') THEN
53            new_output <= <value>;
54         ELSIF (clk'EVENT AND clk='1') THEN --or clk='0'
55            new_output <= output;
56         END IF;
57      END PROCESS;
58   END <architecture_name>;
59   ------------------------------------------------------------
```

FSM Golden Rules

Before we show design examples using the template above, it is very important to consider three fundamental rules in writing good VHDL code for finite state machines.

Rule 1: Fully specify the truth table for the combinational part of the FSM.

The upper section of the FSM is a combinational circuit. Therefore, the complete lookup table should be covered, or the compiler might infer latches to hold a previous value. In other words, in the CASE statement (lines 29–47) of the template above, always specify all conditions for each state (output values and next state). For example, if the output value in a particular state does not matter, express it using the "don't care" condition ('-' or "---...", same as 'X' or "XXX...").

Rule 2: Assignments made in the combinational part are only guaranteed while the FSM is in the state where the assignments were made.

A common mistake is to make an assignment in the combinational part (process of lines 27–48 in the template above) and think that it will be stored. Remember, the upper section is intended to be a purely combinational circuit. As an example, say that one specifies "$y <= x + 1;$" in a certain state; this does not mean that y will be incremented, but simply that the value of y will be $x + 1$ while the machine is in that state.

Rule 3: Anything that must be registered should go in the sequential section of the FSM.

For the upper section to remain purely combinational, all flip-flops must be installed in the lower (sequential) part of the machine (except for the optional glitch-remover section). In other words, the corresponding code should be inserted in the process of lines 18–25 of the template above (or in a separate, complementary process). A very important example of this practice will be seen in section 11.6, where a modification will be made to the template in order to be able to design complex timed machines in a simple, systematic way.

Example 11.1: Vending-Machine Controller

Figure 11.3a shows the top-level diagram of a simplified controller for a vending machine that sells candy bars for 25 cents. The inputs are *nickel_in*, *dime_in*, and *quarter_in*, indicating the type of coin that was deposited, plus clock (*clk*) and reset (*rst*), to which the circuit responds with the outputs *candy_out*, to dispense a candy bar, plus *nickel_out* or *dime_out*, asserted when change is due. Design this circuit using the FSM approach. Also, estimate the number of flip-flops that will be required.

Solution As mentioned earlier, digital controllers are good examples of circuits that can be efficiently implemented using the FSM approach. The state transition diagram for the present example is depicted in figure 11.3b, where the state names (*st0*, *st5*, *st10*, etc.) indicate the total amount of credit (only nickels, dimes, and quarters are accepted). To simplify the diagram, the following notation was used: *ni = nickel_in*, *di = dime_in*,

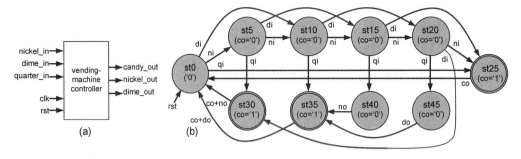

Figure 11.3
Vending machine of example 11.1: (a) Top-level circuit diagram; (b) State transition diagram.

$qi = quarter\_in$, $no = nickel\_out$, $do = dime\_out$, $co = candy\_out$. Note that in states *st25*, *st30*, and *st35* a candy bar is dispensed ($co = $ '1'), accompanied by a nickel or a dime in the last two.

A VHDL code for this circuit is shown below, which obeys the template just presented. The project's name is *vending_machine* (line 5), and the input-output signals (lines 6–8) are from figure 11.3a.

The architecture is in lines 11–123. Its declarative part (lines 12–16) contains FSM-related declarations. An enumerated data type, called *state*, which contains all states of the state transition diagram, was specified in lines 12–13, then, in line 14, the signals *pr_state* and *nx_state* were declared as conforming with that data type. The optional *enum_encoding* attribute (explained in section 11.4) for enumerated data types was used in lines 15–16 to specify the desired encoding scheme for the machine states, which is sequential. Consequently, given that the machine has 10 states, $\lceil \log_2 10 \rceil = 4$ bits (hence 4 flip-flops) are needed to represent them, following the sequence $st0 = $ "0000" (decimal 0), $st5 = $ "0001" (decimal 1), $st10 = $ "0010" (decimal 2),..., $st45 = $ "1001" (decimal 9). In the simulations, these decimal values are expected to be exhibited instead of the given state names.

The lower (sequential) section of the FSM is in the process of lines 19–26, while the upper (combinational) section is in the process of lines 28–122. Note that only the former has clock and reset. Observe also that the optional output section (glitch remover) was not employed.

```
1   ------------------------------------------------------------
2   LIBRARY ieee;
3   USE ieee.std_logic_1164.all;
4   ------------------------------------------------------------
5   ENTITY vending_machine IS
6      PORT (clk, rst: IN STD_LOGIC;
7              nickel_in, dime_in, quarter_in: IN BOOLEAN;
```

```
8              candy_out, nickel_out, dime_out: OUT STD_LOGIC);
9   END vending_machine;
10  ------------------------------------------------------------
11  ARCHITECTURE fsm OF vending_machine IS
12     TYPE state IS (st0, st5, st10, st15, st20, st25, st30,
13        st35, st40, st45);
14     SIGNAL pr_state, nx_state: state;
15     ATTRIBUTE enum_encoding: STRING; --optional attribute
16     ATTRIBUTE enum_encoding OF state: TYPE IS "sequential";
17  BEGIN
18     ----Lower section of FSM:-----------
19     PROCESS (rst, clk)
20     BEGIN
21        IF (rst='1') THEN
22           pr_state <= st0;
23        ELSIF (clk'EVENT AND clk='1') THEN
24           pr_state <= nx_state;
25        END IF;
26     END PROCESS;
27     ----Upper section of FSM:-----------
28     PROCESS (pr_state, nickel_in, dime_in, quarter_in)
29     BEGIN
30        CASE pr_state IS
31           WHEN st0 =>
32              candy_out <= '0';
33              nickel_out <= '0';
34              dime_out <= '0';
35              IF (nickel_in) THEN
36                 nx_state <= st5;
37              ELSIF (dime_in) THEN
38                 nx_state <= st10;
39              ELSIF (quarter_in) THEN
40                 nx_state <= st25;
41              ELSE
42                 nx_state <= st0;
43              END IF;
44           WHEN st5 =>
45              candy_out <= '0';
46              nickel_out <= '0';
47              dime_out <= '0';
48              IF (nickel_in) THEN
49                 nx_state <= st10;
50              ELSIF (dime_in) THEN
```

```
51              nx_state <= st15;
52          ELSIF (quarter_in) THEN
53              nx_state <= st30;
54          ELSE
55              nx_state <= st5;
56          END IF;
57      WHEN st10 =>
58          candy_out <= '0';
59          nickel_out <= '0';
60          dime_out <= '0';
61          IF (nickel_in) THEN
62              nx_state <= st15;
63          ELSIF (dime_in) THEN
64              nx_state <= st20;
65          ELSIF (quarter_in) THEN
66              nx_state <= st35;
67          ELSE
68              nx_state <= st10;
69          END IF;
70      WHEN st15 =>
71          candy_out <= '0';
72          nickel_out <= '0';
73          dime_out <= '0';
74          IF (nickel_in) THEN
75              nx_state <= st20;
76          ELSIF (dime_in) THEN
77              nx_state <= st25;
78          ELSIF (quarter_in) THEN
79              nx_state <= st40;
80          ELSE
81              nx_state <= st15;
82          END IF;
83      WHEN st20 =>
84          candy_out <= '0';
85          nickel_out <= '0';
86          dime_out <= '0';
87          IF (nickel_in) THEN
88              nx_state <= st25;
89          ELSIF (dime_in) THEN
90              nx_state <= st30;
91          ELSIF (quarter_in) THEN
92              nx_state <= st45;
93          ELSE
```

```
 94                    nx_state <= st20;
 95                 END IF;
 96              WHEN st25 =>
 97                 candy_out <= '1';
 98                 nickel_out <= '0';
 99                 dime_out <= '0';
100                 nx_state <= st0;
101              WHEN st30 =>
102                 candy_out <= '1';
103                 nickel_out <= '1';
104                 dime_out <= '0';
105                 nx_state <= st0;
106              WHEN st35 =>
107                 candy_out <= '1';
108                 nickel_out <= '0';
109                 dime_out <= '1';
110                 nx_state <= st0;
111              WHEN st40 =>
112                 candy_out <= '0';
113                 nickel_out <= '1';
114                 dime_out <= '0';
115                 nx_state <= st35;
116              WHEN st45 =>
117                 candy_out <= '0';
118                 nickel_out <= '0';
119                 dime_out <= '1';
120                 nx_state <= st35;
121           END CASE;
122        END PROCESS;
123   END fsm;
124   -----------------------------------------------------------
```

Simulation results are depicted in figure 11.4. First, one nickel is deposited, causing the system to move to state *st5* (decimal 1). Then a dime is entered, moving the machine to state *st15* (decimal 3), followed by a quarter (state *st40*, decimal 8). In this state, a nickel is returned to the customer (note *nickel_out* = '1'), moving the controller automatically to state *st35* (decimal 7), where a candy and a dime are dispensed, with the system returning then to its idle state (*st0*).

One important aspect of this design is that the outputs (generated by combinational circuits), depending on the chosen CPLD/FPGA and on the particular routing inside that device, are subject to glitches, which can indeed be observed in figure 11.4. Since in this type of application (involving money) glitches are definitely not acceptable, the optional output section seen in the FSM template must be included (see next example).

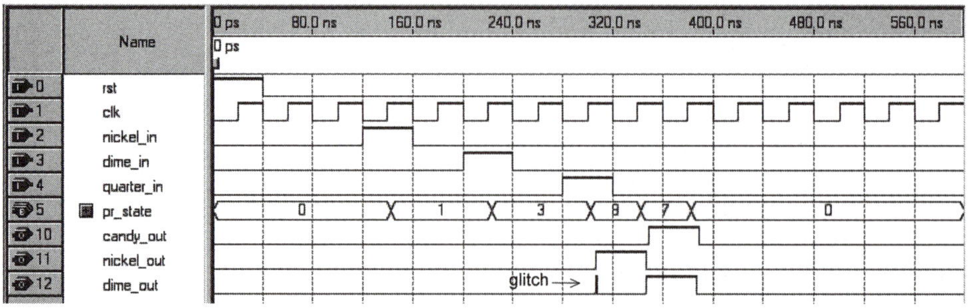

Figure 11.4
Simulation results from the vending-machine controller of example 11.1.

Example 11.2: Glitch-Free Vending-Machine Controller

Add the optional output section seen in the FSM template to the vending-machine controller above to guarantee that its outputs are free from glitches (see glitch in figure 11.4).

Solution First, it is important to mention that when the simulator does not show glitches in the output signals it does not necessarily mean that the outputs are glitch-free (this can only be determined by circuit analysis), because that depends on the cells used in the target device—that is, it might be just a coincidence that the delays in the routings favored a glitch-free response, which might not hold in another routing or another device.

To obtain the glitch-free solution, just add the code below to that in the previous example. The new outputs are called *new_candy_out*, *new_nickel_out*, and *new_dime_out*, which are obviously one clock cycle behind the original outputs (the delay would be one-half of a clock cycle if the negative clock edge were employed). If the original outputs are still kept as output signals in your code, remember to either declare them as BUFFER or to use auxiliary internal signals. Corresponding glitch-free simulation results are depicted in figure 11.5.

```
-----Output section of FSM:-------------
   PROCESS (rst, clk)
   BEGIN
      IF (rst='1') THEN
         new_candy_out <= '0';
         new_nickel_out <= '0';
         new_dime_out <= '0';
      ELSIF (clk'EVENT AND clk='1') THEN
         new_candy_out <= candy_out;
         new_nickel_out <= nickel_out;
         new_dime_out <= dime_out;
      END IF;
   END PROCESS;
-----------------------------------------
```

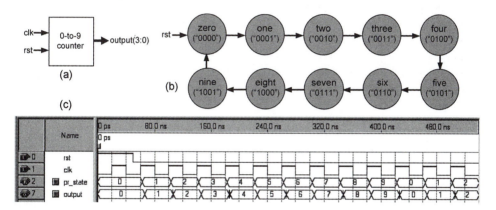

Figure 11.5
Simulation results from the code of example 11.2.

Figure 11.6
Zero-to-nine counter of example 11.3.

Example 11.3: Zero-to-Nine Counter

Figure 11.6a shows the top-level diagram for a sequential binary 0-to-9 counter. The inputs are *clk* and *rst*, and the output is *output(3:0)*. Design this circuit using the FSM approach.

Solution   The corresponding state transition diagram is depicted in figure 11.6b, showing the machine's ten states, with their names and respective output values. Note that this is a Moore machine because there are no control inputs (besides reset, of course), so the next state depends solely on the present state.

A VHDL code for this circuit is presented below, which is again a direct application of the template introduced earlier. The optional output section is not needed here because the outputs come directly from flip-flops, so each bit is automatically glitch-free. Simulation results are depicted in figure 11.6c. Recall that the glitches seen in the figure after some of the transitions are not necessarily glitches; this is because *output* represents a bus, not a single bit, and multiple bits neither change instantaneously (perfectly vertical transitions) nor change all exactly at the same time.

```
1   -----------------------------------------------------------
2   ENTITY counter IS
3      PORT (clk, rst: IN BIT;
4            output: OUT NATURAL RANGE 0 TO 9);
5   END counter;
6   -----------------------------------------------------------
7   ARCHITECTURE fsm OF counter IS
8      TYPE state IS (zero, one, two, three, four, five,
9         six, seven, eight, nine);
10     SIGNAL pr_state, nx_state: state;
11     ATTRIBUTE enum_encoding: STRING; --optional attribute
12     ATTRIBUTE enum_encoding OF state: TYPE IS "sequential";
13  BEGIN
14     ----Lower section of FSM:-----------
15     PROCESS (rst, clk)
16     BEGIN
17        IF (rst='1') THEN
18           pr_state <= zero;
19        ELSIF (clk'EVENT AND clk='1') THEN
20           pr_state <= nx_state;
21        END IF;
22     END PROCESS;
23     ----Upper section of FSM:-----------
24     PROCESS (pr_state)
25     BEGIN
26        CASE pr_state IS
27           WHEN zero =>
28              output <= 0;
29              nx_state <= one;
30           WHEN one =>
31              output <= 1;
32              nx_state <= two;
33           WHEN two =>
34              output <= 2;
35              nx_state <= three;
36           WHEN three =>
37              output <= 3;
```

```
38                  nx_state <= four;
39              WHEN four =>
40                  output <= 4;
41                  nx_state <= five;
42              WHEN five =>
43                  output <= 5;
44                  nx_state <= six;
45              WHEN six =>
46                  output <= 6;
47                  nx_state <= seven;
48              WHEN seven =>
49                  output <= 7;
50                  nx_state <= eight;
51              WHEN eight =>
52                  output <= 8;
53                  nx_state <= nine;
54              WHEN nine =>
55                  output <= 9;
56                  nx_state <= zero;
57          END CASE;
58      END PROCESS;
59  END fsm;
60  ------------------------------------------------------------
```

11.3   Poor FSM Model

Below is a common FSM design template using VHDL. The purpose of this section is to show that this is a poor approach, and so further illustrate the construction of state machines.

```
1   --------------------------------------------------
2   LIBRARY ieee;
3   USE ieee.std_logic_1164.all;
4   --------------------------------------------------
5   ENTITY poor_template IS
6       PORT (clk, rst, input: IN STD_LOGIC;
7               output: OUT STD_LOGIC_VECTOR(2 DOWNTO 0));
8   END poor_template;
9   --------------------------------------------------
10  ARCHITECTURE poor_fsm OF poor_template IS
11      TYPE state_type IS (A, B, C, ...);
12      SIGNAL state: state_type;
13  BEGIN
```

```
14      PROCESS (clk, rst)
15      BEGIN
16         IF (rst='1') THEN
17            state <= A;
18         ELSIF (rising_edge(clk)) THEN
19            CASE state IS
20               WHEN A=>
21                  IF (input='1') THEN
22                     state <= B;
23                  ELSE
24                     state <= A;
25                  END IF;
26               WHEN B=>
27                  IF (input='1') THEN
28                     state <= C;
29                  ELSE
30                     state <= B;
31                  END IF;
32               ...
33            END CASE
34         END IF;
35      END PROCESS;
36      -------------------------------
37      PROCESS (state, input)
38      BEGIN
39         CASE state IS
40            WHEN A=>
41               IF (input='1') THEN
42                  output <= "001";
43               ELSE
44                  output <= "000";
45               END IF;
46            WHEN B=>
47               IF (input='1') THEN
48                  output <= "101";
49               ELSE
50                  output <= "001";
51               END IF;
52            ...
53         END CASE;
54      END PROCESS;
55   END poor_fsm;
56   ---------------------------------------------------
```

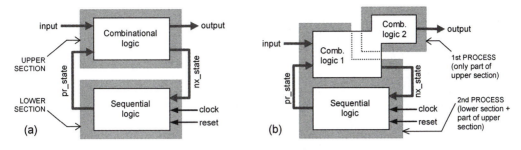

Figure 11.7
(a) Proposed versus (b) poor FSM model (from a hardware point of view).

Comparing the template above with that proposed in section 11.2, one verifies that while the latter directly resembles the FSM model of figure 11.7a (copied from figure 11.2), the former implements the model of figure 11.7b. Consequently, while the former is consistent with the type of logic (sequential versus combinational) employed in the circuits, the other mixes the logic types, losing essential advantages of well-established design techniques. Additionally, there is so little left for the second process that it is then generally not worth having a second part at all.

11.4   FSM Encoding Styles

The states of an FSM can be encoded in several ways, which are described below using the following enumerated data type:

```
TYPE state IS (A, B, C, D, E);
```

▪ *Sequential encoding*:   The minimum number of bits is employed and the states are encoded in ascending order of decimal values. With N bits (N flip-flops), 2^N states can be encoded. For the type above, the following would result: $A = $ "000" ($= 0$ decimal), $B = $ "001" ($= 1$), $C = $ "010" ($= 2$), $D = $ "011" ($= 3$), and $E = $ "100" ($= 4$).

▪ *Gray encoding*:   Again, the minimum number of bits is used, with the states encoded using the Gray code, so neighboring code words differ by just one bit. For the type above, the following would result: $A = $ "000", $B = $ "001", $C = $ "011", $D = $ "010", and $E = $ "110".

▪ *Johnson encoding*:   Like the Gray code, neighboring words differ by just one bit. However, with N bits (N flip-flops), only $2N$ states can be encoded. Each new code word is obtained by circularly shifting the previous one to the right by one position, with the new MSB equal to the reverse of the previous LSB. For the type above, the following would result: $A = $ "000", $B = $ "100", $C = $ "110", $D = $ "111", and $E = $ "011".

▪ *One-hot encoding*:   To encode N states, N flip-flops are needed. Each code word contains only one bit distinct from the others (that is, all bits are '0', except one, or vice versa).

FSM encodings (for 4-bit systems)					
Sequential		Gray		Johnson	One-hot
0000	1000	0000	1100	0000	0001
0001	1001	0001	1101	1000	0010
0010	1010	0011	1111	1100	0100
0011	1011	0010	1110	1110	1000
0100	1100	0110	1010	1111	---
0101	1101	0111	1011	0111	---
0110	1110	0101	1001	0011	---
0111	1111	0100	1000	0001	---
Total=16 states		Total=16 states		Total=8 states	Total=4 states

Figure 11.8
Main FSM encoding styles and respective code words with four bits.

For the type above, the following would result: $A =$ "00001", $B =$ "00010", $C =$ "00100", $D =$ "01000", and $E =$ "10000".

• *User-defined encoding*: This includes any other encoding scheme specified by the designer.

The one-hot style is often used in applications where flip-flops are abundant, like in FPGAs, while in compact ASIC implementations sequential encoding is often employed. Even though the latter requires the minimum number of flip-flops, the former requires the least amount of combinational logic, thus generally resulting the fastest possible circuit.

In VHDL, a special attribute, called *enum_encoding*, is available to specify the encoding style. Its syntax is:

```
ATTRIBUTE enum_encoding: STRING;
ATTRIBUTE enum_encoding OF state: TYPE IS "sequential";
```

The options normally synthesizable without restrictions in the attribute above are *sequential* and *one-hot*. For the others, it may be advisable to set up the compiler directly. For example, if using Quartus II, the setup is the following: Assignments > Settings > Analysis & Synthesis Settings > More Settings > State Machine Processing > (choose encoding option). For example, the default encoding adopted by Quartus II 8.1 is *sequential* for up to 4 states, *one-hot* for 5–49 states, or *gray* otherwise.

To conclude, figure 11.8 illustrates how the states would be encoded with four bits, for all four encoding options described above.

11.5   The State-Bypass Problem in FSMs

Figure 11.9a shows a simplified state transition diagram for a car alarm (Pedroni 2008). It contains three states, called *disarmed*, *armed*, and *intrusion* (in the last state, *siren* = '1'

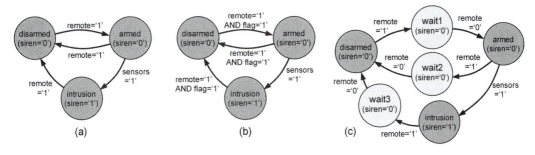

Figure 11.9
(a) Simplified diagram for a car alarm, which exhibits a major flaw (state-bypass); (b) Bypasses prevented with a flag; (c) Bypasses prevented with additional states.

causes the siren to go off). Note that there is a major flaw in this FSM because if a long (several clock cycles) *remote* = '1' command (from the remote control) occurs, the system flips back and forth between the *disarmed* and *armed* states, or even between these states and the *intrusion* state if *sensors* is high. In summary, some (or all) system states are essentially bypassed (undesirable loop operation). Of course, there are applications in which this is indeed the desired behavior (that is, to have the system stay in a certain state during only one clock period), but in other applications (like the car alarm) this is a problem, so it needs to be fixed.

Figures 11.9b–c show two techniques for solving the state-bypass problem, with the former employing a flag while the latter uses additional states.

The solution with a flag consists of resetting the flag when the system enters certain (or all) states, which is then reasserted only if a predetermined condition occurs. In the example of figure 11.9b, this condition is *remote* = '0': that is, once the system enters a state (due to *remote* = '1'), it will only leave that state after *remote* = '0' happens, thus preventing any state from being bypassed.

The solution with additional states consists of including "wait" states in the paths subject to bypass. In the example of figure 11.9c, all paths are subject to bypass, so three extra states are needed. The system enters such states when a *remote* = '1' command is received, leaving only after *remote* returns to '0'. Observe that each wait state should produce the same output value as the *next* state.

In terms of hardware, there generally is not much difference between these two solutions. In a large system with multiple bypass points, the latter tends to require more flip-flops (more states), while the former requires more combinational logic to produce the flag. Both approaches are illustrated in the examples that follow.

Example 11.4: Car Alarm with Bypasses Prevented by a Flag

Design the simplified car alarm of figure 11.9 using a flag to prevent state bypasses (figure 11.9b).

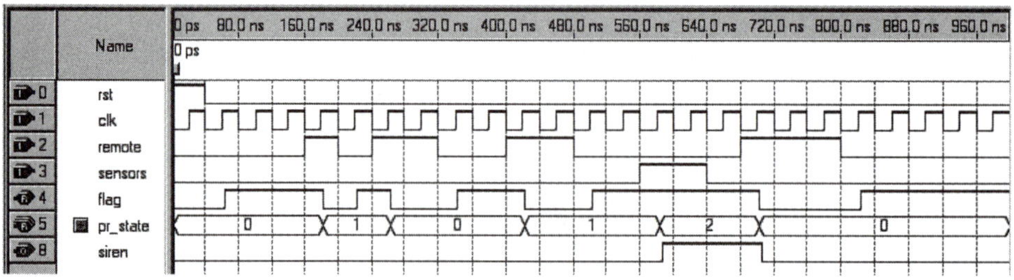

Figure 11.10
Simulation results from the code of example 11.4.

Solution   A VHDL code for this FSM is presented below. Note that an additional process was included to generate the flag, while the rest of the code still complies with the FSM template precisely. In the new process, *remote* = '1' causes the flag to be reset, only being reactivated after *remote* = '0' happens again. Simulation results from this code are shown in figure 11.10.

```
1    ------------------------------------------------------------
2    LIBRARY ieee;
3    USE ieee.std_logic_1164.all;
4    ------------------------------------------------------------
5    ENTITY simple_car_alarm IS
6       PORT (clk, rst, remote, sensors: IN STD_LOGIC;
7             siren: OUT STD_LOGIC);
8    END simple_car_alarm;
9    ------------------------------------------------------------
10   ARCHITECTURE fsm OF simple_car_alarm IS
11      TYPE state IS (disarmed, armed, intrusion);
12      SIGNAL pr_state, nx_state: state;
13      ATTRIBUTE enum_encoding: STRING;
14      ATTRIBUTE enum_encoding OF state: TYPE IS "sequential";
15      SIGNAL flag: STD_LOGIC;
16   BEGIN
17      ----Flag generator:----------------
18      PROCESS (clk, rst)
19      BEGIN
20         IF (rst='1') THEN
21            flag <= '0';
22         ELSIF (clk'EVENT AND clk='1') THEN
23            IF (remote='0') THEN
24               flag <= '1';
25            ELSE
```

```
26              flag <= '0';
27          END IF;
28        END IF;
29      END PROCESS;
30      ----Lower section of FSM:---------
31      PROCESS (clk, rst)
32      BEGIN
33        IF (rst='1') THEN
34            pr_state <= disarmed;
35        ELSIF (clk'EVENT AND clk='1') THEN
36            pr_state <= nx_state;
37        END IF;
38      END PROCESS;
39      ----Upper section of FSM:---------
40      PROCESS (pr_state, remote, sensors, flag)
41      BEGIN
42        CASE pr_state IS
43          WHEN disarmed =>
44              siren <= '0';
45              IF (remote='1' AND flag='1') THEN
46                  nx_state <= armed;
47              ELSE
48                  nx_state <= disarmed;
49              END IF;
50          WHEN armed =>
51              siren <= '0';
52              IF (sensors='1') THEN
53                  nx_state <= intrusion;
54              ELSIF (remote='1' AND flag='1') THEN
55                  nx_state <= disarmed;
56              ELSE
57                  nx_state <= armed;
58              END IF;
59          WHEN intrusion =>
60              siren <= '1';
61              IF (remote='1' AND flag='1') THEN
62                  nx_state <= disarmed;
63              ELSE
64                  nx_state <= intrusion;
65              END IF;
66        END CASE;
67      END PROCESS;
68    END fsm;
69    ------------------------------------------------------------
```

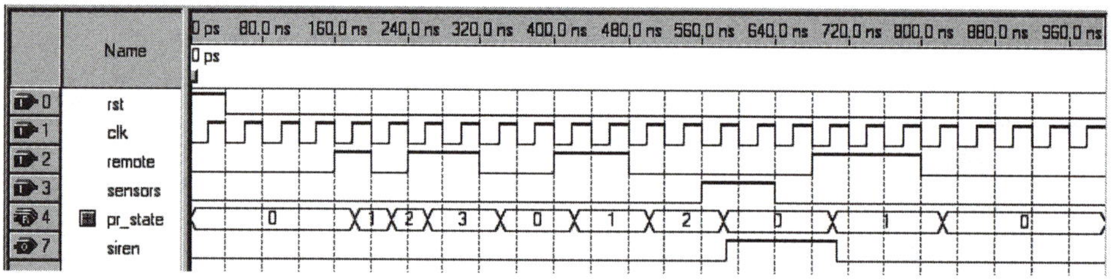

Figure 11.11
Simulation results from the code of example 11.5.

Example 11.5: Car Alarm with Bypasses Prevented by Additional States

Design the simplified car alarm of figure 11.9 using additional states to prevent bypasses (figure 11.9c).

Solution A VHDL code for this FSM is presented below. Note the inclusion of three wait states, so the additional process (for the flag) is no longer needed and the code once again fully complies with the general FSM template of section 11.2. Simulation results are depicted in figure 11.11 (note that *siren* in figures 11.10 and 11.11 are equal).

```
1    ----------------------------------------------------------------
2    LIBRARY ieee;
3    USE ieee.std_logic_1164.all;
4    ----------------------------------------------------------------
5    ENTITY simple_car_alarm IS
6       PORT (clk, rst, remote, sensors: IN STD_LOGIC;
7             siren: OUT STD_LOGIC);
8    END simple_car_alarm;
9    ----------------------------------------------------------------
10   ARCHITECTURE fsm OF simple_car_alarm IS
11      TYPE state IS (disarmed, wait1, armed, wait2, intrusion, wait3);
12      ATTRIBUTE enum_encoding: STRING; --optional attribure
13      ATTRIBUTE enum_encoding OF state: TYPE IS "sequential";
14      SIGNAL pr_state, nx_state: state;
15   BEGIN
16      ----Lower section of FSM:---------
17      PROCESS (clk, rst)
18      BEGIN
19         IF (rst='1') THEN
20            pr_state <= disarmed;
21         ELSIF (clk'EVENT AND clk='1') THEN
22            pr_state <= nx_state;
23         END IF;
```

```
24      END PROCESS;
25      ----Upper section of FSM:---------
26      PROCESS (pr_state, remote, sensors)
27      BEGIN
28        CASE pr_state IS
29          WHEN disarmed =>
30              siren <= '0';
31              IF (remote='1') THEN
32                  nx_state <= wait1;
33              ELSE
34                  nx_state <= disarmed;
35              END IF;
36          WHEN wait1 =>
37              siren <= '0';
38              IF (remote='0') THEN
39                  nx_state <= armed;
40              ELSE
41                  nx_state <= wait1;
42              END IF;
43          WHEN armed =>
44              siren <= '0';
45              IF (sensors='1') THEN
46                  nx_state <= intrusion;
47              ELSIF (remote='1') THEN
48                  nx_state <= wait2;
49              ELSE
50                  nx_state <= armed;
51              END IF;
52          WHEN wait2 =>
53              siren <= '0';
54              IF (remote='0') THEN
55                  nx_state <= disarmed;
56              ELSE
57                  nx_state <= wait2;
58              END IF;
59          WHEN intrusion =>
60              siren <= '1';
71              IF (remote='1') THEN
72                  nx_state <= wait3;
73              ELSE
74                  nx_state <= intrusion;
75              END IF;
76          WHEN wait3 =>
```

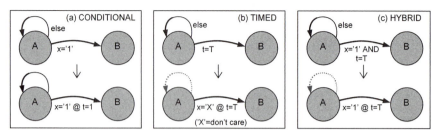

Figure 11.12
(a) Conditional, (b) timed, and (c) hybrid FSM transitions. Any type of FSM transition will fall in one of these three categories.

```
77                  siren <= '0';
78                  IF (remote='0') THEN
79                        nx_state <= disarmed;
80                  ELSE
81                        nx_state <= wait3;
82                  END IF;
83           END CASE;
84       END PROCESS;
85   END fsm;
86   -------------------------------------------------------------------
```

11.6   Systematic Design Technique for Timed Machines

This section deals with the design of large state machines, which normally contain embedded counters, operating as timers. Even though a counter itself is a state machine, it can be designed very easily without the FSM approach, so the counter can be considered as just an auxiliary circuit whose output is used as an input by the main machine, greatly simplifying the overall design.

A systematic design technique for regular FSMs was discussed in section 11.2. Even though the VHDL template seen there can be used to implement *any* machine, the adaptations needed to accommodate the case of timed machines may vary in style and also in correctness. For that reason, an *extended systematic approach* is proposed here, which expands that template to also cover machines with built-in counters, hence contemplating essentially any kind of FSM-based design.

The technique is based on observations shown in figure 11.12, which contains three types of state transitions, identified as (a) *conditional* (only logic involved), (b) *timed* (only time involved), and (c) *hybrid* (logic and time included). The upper part of figure 11.12 shows the following: in (a), the machine moves from state A to state B when $x = $ '1' occurs (at the proper clock transition, of course); in (b), the change occurs when t (the number of clock cycles) reaches T (a predefined positive value), regardless of the circuit's logic values;

finally, in (c), the change only occurs if $x = $ '1' after T clock cycles (hence with logic and time involved).

An equivalent representation is presented in the lower part of figure 11.12. Note $t = 1$ in (a), meaning that the condition is evaluated after one clock cycle (in other words, at *every* clock cycle). In (b), $x = $ 'X' (don't care) is evaluated after T clock cycles (so only time matters). Finally, in (c), the condition $x = $ '1' is evaluated after T clock cycles. Two important aspects of these transformations are that in the timed cases the "else" condition becomes automatically absorbed by the timer, and that any timed situation, even if the machine looks very complex, will fall in one of these categories.

A VHDL template incorporating these conditions is shown next. Observe that in the code for the lower section (process of lines 17–30) the only change is the inclusion of a counter, whose final value, *timer*, is equal to 1 (for conditional transitions) or equal to T (for timed or hybrid transitions). In the code for the upper section (process of lines 32–57), the only change is the inclusion of a value for *timer* in each state (that is, besides *output* and *nx_ state*, *timer* must now also be specified). Examples are presented to illustrate the use of this technique.

```
1    -----------------------------------------------------
2    LIBRARY ieee;
3    USE ieee.std_logic_1164.all;
4    -----------------------------------------------------
5    ENTITY <entity_name> IS
6       PORT (clk, rst: IN STD_LOGIC;
7              input: IN <data_type>;
8              output: OUT <data_type>);
9    END <entity_name>;
10   -----------------------------------------------------
11   ARCHITECTURE <architecture_name> OF <entity_name> IS
12      TYPE state IS (A, B, C, ...);
13      SIGNAL pr_state, nx_state: state;
14      SIGNAL timer: INTEGER RANGE 0 TO max;
15   BEGIN
16      ------Lower section of FSM:------------
17      PROCESS (clk, rst)
18         VARIABLE count: INTEGER RANGE 0 TO max;
19      BEGIN
20         IF (rst='1') THEN
21            pr_state <= A;
22            count := 0;
23         ELSIF (clk'EVENT AND clk='1') THEN
24            count := count + 1;
25            IF (count>=timer) THEN --">=", not "="
```

```
26              pr_state <= nx_state;
27              count := 0;
28          END IF;
29       END IF;
30    END PROCESS;
31    ------Upper section of FSM:------------
32    PROCESS (pr_state, input)
33    BEGIN
34       CASE pr_state IS
35          WHEN A =>
36              output <= <value>;
37              IF (input=<value>) THEN
38                  timer <= <value>;
39                  nx_state <= B;
40                  ...
41              ELSE
42                  timer <= <value>;
43                  nx_state <= A;
44              END IF;
45          WHEN B =>
46              output <= <value>;
47              IF (input=<value>) THEN
48                  timer <= <value>;
49                  nx_state <= C;
50                  ...
51              ELSE
52                  timer <= <value>;
53                  nx_state <= B;
54              END IF;
55          WHEN ...
56       END CASE;
57    END PROCESS;
58    ------Output section (optional):-------
59    PROCESS (clk, rst)
60    BEGIN
61       IF (rst='1') THEN
62          new_output <= <value>;
63       ELSIF (clk'EVENT AND clk='1') THEN --or clk='0'
64          new_output <= output;
65       END IF;
66    END PROCESS;
67 END <architecture_name>;
68 ------------------------------------------------------------
```

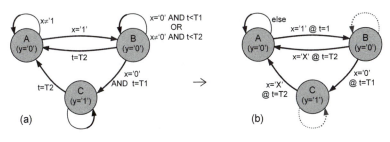

Figure 11.13
(a) Timed FSM of example 11.6; (b) Version with transitions adapted to the models of figure 11.12.

Example 11.6: FSM with Embedded Timer

Figure 11.13a shows a timed FSM, which operates as follows. When in state A, the machine must progress to B if $x = $ '1' occurs (at the proper clock edge, of course), or remain in A otherwise. If in B, it must move to C after $T1$ clock cycles if $x = $ '0' or return to A after $T2$ clock cycles otherwise ($T1 < T2$). Finally, when in C, it must return to A unconditionally after $T2$ clock periods. Design this FSM using the systematic approach described. Note that state A is subject to state-bypass, which should not be considered a problem in this exercise.

Solution First, observe that this FSM exhibits all three types of transitions seen in figure 11.12—that is, *conditional* (from A to B), *timed* (from B to A and C to A), and *hybrid* (from B to C), thus representing an excellent opportunity to illustrate the proposed design technique.

Using the principles of figure 11.12, the diagram of figure 11.13b results, so the application of the VHDL template is straightforward. (Even though it is obviously not necessary to include $x = $ 'X' in the diagram, it was included in this first example for clarity.)

A VHDL code for this FSM is presented below, under the project name *timed_machine* (line 5). It starts with GENERIC definitions for the time limits ($T1$, $T2$) in lines 6–8, followed by the input-output signals in lines 9–12. The architecture is in lines 15–64. Its declarative part (lines 16–20) contains the usual FSM-related specifications, in this example including the optional *enum_encoding* attribute, plus a signal to build the timer. The first process (lines 23–36) implements the lower (sequential) section of the FSM, in which the only modification (compared to the original template of section 11.2) is the inclusion of a counter (timer). The second process (lines 38–63) implements the upper (combinational) part of the FSM, in which the only modification is the inclusion of time values in each state, also according with the design technique introduced above.

In summary, with very simple and *systematic* modifications, the original template can be used to build any type of timed FSM. Simulation results for $T1 = 3$ and $T2 = 4$ clock pulses are depicted in figure 11.14.

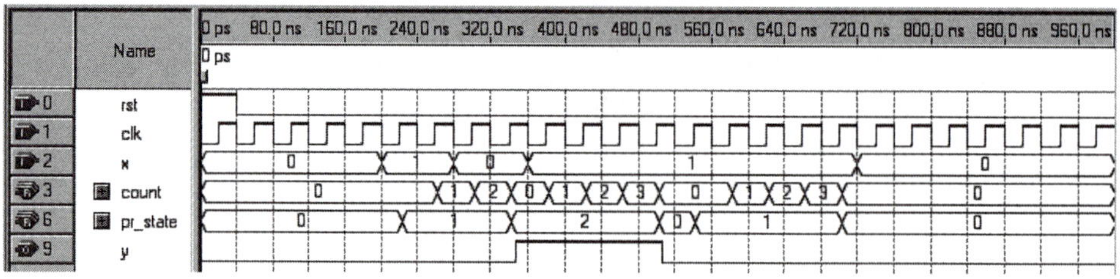

Figure 11.14
Simulation results from the VHDL code for the FSM of figure 11.13.

```
1    ------------------------------------------------------------
2    LIBRARY ieee;
3    USE ieee.std_logic_1164.all;
4    ------------------------------------------------------------
5    ENTITY timed_machine IS
6       GENERIC (
7          T1: INTEGER := 3;     --time limit 1
8          T2: INTEGER := 4;     --time limit 2 (T2>T1)
9       PORT (
10         clk, rst: IN STD_LOGIC;
11         x: IN STD_LOGIC;
12         y: OUT STD_LOGIC);
13   END timed_machine;
14   ------------------------------------------------------------
15   ARCHITECTURE fsm OF timed_machine IS
16      TYPE state IS (A, B, C);
17      SIGNAL pr_state, nx_state: state;
18      ATTRIBUTE enum_encoding: STRING; --optional attribute
19      ATTRIBUTE enum_encoding OF state: TYPE IS "sequential";
20      SIGNAL timer: INTEGER RANGE 0 TO T2;
21   BEGIN
22      ----Lower section of FSM:-----------
23      PROCESS (clk, rst)
24         VARIABLE count: INTEGER RANGE 0 TO T2;
25      BEGIN
26         IF (rst='1') THEN
27            pr_state <= A;
28            count := 0;
29         ELSIF (clk'EVENT AND clk='1') THEN
30            count := count + 1;
31            IF (count>=timer) THEN
32               pr_state <= nx_state;
```

```
33                    count := 0;
34                END IF;
35            END IF;
36        END PROCESS;
37        ----Upper section of FSM:-----------
38        PROCESS (pr_state, x)
39        BEGIN
40            CASE pr_state IS
41                WHEN A =>
42                    y <= '0';
43                    timer <= 1;
44                    IF (x='1') THEN
45                        nx_state <= B;
46                    ELSE
47                        nx_state <= A;
48                    END IF;
49                WHEN B =>
50                    y <= '0';
51                    IF (x='0') THEN
52                        timer <= T1;
53                        nx_state <= C;
54                    ELSE
55                        timer <= T2;
56                        nx_state <= A;
57                    END IF;
58                WHEN C =>
59                    y <= '1';
60                    timer <= T2;
61                    nx_state <= A;
62            END CASE;
63        END PROCESS;
64    END fsm;
65    ------------------------------------------------------------
```

Example 11.7: Traffic-Light Controller

This example presents another timed FSM. Figure 11.15 shows a traffic-light controller (TLC), which must be designed with the following features (see the table in figure 11.15):

1) Three modes of operation: *regular*, *test*, and *standby*.

2) In regular mode: Four states of operation, called *RG* (red in direction 1 and green in direction 2 ON), *RY* (red in direction 1 and yellow in direction 2 ON), *GR* (green in direction 1 and red in direction 2 ON), and *YR* (yellow in direction 1 and red in direction 2 ON), each with an independent time duration.

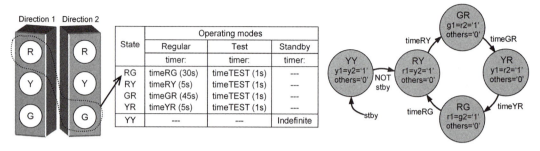

Figure 11.15
Traffic-light controller of example 11.7.

3) In test mode: Allows all preprogrammed times to be overwritten (by a manual switch) with a small value, such that the system can be easily tested during maintenance (1 second per state).

4) In standby mode: If set (by a sensor accusing malfunctioning, for example, or by a manual switch) the system should activate the yellow lights in both directions, remaining so while the standby signal is active.

5) High precision is not required in this kind of application, so assume that the clock is a 60 Hz square wave derived from the power line itself (otherwise, a regular crystal-controlled oscillator could be employed).

Design this circuit using the extended FSM approach introduced in this section. Also, estimate the number of flip-flops that will be required.

Solution As mentioned earlier, digital controllers are good examples of circuits that can be efficiently implemented using the FSM approach. The state transition diagram for the present example is shown on the right of figure 11.15, where the time values change with the state and with the operating mode (*regular* or *test*). Note that all transitions are timed-only. The inputs are *clk*, *stby*, and *test*, while the outputs are *r1*, *y1*, *g1*, *r2*, *y2*, and *g2* (red, yellow, and green lights in directions 1 and 2).

A VHDL code for this FSM, obeying the modified template introduced in this section, is presented below. The time values were specified using GENERIC declarations (lines 6–12), so they can be easily changed. The first process (lines 24–37) again implements the FSM's lower section, while the second process (lines 39–84) implements the upper section. Note in line 32 that *count* ≥ *timer* was used instead of *count* = *timer* to allow the system to switch to test mode immediately when the test switch is activated, even if *count* > *timeTEST* at that moment. (This is indeed necessary in all machines designed with the new template.)

The number of flip-flops depends on the counter (timer) and on the number of FSM states. The former requires $\lceil \log_2 2700 \rceil = 12$ flip-flops, while the latter requires $\lceil \log_2 5 \rceil =$

3 if sequential encoding is used or 5 if one-hot encoding is chosen instead, hence totaling 15 or 17 registers. If, instead of a 60 Hz clock, a 40 MHz clock is employed, then the counter will require $\lceil \log_2 1800M \rceil = 31$ flip-flops.

The reader is invited to compile this code and check these results. Note that this is another interesting circuit to be tested in the FPGA development board (see exercises 11.16 and 11.17).

```
1   ---------------------------------------------------------
2   LIBRARY ieee;
3   USE ieee.std_logic_1164.all;
4   ---------------------------------------------------------
5   ENTITY tlc IS
6      GENERIC (
7          timeRG: POSITIVE := 1800;   --30s with 60Hz clock
8          timeRY: POSITIVE := 300;    --5s with 60Hz clock
9          timeGR: POSITIVE := 2700;   --45s with 60Hz clock
10         timeYR: POSITIVE := 300;    --5s with 60Hz clock
11         timeTEST: POSITIVE := 60;   --1s with 60Hz clock
12         timeMAX: POSITIVE := 2700); --max of all above
13      PORT (
14         clk, stby, test: IN STD_LOGIC;
15         r1, r2, y1, y2, g1, g2: OUT STD_LOGIC);
16   END tlc;
17   ---------------------------------------------------------
18   ARCHITECTURE fsm OF tlc IS
19      TYPE state IS (RG, RY, GR, YR, YY);
20      SIGNAL pr_state, nx_state: state;
21      SIGNAL timer: INTEGER RANGE 0 TO timeMAX;
22   BEGIN
23      ----Lower section of FSM:-----------
24      PROCESS (clk, stby)
25         VARIABLE count : INTEGER RANGE 0 TO timeMAX;
26      BEGIN
27         IF (stby='1') THEN
28          pr_state <= YY;
29          count := 0;
30         ELSIF (clk'EVENT AND clk='1') THEN
31          count := count + 1;
32          IF (count>=timer) THEN
33             pr_state <= nx_state;
34             count := 0;
35          END IF;
36         END IF;
37      END PROCESS;
```

```
38        ----Upper section of FSM:-----------
39        PROCESS (pr_state, test)
40        BEGIN
41           CASE pr_state IS
42              WHEN RG =>
43                    r1<='1'; y1<='0'; g1<='0';
44                    r2<='0'; y2<='0'; g2<='1';
45                    nx_state <= RY;
46                    IF (test='0') THEN
47                        timer <= timeRG;
48                    ELSE
49                        timer <= timeTEST;
50                    END IF;
51              WHEN RY =>
52                    r1<='1'; y1<='0'; g1<='0';
53                    r2<='0'; y2<='1'; g2<='0';
54                    nx_state <= GR;
55                    IF (test='0') THEN
56                        timer <= timeRY;
57                    ELSE
58                        timer <= timeTEST;
59                    END IF;
60              WHEN GR =>
61                    r1<='0'; y1<='0'; g1<='1';
62                    r2<='1'; y2<='0'; g2<='0';
63                    nx_state <= YR;
64                    IF (test='0') THEN
65                        timer <= timeGR;
66                    ELSE
67                        timer <= timeTEST;
68                    END IF;
69              WHEN YR =>
70                    r1<='0'; y1<='1'; g1<='0';
71                    r2<='1'; y2<='0'; g2<='0';
72                    nx_state <= RG;
73                    IF (test='0') THEN
74                        timer <= timeYR;
75                    ELSE
76                        timer <= timeTEST;
77                    END IF;
78              WHEN YY =>
79                    r1<='0'; y1<='1'; g1<='0';
80                    r2<='0'; y2<='1'; g2<='0';
81                    timer <= timeTEST; --to avoid latches
82                    nx_state <= RY;
```

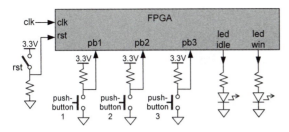

Figure 11.16
Pushbutton sequence detector of example 11.8.

```
83            END CASE;
84        END PROCESS;
85    END fsm;
86    ----------------------------------------------------------
```

Example 11.8: Pushbutton Sequence Detector

Like the previous two examples, this too illustrates the construction of an FSM with an embedded timer, designed according to the modified template introduced in this section.

Figure 11.16 shows three pushbuttons (*pb1*, *pb2*, *pb3*) connected to an FPGA, along with clock and reset signals, to which the FPGA responds with two signals (*led_idle*, *led_win*) that feed two LEDs (see details about LEDs in chapter 12). The purpose of this circuit is to turn *led_win* ON when the correct sequence of pushbuttons is pressed, whereas *led_idle* must be ON when the system is in the idle state. Design this circuit with the following specifications:

1) Turn *led_idle* ON when the system is in the idle state.

2) Turn *led_win* ON when the correct pushbutton sequence is entered.

3) The sequence must consist of exactly three key strokes, with repetitions allowed. Assume that the keys are already debounced.

4) After a key is pressed, only three seconds should be allowed until a new key is pressed, otherwise the sequence must be restarted.

5) When *led_idle* is turned ON, the system must wait three seconds before any keystroke is again considered.

Solution   A progressive construction for the state transition diagram of this FSM is presented in figure 11.17. Initially, the core is shown. The states are *idle* (*led_idle* ON), *key1* (to which the machine moves if the first keystroke is correct), *key2* (is the second key is also correct), and finally *key3* (third key correct, *led_win* ON). The correct sequence of pushbuttons is represented as $a \rightarrow b \rightarrow c$, where *a*, *b*, and *c* can assume any of the values "011" (*pb1* pressed), "101" (*pb2* pressed), or "110" (*pb3* pressed), with repetitions allowed ($a = b$, for example). When none of the pushbuttons is pressed, the input is "111".

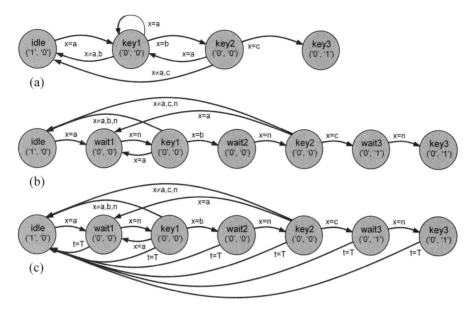

Figure 11.17
State transition diagram for the pushbutton sequence detector of example 11.8. (a) Core; (b) With wait states included to prevent state-bypass; (c) With timed transitions also included.

Analyzing the diagram in figure 11.17a, one verifies that it is subject to state-bypass (section 11.5). For example, if the machine is in state *key1* and *b* (the next input) lasts more than one clock cycle (which it certainly will because it comes from a mechanical switch), the system will go to state *key2* in one clock cycle and then directly back to *idle* in the next. This problem is solved with *wait* states (section 11.5), shown in figure 11.17b (some of the "else" conditions were omitted to preserve drawing clarity). Consequently, now it is necessary to release the pushbutton before a new input can be considered.

Finally, in figure 11.17c, the complete diagram is presented, with the timed transitions also included. If the system stays T clock cycles in any state, it will be forced back to *idle* and the player loses.

A VHDL code for this FSM is presented below, which is a straight application of the VHDL template introduced in this section. The project's name is *pb_sequence_detector* (line 5). The "password" (*abc*) and the delay (*T*) were entered using GENERIC declarations (lines 7–11), so they can be easily modified.

The architecture is in lines 18–125. Its declarative part (lines 19–24) contains FSM-related and system signals. Note *timer* in line 24, which is used to pass to the counter the intended delay before each transition. The code begins with a construction for x (line 27), which reads the three pushbuttons. The two FSM processes then follow. As usual, the first process (lines 29–42) implements the lower (sequential) part of the FSM, while the second one (lines 44–124) implements its upper (combinational) part. Because repetitions in the pushbutton sequence (password) must be allowed, observe that the $x = b$ test (line 69) must come before the $x = a$ test (line 72), the same being true for the $x = c$ (line 95) and $x = a$ (line 98) tests.

The reader is invited to compile this code and test it in the FPGA board (see exercise 11.18).

```
1   ----------------------------------------------------------------------
2   LIBRARY ieee;
3   USE ieee.std_logic_1164.all;
4   ----------------------------------------------------------------------
5   ENTITY pb_sequence_detector IS
6      GENERIC (
7         a: STD_LOGIC_VECTOR(2 DOWNTO 0) := "011"; --"011","101",or"110"
8         b: STD_LOGIC_VECTOR(2 DOWNTO 0) := "101"; --"011","101",or"110"
9         c: STD_LOGIC_VECTOR(2 DOWNTO 0) := "110"; --"011","101",or"110"
10        none: STD_LOGIC_VECTOR(2 DOWNTO 0) := "111";       --always "111"
11        T: INTEGER := 150_000_000); --delay=3s with 50MHz clock
12     PORT (
13        clk, rst: IN STD_LOGIC;
14        pb1, pb2, pb3: IN STD_LOGIC;
15        led_idle, led_win: OUT STD_LOGIC);
16  END pb_sequence_detector;
17  ----------------------------------------------------------------------
18  ARCHITECTURE fsm OF pb_sequence_detector IS
19     TYPE state IS (idle, wait1, key1, wait2, key2, wait3, key3);
20     SIGNAL pr_state, nx_state: state;
21     ATTRIBUTE enum_encoding: STRING;
22     ATTRIBUTE enum_encoding OF state: TYPE IS "sequential";
23     SIGNAL x: STD_LOGIC_VECTOR(2 DOWNTO 0);
24     SIGNAL timer: INTEGER RANGE 0 TO T;
25  BEGIN
26     ----Construction of x:----------------
27     x <= (pb1 & pb2 & pb3);
28     ----Lower section of FSM:-------------
29     PROCESS (clk, rst)
30        VARIABLE count: INTEGER RANGE 0 TO T;
31     BEGIN
32        IF (rst='1') THEN
33           pr_state <= idle;
34           count := 0;
35        ELSIF (clk'EVENT AND clk='1') THEN
36           count := count + 1;
37           IF (count>=timer) THEN
38              pr_state <= nx_state;
```

```
39              count := 0;
40           END IF;
41        END IF;
42     END PROCESS;
43     ----Upper section of FSM:------------
44     PROCESS (pr_state, x)
45     BEGIN
46        CASE pr_state IS
47           WHEN idle =>
48              led_idle <= '1';
49              led_win <= '0';
50              timer <= 1;
51              IF (x=a) THEN
52                 nx_state <= wait1;
53              ELSE
54                 nx_state <= idle;
55              END IF;
56           WHEN wait1 =>
57              led_idle <= '0';
58              led_win <= '0';
59              IF (x=none) THEN
60                 timer <= 1;
61                 nx_state <= key1;
62              ELSE
63                 timer <= T;
64                 nx_state <= idle;
65              END IF;
66           WHEN key1 =>
67              led_idle <= '0';
68              led_win <= '0';
69              IF (x=b) THEN --x=b test must be before x=a test
70                 timer <= 1;
71                 nx_state <= wait2;
72              ELSIF (x=a) THEN
73                 timer <= 1;
74                 nx_state <= wait1;
75              ELSIF (x/=none) THEN
76                 timer <= 1;
77                 nx_state <= idle;
78              ELSE
79                 timer <= T;
80                 nx_state <= idle;
81              END IF;
82           WHEN wait2 =>
```

```
83                  led_idle <= '0';
84                  led_win <= '0';
85                  IF (x=none) THEN
86                      timer <= 1;
87                      nx_state <= key2;
88                  ELSE
89                      timer <= T;
90                      nx_state <= idle;
91                  END IF;
92              WHEN key2 =>
93                  led_idle <= '0';
94                  led_win <= '0';
95                  IF (x=c) THEN --x=c test must be before x=a test
96                      timer <= 1;
97                      nx_state <= wait3;
98                  ELSIF (x=a) THEN
99                      timer <= 1;
100                     nx_state <= wait1;
101                 ELSIF (x/=none) THEN
102                     timer <= 1;
103                     nx_state <= idle;
104                 ELSE
105                     timer <= T;
106                     nx_state <= idle;
107                 END IF;
108             WHEN wait3 =>
109                 led_idle <= '0';
110                 led_win <= '1';
111                 IF (x=none) THEN
112                     timer <= 1;
113                     nx_state <= key3;
114                 ELSE
115                     timer <= T;
116                     nx_state <= idle;
117                 END IF;
118             WHEN key3 =>
119                 led_idle <= '0';
120                 led_win <= '1';
121                 timer <= T;
122                 nx_state <= idle;
123         END CASE;
124     END PROCESS;
125 END fsm;
126 --------------------------------------------------------------------
```

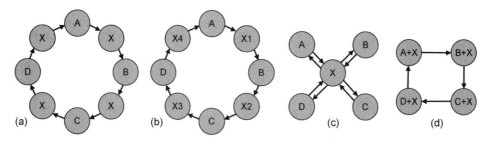

Figure 11.18
(a) FSM with a repetitive state X; (b) Adjusted diagram with the repetitive state distinctly named in each instantiation; (c) Reorganized radial version, which avoids state repetitions, but demands extra registers; (d) Reorganized version with the repetitive state absorbed by the others, again avoiding repetitions.

11.7 FSMs with Repetitive States

Figure 11.18a shows the state transition diagram of an FSM that has a repetitive state X. This kind of situation often arises when dealing with memories. For example, to write data to a memory, after some data manipulations the need for a memory-write pulse is inevitable.

Three solutions for this kind of situation are depicted in figures 11.18b–d. In the first solution, all states are maintained, so the repetitive state must be distinctly named in each instantiation (see $X1$–$X4$). The other two solutions avoid state repetitions. In figure 11.18c, a *radial* architecture is presented, in which the repetitive state appears only once, in the center. In figure 11.18d, a still conventional diagram is presented, but with the repetitive state *incorporated* into the others.

The solution of figure 11.18b is the most straightforward, but has a larger number of states. On the other hand, in the radial solution of figure 11.18c, when the machine goes to X it must *memorize* the state to which it must go *after* that state; for example, if it goes from A to X, then it must remember that B is the next state. Finally, the solution in figure 11.18d maintains the overall FSM architecture, so it is normally relatively simple; in it, the repetitive state is simply absorbed by the preceding state. Regardless of the option chosen, the amount of logic is expected to be nearly the same in all three solutions (note that, even though (b) has more states than (c)–(d), other kinds of information must be stored in the latter, not needed in the former).

11.8 Other FSM Designs

Besides the designs already seen and those proposed in section 11.9, see also the following designs involving finite state machines in other chapters:

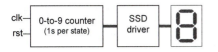

Figure 11.19

Basic LED/SSD/LCD driver (section 12.2)

Counters with SSD and LCD displays (exercises 12.1 and 12.2)

Frequency meter (section 12.4 and exercise 12.4)

I^2C interface for an EEPROM memory (section 14.4)

Playing with an SSD (section 12.3 and exercise 12.3)

SPI interface for an FRAM memory (section 14.5).

11.9   Exercises

Exercise 11.1: Gray-Encoded Counter

Design a 0-to-8 counter with Gray-encoded outputs. (Suggestion: see example 11.3).

a) Draw the state transition diagram.

b) Estimate the number of flip-flops that will be needed.

c) Write the VHDL code, then compile and simulate it.

d) Check whether the number of DFFs inferred by the compiler matches your prediction.

Exercise 11.2: Johnson-Encoded Counter

Repeat exercise 11.1 for a counter with Johnson-encoded output instead of Gray-encoded.

Exercise 11.3: One-Hot-Encoded Counter

Repeat exercise 11.1 for a counter with one-hot output instead of Gray output.

Exercise 11.4: Zero-to-Nine Counter

The purpose of this exercise is to modify the counter designed in example 11.3, such that the circuit stays in each state during $T = 1$ s, with the output displayed by an SSD (seven-segment display—described in section 12.1, see figure 12.2). This arrangement is depicted in figure 11.19, which also shows an SSD driver, which is a combinational circuit that converts the 4-bit output from the counter into a 7-bit signal to feed the SSD's segments. Assume that the clock frequency is 50 MHz. (Suggestion: see section 11.6.)

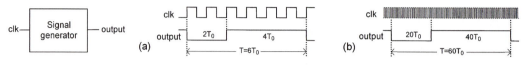

Figure 11.20

a) Draw the new state transition diagram for the counter.

b) Write a VHDL code for this circuit, including in it, besides the FSM, the SSD driver. Enter T using GENERIC, so changes are easy to make. Compile and simulate your code.

c) Finally, download the design to your FPGA board and test its operation. Recall that while $rst = '1'$ the display must remain zeroed.

Suggestion: See a related design, without the FSM approach, in example 6.6.

Exercise 11.5: Signal Generator #1

Figure 11.20a shows the output of a signal generator, which stays low during two clock cycles and high during four cycles (hence $T = 6T_0$, where T_0 is the clock period). Observe that this machine has only six states, and recall that in a signal generator glitches are not acceptable.

a) Draw the state transition diagram.

b) Estimate the number of DFFs that will be needed for sequential and one-hot encodings.

c) Write the VHDL code, then compile and simulate it.

d) Check whether the number of DFFs inferred by the compiler matches your prediction for each encoding option.

Exercise 11.6: Signal Generator #2

Figure 11.20b shows the output of another signal generator, similar to that in the previous exercise, but now the output stays low during 20 clock cycles and high during 40 cycles (hence $T = 60T_0$), instead of two and four cycles, respectively. Note that the machine has now 60 states. Is it fine to use exactly the same design technique employed in the previous problem? Recall that in a signal generator glitches are not acceptable. (Suggestion: see section 11.6.)

a) Draw the state transition diagram.

b) Estimate the number of DFFs that will be needed with sequential and one-hot encodings.

c) Write the VHDL code, then compile and simulate it.

d) Check whether the number of DFFs inferred by the compiler matches your predictions for both encoding options.

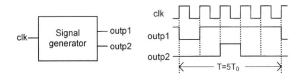

Figure 11.21

Exercise 11.7: Signal Generator #3

This exercise deals with the design of a two-output signal generator, depicted in figure 11.21. Recall that in a signal generator glitches are not acceptable.

a) Draw the state transition diagram.

b) Write the VHDL code, then compile and simulate it.

c) Examine your solution to determine whether it is subject to glitches. If it is, include the optional output section seen in the FSM template of section 11.2.

Exercise 11.8: ASCII Sequence Detector

Using the FSM approach, design a circuit that receives a serial data stream of ASCII characters and asserts a flag when the sequence "VHDL" occurs. Recall that ASCII characters are synthesizable.

a) Draw the state transition diagram.

b) Estimate the number of DFFs that will be needed with sequential encoding. Note that a flag might also be stored.

c) Write the VHDL code, then compile and simulate it.

d) Check whether the number of DFFs inferred by the compiler matches your prediction.

Exercise 11.9: Preventing State-Bypass with a Flag #1

Figure 11.22a shows the state transition diagram for an FSM subject to state-bypass (note that when a long $x = '1'$ occurs the system flips back and forth between A and B). Design the corresponding circuit using a flag to prevent such a problem.

Exercise 11.10: Preventing State-Bypass with Additional States #1

Repeat the exercise above, now using additional (wait) states to prevent the state-bypass problem.

Exercise 11.11: Preventing State-Bypass with a Flag #2

Figure 11.22b shows the state transition diagram for another FSM subject to state-bypass (note that when the system is in state A and $x = '1'$ occurs, lasting two or more clock

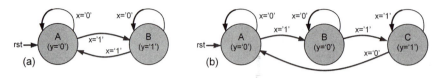

Figure 11.22

cycles, the machine goes directly to C (it stays in B during only one clock period)). Design the corresponding circuit using a flag to prevent such a problem.

Exercise 11.12: Preventing State-Bypass with Additional States #2

Repeat the exercise above, now using additional (wait) states to prevent the state-bypass problem.

Exercise 11.13: Timed FSM #1

A certain machine has two states, called A and B. If in A and $x = \text{'1'}$ occurs, and $x = \text{'1'}$ lasts *T1* clock cycles, then the machine must move to B (x triggers the transition and must remain high during the whole time interval). When in B, it must return to A unconditionally after *T2* clock cycles. Recall that, as in any other design, it might be necessary to add to the original states (A, B) other states (wait states, for example) in order to fulfill the specifications.

a) Draw the corresponding state transition diagram.

b) Classify each transition as logic, timed, or hybrid (see figure 11.12).

c) Design this circuit using VHDL.

Exercise 11.14: Timed FSM #2

Design the FSM of figure 11.23a knowing that *time1* = 3 and *time2* = 5 clock pulses. The possible values for x are '0' and '1'.

Exercise 11.15: FSM with Embedded Timer #2

Design the FSM of figure 11.23b knowing that *time1* = 3 and *time2* = 5 clock pulses. The possible values for x are 0, 1, and 2.

Exercise 11.16: Traffic-Light Controller #1

This exercise concerns the TLC designed in example 11.7.

a) Compile that code and check if the estimated numbers of flip-flops match the predictions.

b) If *fclk* = 50 MHz, can the code be compiled directly? (Hint: think about the range of integers.) If there is a problem, how can it be solved? How many flip-flops are now needed?

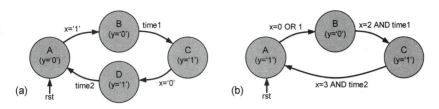

Figure 11.23

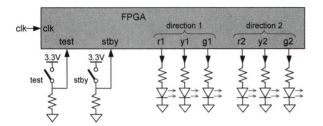

Figure 11.24

c) To make the code truly generic, the following could be done in the generics declaration list (lines 7–12): include *fclk* (in Hz) and specify all the times in seconds instead of in number of clock pulses. Why would it be now truly generic? Make such modifications in the code.

d) Physically implement that circuit in your FPGA board. The overall arrangement is shown in figure 11.24, which exhibits three inputs (*clk* plus two toggle switches for *test* and *stby*) and six outputs (an LED for each light in the two directions). Test its operation for all three operating modes (*regular*, *test*, and *standby*).

Exercise 11.17: Traffic-Light Controller #2

In continuation to the exercise above, modify the design presented in example 11.7 such that the yellow lights, when in standby mode, blink with a 2 Hz frequency instead of remaining statically ON. Download then the design to your FPGA board to test this new feature.

Exercise 11.18: Pushbutton Sequence Detector

This exercise regards the pushbutton sequence detector designed in example 11.8.

a) Estimate the number of flip-flops that will be needed to implement that circuit for a 50MHz clock and *sequential* FSM encoding.

b) What is the number of flip-flops if *one-hot* encoding is used instead?

Figure 11.25

c) Compile the code and check whether the inferred numbers of registers match your predictions (note that the only change needed in the code from one case to the other is in line 22).

d) Physically implement the circuit in your FPGA board and check its operation. Try it with several values for a, b, and c (lines 7–9), including repetitions.

e) Modify your design by including one LED for each state, then retest it in the FPGA board.

Exercise 11.19: Car Speed Monitor

The speed of an automobile is generally measured by a Hall effect sensor installed near the transmission. The signal provided by this sensor has a frequency *fpulse* proportional to the car's speed, normally of about 500–2,000 Hz per miles/hour (assume 1 kHz). This exercise concerns the design of a speed monitor, taking *fpulse* as input, with the following features (a possible view is shown in figure 11.25):

1) Speed selection pushbutton (*speed*): Every time it is pressed, the next speed to be monitored is selected, in miles/hour ($35 \rightarrow 45 \rightarrow 55 \rightarrow 60 \rightarrow 65 \rightarrow 70 \rightarrow 75 \rightarrow 80 \rightarrow 35 \ldots$).

2) LEDs: Total of eight (one for each speed), with only one lit at a time, indicating the speed being monitored. All LEDs must be turned OFF when the power switch is turned OFF.

3) Numeric display: 3-digit display (LCD or SSD) that shows the car's actual speed (based on *fpulse*).

4) Buzzer: Must emit a sound when the car approaches the selected speed. Adopt a 2Hz signal when the speed is 3 miles/hour or less from the selected speed, or a continuous signal when at or above the selected speed. Consider that it is a buzzer with internal oscillator, so only a 2Hz square wave is needed in the former case and a continuous DC voltage in the latter.

a) Draw the state transition diagram.

b) Write the VHDL code, then compile and simulate it.

c) Physically implement the circuit in your FPGA board. Use a signal generator or the board's clock to emulate *fpulse*.

12 VHDL Designs with Basic Displays

12.1 Introduction

Because feedback is always wanted, designs that include a visual device (thus providing a simple physical response) are very attractive, helping immensely to motivate students. For that reason, a series of design examples using VHDL and basic displays are included in this chapter, continuing with other, more complex displays in chapters 15–17.

By "basic displays" we mean LEDs, SSDs, and alphanumeric LCDs.

LED (Light Emitting Diode)

Figure 12.1a shows examples of commercial LEDs. The most common material employed in their fabrication is gallium arsenide, which emits a radiation in the infrared spectrum (a typical application for this is in remote controls for TVs, sound, etc.). To have an LED emit a radiation in the visible spectrum, other materials must be added to gallium arsenide, such as aluminum or phosphorous to yield red LEDs, for example. Many other materials are used to attain higher-frequency radiations, like indium gallium nitride and zinc selenide for blue LEDs.

An LED emits light when traversed by an electric current, so it needs to be forward biased, as illustrated in figure 12.1b, which also includes a current limiting resistor (LEDs normally operate under 20mA).

SSD (Seven Segment Display)

SSDs are just special 7-LED arrangements (8 if a decimal point is also included). This kind of device is illustrated in figure 12.2. In (a), an example of commercial SSD (two digits with decimal points) is shown. In (b), a typical notation for the segment names (*abcdefg*) is included. In (c), the common-anode configuration is presented. Finally, the table in (d) shows which segments must be lit to attain 0-to-F characters (note that it was considered a common-anode SSD, so a low voltage must be applied to light a segment).

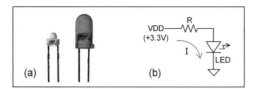

Figure 12.1
(a) Examples of commercial LEDs; (b) Typical usage.

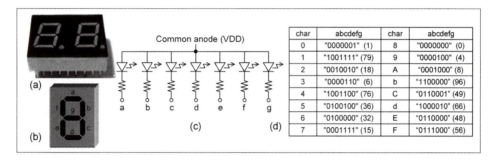

char	abcdefg	char	abcdefg
0	"0000001" (1)	8	"0000000" (0)
1	"1001111" (79)	9	"0000100" (4)
2	"0010010" (18)	A	"0001000" (8)
3	"0000110" (6)	b	"1100000" (96)
4	"1001100" (76)	C	"0110001" (49)
5	"0100100" (36)	d	"1000010" (66)
6	"0100000" (32)	E	"0110000" (48)
7	"0001111" (15)	F	"0111000" (56)

Figure 12.2
(a) Example of commercial SSD; (b) Segment names (*abcdefg*); (c) Common-anode configuration; (d) Segments that must be lit to attain 0-to-F characters.

Alphanumeric LCD (Liquid Crystal Display)

Figure 12.3 shows a popular alphanumeric LCD, which contains two lines of 16 characters each. A picture of the display is shown in (a). The corresponding pinout is exhibited in (b). The internal display layout is illustrated in (c), showing 16×2 dot arrays of size 8×5 each. In (d), its most frequent exhibition mode is depicted, consisting of 8×5-dot arrays, for 7×5 characters. Finally, in (e), its other predefined exhibition mode is depicted, consisting of 11×5-dot arrays, for 10×5 characters.

On the back of the LCD there is a controller (generally HD44780U, from Hitachi, or an equivalent device) that acts as the interface between the LCD and the external world. This controller can be accessed through 16 pins, listed in figure 12.3b, which include power, contrast, control, and data lines.

The predefined characters, stored in the LCD controller's CGROM (character generator ROM) are shown in figure 12.4, consisting of 192 7×5 characters. The system contains also 32 characters of size 10×5. When using the former (hence with 8×5-dot blocks) the LCD can operate with two lines, while in the latter (11×5-dot blocks) only single-line operation is possible. User-defined characters are also allowed, so other exhibition modes are possible, like full-height (16×5-dot) characters.

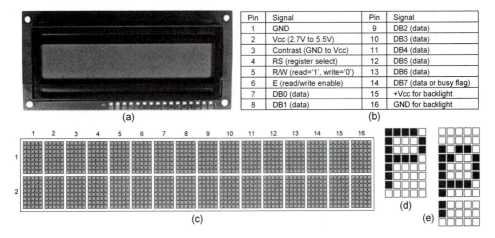

Pin	Signal	Pin	Signal
1	GND	9	DB2 (data)
2	Vcc (2.7V to 5.5V)	10	DB3 (data)
3	Contrast (GND to Vcc)	11	DB4 (data)
4	RS (register select)	12	DB5 (data)
5	R/W (read='1', write='0')	13	DB6 (data)
6	E (read/write enable)	14	DB7 (data or busy flag)
7	DB0 (data)	15	+Vcc for backlight
8	DB1 (data)	16	GND for backlight

Figure 12.3
(a) Picture of a 16×2 alphanumeric LCD; (b) LCD controller's pinout; (c) Internal layout (16×2 8×5-dot blocks); (d) Standard 8×5-dot exhibition mode (two lines of 7×5 characters); (e) Standard 11×5-dot configuration (single line of 10×5 characters).

Figure 12.4
Predefined characters available in the LCD controller's ROM.

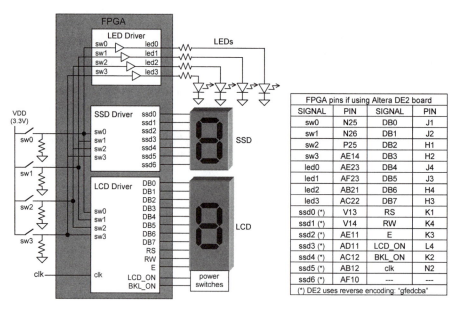

Figure 12.5
Basic LED/SSD/LCD driver designed in section 12.2.

Other details on the operation of this type of display, like controller's instructions, initialization procedure, state-machine diagram, and corresponding VHDL code, can be seen in Pedroni (2008), which was used as a reference for all codes for LCDs presented in this chapter.

A series of designs, using the three basic displays described, are presented next.

12.2   Basic LED/SSD/LCD Driver

This first design concerns the multiple driver of figure 12.5. The input consists of four toggle switches (plus clock), hence encoding values from "0000" to "1111". These values must be exhibited at the output by four LEDs, one SSD, and one alphanumeric LCD. In the last two, hexadecimal representation (0 to F) must be used.

As depicted in figure 12.5, the LED driver is just a buffer, because *led0 = sw0*, *led1 = sw1*, and so on (note that the input switches convey a '0' when open and a '1' when closed). The second driver (SSD) is a fully combinational circuit (so clock is not needed) that simply converts a 4-bit signal from the switches into a 7-bit signal for the SSD, according with the table in figure 12.2d. Finally, the LCD driver is more complex because it must communicate with the LCD controller, to which it provides eight data bits (*DB7-DB0*) and three control bits (*RS*, *RW*, *E*), plus two additional signals (*LCD_ON*, *BKL_ON*) to turn the LCD and its backlight ON. Figure 12.5 also shows the FPGA pins that can be used if implementing this design in Altera's DE2 development board. A VHDL code for this circuit is presented next (divided into two parts).

Part 1: Package with Conversion Functions

To ease the conversion to SSD and LCD characters, two functions were created in the package below, called *my_functions*, which are then used in lines 27 and 30 of the main code. Values in lines 38–53 are from figure 12.4.

```
1    -----Package with functions:-------------------------------------------
2    LIBRARY ieee;
3    USE ieee.std_logic_1164.all;
4    PACKAGE my_functions IS
5       FUNCTION integer_to_ssd (SIGNAL input: NATURAL) RETURN STD_LOGIC_VECTOR;
6       FUNCTION integer_to_lcd (SIGNAL input: NATURAL) RETURN STD_LOGIC_VECTOR;
7    END my_functions;
8    ----------------------------------------------------------------------
9    PACKAGE BODY my_functions IS
10      FUNCTION integer_to_ssd (SIGNAL input: NATURAL) RETURN STD_LOGIC_VECTOR
11         IS VARIABLE output: STD_LOGIC_VECTOR(6 DOWNTO 0);
12      BEGIN
13         CASE input IS
14            WHEN 0 => output:="0000001";     --"0" on SSD
15            WHEN 1 => output:="1001111";     --"1" on SSD
16            WHEN 2 => output:="0010010";     --"2" on SSD
17            WHEN 3 => output:="0000110";     --"3" on SSD
18            WHEN 4 => output:="1001100";     --"4" on SSD
19            WHEN 5 => output:="0100100";     --"5" on SSD
20            WHEN 6 => output:="0100000";     --"6" on SSD
21            WHEN 7 => output:="0001111";     --"7" on SSD
22            WHEN 8 => output:="0000000";     --"8" on SSD
23            WHEN 9 => output:="0000100";     --"9" on SSD
24            WHEN 10 => output:="0001000";    --"A" on SSD
25            WHEN 11 => output:="1100000";    --"b" on SSD
26            WHEN 12 => output:="0110001";    --"C" on SSD
27            WHEN 13 => output:="1000010";    --"d" on SSD
28            WHEN 14 => output:="0110000";    --"E" on SSD
29            WHEN OTHERS=>output:="0111000";  --"F" on SSD
30         END CASE;
31         RETURN output;
32      END integer_to_ssd;
```

```
33   -------------------------------------------------------------------------
34   FUNCTION integer_to_lcd (SIGNAL input: NATURAL) RETURN STD_LOGIC_VECTOR
35      IS VARIABLE output: STD_LOGIC_VECTOR(7 DOWNTO 0);
36   BEGIN
37      CASE input IS
38         WHEN 0 => output:="00110000";    --"0" on LCD
39         WHEN 1 => output:="00110001";    --"1" on LCD
40         WHEN 2 => output:="00110010";    --"2" on LCD
41         WHEN 3 => output:="00110011";    --"3" on LCD
42         WHEN 4 => output:="00110100";    --"4" on LCD
43         WHEN 5 => output:="00110101";    --"5" on LCD
44         WHEN 6 => output:="00110110";    --"6" on LCD
45         WHEN 7 => output:="00110111";    --"7" on LCD
46         WHEN 8 => output:="00111000";    --"8" on LCD
47         WHEN 9 => output:="00111001";    --"9" on LCD
48         WHEN 10 => output:="01000001";   --"A" on LCD
49         WHEN 11 => output:="01000010";   --"B" on LCD
50         WHEN 12 => output:="01000011";   --"C" on LCD
51         WHEN 13 => output:="01000100";   --"D" on LCD
52         WHEN 14 => output:="01000101";   --"E" on LCD
53         WHEN OTHERS=>output:="01000110"; --"F" on LCD
54      END CASE;
55      RETURN output;
56   END integer_to_lcd;
57 END my_functions;
58 -------------------------------------------------------------------------
```

Part 2: Main Code

The main code is shown below, under the project name *led_ssd_lcd_driver* (line 6). The inputs are *clk* and *switches* (lines 9–10), while the outputs are *leds* (line 11 for the LED driver) and *ssd* (line 12 for the SSD driver), plus *RS*, *RW*, *LCD_ON*, *BK_LIGHT*, *E*, and *DB* (lines 13–15 for the LCD driver).

The architecture is divided into three parts: LED driver (line 25), SSD driver (line 27), and LCD driver (rest of the code). The first driver requires just one assignment, while the second is implemented with just a function call (*integer_to_ssd*), which converts 4-bit decimal values into 7-bit values for the SSD. The third driver is more complex because the LCD can only be reached through its microcontroller. The corresponding code is in lines 28–95, based on the state machine developed in Pedroni (2008). As with the SSD driver, a special function (*integer_to_lcd*, line 30) was written, which simplifies the access to the LCD ROM (the function provides an address from which the corresponding character is retrieved and used—see figure 12.4).

```
1   ------Main code:-------------------------------------------------------
2   LIBRARY ieee;
3   USE ieee.std_logic_1164.all;
4   USE work.my_functions.all;
5   ----------------------------------------------------------------------
6   ENTITY led_ssd_lcd_driver IS
7      GENERIC (clk_divider: POSITIVE := 50_000); --50MHz to 500Hz
8      PORT (
9         clk: IN STD_LOGIC;
10        switches: IN NATURAL RANGE 0 TO 15;
11        leds: OUT NATURAL RANGE 0 TO 15;
12        ssd: OUT STD_LOGIC_VECTOR(6 DOWNTO 0);
13        RS, RW, LCD_ON, BKL_ON: OUT STD_LOGIC;
14        E: BUFFER STD_LOGIC;
15        DB: OUT STD_LOGIC_VECTOR(7 DOWNTO 0));
16  END led_ssd_lcd_driver;
17  ----------------------------------------------------------------------
18  ARCHITECTURE led_ssd_lcd_driver OF led_ssd_lcd_driver IS
19     TYPE state IS (FunctionSet1, FunctionSet2, FunctionSet3, FunctionSet4,
20        ClearDisplay, DisplayControl, EntryMode, WriteData, ReturnHome);
21     SIGNAL pr_state, nx_state: state;
22     SIGNAL input_lcd: STD_LOGIC_VECTOR(7 DOWNTO 0);
23  BEGIN
24     -----Part 1: LED driver:----------------------
25     leds <= switches;
26     -----Part 2: SSD driver:----------------------
27     ssd <= integer_to_ssd(switches);
28     -----Part 3: LCD driver (FSM-based):----------
29     --Get LCD character:
30     input_lcd <= integer_to_lcd(switches);
31     --Turn LCD and its backlight ON:
32     LCD_ON <= '1'; BKL_ON <= '1';
33     --Clock generator for LCD(E=500Hz):
34     PROCESS (clk)
35        VARIABLE count: INTEGER RANGE 0 TO clk_divider;
36     BEGIN
37        IF (clk'EVENT AND clk='1') THEN
38           count := count + 1;
39           IF (count=clk_divider) THEN
40              E <= NOT E;
41              count := 0;
42           END IF;
43        END IF;
```

```
44      END PROCESS;
45      --Lower section of FSM:
46      PROCESS (E)
47      BEGIN
48         IF (E'EVENT AND E='1') THEN
49             pr_state <= nx_state;
50         END IF;
51      END PROCESS;
52      --Upper section of FSM:
53      PROCESS (pr_state, input_lcd)
54      BEGIN
55         CASE pr_state IS
56             ---Initialize LCD:
57             WHEN FunctionSet1 =>
58                 RS<='0'; RW<='0';
59                 DB <= "0011XX00";
60                 nx_state <= FunctionSet2;
61             WHEN FunctionSet2 =>
62                 RS<='0'; RW<='0';
63                 DB <= "0011XX00";
64                 nx_state <= FunctionSet3;
65             WHEN FunctionSet3 =>
66                 RS<='0'; RW<='0';
67                 DB <= "0011XX00";
68                 nx_state <= FunctionSet4;
69             WHEN FunctionSet4 =>
70                 RS<='0'; RW<='0';
71                 DB <= "00111000";
72                 nx_state <= ClearDisplay;
73             WHEN ClearDisplay =>
74                 RS<='0'; RW<='0';
75                 DB <= "00000001";
76                 nx_state <= DisplayControl;
77             WHEN DisplayControl =>
78                 RS<='0'; RW<='0';
79                 DB <= "00001100";
80                 nx_state <= EntryMode;
81             WHEN EntryMode =>
82                 RS<='0'; RW<='0';
83                 DB <= "00000110";
84                 nx_state <= WriteData;
85             ---Write data to LCD:
86             WHEN WriteData =>
```

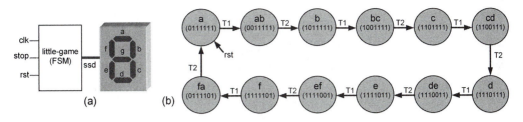

Figure 12.6
Playing with an SSD. (a) Top-level circuit diagram; (b) State transition diagram.

```
87                    RS<='1'; RW<='0';
88                    DB <= input_lcd;
89                    nx_state <=ReturnHome;
90                 WHEN ReturnHome =>
91                    RS<='0'; RW<='0';
92                    DB <= "10000000";
93                    nx_state <= WriteData;
94            END CASE;
95         END PROCESS;
96    END led_ssd_lcd_driver;
97    -----------------------------------------------------------------------
```

12.3 Playing with a Seven-Segment Display

Figure 12.6a shows a finite state machine, called *little-game*, which feeds an SSD. The circuit has three inputs (*clk*, *stop*, *rst*) and one output (*ssd(6:0)*). It must be designed in such a way to produce a continuous clockwise rotating movement of the display segments. To make the movement more realistic, a momentary overlap of neighboring segments should occur—that is, $a \rightarrow ab \rightarrow b \rightarrow bc \rightarrow c \rightarrow \cdots \rightarrow fa \rightarrow a \ldots$, with the overlapping states (*ab*, *bc*, etc.) lasting less ($T2 = 40\text{ms}$) than the others ($T1 = 120\text{ms}$). The *stop* switch must interrupt the movement when asserted, with the movement resumed from the *same* position when *stop* returns to zero. The reset switch, on the other hand, must cause the machine to go to state *a* when asserted, resuming from there when *rst* returns to zero.

The state transition diagram for this FSM is presented in figure 12.6b, with the state names indicating which SSD segments should be ON in each state. The corresponding output (*ssd*) values are shown between parentheses (for a common-anode SSD, as in figure 12.2c).

A VHDL code for this circuit is presented below, which is a direct application of the FSM template for timed machines studied in section 11.6. The clock frequency (*fclk*) and the time delays (*T1*, *T2*) were entered using GENERIC declarations (lines 3–5), so they can be easily modified. The pin names (lines 6–7) obey the diagram in figure 12.6a. A user-defined type for the state machine was created in line 14 and then used in line 15. The lower (sequential) section of the FSM is in the process of lines 18–32, while the upper (combinational) section is in the process of lines 34–96. Note that library declarations were not needed because only data types defined in the package *standard* were employed, which is visible by default. Additional features are added to this machine in exercise 12.3.

```
1   ------------------------------------------------------------
2   ENTITY little_game IS
3      GENERIC (fclk: INTEGER := 50_000;  --clk frequency (kHz)
4                T1: INTEGER := 120;       --long delay (ms)
5                T2: INTEGER := 40);       --short delay (ms)
6      PORT (clk, stop, rst: IN BIT;
7            ssd: OUT BIT_VECTOR(6 DOWNTO 0));
8   END little_game;
9   ------------------------------------------------------------
10  ARCHITECTURE fsm OF little_game IS
11     CONSTANT time1: INTEGER := fclk*T1;
12     CONSTANT time2: INTEGER := fclk*T2;
13     SIGNAL delay: INTEGER RANGE 0 TO time1;
14     TYPE state IS (a, ab, b, bc, c, cd, d, de, e, ef, f, fa);
15     SIGNAL pr_state, nx_state: state;
16  BEGIN
17     -------Lower section of FSM:------------
18     PROCESS (clk, stop, rst)
19        VARIABLE count: INTEGER RANGE 0 TO time1;
20     BEGIN
21        IF (rst='1') THEN
22           pr_state <= a;
23        ELSIF (clk'EVENT AND clk='1') THEN
24           IF (stop='0') THEN
25              count := count + 1;
26              IF (count=delay) THEN
27                 count := 0;
28                 pr_state <= nx_state;
29              END IF;
30           END IF;
31        END IF;
32     END PROCESS;
33     -------Upper section of FSM:------------
34     PROCESS (pr_state)
35     BEGIN
36        CASE pr_state IS
```

```
37            WHEN a =>
38                ssd <= "0111111"; --decimal 63
39                delay <= time1;
40                nx_state <= ab;
41            WHEN ab =>
42                ssd <= "0011111"; --decimal 31
43                delay <= time2;
44                nx_state <= b;
45            WHEN b =>
46                ssd <= "1011111";  --decimal 95
47                delay <= time1;
48                nx_state <= bc;
49            WHEN bc =>
50                ssd <= "1001111";  --decimal 79
51                delay <= time2;
52                nx_state <= c;
53            WHEN c =>
54                ssd <= "1101111";  --decimal 111
55                delay <= time1;
56                nx_state <= cd;
57            WHEN cd =>
58                ssd <= "1100111";  --decimal 103
59                delay <= time2;
60                nx_state <= d;
71            WHEN d =>
72                ssd <= "1110111";  --decimal 119
73                delay <= time1;
74                nx_state <= de;
75            WHEN de =>
76                ssd <= "1110011";  --decimal 115
77                delay <= time2;
78                nx_state <= e;
79            WHEN e =>
80                ssd <= "1111011";  --decimal 123
81                delay <= time1;
82                nx_state <= ef;
83            WHEN ef =>
84                ssd <= "1111001";  --decimal 121
85                delay <= time2;
86                nx_state <= f;
87            WHEN f =>
88                ssd <= "1111101";  --decimal 125
89                delay <= time1;
```

```
90              nx_state <= fa;
91          WHEN fa =>
92              ssd <= "0111101";   --decimal 61
93              delay <= time2;
94              nx_state <= a;
95      END CASE;
96   END PROCESS;
97 END fsm;
98 -------------------------------------------------------------
```

12.4 Frequency Meter (with LCD)

The purpose of this section is to design a frequency meter that measures frequencies in the 00.0 to 99.9 kHz range, displaying the result on an alphanumeric LCD, which is updated once per second. A circuit diagram is suggested in figure 12.7a, with the system divided into two portions: frequency meter and LCD driver (synchronizer not included yet).

The frequency meter receives *input*, which is the signal whose frequency we want to measure, and produces at the output the decimal digits *output1*, *output2*, and *output3*, containing the result of the measurement. These three signals are then passed to the LCD driver, which converts them into the proper form for LCD exhibition. The other input to both blocks is obviously *clk*.

Two construction techniques for frequency meters were introduced in Pedroni (2008), which depend on the frequency being measured and the refresh rate (see also the notes

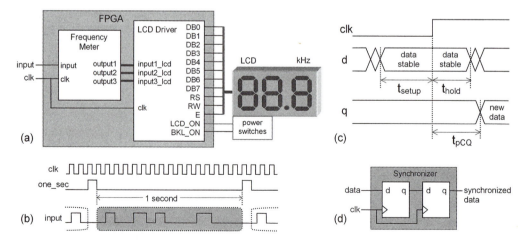

Figure 12.7
(a) Frequency meter of section 12.4; (b) Suggested timing diagram; (c) Flip-flop setup and hold times; (d) Synchronizer to prevent metastable states.

after the code below regarding a test procedure and synchronization). The technique that is appropriate in the present problem is summarized by the timing diagram of figure 12.7b, which consists of creating a one-second time window and then counting the number of input pulses during that time interval, with the result stored by output registers at the end of that time period (see *one_sec* and *input* waveforms in figure 12.7b; the pulses counted in the one-second period are those in the dark area).

A VHDL code for this problem is presented below. Again, to make it easier to retrieve the character codes from the LCD controller's ROM, a function was used. The function, called *integer_to_lcd*, was written in section 12.2 and located in a package named *my_functions* (see package declaration in line 6 and function calls in lines 84–86 of the code below). Consequently, that file must be included in the project.

The project's name is *frequencimeter* (line 8) and the input-output port names (lines 13–17) are from figure 12.7a. Two generic parameters were also included in order to make the code easy to adapt to any other clock frequency. In the declarative part of the architecture (lines 21–30), a series of signals were created, including an enumerated data type (lines 27–30) for a state machine, which will be used to implement the LCD driver.

The code proper (lines 31–164) was divided into two parts (again following the diagram of figure 12.7a), with the frequency meter in lines 32–81 and the LCD driver in lines 82–163. The former follows the diagram of figure 12.7b, while the latter employs the same structure seen in a previous code. Recall that the *next* character entered in an LCD display is shifted to the right, not the one already entered, so the most significant character must be entered first (see the sequence in lines 144, 148, 152, and 156). To test this circuit, see the suggestion after the code below.

Observe in lines 59–74 how the input pulses are counted and the technique used to achieve the proper rounding. Whenever *count1* reaches 50 (line 61), *count2* is incremented, which can cause the other variables to be incremented as well; however, *count1* is only zeroed when it reaches 100 (line 71). To test this circuit and to add a synchronizer, see the suggestions after the code.

```
1   ----------------------------------------------------------------
2   --Remember to include the function "integer-to-lcd" in the project.
3   ----------------------------------------------------------------
4   LIBRARY ieee;
5   USE ieee.std_logic_1164.all;
6   USE work.my_functions.all;
7   ----------------------------------------------------------------
8   ENTITY frequencimeter IS
9      GENERIC (
10        fclk: POSITIVE := 50_000_000;        --clock frequency
11        clk_divider: POSITIVE := 50_000);    --for E=500Hz @LCD
12     PORT (
```

```
13          clk: IN STD_LOGIC;                      --system clock
14          input: IN STD_LOGIC;                    --signal to measure
15          RS, RW, LCD_ON, BKL_ON: OUT STD_LOGIC; --LCD signals
16          E: BUFFER STD_LOGIC;                    --LCD signal
17          DB: OUT STD_LOGIC_VECTOR(7 DOWNTO 0)); --LCD signal
18 END frequencimeter;
19 -------------------------------------------------------------------
20 ARCHITECTURE frequencimeter OF frequencimeter IS
21    SIGNAL one_sec: STD_LOGIC;
22    SIGNAL output1, output2, output3: INTEGER RANGE 0 TO 9;
23    SIGNAL input1_lcd: STD_LOGIC_VECTOR(7 DOWNTO 0);
24    SIGNAL input2_lcd: STD_LOGIC_VECTOR(7 DOWNTO 0);
25    SIGNAL input3_lcd: STD_LOGIC_VECTOR(7 DOWNTO 0);
26    --SIGNAL input: STD_LOGIC;
27    TYPE state IS (FunctionSet1, FunctionSet2, FunctionSet3,
28       FunctionSet4, ClearDisplay, DisplayControl, EntryMode,
29       WriteData1, WriteData2, WriteData3, WriteData4, ReturnHome);
30    SIGNAL pr_state, nx_state: state;
31 BEGIN
32    -------Part 1: Freq. meter:----------------
33    ---Create 1-second time window:
34    PROCESS (clk)
35       VARIABLE count: INTEGER RANGE 0 TO fclk+1;
36    BEGIN
37       IF (clk'EVENT AND clk='1') THEN
38          count := count + 1;
39          IF (count=fclk+1) THEN
40             one_sec <= '1';
41             count := 0;
42          ELSE
43             one_sec <= '0';
44          END IF;
45       END IF;
46    END PROCESS;
47    ---Count input pulses in 1sec and store result:
48    PROCESS (input, one_sec)
49       VARIABLE count1: INTEGER RANGE 0 TO 100;
50       VARIABLE count2: INTEGER RANGE 0 TO 10;
51       VARIABLE count3: INTEGER RANGE 0 TO 10;
52       VARIABLE count4: INTEGER RANGE 0 TO 10;
53    BEGIN
54       IF (one_sec='1') THEN
55          count1 := 0;
```

```
56              count2 := 0;
57              count3 := 0;
58              count4 := 0;
59          ELSIF (input'EVENT AND input='1') THEN
60              count1 := count1 + 1;
61              IF (count1=50) THEN
62                  count2 := count2 + 1;
63                  IF (count2=10) THEN
64                      count2 := 0;
65                      count3 := count3 + 1;
66                      IF (count3=10) THEN
67                          count3 := 0;
68                          count4 := count4 + 1;
69                      END IF;
70                  END IF;
71              ELSIF (count1=100) THEN
72                  count1 := 0;
73              END IF;
74          END IF;
75          ---Store meter outputs:
76          IF (one_sec'EVENT AND one_sec='1') THEN
77              output1 <= count2;
78              output2 <= count3;
79              output3 <= count4;
80          END IF;
81      END PROCESS;
82      --------Part 2: LCD driver:------------------
83      ---Get LCD characters:
84      input1_lcd <= integer_to_lcd(output1);
85      input2_lcd <= integer_to_lcd(output2);
86      input3_lcd <= integer_to_lcd(output3);
87      ---Turn LCD and its backlight ON:
88      LCD_ON <= '1'; BKL_ON <= '1';
89      ---Generate clock for LCD(E=500Hz):
90      PROCESS (clk)
91          VARIABLE count: INTEGER RANGE 0 TO clk_divider;
92      BEGIN
93          IF (clk'EVENT AND clk='1') THEN
94              count := count + 1;
95              IF (count=clk_divider) THEN
96                  E <= NOT E;
97                  count := 0;
98              END IF;
```

```
 99        END IF;
100    END PROCESS;
101    ---Lower section of FSM:
102    PROCESS (E)
103    BEGIN
104       IF (E'EVENT AND E='1') THEN
105           pr_state <= nx_state;
106       END IF;
107    END PROCESS;
108    ---Upper section of FSM:
109    PROCESS (pr_state, input1_lcd, input2_lcd, input3_lcd)
110    BEGIN
111      CASE pr_state IS
112          ---Initialize LCD:
113          WHEN FunctionSet1 =>
114             RS<='0'; RW<='0';
115             DB <= "0011XX00";
116             nx_state <= FunctionSet2;
117          WHEN FunctionSet2 =>
118             RS<='0'; RW<='0';
119             DB <= "0011XX00";
120             nx_state <= FunctionSet3;
121          WHEN FunctionSet3 =>
122             RS<='0'; RW<='0';
123             DB <= "0011XX00";
124             nx_state <= FunctionSet4;
125          WHEN FunctionSet4 =>
126             RS<='0'; RW<='0';
127             DB <= "00111000";
128             nx_state <= ClearDisplay;
129          WHEN ClearDisplay =>
130             RS<='0'; RW<='0';
131             DB <= "00000001";
132             nx_state <= DisplayControl;
133          WHEN DisplayControl =>
134             RS<='0'; RW<='0';
135             DB <= "00001100";
136             nx_state <= EntryMode;
137          WHEN EntryMode =>
138             RS<='0'; RW<='0';
139             DB <= "00000110";
140             nx_state <= WriteData1;
141          ---Write data to LCD:
```

```
142              WHEN WriteData1 =>
143                  RS<='1'; RW<='0';
144                  DB <= input3_lcd; --left digit
145                  nx_state <= WriteData2 ;
146              WHEN WriteData2 =>
147                  RS<='1'; RW<='0';
148                  DB <= input2_lcd; --central digit
149                  nx_state <= WriteData3;
150              WHEN WriteData3 =>
151                  RS<='1'; RW<='0';
152                  DB <= "00101110"; --point
153                  nx_state <= WriteData4;
154              WHEN WriteData4 =>
155                  RS<='1'; RW<='0';
156                  DB <= input1_lcd; --right digit
157                  nx_state <= ReturnHome;
158              WHEN ReturnHome =>
159                  RS<='0'; RW<='0';
160                  DB <= "10000000";
161                  nx_state <= WriteData1;
162          END CASE;
163      END PROCESS;
164  END frequencimeter;
165  ------------------------------------------------------------------
```

Testing the Frequency Meter

A simple way of testing the code above is by creating an input signal derived from the system clock itself. To do so, comment out line 14 and uncomment line 26, causing *input* to be an internal signal. Now add the code below to your code (for example, between lines 81–82). Since in our example $fclk = 50$ MHz, any test parameter *test_par* > 250 can create a useful test signal (because 50MHz/(2*250) = 100 kHz). For example, with *test* = 2041, the resulting input frequency is $fclk/(2*2041) = 12.249$ kHz, while *test* = 2040 gives $fclk/(2*2040) = 12.255$ kHz; consequently, after rounding, the former must produce 12.2 kHz on the display, while the latter must produce 12.3 kHz. The reader is now invited to test this code in the FPGA board. If the DE2 board is used, then the FPGA pins listed on the right-hand part of the table in figure 12.5 can be used.

```
-----Test signal:-----------------------
PROCESS (clk)
   VARIABLE test: INTEGER RANGE 0 TO fclk;
BEGIN
   IF (clk'EVENT AND clk='0') THEN
```

```
        test := test + 1;
        IF (test=2040) THEN --or 2041
            input <= NOT input;
            test := 0;
        END IF;
    END IF;
END PROCESS;
------------------------------------------
```

Observe that the test code above does not bother with the fact that during one clock period after every 50M clock periods the counter is inactive (while *one_sec* = '1'). This can be a problem when generating small frequencies because one of the pulses might fall occasionally in that time interval, so the actual number of pulses in the dark area of figure 12.7b is one less than the expected number. In summary, when generating low frequencies, particularly around the smallest sensitive value, which is 50 Hz (because for 49 Hz the output must be 00.0 KHz, while for 50 Hz it has to be 0.01 kHz), one must know the exact number of pulses being generated between two consecutive *one_sec* pulses in order to conduct a proper test. The generation of such a precise signal can be easily achieved with some modifications in the above test code (exercise 12.11).

Including a Synchronizer in the Design

The design shown in this section has two circuits operating with different clocks: *input* is the clock for the counter that counts the input pulses, while the system clock, *clk*, produces the signal *one_sec* responsible for storing the output of the counter into a register for display exhibition.

Because these signals are *asynchronous* with respect to each other, the output of the counter may change during the *evaluation window* of the output register (that is, during the setup plus hold time interval of its flip-flops), in which case there is no guarantee that the correct value will be stored by the register. For example, a *metastable* state (somewhere between '0' and '1') can occur.

The operation of a DFF is illustrated in figure 12.7c. Observe that the data must remain stable during t_{setup} seconds before and t_{hold} seconds after the clock edge to guarantee that the correct value will be stored.

To (practically) eliminate the occurrence of metastable states, a *synchronizer* is normally employed, which provides a means for synchronization between the two circuits. Note that in the present example, even though we want to prevent the counter output from changing its value while that value is being evaluated (stored) by the output register, that point (interface between the counter and the register) is not the only point where the synchronizer can be inserted.

A popular synchronizer implementation is shown in figure 12.7d, which consists simply of a two-stage shift register. A metastable state can only occur in the first stage; if its dura-

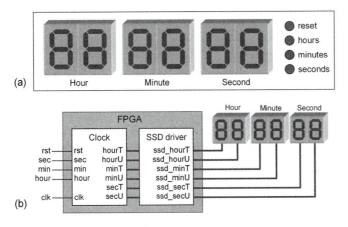

Figure 12.8
(a) Digital clock designed in section 12.5 (with six SSD digits and four pushbuttons); (b) Circuit diagram.

tion is less than the system clock period, a valid value ('0' or '1') will then be passed ahead by the second flip-flop.

For more on this, please refer to exercise 12.12. And for another alternative to reduce errors during the flip-flops' evaluation window, which employs a Gray counter, please refer to exercise 12.13.

12.5   Digital Clock (with SSDs)

Figure 12.8a shows a digital clock that displays hours, minutes, and seconds. The purpose of this section is to design that clock, using SSDs as displays, and four control pushbuttons as follows.

Reset: When asserted, must zero the display, with precedence over any other button.

Seconds: When asserted, must increase the speed of the counter by a factor of 8.

Minutes: When asserted, must increase the speed of the counter by a factor of 256.

Hours: When asserted, must increase the speed of the counter by a factor of 8,192.

Note that powers of two (shifts) were chosen to optimize the amount of hardware. It will be assumed that the clock frequency is 50 MHz, entered using GENERIC, so changing it is easy. The SSDs will be assumed to be of common-anode type (figure 12.2).

In figure 12.8b, a diagram, with the circuit broken into two sections (clock and SSD driver), is suggested, which also includes signal names.

A corresponding VHDL code is shown below. To avoid writing the code for the SSD driver six times (one for each digit), a function named *integer_to_ssd* was constructed (lines

31–48). However, unlike a previous design, this time the function was located in the main code (in the declarative part of the architecture) instead of in a package. The project's name is *clock_with_ssds* (line 5) and *fclk* was entered using a GENERIC declaration (line 6). The input and output signals, obeying the names in figure 12.8b, are declared in lines 8–18. The code proper (architecture) is in lines 21–125, with the clock designed in the first part (lines 50–117) and the SSD driver, using the function mentioned earlier, in the second part (lines 118–124). The reader is again invited to test the design in the FPGA board.

```
1    ------------------------------------------------------------------------
2    LIBRARY ieee;
3    USE ieee.std_logic_1164.all;
4    ------------------------------------------------------------------------
5    ENTITY clock_with_ssds IS
6       GENERIC (fclk: INTEGER := 50_000_000); --50MHz clock
7       PORT (
8          clk: IN STD_LOGIC;
9          rst: IN STD_LOGIC;
10         sec: IN STD_LOGIC;                   --fast adjustment for seconds
11         min: IN STD_LOGIC;                   --fast adjustment for minutes
12         hour: IN STD_LOGIC;                  --fast adjustment for hours
13         ssd_secU: OUT STD_LOGIC_VECTOR(6 DOWNTO 0);   --units of seconds
14         ssd_secT: OUT STD_LOGIC_VECTOR(6 DOWNTO 0);   --tens of seconds
15         ssd_minU: OUT STD_LOGIC_VECTOR(6 DOWNTO 0);   --units of minutes
16         ssd_minT: OUT STD_LOGIC_VECTOR(6 DOWNTO 0);   --tens of minutes
17         ssd_hourU: OUT STD_LOGIC_VECTOR(6 DOWNTO 0);  --units of hours
18         ssd_hourT: OUT STD_LOGIC_VECTOR(6 DOWNTO 0)); --tens of hours
19   END clock_with_ssds;
20   ------------------------------------------------------------------------
21   ARCHITECTURE clock_with_ssds OF clock_with_ssds IS
22      ---Signals needed because variables cannot be passed to functions
23      SIGNAL secUnits: NATURAL RANGE 0 TO 10;
24      SIGNAL secTens: NATURAL RANGE 0 TO 6;
25      SIGNAL minUnits: NATURAL RANGE 0 TO 10;
26      SIGNAL minTens: NATURAL RANGE 0 TO 6;
27      SIGNAL hourUnits: NATURAL RANGE 0 TO 10;
28      SIGNAL hourTens: NATURAL RANGE 0 TO 3;
29      SIGNAL limit: INTEGER RANGE 0 TO fclk;
30      ---Function for SSD driver:
31      FUNCTION integer_to_ssd (SIGNAL input: NATURAL) RETURN STD_LOGIC_VECTOR
32         IS VARIABLE output: STD_LOGIC_VECTOR(6 DOWNTO 0);
33      BEGIN
34         CASE input IS
35            WHEN 0 => output:="0000001";       --"0" on SSD
```

```
36              WHEN 1 => output:="1001111";        --"1" on SSD
37              WHEN 2 => output:="0010010";        --"2" on SSD
38              WHEN 3 => output:="0000110";        --"3" on SSD
39              WHEN 4 => output:="1001100";        --"4" on SSD
40              WHEN 5 => output:="0100100";        --"5" on SSD
41              WHEN 6 => output:="0100000";        --"6" on SSD
42              WHEN 7 => output:="0001111";        --"7" on SSD
43              WHEN 8 => output:="0000000";        --"8" on SSD
44              WHEN 9 => output:="0000100";        --"9" on SSD
45              WHEN OTHERS => output:="0110000";   --"E" on SSD
46          END CASE;
47          RETURN output;
48      END integer_to_ssd;
49  BEGIN
50      --------Part 1: Clock----------------------
51      ---Speed-up factors:
52      limit <= fclk/8192 WHEN hour='1' ELSE
53              fclk/256 WHEN min='1' ELSE
54              fclk/8 WHEN sec='1' ELSE
55              fclk;
56      ---Clock design:
57      PROCESS (clk, rst)
58          VARIABLE one_sec: NATURAL RANGE 0 TO fclk;
59          VARIABLE secU: NATURAL RANGE 0 TO 10;
60          VARIABLE secT: NATURAL RANGE 0 TO 6;
71          VARIABLE minU: NATURAL RANGE 0 TO 10;
72          VARIABLE minT: NATURAL RANGE 0 TO 6;
73          VARIABLE hourU: NATURAL RANGE 0 TO 10;
74          VARIABLE hourT: NATURAL RANGE 0 TO 3;
75      BEGIN
76          IF (rst='1') THEN
77              one_sec := 0; secU := 0; secT := 0;
78              minU := 0; minT := 0;
79              hourU := 0; hourT := 0;
80          ELSIF (clk'EVENT AND clk='1') THEN
81              one_sec := one_sec + 1;
82              IF (one_sec=limit) THEN
83                  one_sec := 0;
84                  secU := secU + 1;
85                  IF (secU=10) THEN
86                      secU := 0;
87                      secT := secT + 1;
88                      IF (secT=6) THEN
```

```
89                    secT := 0;
90                    minU := minU + 1;
91                    IF (minU=10) THEN
92                       minU := 0;
93                       minT := minT + 1;
94                       IF (minT=6) THEN
95                          minT := 0;
96                          hourU := hourU + 1;
97                          IF ((hourT/=2 AND hourU=10) OR
98                              (hourT=2 AND hourU=4)) THEN
99                             hourU := 0;
100                            hourT := hourT + 1;
101                            IF (hourT=3) THEN
102                               hourT := 0;
103                            END IF;
104                         END IF;
105                      END IF;
106                   END IF;
107                END IF;
108             END IF;
109          END IF;
110       END IF;
111       secUnits <= secU;
112       secTens <= secT;
113       minUnits <= minU;
114       minTens <= minT;
115       hourUnits <= hourU;
116       hourTens <= hourT;
117    END PROCESS;
118    --------Part 2: SSD driver----------------
119    ssd_secU <= integer_to_ssd(secUnits);
120    ssd_secT <= integer_to_ssd(secTens);
121    ssd_minU <= integer_to_ssd(minUnits);
122    ssd_minT <= integer_to_ssd(minTens);
123    ssd_hourU <= integer_to_ssd(hourUnits);
124    ssd_hourT <= integer_to_ssd(hourTens);
125 END clock_with_ssds;
126 --------------------------------------------------------------------------
```

12.6   Quick-Finger Game (with LEDs and SSDs)

Figure 12.9a shows the top-level diagram for the quick-finger game to be designed in this section. Its inputs are four pushbutton-type switches (*sw3*, . . . , *sw0*) that will be activated

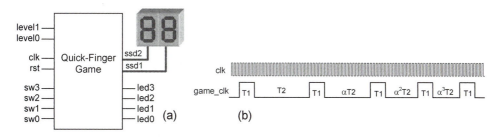

Figure 12.9
(a) Top-level diagram for the quick-finger game; (b) Geometric game clock.

by the player and two toggle-type switches (*level1*, *level0*) to set the game's level of difficulty, plus *clk* and *rst*. The outputs are a set of four LEDs (*led3*, ..., *led0*), plus a pair of SSDs to display the score.

Figure 12.9b depicts the clock waveform needed for the game (*game_clk*), which stays high during a fixed amount of time *T1* and low during a geometrically decreasing time interval $\alpha^i T2$ ($\alpha < 1$, $i = 0, 1, 2, ...$). The player must press, within the time interval $\alpha^i T2$, the switch corresponding to the LED that has been lit during the preceding time interval *T1*. This obviously becomes harder as the game progresses, and the player will win when 15 consecutive correct key presses occur. If the player makes an incorrect choice or does not make any choice within the given time interval, the game must end, so all LEDs must be turned ON indefinitely and the final score must be kept by the SSDs. The same must happen when the player wins the game (score = 15), though having the LEDs blinking in this case would be more interesting (exercise 12.6).

The level of difficulty increases when a smaller α is chosen (*level* switches). The values to be adopted should be such that the reduction of *T2* from one interval to the next falls approximately in the 3% to 10% range. This geometric clock has the advantage of reducing the time interval more in the beginning and less at the end, but a linear clock could also be used, in which case the time reduction from one time interval to the next would be constant (exercise 12.7).

A VHDL code for this circuit is shown below, under the name *quick_finger* (line 2). The clock frequency (*fclk*, line 3) and the desired value for *T1* (0.2s, line 4) were specified using GENERIC so they can be easily changed. The pin names (lines 5–9) are based on figure 12.9a. Note that library declarations were not needed because only data types defined in the package *standard* were employed (visible by default).

The code proper is in lines 15–140, and is organized in four sections. The first section (lines 17–49) generates the waveform *game_clk* depicted in figure 12.9b. Note that $\alpha = factor/128$ (*factor* is defined in lines 23–28, while the denominator, 128, is in the equation of line 40). The way this expression is organized is important for two reasons. First, 128 is a power of two, so only shifting is needed to produce this division (inexpensive hardware);

second, recall that *factor**(time_low*/128) is different from (*factor**time_low*)/128, because we are dealing with integers. Note also that the values of *factor* are never a multiple of four (the "random" counter has four states, needed to feed the four LEDs). With the values given in lines 24–27, the reduction rates are $1 - 123/128 = 0.039$ (3.9%), 5.5%, 7%, and 8.6%.

The second portion is in lines 51–77. It is responsible for obtaining a "random" number between 0 and 3 to feed the LEDs, then storing that value during a whole period of *game_clk*. Such a number is produced by the free-running (no reset) counter in lines 55–57. Because *fclk* = 50 MHz, a whole counter cycle takes only 80 ns. Given that the human finger takes at least a few milliseconds to react, and considering that the results of trying to turn the reset switch ON and OFF with the same time duration is Gaussian, its standard deviation is expected to be also at least in the ms range. Therefore, within a 80 ns interval about the mean, the probability distribution is approximately constant, so the chance of getting any state when the reset switch is turned OFF is approximately uniform ($\sim$25%).

There is a problem though, because the randomness discussed above only applies to the *starting* value. Since the sampling is done in a predetermined fashion (at the rising edge of *game_clk*—see line 59), the sequence that follows is deterministic. To avoid having it repeat itself after only a few draws, it is important that the reduction factor of *T2* be not a multiple of 4, which is the reason why only odd values were chosen in lines 24–27, resulting in a randomness just good enough for the present experiment (more sophisticated techniques can include the use of a linear feedback shift register [Pedroni 2008] and other random values, like the duration of the reset signal).

The third part, responsible for feeding the LEDs, is in lines 79–88. It simply turns all LEDs ON when the game is over, turns just one LED ON according to the random number drawn during *T1* = 0.2 s (observation time), or turns all of them OFF while *game_clk* is low (betting time).

The final part, in lines 90–139, is responsible for reading the pushbutton pressed by the player and displaying the result. It checks for a hit (lines 95–103), updates the score (lines 105–117), and finally sends the score to the SSDs (lines 119–138). As with all previous designs, the reader is now invited to test this code in the FPGA board.

```
1   ------------------------------------------------------------
2   ENTITY quick_finger IS
3     GENERIC (fclk: POSITIVE := 50_000_000;        --50MHz clock
4              time_high: POSITIVE := 10_000_000); --T1=0.2s
5     PORT (clk, rst: IN BIT;
6           level: IN BIT_VECTOR(1 DOWNTO 0);
7           switches: IN BIT_VECTOR(3 DOWNTO 0);
8           leds: OUT BIT_VECTOR(3 DOWNTO 0);
9           ssd1, ssd2: OUT BIT_VECTOR(6 DOWNTO 0));
10  END quick_finger;
```

```
11   ------------------------------------------------------------
12   ARCHITECTURE quick_finger OF quick_finger IS
13      SIGNAL game_clk, game_over: BIT;
14      SIGNAL leds_memory: BIT_VECTOR(3 DOWNTO 0);
15   BEGIN
16      ---------Generate game clock:---------------
17      PROCESS (clk, rst, level)
18         VARIABLE count: INTEGER RANGE 0 TO fclk;
19         VARIABLE time_low: INTEGER RANGE 0 TO fclk;
20         VARIABLE factor: INTEGER RANGE 0 TO 128;
21      BEGIN
22         --Specify difficulty factor:
23         CASE level IS
24            WHEN "00" => factor:=123;     --3.9%
25            WHEN "01" => factor:=121;     --5.5%
26            WHEN "10" => factor:=119;     --7%
27            WHEN OTHERS => factor:=117;   --8.6%
28         END CASE;
29         --Generate game_clk:
30         IF (rst='1') THEN
31            count := 0;
32            game_clk <= '0';
33            time_low := fclk;
34         ELSIF (clk'EVENT AND clk='1') THEN
35            count := count + 1;
36            IF (game_clk='0') THEN
37               IF (count=time_low) THEN
38                  count := 0;
39                  game_clk <= '1';
40                  time_low := factor*(time_low/128);
41               END IF;
42            ELSE
43               IF (count=time_high) THEN
44                  count := 0;
45                  game_clk <= '0';
46               END IF;
47            END IF;
48         END IF;
49      END PROCESS;
50      ---------Get random value:------------------
51      PROCESS (clk, rst, game_clk)
52         VARIABLE random: INTEGER RANGE 0 TO 3;
53      BEGIN
```

```
54          --Get value (no reset here):
55          IF (clk'EVENT AND clk='0') THEN
56              random := random + 1;
57          END IF;
58          --Store value:
59          IF (game_clk'EVENT AND game_clk='1') THEN
60              CASE random IS
71                  WHEN 0 => leds_memory <= "0001";
72                  WHEN 1 => leds_memory <= "0010";
73                  WHEN 2 => leds_memory <= "0100";
74                  WHEN OTHERS => leds_memory <= "1000";
75              END CASE;
76          END IF;
77      END PROCESS;
78      ---------Activate LEDs:--------------------
79      PROCESS (game_clk, game_over, leds_memory)
80      BEGIN
81          IF (game_over='1') THEN
82              leds <= "1111";
83          ELSIF (game_clk='0') THEN
84              leds <= "0000";
85          ELSE
86              leds <= leds_memory;
87          END IF;
88      END PROCESS;
89      -----Read pushbuttons & display result:-----
90      PROCESS (clk, rst, game_clk)
91          VARIABLE count: INTEGER RANGE 0 TO 16;
92          VARIABLE hit: BIT;
93      BEGIN
94          --Check for a hit:
95          IF (rst='1') THEN
96              hit := '0';
97          ELSIF (clk'EVENT AND clk='0') THEN --inverted switches:
98              IF (game_clk='0' AND switches=NOT leds_memory) THEN
99                  hit := '1';
100             ELSIF (game_clk='1') THEN
101                 hit := '0';
102             END IF;
103         END IF;
104         --Increase score or end game:
105         IF (rst='1') THEN
106             count := 0;
```

```
107              game_over <= '0';
108         ELSIF (game_clk'EVENT AND game_clk='1') THEN
109            IF (hit='1') THEN
110               count := count + 1;
111               IF (count=16) THEN
112                  game_over <= '1';
113               END IF;
114            ELSE
115               game_over <= '1';
116            END IF;
117         END IF;
118         --Display result (SSD driver):
119         CASE count IS
120            WHEN 0 => ssd2<="0000001"; ssd1<="0000001";    --"00"
121            WHEN 1 => ssd2<="0000001"; ssd1<="0000001";    --"00"
122            WHEN 2 => ssd2<="0000001"; ssd1<="1001111";    --"01"
123            WHEN 3 => ssd2<="0000001"; ssd1<="0010010";    --"02"
124            WHEN 4 => ssd2<="0000001"; ssd1<="0000110";    --"03"
125            WHEN 5 => ssd2<="0000001"; ssd1<="1001100";    --"04"
126            WHEN 6 => ssd2<="0000001"; ssd1<="0100100";    --"05"
127            WHEN 7 => ssd2<="0000001"; ssd1<="0100000";    --"06"
128            WHEN 8 => ssd2<="0000001"; ssd1<="0001111";    --"07"
129            WHEN 9 => ssd2<="0000001"; ssd1<="0000000";    --"08"
130            WHEN 10 => ssd2<="0000001"; ssd1<="0000100";   --"09"
131            WHEN 11 => ssd2<="1001111"; ssd1<="0000001";   --"10"
132            WHEN 12 => ssd2<="1001111"; ssd1<="1001111";   --"11"
133            WHEN 13 => ssd2<="1001111"; ssd1<="0010010";   --"12"
134            WHEN 14 => ssd2<="1001111"; ssd1<="0000110";   --"13"
135            WHEN 15 => ssd2<="1001111"; ssd1<="1001100";   --"14"
136            WHEN 16 => ssd2<="1001111"; ssd1<="0100100";   --"15"
137            WHEN OTHERS=> ssd2<="0110000"; ssd1<="0110000";--"EE"
138         END CASE;
139      END PROCESS;
140  END quick_finger;
141  -------------------------------------------------------------
```

12.7   Other Designs with Basic Displays

See also the following designs using basic displays in other chapters:

Car speed monitor (exercise 11.19).

I^2C interface for an EEPROM memory (section 14.4.3).

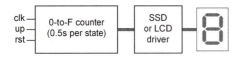

Figure 12.10

PS2 keyboard interface (section 14.3 and exercises 14.8).

Pushbutton sequence detector (example 11.8 and exercise 11.18).

SPI interface for a FRAM memory (section 14.5).

Timer (example 6.6 and exercise 6.10).

Traffic-light controller (example 11.7 and exercises 11.16 and 11.17).

Zero-to-nine counter (example 6.6 and exercise 11.4).

12.8   Exercises

Exercise 12.1: Counter with SSD Display

Design the 0-to-F counter depicted in figure 12.10 using an SSD to display the result. It must count upward when $up = \text{'1'}$ or downward otherwise, remaining in each state for 0.5 seconds and stopping when F or 0 is reached. If *rst* is activated, then it must return to zero (if counting upward) or F (if counting downward) and resume counting from there (when *rst* is deactivated, of course).

Exercise 12.2: Counter with LCD Display

Repeat the design in exercise 12.1, this time with an LCD displaying the result.

Exercise 12.3 Playing with a Seven-Segment Display

Consider the design developed in section 12.3. Redesign it with the three pushbuttons as shown in figure 12.11, which add the following features to the circuit:

1) The *stop* button, when pressed, must cause the movement to stop if running, or resume running if stopped. If pressed for longer than 2 s, it must reset the system (that is, return to state *a*).

2) The *dir* button must reverse the direction of the movement (counterclockwise if going clockwise or vice versa) every time it is pressed.

3) The *speed* button must change the speed of the movement every time it is pressed, with the following sequence of four values for *T1* (*T2* must remain at 40ms): 80ms → 140ms → 200ms → 260ms → 80ms . . . .

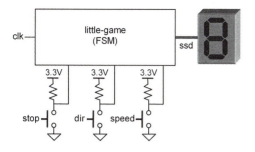

Figure 12.11

Figure 12.12

Exercise 12.4: Frequency Meter (with SSDs)

Modify the design of section 12.4 for the circuit to operate with SSDs instead of an LCD.

Exercise 12.5: Digital Clock (with LCD)

Modify the design of section 12.5 for the circuit to operate with an LCD instead of SSDs. Include colons to separate the pairs of digits, as shown in figure 12.12.

Exercise 12.6: Quick-Finger Game with Blinking LEDs

Modify the design of section 12.6 such that the LEDs, instead of being just ON, blink with a frequency of 2 Hz when the player wins the game.

Exercise 12.7: Quick-Finger Game with Linear Time Reduction

Modify the design of section 12.6 such that the game clock (*game_clk*) has *T2* reduced linearly instead of geometrically (keep *T1* = 0.2 s). Consider the initial value as *T2* = 1 s and adopt the following reduction values: *level* = "00" → 25ms, "01" → 35ms, "10" → 45ms, and "11" → 55ms.

Exercise 12.8: Timer (with LED and SSDs)

Design the timer of figure 12.13, with the following features:

1) The circuit must operate with two SSD displays, which display seconds in the 00 to 60 range.

Figure 12.13

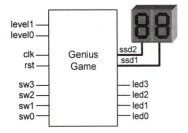

Figure 12.14

2) The timer must count downward, from an adjustable initial value (*max* = 60) down to 00.

3) A pushbutton (*time*) must be included to adjust the starting value, which should occur in steps of one second every time the button is pressed. If the pushbutton remains pressed for a long time, it should not be interpreted as more than one command.

4) Another pushbutton (*stop*) must cause the timer to stop when pressed (and released). If pressed again, the timer should resume counting.

5) If pressed for longer than 2 seconds, *stop* must reset the counter (00).

6) Finally, an LED must be included, which remains lit while the timer is down counting. If the timer is stopped, the LED must be turned OFF, returning to the ON state when the timer resumes counting. When the timer reaches its final state (00), the LED must blink with a frequency of 2 Hz. Only after the reset button is pressed the LED should be turned OFF.

Exercise 12.9: Timer (with LED and LCD)

Repeat the design in exercise 12.8, this time with an LCD instead of SSDs.

Exercise 12.10: Genius Game (with LEDs and SSDs)

The purpose of this exercise is to design the well-known *genius game*. The top-level diagram is shown in figure 12.14, containing at the input four pushbutton-type switches (*sw3*, . . . , *sw0*) that will be activated by the player and two toggle-type switches (*level1*, *level0*) to set the game's level of difficulty, plus *clk* and *rst*. The outputs are a set of four LEDs (*led3*,..., *led0*), plus a pair of SSDs to display the score. The game consists of the following: a random number (between 0 and 3) is drawn and the corresponding LED is lit momentarily, after which the player must press the corresponding pushbutton. If the result is correct, the score is incremented and a new round starts, which consists of lighting the same LED, followed by a new one (repetitions are allowed), so the player is required to make two correct key presses this time. The game ends when the player either makes a mistake or reaches a predefined sequence size (say, 12). Here are some suggestions for the timing: light each LED for 0.3s; the time separation between LEDs must be set by the *level* switches, say "00" → 0.8s, "01" → 0.7s, "10" → 0.6s, and "11" → 0.5s. After the sequence of LEDs is finished, allow a maximum of about 2s between pushbutton depressions.

Exercise 12.11: Frequency Meter Testing Procedure

For the frequency meter designed in section 12.4 or for that in exercise 12.4, do the following:

a) Add to the original code the test code shown at the end of section 12.4, and check whether the results for the two values of *test* (2040 and 2041) match those presented in the text.

b) Modify the test code as needed in order to test the circuit with frequencies as close as possible to 49 Hz (00.0 kHz on the display) and 50 Hz (00.1 kHz on the display).

c) Further test the circuit by including a frequency where the increment of the other digits can also be verified (for example, for the following pair of points: 00.9 and 01.0; 09.9 and 10.0; 89.9 and 90.0).

Exercise 12.12: Frequency Meter with a Synchronizer

For the frequency meter designed in section 12.4 or for that in exercise 12.4, do the following:

a) Draw a more detailed block diagram (as in figure 22.9a of Pedroni [2008]) and briefly explain how it works.

b) In which points of this circuit can a synchronizer be included? Which one requires fewer flip-flops?

c) Draw the new circuit, with the synchronizer included.

d) Design this new circuit using VHDL and repeat the tests of exercise 12.11 to check if the same results are obtained.

Exercise 12.13: Frequency Meter with Gray Counter

The frequency meter of section 12.4 (or of exercise 12.4) operates with two asynchronous signals (*input* and *clk*). One way of preventing the output register from storing an incorrect value was shown in section 12.4 and further discussed in exercise 12.12. Another way is to use a Gray counter (that is, a counter whose output value follows the Gray code) instead of a regular sequential counter.

a) Explain why a Gray counter can be helpful (Hint: Think of how many bits change at a time.)

b) Is this solution less expensive (in terms of hardware) than that with the synchronizer?

13 VHDL Design of Memory Circuits

13.1 Introduction

Figure 13.1 shows the most basic ROM (read-only memory) and RAM (random access memory) configurations. The ROM has as input the signal *address* (*M* bits) and as output the signal *data_out* (*N* bits), with the latter exhibiting the contents in the specified memory address. Contrary to a ROM, the contents of a RAM can be modified freely, so the RAM has two additional inputs, called *data_in* (data to be stored in the memory) and *we* (write enable); the latter, when asserted, causes *data_in* to be stored in the specified address. This RAM arrangement is called *read-on-write*, because during the writing process the output reads the value that is being stored in the memory. Both circuits in figure 13.1 have $depth = 2^M$ (number of words) and $width = N$ (number of bits in each word).

Several variations of the architectures above exist. For example, in high-performance systems, memories are normally *synchronous* to allow the system clock to control their operation. This means that the memory inputs and/or outputs are *registered* (that is, the address and/or the data buses are stored by flip-flops).

An aspect that is sometimes relevant when writing VHDL code to implement memory in CPLD/FPGA devices is the *location* where such memories will be built. Because FPGAs normally have built-in user SRAM blocks, ROMs and RAMs can be implemented in them. But because logic cells and LUTs (lookup tables) can also emulate memory, that is another possible location for such memories. Consequently, if one wants, for example, to force the compiler to use the SRAM blocks, special synthesis attributes or special functions must be used. Moreover, when using such built-in SRAM blocks, one must remember that they usually already have flip-flops at all inputs and, optionally, also at the outputs, so no additional flip-flops are needed to turn the memory into a synchronous one (just make sure that the proper setup is chosen when instantiating these blocks). On the other hand, when using CPLDs, such user SRAM blocks normally do not exist, so regular logic cells will be used to implement the desired memory functionality.

In the sections that follow, several of the cases described above will be examined and designed. Additionally, the design of memory interfaces, needed when the memory is not located inside the FPGA but rather in an external memory chip, will also be discussed.

Figure 13.1
Basic asynchronous ROM and RAM configurations.

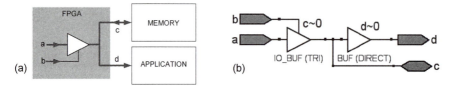

Figure 13.2
(a) System with a bidirectional data bus; (b) Synthesis result.

13.2 Implementing Bidirectional Buses

Before we start discussing memory implementations, let us see how the INOUT mode works (recall from section 2.4 that the VHDL modes are IN, OUT, BUFFER, and INOUT). The reason for this analysis is that many memories operate with a single data bus, which must allow data to be entered (written) into the memory as well as data to be retrieved (read) from the memory, being therefore bidirectional.

As an example, consider the situation depicted in figure 13.2a, which shows an FPGA providing interface between an external memory and an application circuit that uses data stored in that memory (only the data bus is shown; the address bus and the control signals are not relevant for the present discussion). Because this memory is assumed to have just one data bus (signal c), such a bus has to be bidirectional. The FPGA must be able to store data in the memory and also to retrieve data from it, passing the data to the application circuit (signal d). A tristate buffer, with input a and control b, is also employed in these operations.

A VHDL code for the bidirectional part of the memory interface above is shown below. Note in the entity (lines 5–9) that the inputs are a and b, while c is bidirectional and d is an output, all of type STD_LOGIC (single-bit was assumed for simplicity).

Two equivalent architectures are presented. The first (*arch1*, lines 11–15) employs concurrent code (WHEN statement), while the second (*arch2*, lines 17–26) uses sequential code (IF statement, in a process). Recall from section 8.6 that to compile this code a CONFIGURATION declaration is needed or one of the architectures must be commented out.

```
1   -----------------------------
2   LIBRARY ieee;
3   USE ieee.std_logic_1164.all;
4   -----------------------------
5   ENTITY bidir IS
6      PORT (a, b: IN STD_LOGIC;
7            c: INOUT STD_LOGIC;
8            d: OUT STD_LOGIC);
9   END ENTITY;
10  -----------------------------
11  ARCHITECTURE arch1 OF bidir IS
12  BEGIN
13     c <= a WHEN b='1' ELSE 'Z';
14     d <= c;
15  END ARCHITECTURE;
16  -----------------------------
17  ARCHITECTURE arch2 OF bidir IS
18  BEGIN
19     PROCESS (a, b)
20     BEGIN
21        d <= c;
22        IF (b='1') THEN c <= a;
23        ELSE c <= 'Z';
24        END IF;
25     END PROCESS;
26  END ARCHITECTURE;
27  -----------------------------
```

The crucial points to be observed in either architecture are:

1) *d* is always equal to *c* (line 14 or 21);

2) *c* is equal to *a* when sending data to the memory, or equal to 'Z' when receiving data from it (line 13 or 22–23).

The RTL view produced by the compiler (from either code) is shown in figure 13.2b, which resembles the circuit of figure 13.2a. The points mentioned will be useful later, when designing interfaces for single data bus memories.

13.3 Memory Initialization Files

Another important topic to be examined before we start discussing memory implementations is *memory initialization files*. The following three file formats are described in this section:

```
%----------------%   %----------------%   %---------------%   %---------------%
WIDTH=8;             WIDTH=8;             WIDTH=8;            WIDTH=8;
DEPTH=16;            DEPTH=16;            DEPTH=16;           DEPTH=16;
ADDRESS_RADIX=UNS;   ADDRESS_RADIX=UNS;   ADDRESS_RADIX=UNS;  ADDRESS_RADIX=HEX;
DATA_RADIX=UNS;      DATA_RADIX=BIN;      DATA_RADIX=HEX;     DATA_RADIX=HEX;
CONTENT BEGIN        CONTENT BEGIN        CONTENT BEGIN       CONTENT BEGIN
0 : 0;               0 : 00000000;        0 : 00;            [0..F] :00;
1 : 0;               1 : 00000000;        1 : 00;            2 : FF;
2 : 255;             2 : 11111111;        2 : FF;            3 : 1A;
3 : 26;              3 : 00011010;        3 : 1A;            4 : 05;
4 : 5;               4 : 00000101;        4 : 05;            5 : 50;
5 : 80;              5 : 01010000;        5 : 50;            6 : B0;
6 : 176;             6 : 10110000;        6 : B0;            F : 11;
[7..14] :0;          [7..14] : 00000000;  [7..14] :00;       END;
15 : 17;             15: 00010001;        15 : 11;           %---------------%
END;                 END;                 END;
%---------------%    %---------------%    %---------------%
```

Figure 13.3
Four equivalent representations for a MIF file.

- MIF (memory initialization file), from Altera.
- COE (derived from CGF—core generator file), from Xilinx.
- HEX (Intel hexadecimal file), with general support.

Another standard file format is RIF (RAM initialization file), used in other vendors' EDA software.

MIF File

MIF is a file format that can be used to initialize ROM, RAM, and CAM contents in Altera devices. Four equivalent examples are presented in figure 13.3. The file starts with a declaration regarding the *width* (number of bits) of the stored words, followed by the memory *depth* (number of words). The address radix can be binary (BIN), octal (OCT), hexadecimal (HEX), or unsigned decimal (UNS), while the memory radix can also be signed decimal (DEC). The default for both is HEX. "%" is used for comments. Note that in the last case of figure 13.3 the entire memory was initialized to 00h, then some of the values were overwritten.

COE File

COE is a Xilinx file format whose main purpose is the same as Altera's MIF format (memory initialization). To create a COE file, a CGF (core generator file) must be entered by the user in the memory editor, from which ISE automatically creates the corresponding COE file. Multiple memory blocks can be entered in a single CGF, from which multiple COEs result (one for each memory block).

An example of CGF is shown below (called *example.cgf*). The block depth can be any value in the 1 to 1,048,575 range (default = 256). The default for the data width is 16. The pad bit (used to fill the provided words when they are shorter than the specified data width) can be '0' (default) or '1'. The padding direction can be left (default) or right. The data radix can be 2, 8, 10 signed, 10 unsigned, or 16 (default). Finally, the address radix can be 2, 8, 10 (default), or 16. The symbol "@" indicates the initial address of a memory segment. All unspecified words are automatically filled with the default word.

```
-------------------------------------
#CGF file "example.cgf"
#memory_block_name=block1
#block_depth=8
#data_width=8
#default_word=0
#default_pad_bit_value=0
#pad_direction=left
#data_radix=2
#address_radix=10
#coe_radix=MEMORY_INITIALIZATION_RADIX
#coe_data=MEMORY_INITIALIZATION_VECTOR
#data=
@0
18
3
@5
11
255
#end
-------------------------------------
```

The COE file (*block1.coe*) generated from this CGF file is shown below. For example, at address = 0, the data is 18, which is "10010" in binary form, resulting in "00010010" after left padding. Likewise, at address = 1, the data is 3, which is "11" in binary form, resulting in "00000011" after left padding, and so on.

```
---------------------------------------------------
#COE file "block1.coe" generated from "example.cgf"
MEMORY_INITIALIZATION_RADIX=2;
MEMORY_INITIALIZATION_VECTOR=
00010010,
00000011,
00000000,
```

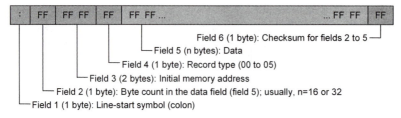

Figure 13.4
Line structure in HEX files.

```
00000000,
00000000,
00001011,
11111111,
00000000;
-----------------------------------------------------
```

HEX File

HEX files (Intel format) can also be used for memory initialization. As figure 13.4 shows, each data line is divided into six fields, which contain the following:

Field 1 (1 byte): Always contains a colon, indicating the beginning of a line.

Field 2 (1 byte): Number of (data) bytes in field 5. Usually, $n = 16$ or 32 bytes.

Field 3 (2 bytes): Initial memory address.

Field 4 (1 byte): Record type, with the following options:

00 = Data line (usual case). A regular data line then is ": xx xxxx 00 xx...xx".

01 = End-of-file (indicates that it is the last line). The last line then is ": 00 0000 01 FF".

02 to 05 = Additional addressing modes.

Field 5 (n bytes): Data field (usually, $n = 16$ or 32).

Field 6 (1 byte): Check sum, obtained by adding all hexa values in fields 2 to 5, then taking the two's complement of this sum's least significant byte.

An example is shown below whose first line contains the same data as the MIF file in figure 13.3. Because the sum of fields 2 to 5 is 23F, C1 (two's complement of 3F, which is the sum's LSByte) was entered as the check sum (field 6). The last line is always ": 00 0000 01 FF".

```
: 10 0000 00 00 00 FF 1A 05 50 B0 00 00 00 00 00 00 00 00 11 C1
: ...
: 00 0000 01 FF
```

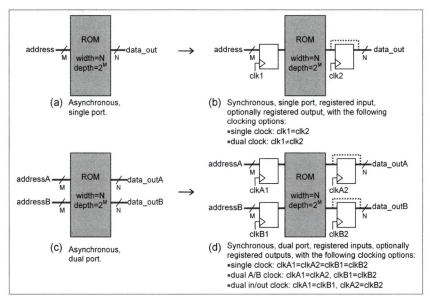

Figure 13.5
ROM configurations.

13.4 ROM Design

Figure 13.5 shows four classic ROM architectures. The circuits in (a) and (b) are similar, both being single-ported (see the note below), but while the former is asynchronous, the latter is synchronous, with registers (D-type flip-flops (DFFs)) installed at the input and also, optionally, at the output. The clocks in (b) are separated, so two clocking options are available (see caption of figure 13.5b). The circuits in (c) and (d) are also similar, both being dual-ported, but while the former is asynchronous, the latter is again synchronous, with registers at the inputs and, optionally, at the outputs. The clocks in (d) are again separated, with three clocking options shown in the figure. Several design techniques for ROMs are described in the sections that follow.

Note: There are several (confusing) ways of determining the number of ports of a memory. Two are described below.

Criterion 1 (cr1): The number of ports is equal to the number of address buses.

Criterion 2 (cr2): A pair composed of read plus write addresses is a *simple* (simple, not single) dual port, while two of such pairs is a *true* dual port.

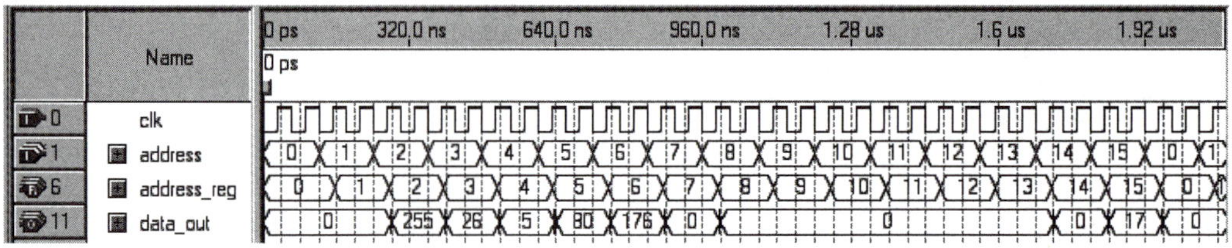

Figure 13.6
Simulation results from the ROM implemented with regular VHDL code in section 13.4.

Examples

Memory with just one address (for reading and writing): Single port (by both criteria).

Memory with one address for reading and another for writing: Dual port (cr1), simple dual port (cr2).

Memory with two addresses for reading and another for writing: Tree ports (cr1), no equivalent in cr2.

Memory with two addresses for reading and two for writing: Four ports (cr1), true dual port (cr2).

ROM Implemented with Regular VHDL Code

In this case, straight VHDL code is used. There is no concern regarding particular encoding styles to help the compiler understand that memory is wanted, as there are no synthesis attributes or special functions to force the compiler to adopt, for example, on-chip SRAM memory blocks (if available). This means that the compiler might use them or not (in general, regular logic cells will be used) and memory will be created independently from the existence or not of such blocks.

An example is shown below whose contents are those of figure 13.3, entered using CONSTANT (lines 14–21). Note that a process (lines 24–29) was used to register the address, so the inferred circuit is that in figure 13.5b, without the output register. Simulation results are depicted in figure 13.6 (compare the memory contents against those in figure 13.3).

```
1   ----------------------------------------------------------------
2   LIBRARY ieee;
3   USE ieee.std_logic_1164.all;
4   ----------------------------------------------------------------
5   ENTITY rom IS
6      PORT (clk: IN STD_LOGIC;
7            address: IN INTEGER RANGE 0 TO 15;
8            data_out: OUT STD_LOGIC_VECTOR(7 DOWNTO 0));
9   END rom;
10  ----------------------------------------------------------------
11  ARCHITECTURE rom OF rom IS
```

```
12      SIGNAL reg_address: INTEGER RANGE 0 TO 15;
13      TYPE memory IS ARRAY (0 TO 15) OF STD_LOGIC_VECTOR(7 DOWNTO 0);
14      CONSTANT myrom: memory := (
15          2 => "11111111", --255
16          3 => "00011010", --26
17          4 => "00000101", --5
18          5 => "01010000", --80
19          6 => "10110000", --176
20          15=> "00010001", --17
21          OTHERS => "00000000");
22  BEGIN
23      --Register the address:----------
24      PROCESS (clk)
25      BEGIN
26          IF (clk'EVENT AND clk='1') THEN
27              reg_address <= address;
28          END IF;
29      END PROCESS;
30      --Get unregistered output:-------
31      data_out <= myrom(reg_address);
32  END rom;
33  -------------------------------------------------------------------
```

Altera proposes, in its VHDL templates for ROMs in Quartus II, the following code (only the part needed here is shown):

```
1   ----------------------------------------------------------
2   ENTITY single_port_rom IS
3       ...
4       PORT (clk: IN STD_LOGIC;
5             addr: IN NATURAL RANGE 0 TO 2**ADDR_WIDTH - 1;
6             q: OUT STD_LOGIC_VECTOR(DATA_WIDTH -1 DOWNTO 0));
7   END single_port_rom;
8   ----------------------------------------------------------
9   ARCHITECTURE rtl OF single_port_rom IS
10  ...
11  BEGIN
12      PROCESS(clk)
13      BEGIN
14          IF (rising_edge(clk)) THEN
15              q <= rom(addr);
16          END IF;
17      END PROCESS;
18  END rtl;
19  ----------------------------------------------------------
```

Note that this code differs from that just presented with respect to the *location* of the register. The reader is invited to determine which ROM of figure 13.5 (if any) this code will implement (exercise 13.4).

ROM Implemented with an Initialization File

In the previous example, the ROM was initialized using CONSTANT, which is fine only for small memories. Data for a large memory is often stored in a file, so the VHDL code must be able to read it.

Taking Altera as an example, files of type MIF and HEX are supported without restrictions. To read them, the synthesis attribute *ram_init_file* (also called *syn_ram_init_file*) can be used. Consequently, if the file in figure 13.3 is saved under the name *rom_contents.mif*, for example, the only changes needed in the previous code are in the architecture declarations, replacing the original text (which employed CONSTANT) with that below (see also exercise 13.5).

```
------------------------------------------------------------------
ARCHITECTURE rom OF rom IS
   SIGNAL reg_address: INTEGER RANGE 0 TO 15;
   TYPE memory IS ARRAY (0 TO 15) OF STD_LOGIC_VECTOR(7 DOWNTO 0);
   SIGNAL rom: memory;
   ATTRIBUTE ram_init_file: STRING;
   ATTRIBUTE ram_init_file OF rom: SIGNAL IS "rom_contents.mif";
BEGIN...
------------------------------------------------------------------
```

Another option, even simpler but *asynchronous*, is shown below.

```
1  ------------------------------------------------------------------
2  LIBRARY ieee;
3  USE ieee.std_logic_1164.all;
4  ------------------------------------------------------------------
5  ENTITY rom IS
6     PORT (address: IN INTEGER RANGE 0 TO 15;
7           data_out: OUT STD_LOGIC_VECTOR(7 DOWNTO 0));
8  END rom;
9  ------------------------------------------------------------------
10 ARCHITECTURE rom OF rom IS
11    SIGNAL reg_address: INTEGER RANGE 0 TO 15;
12    TYPE memory IS ARRAY (0 TO 15) OF STD_LOGIC_VECTOR(7 DOWNTO 0);
13    SIGNAL myrom: memory;
14    ATTRIBUTE ram_init_file: STRING;
15    ATTRIBUTE ram_init_file OF myrom: SIGNAL IS "rom_contents.mif";
16 BEGIN
```

```
17        data_out <= myrom(address);
18   END rom;
19   -------------------------------------------------------------
```

ROM Implemented with a Vendor-Specific Function

In this case, a prewritten VHDL code is employed. Again taking Altera as an example, prebuilt units, for either HDL or graphical entry, are called *macrofunctions*. The *parameterized* code to be used here is part of a subset of the macrofunction collection, called *library of parameterized modules* (LPM). Any LPM cell can be easily instantiated in the main code using the MegaWizard Plug-In Manager unit of Quartus II (see appendix F) (another option is to edit the LPM file directly, making the proper selection of parameters).

The LPM component of interest in this example is *lpm_rom*, which is simply a VHDL code for a ROM similar to that in figure 13.5b, whose parameters (number of data and address bits plus the use or not of registers) can be specified by the user. This component is used in the code below.

Note in the library/package declarations (lines 2–5) the inclusion of the package *lpm_components* from the library *lpm* (from Altera), so the component does not need to be declared in the main code. In the entity, a clock was included because the input was chosen to be registered (the output is not). Observe that all I/Os (lines 8–10) are of type STD_LOGIC(_VECTOR). Finally, in the architecture, the component *lpm_rom* is instantiated (lines 16–22) under the label *myrom*. The same file used before (*rom_contents.mif*) was employed to initialize the memory (line 20). The HEX file in section 13.3 (only two lines) would produce the same results (exercise 13.6).

```
 1   -------------------------------------------------------------
 2   LIBRARY ieee;
 3   USE ieee.std_logic_1164.all;
 4   LIBRARY lpm;
 5   USE lpm.lpm_components.all;
 6   -------------------------------------------------------------
 7   ENTITY rom IS
 8      PORT (clk: IN STD_LOGIC;
 9            address: IN STD_LOGIC_VECTOR(3 DOWNTO 0);
10            data_out: OUT STD_LOGIC_VECTOR(7 DOWNTO 0));
11   END rom;
12   -------------------------------------------------------------
13   ARCHITECTURE rom OF rom IS
14      --Component declaration not needed
15   BEGIN
16      myrom:lpm_rom
17         GENERIC MAP (lpm_widthad => 4,
```

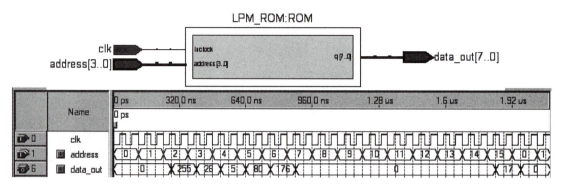

Figure 13.7
RTL view and simulation results from the ROM implemented with a vendor-specific function in section 13.4.

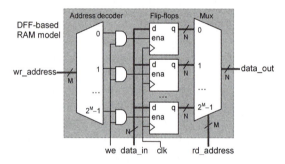

Figure 13.8
Flip-flop-based RAM model.

```
18                         lpm_outdata => "UNREGISTERED",
19                         lpm_address_control => "REGISTERED",
20                         lpm_file => "rom_contents.mif",
21                         lpm_width => 8)
22         PORT MAP (inclock=>clk, address=>address, q=>data_out);
23  END rom;
24  --------------------------------------------------------------
```

Simulation results and the diagram produced by the RTL viewer are presented in figure 13.7. Note that the values stored in this memory do coincide with those in figure 13.3.

13.5 RAM Design

To better understand the construction of RAMs, a RAM model based on DFF blocks is presented in figure 13.8. The memory-write address (*wr_address*) is processed by the address decoder, which produces only one output high, corresponding to the present address value. If write-enable (*we*) is high, then the corresponding DFF block is enabled to store *data_in* at the next positive clock transition. The output is controlled by a multiplexer, which allows only the DFF block selected by the memory-read address (*rd_address*) to be connected to *data_out*.

In the model above, the input and output addresses are not registered, and multiple simultaneous accesses are not included. Consequently, several extensions of this model can be obtained if such alternatives are also considered.

The main cases (besides that in figure 13.8) are depicted in figure 13.9. The case in (a) is asynchronous and equivalent to the model in figure 13.8 (but with a single address) or to the RAM in figure 13.1 The circuit in (b) is the synchronous counterpart of that in (a), with registers at all inputs and optionally also at the output; its separate clocks allow two clocking alternatives. Another asynchronous implementation is shown in (c), which employs a bidirectional bus. A more elaborate access is shown in (d), with separate read/write and in/out addresses, so it is a dual port (cr1) or simple dual port (cr2) RAM. Finally, a circuit with two complete access sections is depicted in (e), again with separate read/write and in/out addresses, so it is a four port (cr1) or true dual port (cr2) RAM. See the clocking options for all cases in figure 13.9.

RAM Implemented with Regular VHDL Code

Straight VHDL code is used here, with no concerns regarding particular encoding styles or synthesis attributes to help the compiler understand that memory is wanted. This means that the compiler might use on-chip SRAM blocks (if available) or might not (in general, regular logic cells are used).

An example is shown below. Because regular code is used, a regular circuit (equivalent to that in figure 13.8) is expected, hence with $8 \times 16 = 128$ flip-flops. After compiling this code, the reader is invited to check that fact in the compilation reports. Recall also that a RAM does not need to be initialized, but if one wants to do so, the same procedure seen for ROMs implemented with an initialization file can be used, which employs the *ram_init_file* attribute (this attribute is used in the code below to load a file called *ram_contents.mif*, containing again the data of figure 13.3). The RTL view obtained from this code is shown in figure 13.10.

```
1   ----------------------------------------------------------------
2   ENTITY ram IS
3      PORT (clk: IN STD_LOGIC;
4             we: IN STD_LOGIC;
5             address: IN INTEGER RANGE 0 TO 15;
6             data_in: IN STD_LOGIC_VECTOR(7 DOWNTO 0);
7             data_out: OUT STD_LOGIC_VECTOR(7 DOWNTO 0));
8   END ram;
```

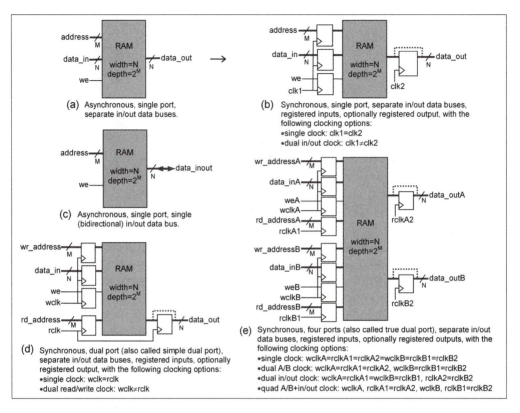

Figure 13.9
RAM configurations.

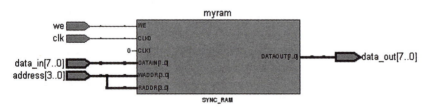

Figure 13.10
RTL view relative to the RAM implemented with regular VHDL code in section 13.5.

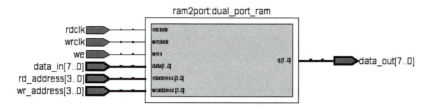

Figure 13.11
RTL view relative to the RAM implemented with a vendor-specific function in section 13.5.

```
 9   --------------------------------------------------------------------
10   ARCHITECTURE ram OF ram IS
11      TYPE memory IS ARRAY (0 to 15) OF STD_LOGIC_VECTOR(7 DOWNTO 0);
12      SIGNAL myram: memory;
13      ATTRIBUTE ram_init_file: STRING;
14      ATTRIBUTE ram_init_file OF myram: SIGNAL IS "ram_contents.mif";
15   BEGIN
16      PROCESS (clk)
17      BEGIN
18         IF (clk'EVENT AND clk='1') THEN
19            IF (we='1') THEN
20               myram(address) <= data_in;
21            END IF;
22         END IF;
23      END PROCESS;
24      data_out <= myram(address);
25   END ram;
26   --------------------------------------------------------------------
```

RAM Implemented with a Vendor-Specific Function

Here a prewritten VHDL code is employed. Taking again Altera as an example, prebuilt units, for either HDL or graphical entry, are called *macrofunctions*, of which the LPM is part. The LPM component used in this example is called *RAM:2-PORT*, accessed using the MegaWizard Plug-In Manager unit of Quartus II (Tools > MegaWizard Plug-In Manager > Installed Plug-Ins > Memory Compiler > RAM:2-PORT). While generating this component, one can choose its location in logic cells or in SRAM blocks.

The corresponding VHDL code is shown below. Note that, because the name chosen for the component (*ram2port*) does not coincide with any of the component names declared in the package *lpm_components*, an explicit component declaration is needed in the main code (lines 14–22). Again, though optional, the RAM was initialized using the *ram_init_ file* attribute (lines 25–26), which loads the contents of *ram_contents.mif*. Observe that all I/Os are of type STD_LOGIC(_VECTOR) (lines 6–10). The RTL view obtained from this code is shown in figure 13.11.

```
1   ------------------------------------------------------------------
2   LIBRARY ieee;
3   USE ieee.std_logic_1164.all;
4   ------------------------------------------------------------------
5   ENTITY ram IS
6      PORT (wrclk, rdclk: IN STD_LOGIC;
7            we: IN STD_LOGIC;
8            wr_address, rd_address: STD_LOGIC_VECTOR(3 DOWNTO 0);
9            data_in: IN STD_LOGIC_VECTOR(7 DOWNTO 0);
10           data_out: OUT STD_LOGIC_VECTOR(7 DOWNTO 0));
11  END ram;
12  ------------------------------------------------------------------
13  ARCHITECTURE ram OF ram IS
14     COMPONENT ram2port IS
15        PORT(data: IN STD_LOGIC_VECTOR (7 DOWNTO 0);
16              rdaddress: IN STD_LOGIC_VECTOR (3 DOWNTO 0);
17              rdclock: IN STD_LOGIC ;
18              wraddress: IN STD_LOGIC_VECTOR (3 DOWNTO 0);
19              wrclock: IN STD_LOGIC ;
20              wren: IN STD_LOGIC := '1';
21              q: OUT STD_LOGIC_VECTOR (7 DOWNTO 0));
22     END COMPONENT;
23     TYPE memory IS ARRAY (0 to 7) OF STD_LOGIC_VECTOR(3 DOWNTO 0);
24     SIGNAL myram: memory;
25     ATTRIBUTE ram_init_file: STRING;
26     ATTRIBUTE ram_init_file OF myram: SIGNAL IS "ram_contents.mif";
27  BEGIN
28     dual_port_ram: ram2port PORT MAP (
29        data => data_in,
30        rdaddress => rd_address,
31        rdclock => rdclk,
32        wraddress => wr_address,
33        wrclock => wrclk,
34        wren => we,
35        q => data_out);
36  END ram;
37  ------------------------------------------------------------------
```

RAM Implemented in a User SRAM Block

If the designer wants to force the compiler to implement a RAM with SRAM blocks, a special synthesis attribute should be used. Altera, for example, calls such an attribute *ram-style* (or *syn_ramstyle*), whose allowed SRAM block options are "M512", "M4K", "M-RAM", "M9K", "M144K", and "MLAB". However, this attribute is only recognized if the code is written in a certain predefined style (style options are available in the proper

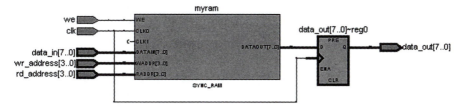

Figure 13.12
RTL view relative to the RAM implemented in a user SRAM block in section 13.5.

documentation, with one example given below). Another way of specifying SRAM blocks is during the instantiation of LPM cells with the MegaWizard Plug-In Manager.

The code below is for a RAM with separate buses for reading and writing and with registered output (thus complying with figure 13.9d, operating with a single clock). The Cyclone II FPGA available in the DE2 board was employed in this design, which contains 105 SRAM blocks of type M4K (~4k bits per block), so "M4K" was declared as the string value.

After compiling this code, the reader is invited to check in the resource section of the fitter whether M4K blocks were indeed used to construct this RAM. Also observe in the code that, as before, the RAM can be optionally initialized using the *ram_init_file* attribute. The RTL view obtained from this code is shown in figure 13.12.

```
1   --------------------------------------------------------------------
2   LIBRARY ieee;
3   USE ieee.std_logic_1164.all;
4   --------------------------------------------------------------------
5   ENTITY ram IS
6      PORT (clk: IN STD_LOGIC;
7            we: IN STD_LOGIC;
8            wr_address: IN INTEGER RANGE 0 TO 15;
9            rd_address: IN INTEGER RANGE 0 TO 15;
10           data_in: IN STD_LOGIC_VECTOR(7 DOWNTO 0);
11           data_out: OUT STD_LOGIC_VECTOR(7 DOWNTO 0));
12  END ram;
13  --------------------------------------------------------------------
14  ARCHITECTURE ram OF ram IS
15     TYPE memory IS ARRAY (0 to 15) OF STD_LOGIC_VECTOR(7 DOWNTO 0);
16     SIGNAL myram: memory;
17     --ATTRIBUTE ram_init_file: STRING;
18     --ATTRIBUTE ram_init_file OF my_ram: SIGNAL IS "ram_contents.mif";
19     ATTRIBUTE ramstyle: STRING;
20     ATTRIBUTE ramstyle OF myram: SIGNAL IS "M4K";
21  BEGIN
```

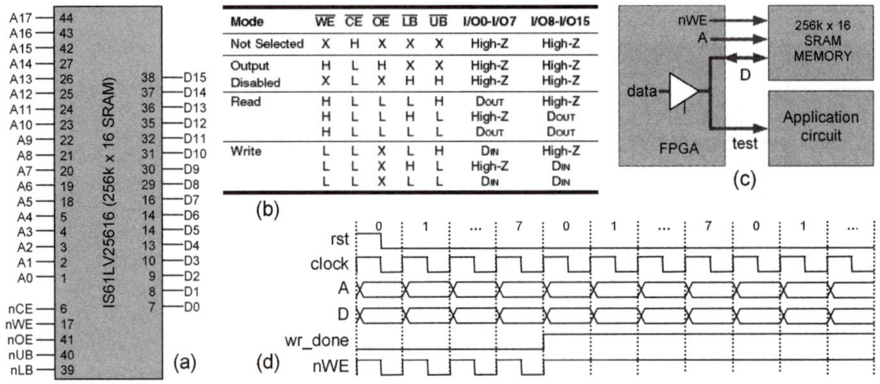

Figure 13.13
(a) $256k \times 16$ SRAM chip; (b) Corresponding truth table; (c) Simplified model; (d) Waveforms for the experiments.

```
22      PROCESS (clk)
23      BEGIN
24          IF (clk'EVENT AND clk='1') THEN
25              IF (we='1') THEN
26                  myram(wr_address) <= data_in;
27              END IF;
28              data_out <= myram(rd_address);
29          END IF;
30      END PROCESS;
31  END ram;
32  --------------------------------------------------------------------
```

As a final note, it is important to mention that a synthesis attribute called *romstyle* also exists, which is equivalent to *ramstyle*, but for ROM implementation (see exercise 13.8).

13.6   External Memory Interfaces

All memory circuits designed so far in this chapter are for memories implemented *inside* the FPGA. A different situation is examined in this section, which consists of having an *external* memory chip so the FPGA does not have to provide memory circuits, but rather an *interfacing* circuit between such a memory and an application circuit that makes use of data stored in the memory. Besides reading data from the memory, the interface must also be able to write data to the memory.

The memory chosen to illustrate this type of design is depicted in figure 13.13a. It is the IS61LV25616 SRAM device from ISSI (available, for example, in Altera's DE2 board), which can store 256 kwords of 16 bits each, hence requiring a 18-bit address bus (*A(17:0)* in the figure) and a 16-bit data bus (*D(15:0)* in the figure). It also contains five control signals, all active low, called *nCE* (chip enable), *nWE* (write enable), *nOE* (output enable), *nLB* (lower byte enable), and *nUB* (upper byte enable). The corresponding truth table is shown in figure 13.13b.

Because this memory has just one data bus, that bus must be bidirectional. Consequently, the analysis presented in section 13.2 will now be helpful.

A simplified diagram (based on section 13.2) for the interface designed in this section is shown in figure 13.13c, where only *nWE* is used (the other four control inputs are enabled permanently with a '0'—see truth table in figure 13.13b). When *nWE* = '0', *D* is written into the memory, while *nWE* = '1' causes data to be read from it. The data read from the memory is passed to the application circuit (signal *test* in the figure).

The experiment is as follows. A ROM is created inside the FPGA containing just eight 8-bit vectors. This data is copied to the first eight addresses of the SRAM, with the upper byte filled with '0's. Next, the data from these eight addresses is read and passed continuously to the test circuit, which displays the received data in eight LEDs, with each data vector lasting 0.5 seconds. Additionally, *nWE* is also displayed by an LED, producing a low brightness while *nWE* is pulsing or full brightness when *nWE* = '1'.

This sequence of events can be observed in the waveforms of figure 13.13d. A control signal called *wr_done* is created, which stays low during eight clock cycles, then is raised permanently to '1'. Writing (that is, *nWE* = '0' pulses) must only occur while *wr_done* = '0'.

A VHDL code for this interface is presented below. Note in the entity (lines 6–14) that the frequency of the system clock was entered using a GENERIC declaration (so the code can be easily adjusted to any clock frequency) and that all ports are of type STD_LOGIC(_VECTOR) (industry standard). In the declarative part of the architecture (lines 17–26), an array type is created, then a ROM conforming with that data type is produced, with a total of eight 8-bit vectors.

The code proper (after line 27) is divided into several small sections. In line 29, static values (= '0') are created for four SRAM control pins. In the first part of the subsequent process, two counters (*count* and *i*) are used in order to produce the memory address (*A*); *count* reduces the actual clock frequency (50 MHz, in this example) down to 2 Hz, while *i* is a 0-to-7 counter that produces the actual address. Note that *A* (line 53) uses a type-conversion function to convert *i* from INTEGER to STD_LOGIC_VECTOR, with 18 bits (see this function in figure 3.10), so the proper package must be declared in the code (*std_logic_arith*, line 4).

The next part of the process (lines 55–61) creates the *nWE* signal (and also *we_test*), following exactly the waveforms in figure 13.13d. Finally, the last part (lines 63–68) creates the bidirectional bus (*D*) between the FPGA and the memory, and the output bus (*test*) between the FPGA and the test circuit. Note that this part obeys the two recommendations introduced in section 13.2. Extensions to this design will be seen in the exercises section.

```
1   ------------------------------------------------------------------
2   LIBRARY ieee;
3   USE ieee.std_logic_1164.all;
4   USE ieee.std_logic_arith.all;
5   ------------------------------------------------------------------
6   ENTITY sram IS
7      GENERIC (fclk: NATURAL := 50_000_000);
8      PORT (clk, rst: IN STD_LOGIC;
9            nCE, nWE, nOE, nUB, nLB: OUT STD_LOGIC;
10           A: OUT STD_LOGIC_VECTOR(17 DOWNTO 0);
11           D: INOUT STD_LOGIC_VECTOR(15 DOWNTO 0);
12           test: OUT STD_LOGIC_VECTOR(7 DOWNTO 0);
13           we_test: OUT STD_LOGIC);
14  END ENTITY;
15  ---------------------------- --------------------------------
16  ARCHITECTURE sram OF sram IS
17     TYPE memory IS ARRAY (0 TO 7) OF STD_LOGIC_VECTOR(7 DOWNTO 0);
18     CONSTANT rom: memory := (
19        0 => "00000000",
20        1 => "00000001",
21        2 => "00000011",
22        3 => "00000111",
23        4 => "00001111",
24        5 => "00011111",
25        6 => "00111111",
26        7 => "01111111");
27  BEGIN
28     --Feed SRAM static pins:--------------
29     nCE<='0'; nOE<='0'; nUB<='0'; nLB<='0';
30     ------------------------------------
31     PROCESS (clk, rst)
32        VARIABLE wr_enable: STD_LOGIC;
33        VARIABLE wr_done: STD_LOGIC;
34        VARIABLE count: NATURAL RANGE 0 TO fclk;
35        VARIABLE i: NATURAL RANGE 0 TO 15;
36     BEGIN
37        --Create address and wr_done:-----
38        IF (rst='1') THEN
39           count := 0;
40           i := 0;
41           wr_done := '0';
42        ELSIF (clk'EVENT AND clk='1') THEN
43           count := count + 1;
44           IF (count=fclk/2) THEN
```

```
45                  count := 0;
46                  i := i + 1;
47                  IF (i=8) THEN
48                      i := 0;
49                      wr_done := '1';
50                  END IF;
51              END IF;
52          END IF;
53          A <= conv_std_logic_vector(i, 18);
54          --Create nWE:--------------------
55          IF (wr_done='0') THEN
56              wr_enable := clk;
57          ELSE
58              wr_enable := '1';
59          END IF;
60          nWE <= wr_enable;
61          we_test <= wr_enable;
62          --Bidirectional bus:--------------
63          test <= D(7 DOWNTO 0);
64          IF (wr_enable='0') THEN
65              D <= "00000000" & rom(i);
66          ELSE
67              D <= (OTHERS => 'Z');
68          END IF;
69      END PROCESS;
70  END ARCHITECTURE;
71  -----------------------------------------------------------------
```

Two other, modern interfaces, called I^{2}C and SPI, will be described in the next chapter. These are serial data interfaces for communication with memories and a variety of other chip families.

13.7   Exercises

Exercise 13.1: Bidirectional Bus

A circuit with a bidirectional (INOUT) bus is shown in figure 13.14. Following the procedure seen in section 13.2, write a VHDL code from which this circuit is inferred. Check the RTL view and also the equations produced by the compiler.

Exercise 13.2: INOUT versus BUFFER Mode

The BUFFER mode is rarely used in VHDL code (an auxiliary internal signal can be used instead, so the actual signal, which goes out of the design, can be declared as OUT instead

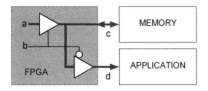

Figure 13.14

of BUFFER). Just as an exercise, consider that line 7 in the code in section 13.2 has the word INOUT replaced with BUFFER.

a) Draw the circuit (as in figure 13.2a) that you believe the compiler will infer.

b) Compile the code and check the circuit (as in figure 13.2b) that the compiler actually implemented.

Exercise 13.3: COE versus MIF Files

Write a CGF and then a COE file with the same contents as the MIF file of figure 13.3.

Exercise 13.4: Synchronous ROM

a) Analyze the code presented in section 13.4 for Altera's ROM template and determine which ROM of figure 13.5 (if any) this code will implement.

b) Compile this code using your synthesis tool and check if the inferred circuit coincides with your prediction.

Exercise 13.5: ROM Implemented with a HEX File #1

Repeat the design of ROM implemented with an initialization file in section 13.4, entering the data with a HEX file instead of a MIF file. Use the HEX file of section 13.3.

Exercise 13.6: ROM Implemented with a HEX File #2

Repeat the design of ROM implemented with a vendor-specific function in section 13.4, entering the data using a HEX file instead of a MIF file. Use the HEX file of section 13.3.

Exercise 13.7: RAM Implemented with a HEX File

Repeat the design of RAM implemented with regular VHDL code in section 13.5, but enter the data using an HEX file instead of a MIF file. Use the HEX file of section 13.3.

Exercise 13.8: ROM Implemented in an User SRAM Block

Make the changes needed for the code of RAM implemented in a user SRAM block in section 13.5 to produce a ROM in an M4K user SRAM block instead of a RAM. Recall that the counterpart of the synthesis attribute *ramstyle* is *romstyle*.

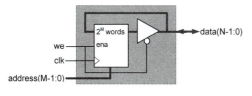

Figure 13.15

Exercise 13.9: Synchronous RAM

Figure 13.15 shows a synchronous DFF-based RAM with a bidirectional data bus. Design this circuit using VHDL. Enter the number of address bits (M) and the number of bits per word (N) as GENERIC declarations.

Exercise 13.10: External Memory Interface

Devise an experiment similar to that in section 13.6, but with the data located in a file rather than in a ROM memory in the FPGA.

14 VHDL Design of Serial Communications Circuits

14.1 Introduction

Figure 14.1 shows a typical arrangement for the physical layer of a modern serial data communications system that employs a high-performance line encoder/decoder plus a high-speed I/O circuit. This structure is employed in basically all high-speed serial data transmission media, like internet connections (through twisted pairs of wires or optical fiber) and DVI (digital visual interface) cables connecting computers to LCD monitors.

The purpose of a line code is to modify the data sequence (either in terms of shape or in terms of content) in order to produce a more robust physical data stream (less noise and EMI effects and better DC balance). On the other hand, the purpose of the I/O circuit is to provide adequate voltage (or current) levels to further improve the physical signal's robustness (using low-swing differential voltages, for example, instead of large, regular voltages). Manchester, MLT-3, 8B/10B, and 4D-PAM5 are examples of line codes, while TTL, LVCMOS, PECL, CML, and LVDS are examples of I/Os (Pedroni 2008).

The communication in figure 14.1 can be classified as *synchronous* or *asynchronous*. In the most general definition, it is said to be synchronous when Rx and Tx operate with the *same* clock, so the clock signal is either transmitted in a separate wire (assuming that it is a wired channel) or the transmitted data contains enough transitions or some kind of embedded bit pattern that allows the receiving end to retrieve the actual clock from the received data. On the other hand, the communication is said to be asynchronous when no clock information is transmitted, in which case start and stop bits are inserted before and after *short* bit streams (typically, one byte), allowing the receiving end to correctly decode the received data even if the Tx and Rx clock frequencies (phases) are not exactly equal (some of the synchronous interfaces also employ start/stop bits). I²C, SPI, and TMDS (described later) are examples of synchronous interfaces, while the RS232 UART is truly asynchronous. Most modern high-performance interfaces are synchronous.

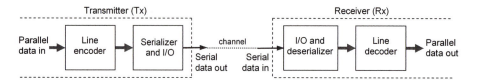

Figure 14.1
Typical high-performance serial data communications link (line encoder plus high-speed I/O).

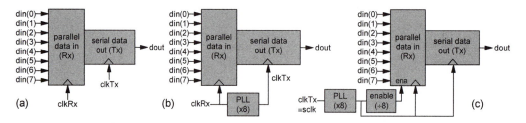

Figure 14.2
Data serializers. (a) General principle; (b) With a single clock; (c) With a single clock and provision for synchronism.

14.2   Data Serializers/Deserializers

Several interfaces for serial data communications are described in this and in following chapters. In some cases, the transformation of data from its usual form (parallel) to serial (time-multiplexed single bits) is part of the interface itself (as in the video interfaces DVI and FPD-Link, studied in chapters 16 and 17). The circuit responsible for such a transformation is the *serializer*, while the circuit that returns the data to its original format is a *deserializer*.

Figure 14.2a symbolically illustrates the data serializing principle. The circuit contains a receiving (Rx) and a transmitting (Tx) section. Rx receives *din* (from some other circuit) in parallel, and passes it to Tx, which sends it out serially, producing *dout*.

Observe in figure 14.2a the presence of two clocks, where *clkTx* must be faster than *clkRx* by a factor *N* (for unbuffered, constant-speed systems), where *N* is the number of bits in *din*.

In figure 14.2b, clock details are included. It is assumed that the transmission clock's frequency is higher than the system clock's, so a PLL (Pedroni 2008) is needed. In summary, the low-frequency clock (*clkRx*) is derived from the system clock, and is then multiplied by *N* to get the high-frequency clock (*clkTx*).

An important consideration about serializers is that some kind of synchronism between *clkRx* and *clkTx* is needed, because if the system does not possess any provision for reset or content initialization, the states of its flip-flops on power up are uncertain, so the

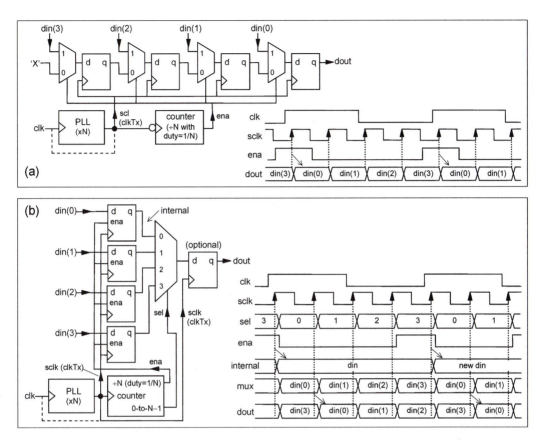

Figure 14.3
Data serializer circuits.

counter that indexes the transmissions of *din* might start from an arbitrary position rather than from the LSB or MSB. Consequently, unless one knows precisely the phase relationship in the PLL (which, by the way, can vary), this arrangement must be improved.

A solution is depicted in figure 14.2c, where the *same* clock feeds both sections. The role of *clkRx* is performed by an enable (*ena*) signal, which stays high during only one out of *N* clock periods. This single clock (*clkTx*) will be represented by *sclk* (serializer clock) in the design ahead.

Serializer Circuits

Figure 14.3 shows two options for implementing a serializer. Note the PLL at the clock input in each case, which is necessary when the required transmission frequency (*sclk*, same as *clkTx*) is higher than that of the system clock (*clk*).

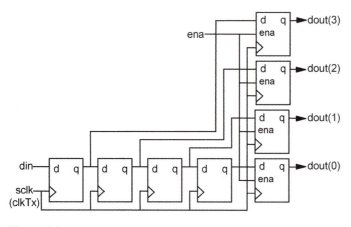

Figure 14.4
Deserializer circuit.

The option in figure 14.3a is constructed with a shift register with data-load capability (Pedroni 2008), plus a counter to generate the load-enable (*ena*) signal and a PLL, if necessary. As shown in the timing diagram, *ena* stays high during only one out of every *N* clock periods, allowing *din* to be loaded into the shift register, which shifts the data out bit by bit subsequently. The counter can operate either at the same clock edge as the shift register or at the other edge.

The option in figure 14.3b is constructed with an input register and a multiplexer, plus a counter to generate the register-enable (*ena*) and port-select (*sel*) signals, and a PLL, if necessary. The output flip-flop is optional; it can be removed if glitches during multiplexer transitions are not a problem in that particular application. As shown again in the timing diagram, *ena* stays high during only one out of every *N* clock periods, allowing *din* to be loaded into the input register (*N* flip-flops), being then passed to the output register (single flip-flop) one bit at a time by the multiplexer.

The two options in figure 14.3 are relatively similar in terms of hardware size, but the latter might be advantageous if *din* has already been stored elsewhere, so the input register can be removed. However, due to the slightly higher propagation delay of its larger multiplexer (*N* inputs, against two inputs in the former), the latter might be slightly slower.

Deserializer Circuit

Once one knows how to construct a serializer, building a deserializer is straightforward because one is just the opposite of the other. An example is depicted in figure 14.4, showing a shift register followed by an (optional) output register. Note that again a single clock (*sclk*, same as *clkTx*) is used, with the slow section controlled by a store-enable (*ena*) signal derived from that clock. (See more on this in exercise 14.5.)

Design of a Fast Serializer

The code below implements the serializer in figure 14.3b, under the project name *fast_serializer* (line 5). It was assumed that $N = 4$ bits, that the system clock frequency is 10 MHz, and that the serializer must operate at full speed—that is, with *din* produced at a rate of 10 MHz. Consequently, a PLL is required in order to generate the actual serializer clock (*sclk*) of 40 MHz. *N* was entered using a GENERIC declaration (line 6), so the code can be easily adjusted to any serializer size (of course, then the PLL multiply and divide factors must also be adjusted). The PLL (see note below) is declared in lines 15–20, then used in line 23.

The code proper is in lines 21–39. In its first part (line 23), it infers the PLL, thus producing *sclk*. In its second part (process of lines 26–38), the serializer is built, based on the diagram in figure 14.3b.

Note: The PLL cell can be instantiated, for example, with the MegaWizard Plug-In Manager of Quartus II (for Altera devices), which was the case in the code below. This creates the additional file *altera_pll.vhd*, which must be included in the project. Further details can be seen in appendix G.

```
1   ------------------------------------------------------------
2   LIBRARY ieee;
3   USE ieee.std_logic_1164.all;
4   ------------------------------------------------------------
5   ENTITY fast_serializer IS
6      GENERIC (N: INTEGER := 4); --number of bits
7      PORT (clk: IN STD_LOGIC;   --10MHz system clock
8            din: IN STD_LOGIC_VECTOR(N-1 DOWNTO 0);
9            dout, sclk_test: OUT STD_LOGIC);
10  END ENTITY;
11  ------------------------------------------------------------
12  ARCHITECTURE fast_serializer OF fast_serializer IS
13     SIGNAL sclk: STD_LOGIC;
14     SIGNAL internal: STD_LOGIC_VECTOR(N-1 DOWNTO 0);
15     COMPONENT altera_pll IS
16        PORT(areset: IN STD_LOGIC := '0';
17             inclk0: IN STD_LOGIC := '0';
18             c0: OUT STD_LOGIC;
19             locked: OUT STD_LOGIC);
20     END COMPONENT;
21  BEGIN
22     -----Get sclk (with PLL):------------
23     pll_circuit: altera_pll PORT MAP ('0', clk, sclk, OPEN);
24     sclk_test <= sclk;
```

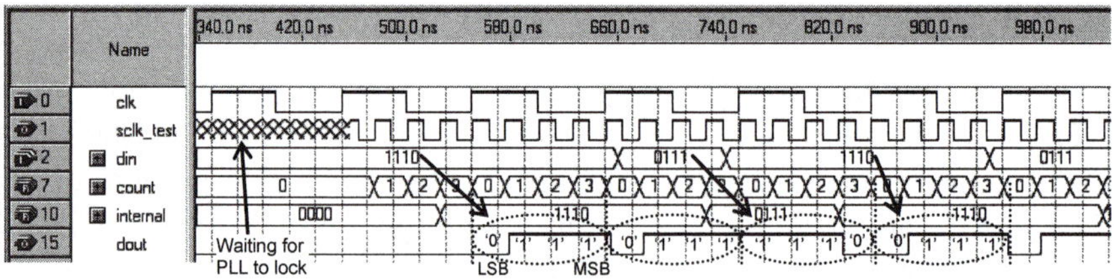

Figure 14.5
Simulation results from the *fast_serializer* code (circuit of figure 14.3b).

```
25      -----Build serializer:---------------
26      PROCESS (sclk)
27         VARIABLE count: INTEGER RANGE 0 TO N;
28      BEGIN
29         IF (sclk'EVENT AND sclk='1') THEN
30            count := count + 1;
31            IF (count=N-1) THEN   --enabled to update "internal"
32               internal <= din;
33            ELSIF (count=N) THEN --counter is 0-to-(N-1)
34               count := 0;
35            END IF;
36            dout <= internal(count); --continuous serial output
37         END IF;
38      END PROCESS;
39   END ARCHITECTURE;
40   ------------------------------------------------------------
```

Simulation results are depicted in figure 14.5, with several values highlighted. It is important to remember that the circuit must be simulated with a clock frequency equal to or near that entered in the PLL file (*altera_pll.vhd*, 10 MHz in this example). Note that in this case the LSB is transmitted first (see exercise 14.2).

14.3   PS2 Interface

The PS2 interface was introduced by IBM in 1987 for connecting computers to keyboard and mouse devices. Even though it is now giving place to the USB (universal serial bus) interface, its study helps understand how computers communicate with peripherals, also allowing interesting designs to be developed in the lab sections.

PS2 (or PS/2, personal system version 2) is a serial, 8-bit oriented bus for communication between a computer and its mouse and keyboard peripherals. As illustrated in figure

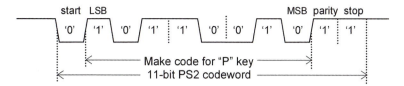

Figure 14.6
PS2 code word, showing the *make* code for the "p" key (4Dh, from right to left).

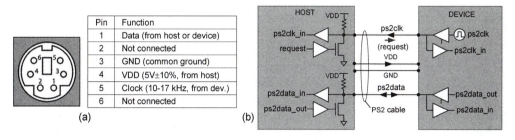

Pin	Function
1	Data (from host or device)
2	Not connected
3	GND (common ground)
4	VDD (5V±10%, from host)
5	Clock (10-17 kHz, from dev.)
6	Not connected

Figure 14.7
(a) Six-pin mini-DIN connector (female part) for PS2 devices; (b) Internal interface details.

14.6, a PS2 code word consists of one data byte, to which *start*, (odd) *parity*, and *stop* bits are added. In this example, the *make* code (explained later) for the "P" keyboard key is shown (this code word is for both "p" and "P", which are differentiated by including or not including the SHIFT key in the operation). The transmission starts from the left, so the code word is 4Dh (hexadecimal form, from right to left). As will be seen, several such code words are required each time a PS2 mouse or keyboard needs to communicate with the host computer.

Host-Device Communication

PS2 devices are normally connected using 6-pin mini-DIN (Deutsches Institut für Normung) connectors. Its female part is depicted in figure 14.7a, along with the respective pin functions. Note the presence of clock in pin 5, so this is a synchronous interface.

Even though the transmission protocol (figure 14.6) is the same for mouse and keyboard communications, the data packets are assembled differently. So to avoid confusion (that is, a mouse connected into a keyboard receptacle or vice versa), different colors are employed in the fabrication of this connector, with green used for the mouse and purple for the keyboard.

Figure 14.7b shows a simplified view of a PS2 host-device physical connection. Both lines (*ps2clk* and *ps2data*) are open-drain, so pull-up resistors are needed, and when

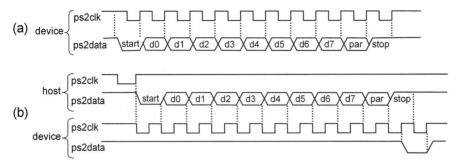

Figure 14.8
(a) Device to host communication; (b) Host to device communication.

these lines are inactive their logic level is '1'. Control always belongs to the host, which can inhibit the communication at any time by pulling the clock line low (see the *request* signal on the host side). On the other hand, the clock is always generated by the device, with a period in the range 60–100 μs, hence a frequency between 10 and 16.7 kHz. Power (V_{DD}) is supplied by the host.

A timing diagram for the device-to-host communication is depicted in figure 14.8a. The device sends new data at the *positive* clock edge, so the receiver (host) samples it at the *negative* clock transition. The communication starts with the device lowering the data line (this line must have been high for at least 50μs), which constitutes the *start* bit, so the host gets ready to receive the subsequent bit stream. The system is busy from the moment when the device lowers the *ps2data* wire to '0' until when *ps2clk* stays high for more than one-half of its time period (recall that this period is in the 60–100 μs range).

A timing diagram for the host-to-device communication is depicted in figure 14.8b. The host lowers the clock line for at least 100μs (see *request* signal in figure 14.7b), informing the device that it wants to send data. The device then releases the clock and data transfer proceeds until the *stop* bit is received, when the device produces a '0' bit in acknowledgement.

At power up, a relatively complex sequence of communications between the host and the devices (mouse and keyboard) takes place, during which the devices inform their IDs, run a self-test (called BAT—basic assurance test) whose result is communicated to the host, and also set up some default parameters.

Two such parameters in the case of a keyboard are called *typematic delay* and *typematic rate*. These parameters establish how the host must behave when a key is pressed during a long time. The typematic delay determines the time interval that the host must wait until it considers a long press to be a *sequence* of key presses rather than just a single press. Its default value is 0.5 seconds. The typematic rate is the number of characters per second produced after the typematic delay has occurred, with a default value normally of 10.

Key	Make code	Break code	Key	Make code	Break code	Key	Make code	Break code
A	1C	F0,1C	0	45	F0,45	ESC	76	F0,76
B	32	F0,32	1	16	F0,16	SPACE	29	F0,29
C	21	F0,21	2	1E	F0,1E	BKSP	66	F0,66
D	23	F0,23	3	26	F0,26	CAPS	58	F0,58
E	24	F0,24	4	25	F0,25	L SHIFT	12	FO,12
F	2B	F0,2B	5	2E	F0,2E	R SHIFT	59	F0,59
G	34	F0,34	6	36	F0,36	L CTRL	14	FO,14
H	33	F0,33	7	3D	F0,3D	R CTRL	E0,14	E0,F0,14
P	4D	F0,4D	F7	83	F0,83	PG DN	E0,7A	E0,F0,7A

Figure 14.9
A portion of the Scan Code Set 2 for keyboard encoding (in hexadecimal form).

PS2 Keyboard Encoding

When a key is pressed, the keyboard processor issues a *make* code, while a *break* code is issued when the key is released (these codes have nothing to do with the ASCII table; if ASCII outputs are needed, the translation must be performed by the host).

The set of make/break code words constitutes a *Scan Code Set*. Even though there are three of such sets, Scan Code Set 2 is practically the only one used (a portion of it is depicted in figure 14.9). Note that the break code is the same as the make code, just preceded by F0h. Observe also that even though most keys are represented (make code) with one byte, some are represented with two bytes. Finally, note that the code corresponding to the "P" key (4Dh) is indeed that used in the example of figure 14.6.

As an example, say that one wants to type the letter "a" (recall that the keyboard shows only capital letters, so to get "A" one of the SHIFT keys would need to be pressed too). The following code words are transmitted by the keyboard controller to the host computer (*start*, *parity*, and *stop* bits not included yet):

First code word: 1Ch when the key is pressed (make code for "a").

Second code word: F0h when the key is released (first part of the break code).

Third code word: 1Ch right after F0h (second part of the break code for "a").

Other cases are treated in exercise 14.6.

PS2 Mouse Encoding

One of the most common mouse constructions is with three pushbuttons and a scrolling wheel. In this case, the data packet transmitted from the mouse to the host when they communicate contains four bytes, as shown in figure 14.10 (hence a 44-bit packet results after PS2 encoding). Note that nine bits are used to represent the X movement (eight bits in byte 2 plus the sign bit in *bit4* of byte 1), hence encompassing the −255 to +255 interval. The same is true for the Y movement, represented in byte 3 and *bit5* of byte 1. The scrolling

byte 1	bit7	bit6	bit5	bit4	bit3	bit2	bit1	bit0
	Y overflow	X overflow	Y sign	X sign	always '1'	middle button	right button	left button
byte 2	X movement (horizontal)							
byte 3	Y movement (vertical)							
byte 4	Z movement (scrolling wheel)							

Figure 14.10
PS2 mouse encoding (for three buttons plus scroll).

Figure 14.11
Experiment with a PS2 keyboard.

wheel (called Z) takes only eight bits (byte 4). The other bits in byte 1 represent the status of the three pushbuttons and possible overflow in X and Y.

Design of a PS2 Keyboard Interface

We conclude this section with the design (using VHDL) of a simplified keyboard interface. The system is depicted in figure 14.11, which shows a PS2 keyboard connected to an FPGA, which in turn feeds an SSD. We want to design a circuit capable of reading the numeric (0, 1,..., 9) keyboard keys and display them on the SSD when such keys are pressed. If any other keyboard key is pressed, then "e" (error) must be displayed. On the other hand, if the error occurs in the read key (that is, if the *start* bit is not '0', the *stop* bit is not '1', or if the parity is not odd), then "E" must be displayed.

A VHDL code for this problem is presented below. Because both signals from the keyboard (*ps2clk* and *ps2data*) will be debounced, a debouncing period (*deb_cycles*) was specified in line 4, using a GENERIC declaration, so this value can be easily changed. The same occurs with the time interval for detection of the idle state (*idle_cycles*, line 5), which has to be larger than one-half the period of *ps2clk* (since this is ≤100 μs, 60 μs was adopted). The input and output signal names (lines 7–10) are from figure 14.11.

The keyboard signals, *ps2clk* and *ps2data*, are debounced in the processes of lines 21–35 and 37–51, respectively. The idle state is detected in the process of lines 53–69; the system is considered busy (not idle) from the moment when the device lowers the voltage of *ps2data* until when *ps2clk* stays high for at least 60 μs.

The data from the keyboard is recorded in the process of lines 71–86. Note that the serial data is stored sequentially by *data* (line 78), then passed in parallel to *dout* (line 82), so the output to the SSD will not be disturbed while the next character is being received.

Because each character is composed of at least three code words (make code plus break code, figure 14.9), it is important to observe that the VHDL code below sequentially detects all such codewords. For example, if "0" is pressed, 45h will be transmitted when the key is pressed, then F0h and 45h will follow when the key is released, all of which are detected by the circuit. The (interesting) consequence of this is discussed in exercise 14.7.

Errors are checked in the process of lines 88–98. Note that this is a purely combinational circuit, so concurrent code (with WHEN) could have been used. Finally, the SSD driver is in lines 102–122, which is a MAKE/BREAK-to-SSD code conversion (again, purely combinational). *ssd* does not need to be stored because *dout* has already been registered.

The reader is now invited to test this design with the FPGA board. For instance, what is expected to happen if line 116 is commented out (exercise 14.7)?

```
1    ---------------------------------------------------------------
2    ENTITY ps2_keyboard IS
3       GENERIC (
4          deb_cycles: INTEGER := 200;      --4us for debouncer (@50 MHz)
5          idle_cycles: INTEGER := 3000);   --60us (>1/2 period ps2_clk)
6       PORT (
7          clk: IN BIT;                     --system clock (50 MHz)
8          ps2clk: IN BIT;                  --clk from keyboard (10-17 kHz)
9          ps2data: IN BIT;                 --data from keyboard
10         ssd: OUT BIT_VECTOR(6 DOWNTO 0));--data out to SSD
11   END ps2_keyboard;
12   ---------------------------------------------------------------
13   ARCHITECTURE ps2_keyboard OF ps2_keyboard IS
14      SIGNAL deb_ps2clk: BIT;     --debounced ps2_clk
15      SIGNAL deb_ps2data: BIT;    --debounced ps2_data
16      SIGNAL data, dout: BIT_VECTOR(10 DOWNTO 0);
17      SIGNAL idle: BIT;    --'1' means data line is idle
18      SIGNAL error: BIT;   --'1' when start, stop, or parity wrong
19   BEGIN
20      ---------Debouncer for ps2clk:---------------
21      PROCESS (clk)
22         VARIABLE count: INTEGER RANGE 0 TO deb_cycles;
23      BEGIN
24         IF (clk'EVENT AND clk='1') THEN
25            IF (deb_ps2clk=ps2clk) THEN
26               count := 0;
27            ELSE
28               count := count + 1;
29               IF (count=deb_cycles) THEN
```

```
30                      deb_ps2clk <= ps2clk;
31                      count := 0;
32                  END IF;
33              END IF;
34          END IF;
35      END PROCESS;
36      ---------Debouncer for ps2data:--------------
37      PROCESS (clk)
38          VARIABLE count: INTEGER RANGE 0 TO deb_cycles;
39      BEGIN
40          IF (clk'EVENT AND clk='1') THEN
41              IF (deb_ps2data=ps2data) THEN
42                  count := 0;
43              ELSE
44                  count := count + 1;
45                  IF (count=deb_cycles) THEN
46                      deb_ps2data <= ps2data;
47                      count := 0;
48                  END IF;
49              END IF;
50          END IF;
51      END PROCESS;
52      ---------Detection of idle state:-----------
53      PROCESS (clk)
54          VARIABLE count: INTEGER RANGE 0 TO idle_cycles;
55      BEGIN
56          IF (clk'EVENT AND clk='0') THEN
57              IF (deb_ps2data='0') THEN
58                  idle <= '0';
59                  count := 0;
60              ELSIF (deb_ps2clk='1') THEN
61                  count := count + 1;
62                  IF (count=idle_cycles) THEN
63                      idle <= '1';
64                  END IF;
65              ELSE
66                  count := 0;
67              END IF;
68          END IF;
69      END PROCESS;
70      ---------Receiving data from keyboard:-------
71      PROCESS (deb_ps2clk)
72          VARIABLE i: INTEGER RANGE 0 TO 15;
```

```
73    BEGIN
74       IF (deb_ps2clk'EVENT AND deb_ps2clk='0') THEN
75          IF (idle='1') THEN
76             i:=0;
77          ELSE
78             data(i) <= deb_ps2data;
79             i := i + 1;
80             IF (i=11) THEN
81                i:=0;
82                dout <= data;
83             END IF;
84          END IF;
85       END IF;
86    END PROCESS;
87    ---------Checking for errors:----------------
88    PROCESS (dout)
89    BEGIN
90       IF (dout(0)='0' AND dout(10)='1' AND (dout(1) XOR
91             dout(2) XOR dout(3) XOR dout(4) XOR dout(5)
92          XOR   dout(6) XOR dout(7) XOR dout(8)
93          XOR dout(9))='1') THEN
94          error <= '0';
95       ELSE
96          error <= '1';
97       END IF;
98    END PROCESS;
99    ---------SSD driver:-----------------------
100   --This process is a MAKE-code to SSD-code conversion.
101   --No need to store "ssd" because "dout" is already registered.
102   PROCESS (dout, error)
103   BEGIN
104      IF (error='0') THEN
105         CASE dout(8 DOWNTO 1) IS
106            WHEN "01000101" => ssd <= "0000001";  --"0" on SSD
107            WHEN "00010110" => ssd <= "1001111";  --"1" on SSD
108            WHEN "00011110" => ssd <= "0010010";  --"2" on SSD
109            WHEN "00100110" => ssd <= "0000110";  --"3" on SSD
110            WHEN "00100101" => ssd <= "1001100";  --"4" on SSD
111            WHEN "00101110" => ssd <= "0100100";  --"5" on SSD
112            WHEN "00110110" => ssd <= "0100000";  --"6" on SSD
113            WHEN "00111101" => ssd <= "0001111";  --"7" on SSD
114            WHEN "00111110" => ssd <= "0000000";  --"8" on SSD
115            WHEN "01000110" => ssd <= "0000100";  --"9" on SSD
```

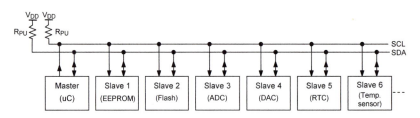

Figure 14.12
I²C bus (two-wire, synchronous, 8-bit oriented, master-slave type).

```
116                  WHEN "11110000" => ssd <= "1111111";   -- blank
117                  WHEN OTHERS => ssd <= "0010000";        --"e" on SSD
118              END CASE;
119          ELSE
120              ssd <= "0110000"; --"E" on SSD
121          END IF;
122      END PROCESS;
123  END ps2_keyboard;
124  -------------------------------------------------------------------
```

14.4   I²C Interface

I²C (inter integrated circuit) is a synchronous 8-bit-oriented serial bus for communication between integrated circuits installed next to each other (normally on the same board). It employs just two wires (plus a common GND) and has four standardized speed modes, called *standard* (100 kbps), *fast* (400 kbps), *fast-plus* (1 Mbps), and *high-speed* (3.3 Mbps), though in practice other, higher speeds are also used.

The original document was released by Philips in 1982, followed by versions 1.0 (1992), 2.0 (1998), 2.1 (2000), and 3.0 (2007).

As depicted in figure 14.12, the I²C bus consists of two wires, called *SCL* (serial clock) and *SDA* (serial data), which interconnect a master unit to a number of slave units. A common ground wire (not shown) is obviously also needed for the system to function. The clock (*SCL*) is unidirectional, always generated by the master (often a microcontroller), while data (*SDA*) is bidirectional. Examples of IC families currently fabricated with I²C support are also shown in figure 14.12; these include microcontrollers, EEPROM and Flash memories, A/D and D/A converters, RTC (real time clock) circuits, and temperature sensors.

The *SCL* and *SDA* chip outputs are open-drain, so external pull-up resistors (R_{PU}), typically in the 1.5 to 33 kΩ range, are connected between them and V_{DD}. The value of R_{PU} depends on the total node capacitance; if it is large (long bus, with many slaves), then the resistor must be small to achieve the minimum rise time defined in the I²C specifications.

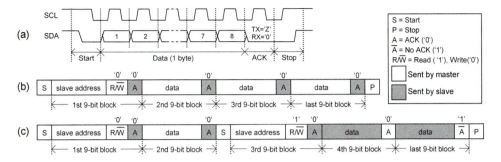

Figure 14.13

I²C communications: (a) Details of *start* and *stop* sequences plus 9-bit block construction; (b) Master writing to slave; (c) Master reading from slave.

The value of V_{DD} was 5V in initial I²C-driven devices, but voltages as low as 1.8V are now common.

The number of devices sharing the same bus can be up to 128 (7-bit address) or 1024 (10-bit address). More than one master is allowed, in which case the I²C protocol provides bus arbitration (the first master to lower the voltage of *SDA* is the current master). I²C provides also other advanced features, such as clock stretching, general call, and reset by software.

Operation of the I²C Interface

The overall operation of the I²C interface is illustrated in figure 14.13. Figure 14.13a shows details of the *start* and *stop* sequences, plus the construction of the 9-bit block. A high-to-low transition of *SDA* while *SCL* is high constitutes a *start* condition, while a low-to-high transition of *SDA* while *SCL* is high is a *stop* condition. Each transmission consists of an 8-bit block, followed by an acknowledgement bit (*ACK* = '0') issued by the corresponding receiver when proper data is identified. Because *SDA* is a bidirectional line, during the *ACK* bit the transmitter must free the line (that is, go into the high-impedance state, 'Z') to allow the receiver to place a '0' on that wire. The specifications require that the data remain stable while *SCL* is high, so changes in *SDA* are only allowed when the clock is low. (Strictly speaking, this specification might not be precise, because flip-flops are used to build these circuits, and they are transparent only at the clock transitions, so the data only need to remain stable during the setup plus hold times of the flip-flops).

Figures 14.3b–c show a simplified view of communications over the I²C bus. The white rectangles correspond to transmissions made by the master, while the gray ones represent transmissions made by the slave (because they share a single data line, while one is transmitting the output of the other must be 'Z').

In figure 14.13b, the master sends data to the slave. It begins with a *start* sequence (S), to which the first 9-bit block follows, consisting of a slave address (seven bits, in general) plus

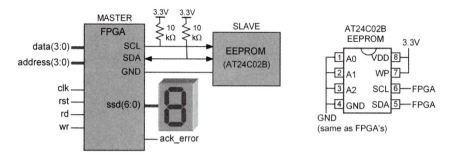

Figure 14.14
I²C interface for an EEPROM (FPGA implements the master, having the memory as its only slave). The figure on the right shows how the EEPROM was wired.

a read/write bit ('0' for writing), causing the corresponding slave to respond with $ACK = $ '0' if that address is identified. The next blocks contain eight bits of data (can also be a memory address, for example), to which the slave again responds with $ACK = $ '0'. When the master is done transmitting, it executes the *stop* sequence (P).

In figure 14.13c, the master reads data from the slave. The initial two 9-bit blocks are exactly the same as those described above (sometimes referred to as "dummy write"), followed by another *start* operation and slave address, but now with R/W = '1' (read), so in the next blocks the slave transmits data to the master, so now the master issues the ACK bit. When the master is done receiving all the data that it wanted, it issues a "no acknowledge" bit ($NoACK = $ '1'), followed by the *stop* sequence.

The design of an actual circuit using the I²C bus is illustrated next, which deals with an EEPROM memory. It is important to mention, however, that for each device category (EEPROM, Flash, ADC, etc.) with I²C support, specific data sequences and address values are employed, so it is indispensable to study the device's datasheets.

Design of an I²C Interface for an EEPROM Memory

As shown in figure 14.12, EEPROM memories are among the several IC categories fabricated with I²C support. The purpose of this exercise is to design a circuit capable of writing data to and reading data from an EEPROM that communicates using the I²C bus. Hence our circuit will play the role of master, while the memory will be its (only) slave. The chosen device is AT24C02B, from Atmel, which is a 256-byte memory (hence both address and data are eight bits wide).

The experiment is depicted in figure 14.14. The inputs are *data* (4-bit data, from four toggle switches), *address* (4-bit address, from other four switches), *clk* (system clock, assumed to be 50 MHz), plus *rst* (optional reset), *rd* (read), and *wr* (write) commands (the last three are also from switches). The outputs are the I²C signals (*SCL* and *SDA*— the latter is in fact bidirectional), with the respective pull-up resistors, plus *ssd*, which feeds an SSD display (chapter 12) to exhibit the data retrieved from the EEPROM, and *ack_error*, which checks whether the three acknowledgement bits (*ACK*) issued during writing are all correct ('0').

Figure 14.14 also shows how the device was wired. Note that besides *SCL* and *SDA* it has also four other programmable pins, labeled A_0, A_1, A_2, and *WP*. The device is fabricated with seven address bits, of which four (the MSBs) are fixed and equal to "1010", hence leaving only three ($A_2A_1A_0$) to be programmed by the designer. For example, if $A_2A_1A_0$ are hardwired to GND, then the device's address is "1010000". The other pin, *WP* (write protect), serves to either block (when '1') or allow (if '0') writing to the device.

There are two options for writing data to this EEPROM, called *byte write* and *page write*. In the former, a single byte is written to the chosen address, while in the latter a set of eight bytes are written sequentially, starting from the given address.

To read from this memory, there are three options, called *current address read*, *random read*, and *sequential read*. In current address read, the last accessed address is read. In random read, any address can be chosen. Finally, in sequential read, it will keep reading until the master stops it with a *NoACK* = '1' bit (followed by a stop sequence). Additional details are available in the device's datasheets.

The recommended approach to solve this type of problem is to use the finite state machine (FSM) model (chapter 11). However, since this is not a trivial circuit, detailed information is needed before we draw the machine's state transition diagram. Two ways of gathering such information are to draw the circuit's flowcharts or to draw its timing diagrams (the former is more behavioral, while the latter is more hardware oriented). Both of these approaches are illustrated below.

Studying the device's datasheets, we get the sequence of events for writing to this memory depicted in figure 14.15a, where the white rectangles represent transmissions made by the master, while the dark ones represent responses issued by the slave. It begins with the master executing the *start* operation (if a *write* command, *wr* = '1', is received), followed by the device address with a *R/W* = '0' bit ('0' is for writing, '1' is for reading), which constitutes the first byte, to which the receiver responds with *ACK* = '0' (if the address is correct). Next, the master sends the intended memory address (again, eight bits), to which the slave again responds with *ACK* = '0'. The third byte contains the data to be stored at the address just sent, followed again by *ACK* = '0' from the slave. If *byte write* is the intended mode, then the dashed arrow in figure 14.15a is the right path; that is, the master simply executes the *stop* sequence and returns to the idle state. On the other hand, if *page write* is the intend mode, then *wr* must remain at '1' for seven more data cycles. Upon returning to the idle state, the *wr* and *rd* inputs are checked, and a new procedure will start if one or both are asserted. In the present example, *byte write* will be implemented.

The same type of reasoning, now for reading, leads to the flowchart of figure 14.15b. To read data, the initial part of the sequence is the same as that for writing (*start*, device address, memory address). In the second part, the *start* operation is repeated and the

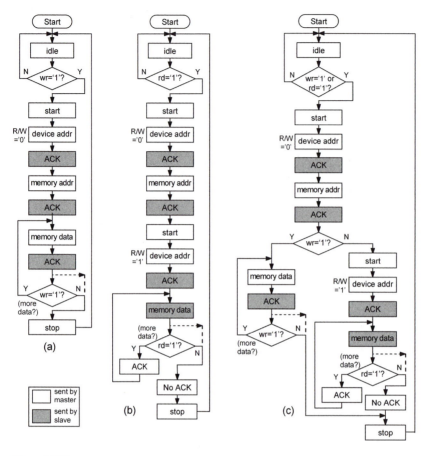

Figure 14.15
Flowcharts for (a) writing to and (b) reading from the EEPROM. The combined procedure is shown in (c).

device address is transmitted, but now with $R/W = $ '1', to which the slave responds with $ACK = $ '0'. After this point, the slave is who transmits the data, with the master then responding with $ACK = $ '0'. If the intended mode is *random read*, then the dashed arrow is the right path—that is, the master sends $NoACK = $ '1', followed by the *stop* sequence, causing the system to return to the idle state. On the other hand, if *sequential read* is the desired mode, then the solid arrow on the left represents the proper path, which is only interrupted when $rd = $ '0', after which $NoACK = $ '1' and *stop* are provided by the master. In figure 14.15c, the combined procedure (write + read) is presented.

As mentioned earlier, another way of gathering information for the state transition diagram is by drawing the system's timing diagram. Based on the datasheets, the timings of figure 14.16 result. The advantage of timing diagrams is that they convey more detailed

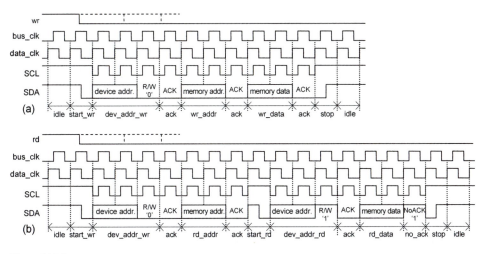

Figure 14.16
EEPROM (a) writing and (b) reading timing diagrams.

information than the flowcharts, but with the drawback that multiple alternatives cannot be expressed (or are difficult to express) (the diagrams of figure 14.16 are for *byte write* and *random read*).

We can now turn to the state transition diagram. Using the information in figure 14.15 or 14.16 (or both), the machine of figure 14.17 results. The upper branch represents the six states common for writing and reading. The central branch represents the continuation for writing and the lower branch is the continuation for reading. The two start sequences are equal, so to differentiate them in the diagram they were called *start_wr* (for writing) and *start_rd* (for reading). The device address is also issued twice, but with the eighth bit (LSB) equal to '0' for writing and '1' for reading, so the corresponding states were named *dev_addr_wr* and *dev_addr_rd*. *ACK* is another state that appears multiple times, so they were called *ACK1*, *ACK2*, and so on. Recall that this machine outputs are the I²C signals (*SCL*, *SDA*), so their values were listed under every state. Because this is an FSM with an embedded timer, the value for *timer* was also listed in each state (see details on how to design this type of machine in section 11.6).

A VHDL code for this FSM is presented below, under the project name *eeprom_i2c* (line 8). Initially, GENERIC was used to enter time-related parameters (lines 10–12), so the code can be easily adapted to any clock frequency, desired speed, and EEPROM write latency. In the present example, the system clock is 50MHz, the desired data rate is 100kbps (standard mode), and the maximum device's write time (t_{wr}) is 5ms (see datasheets). The pin names (lines 13–23) are from figure 14.14.

The code proper (architecture) is in lines 26–199. It starts with a series of signal declarations, which include the device addresses for writing and reading (lines 30–31) and also the

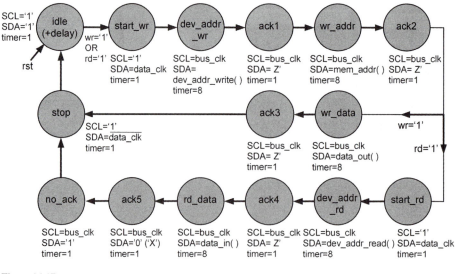

Figure 14.17
Complete finite state machine for EEPROM writing and reading.

FSM-related declarations (lines 40–42). The device's address, with $A_2A_1A_0$ connected to GND (figure 14.14), is "1010000", to which the R/W bit is appended to the right. Note in line 38 that i was declared as a shared variable, so it can be used by more than one process.

Before the processes start, general input-output signals are configured in lines 45–48. Because each memory location is eight bits wide, we can store the SSD value (seven bits) corresponding to each data input value (*data*) directly into the memory, with a '0' appended to its left (line 45). To do so, note in line 45 that, as in previous designs, a function called *integer_to_ssd* (seen in section 12.2, for example) is employed to make the conversion from a 4-bit integer (*data*) to the SSD format (this function was assumed to be in a package called *my_functions*, declared in line 6; the corresponding file must obviously be included in the project). Because the memory addresses are also eight bits wide, four zeros were appended to the left of *address* (line 46). The SSD output (line 47) is a signal read from the memory (*data_in*), with the MSB discarded. Finally, the acknowledge error (*ack_error*, line 48) checks whether the three *ACK* bit received from the slave during the memory-write procedure are correct.

The first process (lines 51–61) derives, from *clk*, an auxiliary clock (*aux_clk*) whose frequency is four times that of the desired data rate. This signal is then used in the next process (lines 64–79) to construct the two clocks seen in the timing diagrams of figure 14.16 (*bus_clk*, *data_clk*), which have phases 90 degrees apart. *bus_clk* is used to create *SCL*, while *data_clk* controls *SDA*.

The other two processes are for the FSM and are direct applications of the general template studied in chapter 11 (see timed FSMs in section 11.6). The lower (sequential) section is in the process of lines 81–113, while the upper (combinational) section is in the process of lines 115–198.

Note: Writing to an EEPROM cell is slow (a few ms/word). For that reason, the chip initially places the data in a regular register (which is fast), to subsequently store it in the EEPROM array. During this operation, the memory ignores any input event. In exercise 14.9, an interesting aspect concerning this latency and the VDHL code below is explored.

```
1   -----------------------------------------------------------------------
2   --Remember to include function "integer_to_ssd" in the project.
3   -----------------------------------------------------------------------
4   LIBRARY ieee;
5   USE ieee.std_logic_1164.all;
6   USE work.my_functions.all; --package with "integer_to_ssd" function.
7   -----------------------------------------------------------------------
8   ENTITY eeprom_i2c IS
9      GENERIC (
10        fclk: POSITIVE := 50_000;      --Freq. of system clock (in kHz)
11        data_rate: POSITIVE := 100;    --Desired I2C bus speed (in kbps)
12        write_time: POSITIVE := 5);    --EEPROM max write time (in ms)
13     PORT (
14        --System signals:
15        clk, rst: IN STD_LOGIC;
16        rd, wr: IN STD_LOGIC;
17        data: IN INTEGER RANGE 0 TO 15;
18        address: IN STD_LOGIC_VECTOR(3 DOWNTO 0);
19        ssd: OUT STD_LOGIC_VECTOR(6 DOWNTO 0);
20        ack_error: OUT STD_LOGIC;
21        --I2C signals:
22        SCL: OUT STD_LOGIC;
23        SDA: INOUT STD_LOGIC);
24  END eeprom_i2c;
25  -----------------------------------------------------------------------
26  ARCHITECTURE fsm OF eeprom_i2c IS
27     --General constants and signals:
28     CONSTANT divider: INTEGER := (fclk/8)/data_rate;
29     CONSTANT delay: INTEGER := write_time*data_rate;
30     CONSTANT dev_addr_write: STD_LOGIC_VECTOR(7 DOWNTO 0):= "10100000";
31     CONSTANT dev_addr_read: STD_LOGIC_VECTOR(7 DOWNTO 0) := "10100001";
32     SIGNAL aux_clk, bus_clk, data_clk: STD_LOGIC;
33     SIGNAL data_in, data_out: STD_LOGIC_VECTOR(7 DOWNTO 0);
34     SIGNAL wr_flag, rd_flag: STD_LOGIC;
35     SIGNAL mem_addr: STD_LOGIC_VECTOR(7 DOWNTO 0);
36     SIGNAL ack: STD_LOGIC_VECTOR(2 DOWNTO 0);
```

```vhdl
37    SIGNAL timer: NATURAL RANGE 0 TO delay;
38    SHARED VARIABLE i: NATURAL RANGE 0 TO delay;
39    --State machine signals:
40    TYPE state IS (idle, start_wr, start_rd, dev_addr_wr, dev_addr_rd,
41       wr_addr, wr_data, rd_data, stop, no_ack, ack1, ack2, ack3, ack4);
42    SIGNAL pr_state, nx_state: state;
43 BEGIN
44    -------General signals:---------------------
45    data_out <= '0' & integer_to_ssd(data);
46    mem_addr <= "0000" & address;
47    ssd <= data_in(6 DOWNTO 0);
48    ack_error <= ack(0) OR ack(1) OR ack(2);
49    -------Auxiliary clock:---------------------
50    --freq=4*data_rate=400kHz for given parameters
51    PROCESS (clk)
52       VARIABLE count: INTEGER RANGE 0 TO divider;
53    BEGIN
54      IF (clk'EVENT AND clk='1') THEN
55         count:= count + 1;
56         IF (count=divider) THEN
57            aux_clk <= NOT aux_clk;
58            count := 0;
59         END IF;
60      END IF;
61    END PROCESS;
62    -------Bus & data reference clocks:----------
63    --freq=data_rate=100kHz for given parameters
64    PROCESS (aux_clk)
65       VARIABLE count: INTEGER RANGE 0 TO 3;
66    BEGIN
67      IF (aux_clk'EVENT AND aux_clk='1') THEN
68         count:= count + 1;
69         IF (count=0) THEN
70            bus_clk <= '0';
71         ELSIF (count=1) THEN
72            data_clk <= '1';
73         ELSIF (count=2) THEN
74            bus_clk <= '1';
75         ELSE
76            data_clk <= '0';
77         END IF;
78      END IF;
79    END PROCESS;
```

```
80      -------Lower section of FSM:----------------
81      PROCESS (data_clk, rst)
82      BEGIN
83        IF (rst='1') THEN
84            pr_state <= idle;
85            i := 0;
86        --Enter data for I2C bus:
87        ELSIF (data_clk'EVENT AND data_clk='1') THEN
88            IF (i=timer-1) THEN
89                pr_state <= nx_state;
90                i := 0;
91            ELSE
92                i := i + 1;
93            END IF;
94        ELSIF (data_clk'EVENT AND data_clk='0') THEN
95            --Store write/read flags:
96            IF (pr_state=idle) THEN
97                wr_flag <= wr;
98                rd_flag <= rd;
99            END IF;
100           --Store ACK signals during writing:
101           IF (pr_state=ack1) THEN
102               ack(0)<=SDA;
103           ELSIF (pr_state=ack2) THEN
104               ack(1)<=SDA;
105           ELSIF (pr_state=ack3) THEN
106               ack(2)<=SDA;
107           END IF;
108           --Store data read from memory:
109           IF (pr_state=rd_data) THEN
110               data_in(7-i) <= SDA;
111           END IF;
112       END IF;
113     END PROCESS;
114     -------Upper section of FSM:----------------
115     PROCESS (pr_state, bus_clk, data_clk, wr_flag,
116         rd_flag, data_out, mem_addr, SDA)
117     BEGIN
118       CASE pr_state IS
119         WHEN idle =>
120             SCL <= '1';
121             SDA <= '1';
122             timer <= delay; --max write time=5ms
```

```
123            IF (wr_flag='1' OR rd_flag='1') THEN
124               nx_state <= start_wr;
125            ELSE
126               nx_state <= idle;
127            END IF;
128         WHEN start_wr =>
129            SCL <= '1';
130            SDA <= data_clk;
131            timer <= 1;
132            nx_state <= dev_addr_wr;
133         WHEN dev_addr_wr =>
134            SCL <= bus_clk;
135            SDA <= dev_addr_write(7-i);
136            timer <= 8;
137            nx_state <= ack1;
138         WHEN ack1 =>
139            SCL <= bus_clk;
140            SDA <= 'Z';
141            timer <= 1;
142            nx_state <= wr_addr;
143         WHEN wr_addr =>
144            SCL <= bus_clk;
145            SDA <= mem_addr(7-i);
146            timer <= 8;
147            nx_state <= ack2;
148         WHEN ack2 =>
149            SCL <= bus_clk;
150            SDA <= 'Z';
151            timer <= 1;
152            IF (wr_flag='1') THEN
153               nx_state <= wr_data;
154            ELSE
155               nx_state <= start_rd;
156            END IF;
157         WHEN wr_data =>
158            SCL <= bus_clk;
159            SDA <= data_out(7-i);
160            timer <= 8;
161            nx_state <= ack3;
162         WHEN ack3 =>
163            SCL <= bus_clk;
164            SDA <= 'Z';
165            timer <= 1;
```

```
166                  nx_state <= stop;
167             WHEN start_rd =>
168                  SCL <= '1';
169                  SDA <= data_clk;
170                  timer <= 1;
171                  nx_state <= dev_addr_rd;
172             WHEN dev_addr_rd =>
173                  SCL <= bus_clk;
174                  SDA <= dev_addr_read(7-i);
175                  timer <= 8;
176                  nx_state <= ack4;
177             WHEN ack4 =>
178                  SCL <= bus_clk;
179                  SDA <= 'Z';
180                  timer <= 1;
181                  nx_state <= rd_data;
182             WHEN rd_data =>
183                  SCL <= bus_clk;
184                  SDA <= 'Z';
185                  timer <= 8;
186                  nx_state <= no_ack;
187             WHEN no_ack =>
188                  SCL <= bus_clk;
189                  SDA <= '1';
190                  timer <= 1;
191                  nx_state <= stop;
192             WHEN stop =>
193                  SCL <= '1';
194                  SDA <= NOT data_clk;
195                  timer <= 1;
196                  nx_state <= idle;
197          END CASE;
198       END PROCESS;
199   END fsm;
200   -----------------------------------------------------------------------
```

14.5 SPI Interface

SPI (serial peripheral interface) is another synchronous serial bus for communication between integrated circuits (installed next to each other, normally on the same board). Like I^2C, it operates in a master-slave architecture, but is simpler to implement and generally operates at higher speeds.

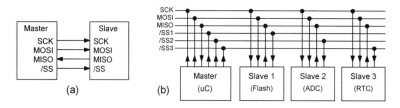

Figure 14.18
SPI bus with (a) single and (b) multiple slaves.

SPI was developed by Motorola for its 68HC family of microcontrollers, now in wide-spread use. Compared to I²C, it has the advantages of requiring a simpler hardware because there is no bidirectional line and making the device selection with a separate wire for each slave rather than a transmitted address, in general operating at higher speeds (multi-MHz range), allowing also duplex mode and flexible message formats. On the other hand, SPI demands more I/O pins, can operate with only one master, has no message acknowledgement, and because there is no standard message format, validation would be more difficult. SPI is said to be a four-wire bus (though that is indeed the least number of wires), while I²C is truly two wires. In some applications, a bidirectional line is used for *MOSI* and *MISO* together, resulting in a three-wire bus.

The SPI bus is illustrated in figure 14.18. In (a), a single slave is shown (normally, the master is a microcontroller), so four wires are needed, called *SCK* (serial clock, always generated by the master), *MOSI* (master out slave in), *MISO* (master in slave out), and *SS* (slave select). When *SS* is low, the slave is selected, to/from which the master sends/receives messages through the *MOSI/MISO* wire. In (b), a multislave system is depicted, so multiple *SS* wires are needed. Examples of ICs with SPI support are also shown in the figure, which are essentially the same categories as for I²C—that is, microcontrollers, EEPROM and Flash memories, A/D and D/A converters, RTCs, and so on.

Operation of the SPI Interface

Figure 14.19 shows the SPI operating modes, which are determined by the clock phase (*CPHA*) and clock polarity (*CPOL*). They are called *mode 00* (*CPHA* = 0, *CPOL* = 0), *mode 01* (*CPHA* = 0, *CPOL* = 1), *mode 10* (*CPHA* = 1, *CPOL* = 0), and *mode 11* (*CPHA* = 1, *CPOL* = 1). In modes 00 and 01, the receiver samples the data at the first clock edge following the lowering of *SS*, so the data must be ready before the first clock transition. In modes 10 and 11, the data is sampled at the second clock edge after *SS* is lowered, so the preceding transition can be used to enter the data. Modes 00 and 11 (positive clock edge) are the most common.

Part of the communication between master and slaves is ruled by information stored in 8-bit registers at both ends. These registers are not standardized, neither in number (≤ 5) nor in content. For example, the SPI in the Motorola MC68HC908GT microcontroller

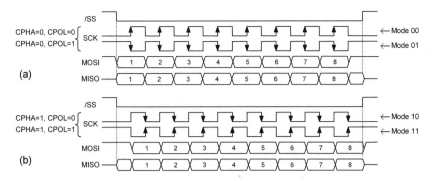

(a)

(b)

Figure 14.19
SPI operating modes (based on clock phase, *CPHA*, and clock polarity, *CPOL*).

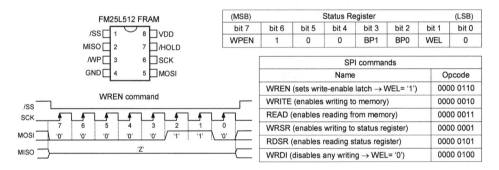

Figure 14.20
SPI-driven FM25L512 FRAM memory (pinout, contents of the status register, timing for the WREN command, and list of SPI commands).

contains three registers (for status, called SPSCR, for control, SPCR, and for data, SPDR), while the SPI in the Maxim DS1306 RTC has two registers (for status and control), and the SPI in the Ramtron FM25L512 FRAM memory contains only one (for status).

To illustrate the use of SPI, the FRAM will be used as an example. It is a 512 kbyte nonvolatile memory with serial access through an SPI bus. Its pinout, contents of the status register, timing for the WREN command, and the set of SPI commands are shown in figure 14.20. An important feature of this new technology (FRAM [Pedroni 2008]) is that data can be written to it at high speed (20MHz in the present example), contrasting with EEPROM (previous section), which generally takes a few ms/word.

Note in figure 14.20 that besides the SPI pins (*SCK*, *MOSI*, *MISO*, and *SS*), the chip contains also two other control pins, called *WP* (write protect) and *HOLD*. The purpose of *WP* is, together with bits 7 (*WPEN*), 3 (*BP1*), and 2 (*BP0*) of the status register, to allow several protection options against writings to both the memory and the status register.

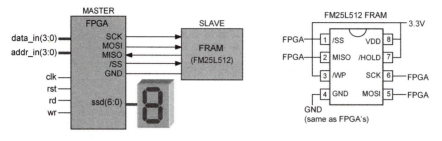

Figure 14.21
SPI interface for a FRAM memory (FPGA implements the master, having the memory as its only slave). The figure on the right shows how the FRAM was wired.

For example, with *WP* high (note that *WP* is active low—we are relaxing the notation, suppressing the negation symbol in the pin names) and $WPEN = BP1 = BP0 = \text{'0'}$, all writings are allowed (see other protection options in the device's datasheets). The role of *HOLD* (also active low) is to handle interrupts.

There is another programmable bit in the status register, called *WEL* (write enable latch), which determines whether writing is allowed (when '1') or not (when '0'). Only when $WEL = \text{'1'}$ are the protection options mentioned earlier in place (any writing is forbidden while $WEL = \text{'0'}$). Because this bit is automatically cleared at power up or at the upward transition of *SS* after a WRITE, WRSR, or WRDI command, any write action must start with the WREN command, because that is the only way of setting *WEL* to '1' (writing to the status register does not affect this bit). Such a command is also illustrated in figure 14.20 (for mode 00), which consists of transmitting the byte "00000110" to the slave (through the *MOSI* port). Note that the slave is selected ($SS = \text{'0'}$) before the command starts and must be unselected ($SS = \text{'1'}$) after it ends. During the command, the slave maintains its output (*MISO*) at high impedance ('Z'). Any SPI communication must start with the MSB.

Additional details about SPI are shown below, where an actual SPI interface is designed using VHDL.

Design of an SPI Interface for an FRAM Memory

The purpose of this exercise is to design a circuit capable of writing data to and reading data from a FRAM that communicates using the SPI bus. Hence our circuit (in the FPGA) will play the role of master, while the memory will be its (only) slave. The chosen device is the FM25L512 FRAM, which supports modes 00 and 11 (automatically selected).

Figure 14.21 shows the setup for the experiment, which is similar to that in the previous section for the I^2C bus and an EEPROM. The inputs are *data* (4-bit data, from four toggle switches), *address* (4-bit address, from another four switches), and *clk* (system clock,

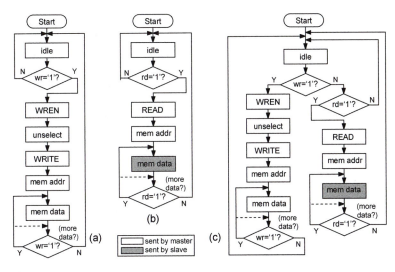

Figure 14.22
Flowcharts for (a) writing to and (b) reading from the FRAM. The combined procedure is shown in (c).

assumed to be 50 MHz), plus *rst* (optional reset), *rd* (read), and *wr* (write) commands (the last three are also from switches). The outputs are the SPI signals (*SCK*, *MOSI*, *MISO*, and *SS—MISO* is in fact an input), connected to the slave, plus *ssd*, which feeds an SSD display (chapter 12) to exhibit the data retrieved from the FRAM. The figure also shows how the device was wired.

Contrary to EEPROMs (previous section), FRAMs can operate in *sequential write* mode because writing to them is fast (20 Mbps in the present example, against a few ms/ word for EEPROMs). In summary, both *sequential write* and *sequential read* are allowed.

The recommended approach to solve this type of problem is the same as that in the previous section—the finite state machine (FSM) model (chapter 11). However, because this too is not a trivial circuit, detailed information is needed before we draw its state transition diagram. As seen before, two ways of gathering such information are to draw the circuit's flowcharts or to draw its timing diagrams. Both of these approaches are illustrated below.

Studying the device's datasheets, the sequence of events that we get for writing is that depicted in figure 14.22a (note that all transmissions are made by the master). It consists of a WREN command (to set *WEN* = '1'), followed by a write enable (WRITE) command, then the memory address, and finally the data to be written to that address. The dashed line in figure 14.22a indicates two options for writing: if the solid arrow is used, then data keeps being written until *wr* returns to zero (*sequential write*; the address is incremented automatically by the slave); otherwise, if the dashed arrow is used, then only one byte is written at a time, with the system again returning to the idle state after *wr* = '0' occurs. The latter will be implemented here.

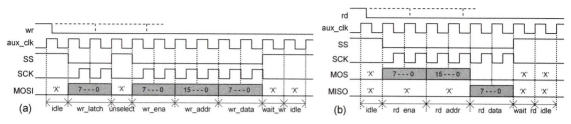

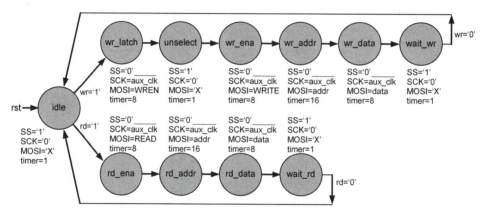

Figure 14.23
FRAM (a) writing and (b) reading timing diagrams.

Figure 14.24
Complete finite state machine for FRAM writing and reading.

The same type of reasoning, now for reading, leads to the flowchart of figure 14.22b. To read data, first the read enable command (READ) is applied, followed by the intended address, after which the slave responds (through the *MISO* wire) with the data stored at that address (the dark rectangle indicates a transmission made by the slave). Here too the dashed arrow indicates two options for reading; *sequential read* if the solid arrow is adopted, or single word reading if the dashed arrow is used instead (the latter will be implemented here). By combining this flowchart with the previous one, the complete write-read process results (figure 14.22c).

As mentioned earlier, another way of gathering information for the state transition diagram is by drawing the system's timing diagram. Based on the datasheets, the timings of figure 14.23 result. The advantage of timing diagrams is that they convey more detailed information than the flowcharts, but with the drawback that multiple alternatives cannot be expressed (or are difficult to express).

We can now turn to the state transition diagram. Using the information in figure 14.22 or 14.23 (or both), the machine of figure 14.24 results, where the upper branch is for writing and the lower one is for reading. The states for writing are *wr_latch* (WREN command), *unselect* (*SS* = '1' to end the WREN command), *wr_enable* (WRITE command), *wr_addr* (memory address where data must be written), *wr_data* (data to be written), and *wait_wr* (waits for *wr* to return to zero). The states for reading are *rd_enable* (READ command), *rd_addr* (memory address from which data must be read), *rd_data* (data read), and finally *wait_rd* (waits for *rd* to return to zero). Recall that this machine must output the master's SPI signals (*SS*, *SCK*, *MOSI*), so their values were listed under every state. Because this is an FSM with an embedded timer, the value for *timer* was also listed in each state (see details on how to design this type of machine in section 11.6).

A VHDL code for this FSM is presented below, under the project name *fram_spi* (line 8). The pin names (lines 9–20) are from figure 14.21. The code proper (architecture) is in lines 23–162. It starts with a series of signal declarations, which include the WREN, WRITE, and READ opcodes (lines 25–27), followed by the FSM-related declarations (lines 29–31), then other system signals (lines 33–37). Note that a shared variable was declared in line 37, so it can be used by more than one process.

Before the processes start, general input-output signals are configured in lines 40–42. Note in line 40 that, as in previous designs, a function called *integer_to_ssd* (seen in section 12.2, for example) was employed to make the conversion from a 4-bit integer (*data*) to the SSD format (this function was assumed to be in a package called *my_functions*, declared in line 6; the corresponding file must obviously be included in the project).

The first process (lines 44–54) derives, from *clk* (50 MHz), an auxiliary clock (*aux_clk*) whose frequency is 1 MHz (the frequency chosen for the present design is 1 Mbps, though this FRAM can be written to at 20 Mbps). As seen in figure 14.23, this clock will define the pace of the SPI signals.

The other two processes are for the FSM and are direct applications of the general template studied in chapter 11 (see timed FSMs in section 11.6) to the diagram of figure 14.24. The lower (sequential) section is in the process of lines 56–75, while the upper (combinational) section is in the process of lines 77–161.

Finally, note that, contrary to the design in the previous section, in the code below there are no GENERIC parameters, which were left out intentionally (see exercise 14.14)

```
1   -----------------------------------------------------------------
2   --Remember to include function "integer_to_ssd" in the project.
3   -----------------------------------------------------------------
4   LIBRARY ieee;
5   USE ieee.std_logic_1164.all;
6   USE work.my_functions.all; --package w/"integer_to_ssd" function.
7   -----------------------------------------------------------------
8   ENTITY fram_spi IS
9      PORT (
10        --System signals:
```

```
11        clk, rst: IN STD_LOGIC;
12        rd, wr: IN STD_LOGIC;
13        data: IN INTEGER RANGE 0 TO 15;
14        address: IN STD_LOGIC_VECTOR(3 DOWNTO 0);
15        ssd: OUT STD_LOGIC_VECTOR(6 DOWNTO 0);
16        --SPI signals:
17        SCK: OUT STD_LOGIC;
18        MOSI: OUT STD_LOGIC;
19        MISO: IN STD_LOGIC;
20        SS: OUT STD_LOGIC);
21   END fram_spi;
22   -------------------------------------------------------------------
23   ARCHITECTURE fsm OF fram_spi IS
24      --SPI commands:
25      CONSTANT WREN: STD_LOGIC_VECTOR(7 DOWNTO 0) := "00000110";
26      CONSTANT WRITEx: STD_LOGIC_VECTOR(7 DOWNTO 0) := "00000010";
27      CONSTANT READx: STD_LOGIC_VECTOR(7 DOWNTO 0) := "00000011";
28      --State machine signals:
29      TYPE state IS (idle, wr_latch, unselect, wr_ena, wr_addr,
30         wr_data, wait_wr, rd_ena, rd_addr, rd_data, wait_rd);
31      SIGNAL pr_state, nx_state: state;
32      --General signals:
33      SIGNAL data_out, data_in: STD_LOGIC_VECTOR(7 DOWNTO 0);
34      SIGNAL mem_addr: STD_LOGIC_VECTOR(15 DOWNTO 0);
35      SIGNAL aux_clk: STD_LOGIC;
36      SIGNAL timer: NATURAL RANGE 0 TO 16;
37      SHARED VARIABLE i: NATURAL RANGE 0 TO 16;
38   BEGIN
39      ------Address, data, and control:--------
40      data_out <= '0' & integer_to_ssd(data);
41      mem_addr <= "000000000000" & address;
42      ssd <= data_in(6 DOWNTO 0);
43      ------Auxiliary clock (1MHz):------------
44      PROCESS (clk)
45         VARIABLE count: NATURAL RANGE 0 TO 25;
46      BEGIN
47         IF (clk'EVENT AND clk='1') THEN
48            count := count + 1;
49            IF (count=25) THEN
50               aux_clk <= NOT aux_clk;
51               count := 0;
52            END IF;
53         END IF;
```

```
54      END PROCESS;
55      ------Lower section of FSM:--------------
56      PROCESS (aux_clk, rst)
57      BEGIN
58         IF (rst='1') THEN
59            pr_state <= idle;
60            i := 0;
61         --Send data to SPI bus:
62         ELSIF (aux_clk'EVENT AND aux_clk='1') THEN
63            IF (i=timer-1) THEN
64               pr_state <= nx_state;
65               i := 0;
66            ELSE
67               i := i + 1;
68            END IF;
69         --Read data from SPI bus:
70         ELSIF (aux_clk'EVENT AND aux_clk='0') THEN
71            IF (pr_state=rd_data) THEN
72               data_in(7-i) <= MISO;
73            END IF;
74         END IF;
75      END PROCESS;
76      -----Upper section of FSM:----------
77      PROCESS (pr_state, aux_clk, wr, rd, data_out, mem_addr)
78      BEGIN
79         CASE pr_state IS
80            WHEN idle =>
81               SS <= '1';
82               SCK <= '0';
83               MOSI <= 'X';
84               timer <= 1;
85               IF (wr='1') THEN
86                  nx_state <= wr_latch;
87               ELSIF (rd='1') THEN
88                  nx_state <= rd_ena;
89               ELSE
90                  nx_state <= idle;
91               END IF;
92            WHEN wr_latch =>
93               SS <= '0';
94               SCK <= NOT aux_clk;
95               MOSI <= WREN(7-i);
96               timer <= 8;
```

```
97              nx_state <= unselect;
98          WHEN unselect =>
99              SS <= '1';
100             SCK <= '0';
101             MOSI <= 'X';
102             timer <= 1;
103             nx_state <= wr_ena;
104         WHEN wr_ena =>
105             SS <= '0';
106             SCK <= NOT aux_clk;
107             MOSI <= WRITEx(7-i);
108             timer <= 8;
109             nx_state <= wr_addr;
110         WHEN wr_addr =>
111             SS <= '0';
112             SCK <= NOT aux_clk;
113             MOSI <= mem_addr(15-i);
114             timer <= 16;
115             nx_state <= wr_data;
116         WHEN wr_data =>
117             SS <= '0';
118             SCK <= NOT aux_clk;
119             MOSI <= data_out(7-i);
120             timer <= 8;
121             nx_state <= wait_wr;
122         WHEN wait_wr =>
123             SS <= '1';
124             SCK <= '0';
125             MOSI <= 'X';
126             timer <= 1;
127             IF (wr='0') THEN
128                 nx_state <= idle;
129             ELSE
130                 nx_state <= wait_wr;
131             END IF;
132         WHEN rd_ena =>
133             SS <= '0';
134             SCK <= NOT aux_clk;
135             MOSI <= READx(7-i);
136             timer <= 8;
137             nx_state <= rd_addr;
138         WHEN rd_addr =>
139             SS <= '0';
```

```
140               SCK <= NOT aux_clk;
141               MOSI <= mem_addr(15-i);
142               timer <= 16;
143               nx_state <= rd_data;
144            WHEN rd_data =>
145               SS <= '0';
146               SCK <= NOT aux_clk;
147               MOSI <= 'X';
148               timer <= 8;
149               nx_state <= wait_rd;
150            WHEN wait_rd =>
151               SS <= '1';
152               SCK <= '0';
153               MOSI <= 'X';
154               timer <= 1;
155               IF (rd='0') THEN
156                  nx_state <= idle;
157               ELSE
158                  nx_state <= wait_rd;
159               END IF;
160         END CASE;
161      END PROCESS;
162 END fsm;
163 -----------------------------------------------------------------
```

14.6 TMDS Interface

TMDS (transition minimized differential signaling) is a modern line code for the serial transmission of video signals. It is used as part of the DVI (digital visual interface) video interface employed for interconnecting desktop computers to LCD monitors, and also as part of the HDMI (high-definition multimedia interface) circuit that connects high-definition camcorders, videogame consoles, set-top boxes, and the like to HDTV (high-definition television) monitors and video projectors. In chapter 16, TMDS will be used as part of a complete DVI circuit designed using VHDL.

The TMDS interface was created by Silicon Image in 1999 with the purpose of connecting computers to flat panel (LCD, in general) monitors. The work was done as part of DDWG (Digital Display Working Group), an industry consortium responsible for the creation of DVI (studied in chapter 16), with TMDS also extended to the HDMI interface.

Its general architecture is depicted in figure 14.25. The TMDS transmitter contains a type of 8B/10B encoder (converts 8 bits into 10 bits, described ahead), followed by a serializer (section 14.2) and a CML (current-mode logic, also described ahead) I/O circuit. The TMDS receiver must then have a CML I/O, a deserializer, and a 10B/8B decoder.

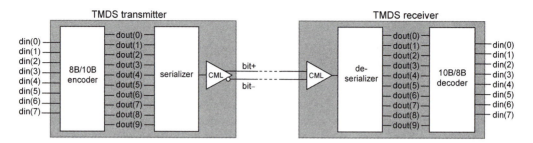

Figure 14.25
TMDS interface.

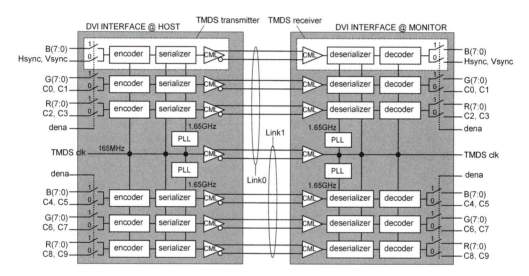

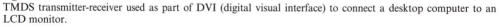

Figure 14.26
TMDS transmitter-receiver used as part of DVI (digital visual interface) to connect a desktop computer to an LCD monitor.

It is important to mention that the 8B/10B code used in TMDS is not equal to the original 8B/10B code introduced by IBM in 1983 (Pedroni 2008). Moreover, in this particular standard (TMDS) the type of I/O logic, which is CML (a type of differential signaling), was also included in the specifications, so the "TM" part of the title is due to the encoder/decoder, while the "DS" part relates to the I/O circuit.

Circuit Details

As mentioned, TMDS is used, for example, as part of the DVI interface between a computer and its LCD monitor. This situation is depicted in figure 14.26.

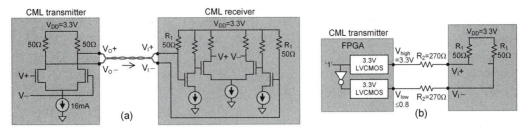

Figure 14.27
(a) CML circuits; (b) CML transmitter emulated with regular 3.3V LVCMOS pads plus resistors.

DVI can contain one or two links, called *link*0 (mandatory) and *link*1 (optional), with each link formed by three TMDS channels. Moreover, each channel can transmit two kinds of data, called *pixel* data ($R =$ red, $G =$ green, and $B =$ blue colors) and *control* data (synchronism and other control signals). The former occurs most of the time and is responsible for creating the images on the screen, while the latter serves to control the monitor, occurring only during the blanking/retrace intervals.

As shown in figure 14.26, the pixel data is eight bits wide, while the control data is just two bits wide. Since the TMDS encoder must produce 10-bit words from 8-bit words, the regular encoding procedure described in figure 14.28 occurs only for pixel data. For control data, fixed 10-bit words are generated (there are only four 10-bit words in this case, because there are only four 2-bit combinations). The resulting 10-bit control words have at least seven transitions, while the 10-bit pixel words have at most five transitions, allowing for safe synchronization.

Note also in figure 14.26 that the typical clock (pixel) frequency is 165 MHz, so because for each pixel a 10-bit vector must be transmitted serially, the transmission clock must be 1.65 GHz. This frequency is produced by a PLL, which multiplies the clock frequency by ten. A similar procedure occurs at the receiving end (obviously a single PLL at each end would suffice).

Observe in figure 14.26 that the I/O circuit is differential, so two wires (plus a global GDN) are required for each channel. This I/O is called CML (current-mode logic) and normally operates with $V_{DD} = 3.3$ V.

A CML Tx-Rx pair is shown in figure 14.27a. The TMDS specifications determine that the two output voltages delivered to the pair of wires that go to the monitor must be V_{DD} (high) and $V_{DD} - 0.4$ V (low). There are 50 Ω resistors at both ends of the wires, so given that only one transistor in the differential pair that composes the CML transmitter is ON at a time, its output voltage is $(50//50)16m = 0.4$ V below V_{DD}, while the other remains at V_{DD}.

Because CML I/Os are not available in FPGAs, in the design examples to come (DVI interface, chapter 16), the circuit of figure 14.27b will be used. In this case, only a pair of

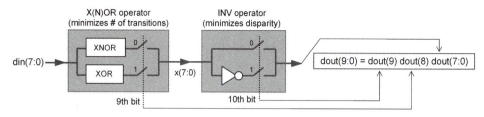

Figure 14.28
The 8B/10B encoder of TMDS.

external resistors (R_2) is needed to emulate a CML pad when regular 3.3V LVTTL or 3.3V LVCMOS I/Os are used instead of CML.

In figure 14.27b, V_{high} and V_{low} represent the '1' and '0' logic voltages produced by the output pad, respectively, while V_{I+} and V_{I-} represent the voltages that actually reach the CML receiver. Assuming that the upper pin is high ($V_{high} = V_{DD}$), then $V_{I+} = V_{DD}$, which obeys the specifications. However, the situation with V_{I-} needs to be examined more closely. Given the arrangement of figure 14.27b, $V_{I-} = (V_{DD} - V_{low})R_2/(R_1 + R_2)$ + V_{low} results. Therefore, $R_2 = [(V_{I-} - V_{low})/(V_{DD} - V_{I-})]R_1$. For $V_{DD} = 3.3$ V, $V_{I-} = 2.9$ V, $V_{low} = 0.8$ V (worst case; see [Pedroni 2008]), and $R_1 = 50$ Ω, $R_2 \approx 270$ Ω is obtained (closest commercial value). As already mentioned, this circuit will be needed in chapter 16.

The TMDS Encoding Algorithm

TMDS coverts an 8-bit word into a 10-bit word with fewer (or at most the same number of) internal transitions, thus reducing high-frequency emissions (less interference between adjacent channels). It also provides a near-perfect DC balance on the communication wires, thus improving the noise margin. However, it is important to remember that a long period without any transitions must be avoided, because it would make clock recovery (for synchronization) more difficult. Even though most TMDS code words have at least one transition, the rare cases when no transitions occur are compensated automatically by the DC balancing technique, which forces transitions to happen (shown later).

The 8/10 encoder of TMDS is depicted in figure 14.28. The first stage is an XOR/ XNOR operator and the second stage is an inverter/noninverter operator. The former minimizes the number of transitions (less EMI), while the latter minimizes the disparity (for DC balance). (*Disparity* is the total number of '1's minus the total number of '0's transmitted in a certain time period, which should remain as near zero as possible.)

The input data byte, *din*(7:0), is passed through an XOR or XNOR gate, producing *x*(7:0). The gate that produces fewer transitions in *x*(7:0) is chosen, and a ninth bit, *x*(8), is added to indicate the choice made ('0' for XNOR, '1' for XOR). Next, *x*(7:0) is inverted or not, producing *dout*(7:0), to which a tenth bit, *dout*(9), is also added to indicate the new choice ('1' when inverted, '0' otherwise). This choice is a function of the accumulated disparity, and it is made as to minimize it. The ninth bit is not affected—that is, $dout(8) = x(8)$. Even though there are only 256 possible values for the pixel data, a total of 460 distinct 10-bit words can result, because some inputs can produce two distinct code words, depending on the disparity.

In the case of video systems using DVI or HDMI (of which TMDS is part), during the transmission of *pixel* data, *dena* (display enable) is asserted, whereas during the transmission of *control* data it is unasserted. In summary:

For *pixel* data: Input = 8 bits, 256 possible values; Output = 10 bits, 460 possible values; *dena* = '1'.

For *control* data: Input = 2 bits, 4 possible values; Output = 10 bits, 4 possible values; *dena* = '0'.

A detailed flowchart for the TMDS algorithm is presented in figure 14.29. This is indeed a modified (by the author) version relative to that from DDWG, introduced here to provide a more hardware-oriented flow, thus helping and optimizing the implementation. The following is employed in the figure (see inset on the upper left corner):

$din(7:0)$ = Input vector with pixel data (eight bits)

$contol(1:0)$ = Input vector with control data (two bits), which can be any of the following pairs: {*Hsync*, *Vsync*}, {*C0*, *C1*}, ..., {*C8*, *C9*}.

$x(8:0)$ = Internal vector after XOR or XNOR ($x(8)$ = '0' for XNOR, or '1' for XOR)

$dout(9:0)$ = Output word (10 bits), which encodes either $din(7:0)$ or $control(1:0)$

dena = Display enable (when '1', *din* is encoded; when '0', *control* is encoded)

$zerosD$ = Number of '0's in $din(7:0)$

$onesD$ = Number of '1's in $din(7:0)$

$zerosX$ = Number of '0's in $x(7:0)$ ($x(8)$ not included)

$onesX$ = Number of '1's in $x(7:0)$ ($x(8)$ not included)

$disp$ = Accumulated disparity ($\sharp$ of '1's minus $\sharp$ of '0's transmitted in a certain period)

Obviously, $zerosD + onesD = 8$, so testing $onesD > zerosD$ is the same as testing $onesD > 4$.

A series of examples are shown in figure 14.30. Note that most inputs can produce two distinct outputs (depending on the disparity), but some can only produce one. Note also that all ten output bits are considered when computing the accumulated disparity.

From figure 14.30, we conclude, for example, that when a long transitionless series occurs (a long series of zeros or of ones), transitions are automatically introduced by the encoder, as shown in the table below.

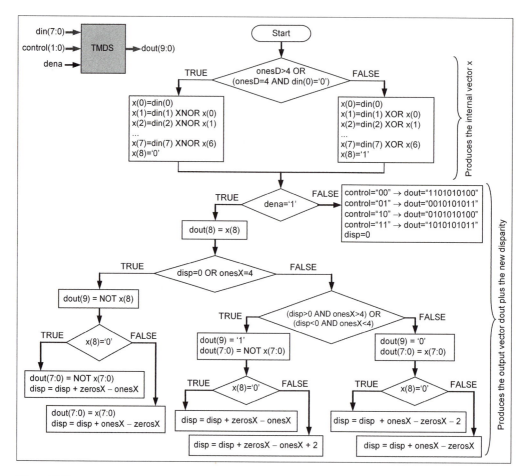

Figure 14.29
Detailed TMDS encoding algorithm.

input ->	1111 1111	1111 1111	1111 1111	1111 1111	1111 1111	1111 1111
disparity ->	0	−8	−2	+4	−4	+2
output ->	10 0000 0000	00 1111 1111	00 1111 1111	10 0000 0000	00 1111 1111	10 0000 0000

Design of a TMDS Encoder

Figure 14.31 shows, on the left, a single TMDS channel (for video applications), and on the right its encoding portion (plus the input multiplexer). This is the circuit that will be designed in this section, using VHDL, then used later (chapter 16) to implement a complete DVI interface.

Input din(7:0)	onesD	Function	x(8:0)	onesX	Assumed disparity	Output dout(9:0)	New disparity
0000 0000	0	XOR	1 0000 0000	0	+2	01 0000 0000	–6
					0	01 0000 0000	–8
					–2	11 1111 1111	+8
1111 1111	8	XNOR	0 1111 1111	8	+2	10 0000 0000	–6
					0	10 0000 0000	–8
					–2	00 1111 1111	+4
0101 0101	4	XOR	1 0011 0011	4	+2	01 0011 0011	+2
					0	01 0011 0011	0
					–2	01 0011 0011	–2
1010 1010	4	XNOR	0 1100 1100	4	+2	10 0011 0011	+2
					0	10 0011 0011	0
					–2	10 0011 0011	–2
0101 0000	2	XOR	1 0011 0000	2	+2	01 0011 0000	–2
					0	01 0011 0000	–4
					–2	11 1100 1111	+4
1010 1111	6	XNOR	0 1100 1111	6	+2	10 0011 0000	–2
					0	10 0011 0000	–4
					–2	00 1100 1111	0

Figure 14.30
Examples of TMDS encodings for several input values and several disparity values.

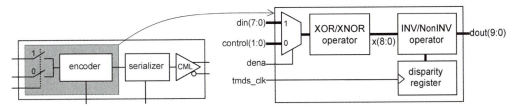

Figure 14.31
Single TMDS channel and its encoding portion (plus mux).

A VHDL code for this circuit is presented below, under the title *tmds* (line 5). The input-output signals (lines 7–11) are from figure 14.31. Several test ports were also included (lines 13–16). The code is divided into four parts. The first part (process of lines 26–37) counts the number of '1's in *din*. Based on this value, the second part (process of lines 39–62) determines *x*(8:0). The third part (process of lines 64–75) computes the number of '1's in *x*(7:0), so the last part (process of lines 78–126) can complete the algorithm. Note that the overall code follows the algorithm of figure 14.29. Simulation results are depicted in figure 14.32.

```
1    --------------------------------------------------------------------
2    LIBRARY ieee;
3    USE ieee.std_logic_1164.all;
4    --------------------------------------------------------------------
5    ENTITY tmds IS
```

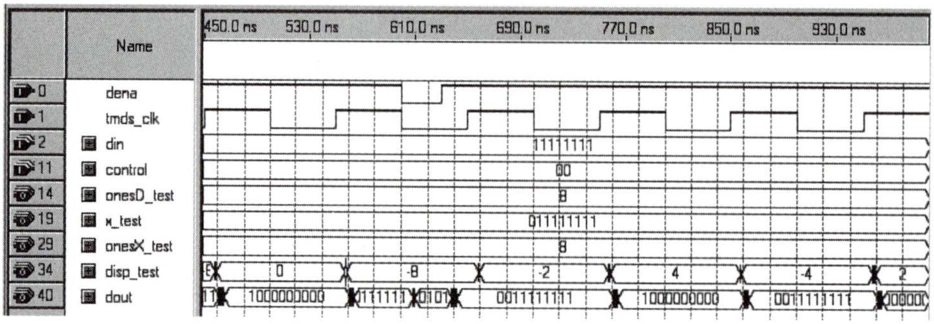

Figure 14.32
Simulation results from the VHDL code for the TMDS encoder. Note that some of the values are identical to those in figure 14.30.

```
 6    PORT (
 7        tmds_clk: IN STD_LOGIC;                    --TMDS clock
 8        dena: IN STD_LOGIC;                        --display enable
 9        din: IN STD_LOGIC_VECTOR(7 DOWNTO 0);      --pixel data
10        control: IN STD_LOGIC_VECTOR(1 DOWNTO 0);--control data
11        dout: OUT STD_LOGIC_VECTOR(9 DOWNTO 0);   --output data
12        ---------------------
13        onesD_test: OUT INTEGER RANGE 0 TO 8;      --test onesD
14        x_test: OUT STD_LOGIC_VECTOR(8 DOWNTO 0);--test x
15        onesX_test: OUT INTEGER RANGE 0 TO 8;      --test onesX
16        disp_test: OUT INTEGER RANGE -16 TO 15); --test disp
17    END tmds;
18    ----------------------------------------------------------------
19    ARCHITECTURE tmds OF tmds IS
20        SIGNAL x: STD_LOGIC_VECTOR(8 DOWNTO 0);   --internal vector
21        SIGNAL onesX: INTEGER RANGE 0 TO 8;        --# of '1's in x(7:0)
22        SIGNAL onesD: INTEGER RANGE 0 TO 8;        --# of '1's in din(7:0)
23        SIGNAL disp: INTEGER RANGE -16 TO 15;      --disparity
24    BEGIN
25        ---Compute number of '1's in din:----------
26        PROCESS (din)
27            VARIABLE counterD: INTEGER RANGE 0 TO 8;
28        BEGIN
29            counterD := 0;
30            FOR i IN 0 TO 7 LOOP
31                IF (din(i)='1') THEN
32                    counterD := counterD + 1;
33                END IF;
34            END LOOP;
```

```
35       onesD <= counterD;
36       onesD_test <= counterD;
37    END PROCESS;
38    ---Produce internal vector x:--------------
39    PROCESS (din, onesD)
40    BEGIN
41       x(0) <= din(0);
42       IF (onesD>4 OR (onesD=4 AND din(0)='0')) THEN
43          x(1) <= din(1) XNOR x(0);
44          x(2) <= din(2) XNOR x(1);
45          x(3) <= din(3) XNOR x(2);
46          x(4) <= din(4) XNOR x(3);
47          x(5) <= din(5) XNOR x(4);
48          x(6) <= din(6) XNOR x(5);
49          x(7) <= din(7) XNOR x(6);
50          x(8) <= '0';
51       ELSE
52          x(1) <= din(1) XOR x(0);
53          x(2) <= din(2) XOR x(1);
54          x(3) <= din(3) XOR x(2);
55          x(4) <= din(4) XOR x(3);
56          x(5) <= din(5) XOR x(4);
57          x(6) <= din(6) XOR x(5);
58          x(7) <= din(7) XOR x(6);
59          x(8) <= '1';
60       END IF;
61       x_test <= x;
62    END PROCESS;
63    ---Compute number of '1's in x:------------
64    PROCESS (x)
65       VARIABLE counterX: INTEGER RANGE 0 TO 8;
66    BEGIN
67       counterX := 0;
68       FOR i IN 0 TO 7 LOOP
69          IF (x(i)='1') THEN
70             counterX := counterX + 1;
71          END IF;
72       END LOOP;
73       onesX <= counterX;
74       onesX_test <= counterX;
75    END PROCESS;
76    ---Produce output and new disparity:-------
77    ---(note that only disp requires storage)--
78    PROCESS (disp, x, onesX, dena, control, tmds_clk)
```

```
79          VARIABLE disp_new: INTEGER RANGE -31 TO 31;
80      BEGIN
81        IF (dena='1') THEN
82          dout(8) <= x(8);
83          IF (disp=0 OR onesX=4) THEN
84            dout(9) <= NOT x(8);
85            IF (x(8)='0') THEN
86              dout(7 DOWNTO 0) <= NOT x(7 DOWNTO 0);
87              disp_new := disp - 2*onesX + 8;
88            ELSE
89              dout(7 DOWNTO 0) <= x(7 DOWNTO 0);
90              disp_new := disp + 2*onesX - 8;
91            END IF;
92          ELSE
93            IF ((disp>0 AND onesX>4) OR (disp<0 AND onesX<4)) THEN
94              dout(9) <= '1';
95              dout(7 DOWNTO 0) <= NOT x(7 DOWNTO 0);
96              IF (x(8)='0') THEN
97                disp_new := disp - 2*onesX + 8;
98              ELSE
99                disp_new := disp - 2*onesX + 10;
100             END IF;
101           ELSE
102             dout(9) <= '0';
103             dout(7 DOWNTO 0) <= x(7 DOWNTO 0);
104             IF (x(8)='0') THEN
105               disp_new := disp + 2*onesX - 10;
106             ELSE
107               disp_new := disp + 2*onesX - 8;
108             END IF;
109           END IF;
110         END IF;
111       ELSE
112         disp_new := 0;
113         IF (control="00") THEN
114           dout <= "1101010100";
115         ELSIF (control="01") THEN
116           dout <= "0010101011";
117         ELSIF (control="10") THEN
118           dout <= "0101010100";
119         ELSE
120           dout <= "1010101011";
121         END IF;
122       END IF;
```

```
123          IF (tmds_clk'EVENT AND tmds_clk='1') THEN
124             disp <= disp_new;
125          END IF;
126       END PROCESS;
127       disp_test <= disp;
128    END tmds;
129    ------------------------------------------------------------------
```

14.7 Video Interfaces: VGA, DVI, and FPD-Link

Special attention will be given in successive chapters to video interfaces. Even though that material also belongs to the overall serial data communications category, the length of the material calls for each video interface to be treated in a specific chapter, as follows.

VGA Video Interface: chapter 15.

DVI Video Interface: chapter 16.

FPD-Link Video Interface: chapter 17.

14.8 Exercises

Note: For exercise solutions, please consult the book website.

Exercise 14.1: Synchronous versus Asynchronous Communication

Based on the definition of *synchronous* versus *asynchronous* communication in section 14.1, in which of them (if any) can the communication be stopped during the idle intervals (that is, while there is no data to be sent)? Why?

Exercise 14.2: Generic Serializer

a) The serializer designed in section 14.2 starts the transmissions with the LSB. What needs to be changed in that code for it to start with the MSB?

b) That code has already a GENERIC parameter (called N) that specifies the number of bits in the input word. To make the code even more generic, add a control input to specify from which end of the input word the data transmissions should start (LSB-first when *direction* = '0', MSB-first otherwise).

c) How many flip-flops do you expect this circuit to need (as a function of N)?

Exercise 14.3: FSM-Based Serializer

Design a serializer using the finite state machine approach (chapter 11). Assume that *sclk* is already available. Include in it a *direction* control, as in exercise 14.2.

Exercise 14.4: Design of a Shift-Register-Based Serializer

a) The serializer designed in section 14.2 was based on the circuit of figure 14.3b. Develop a similar design (with the same frequencies), now based on the circuit of figure 14.3a.

b) How many flip-flops do you expect this circuit will need (for $N = 4$)?

Exercise 14.5: Deserializer

a) Describe the operation of the deserializer of figure 14.4. Pay particular attention to the clock and enable signals (describe the relationship between them and how they can be obtained).

b) Design this circuit using VHDL. Assume that *sclk* is already available.

Exercise 14.6: Code Words for PS2 Keyboard

What are the code words transmitted by a PS2 keyboard controller to the host computer to encode the following information:

a) Lowercase "b"

b) Uppercase "B"

c) Page Down key.

Exercise 14.7: Analysis of PS2 Keyboard Interface

In the design of a PS2 keyboard interface in section 14.3, the circuit detects all serial code words related to the same key press. In such a case, line 116 of the code, though not indispensable, is highly desirable. Explain why, then test the code with and without that line to confirm your answer.

Exercise 14.8: PS2 Keyboard Interface with LCD

Redo the design of a PS2 keyboard interface in section 14.3, this time with an LCD display instead of an SSD.

Exercise 14.9: Measuring EEPROM Latency with I^2C Interface

In section 14.4, an I^2C interface for an EEPROM was constructed. In it, acknowledgement error (*ack_error*) was also included, which checks whether all three ACK bits issued by the memory during the data-write operation are low.

a) What actually happens if the write time (line 12) is lowered to, say, 0.5 ms?

b) How can this bit be used to measure the actual value of the EEPROM's write time (t_{wr})?

c) Can you do it by changing just one value (number) in the VHDL code given?

Exercise 14.10: I^{2}C Interface for an RTC

Design an application for a circuit capable of interfacing using the I^{2}C bus with a RTC (real time clock). The device can be, for example, IPCA8563, from NXP, or DS3231, from Maxim.

Exercise 14.11: I^{2}C Interface for an ADC

Design an application for a circuit capable of interfacing using the I^{2}C bus with an ADC. The device can be, for example, AD7991, from Analog Devices, or PCF8591, from NXP.

Exercise 14.12: I^{2}C Interface for a Temperature Sensor

Design an application for a circuit capable of interfacing using the I^{2}C bus with a temperature sensor. The device can be, for example, LM75A, from NXP, or AD7416, from Analog Devices.

Exercise 14.13: I^{2}C versus SPI

a) Make a table comparing the main features of the I^{2}C and SPI interfaces. Include (at least) the following topics in your analysis: synchronous or asynchronous, number of wires, duplex or simplex, single- or multimaster, with data acknowledgement or not, which hardware is simpler and why, who generates clock and data, and who operates at higher speed.

b) Make a list of device categories (EEPROM, ADC, DAC, etc.) of ICs currently fabricated with I^{2}C support. For each category, include at least one commercial part number.

c) Repeat part (b) above for SPI.

Exercise 14.14: Generic SPI Interface for a FRAM

Contrary to other designs, the code for an SPI interface in section 14.5 does not contain generic parameters. Improve that code by making it as generic as possible (see, for example, the design of an I^{2}C interface for an EEPROM memory in section 14.4).

Exercise 14.15: SPI Interface for a Flash Memory

Design an application for a circuit capable of interfacing using the SPI bus with a Flash memory. The device can be, for example, S25FL008A, from Spansion.

Exercise 14.16: SPI Interface for an ADC

Design an application for a circuit capable of interfacing using the SPI bus with an ADC. The device can be, for example, MAX1242, from Maxim.

Exercise 14.17: SPI Interface for an RTC

Design an application for a circuit capable of interfacing using the SPI bus with an RTC. The device can be, for example, DS1306, from Maxim.

Exercise 14.18: TMDS Encoder

a) Can the disparity value during TMDS encoding be odd?

b) Apply the values shown in the first column of figure 14.30, along with the assumed disparity, to the algorithm of figure 14.29 and check whether the values listed in figure 14.30 indeed result.

15 VHDL Design of VGA Video Interfaces

15.1 Introduction

VGA (video graphics array) is a standard interface introduced by IBM in 1987 for connecting computers to analog video monitors (figure 15.1a). The circuits responsible for generating, processing, and storing the video signals are called *graphics controller* (computer side) and *display controller* (monitor side) (figure 15.1b).

VGA monitors are CRTs (cathode ray tubes). Their current most basic resolution is $640 \times 480 \times 60$Hz VGA; it consists of 640 columns by 480 lines of pixels (picture elements), refreshed 60 times per second (see list in figure 15.2b). It employs three colors ($R = $ red, $G = $ green, $B = $ blue) per pixel, with the intensity of each color determined by an analog voltage in the 0V-to-0.7V range (indeed, for green, it can be either the regular 0V-to-0.7V range or 0.3 V higher; that is, 0.3V-to-1V when sync-on-green is used—explained later). To generate the colors, six bits per color were initially employed, thus allowing a total of $(2^6)^3 = 266,144$ distinct colors or shades (though only 16 or 256 of them were made available in the original VGA palette).

Other, higher-resolution standards succeeded VGA, such as SVGA (super VGA), XGA (extended graphics array), SXGA (super extended graphics array), and more (see their resolutions in figure 15.2). Such analog interfaces are *collectively* referred to as *VGA modes*, with the original $640 \times 480 \times 60$Hz version still being the default mode for most analog computer monitors (that is the mode in which PCs operate at the beginning of the start-up sequence).

VGA monitors are now being replaced with LCDs (liquid crystal displays), which operate differently from CRTs. They are fully digital, and their standard interface is called DVI (digital visual interface). Despite its digital behavior, in many cases the graphics controller also contains a VGA section, so VGA monitors can still be used. This option (analog + digital) is called DVI-I (integrated DVI), while the digital-only version is called DVI-D. Details on the theory and design of DVI systems will be given in chapter 16.

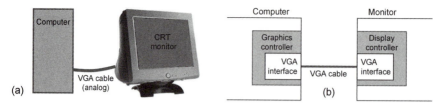

Figure 15.1
(a) Computer connected to an analog VGA monitor; (b) Circuits responsible for the video signals (*graphics controller* on the host side, *display controller* on the device side).

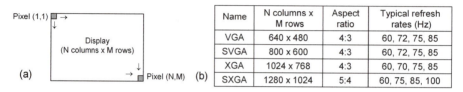

Figure 15.2
(a) Pixel count; (b) Examples of display resolutions (collectively called VGA modes).

As in chapters 16 and 17, which also deal with video interfaces, here we focus on the following fundamental aspects:

1) Operation of the VGA interface.

2) How the circuit should be divided to make the design as simple and as standard as possible.

3) How the control signals operate and how they should be generated.

4) How images can be generated (from local hardware, external memory, file, etc.), rather than focusing on images themselves (software).

15.2   VGA Connector

Figure 15.3 shows a VGA connector, which is a 15-pin connector called DB15. The figure also shows the corresponding pinout, defined by the VESA (Video Electronics Standards Association) through a document called DDC (display data channel).

Observe the following in figure 15.3:

1) The table is separated into two parts, with the upper part containing the standard signals and the lower part containing the signals used for monitor identification.

2) In the upper part, three wires are employed for the colors (R, G, B), which are analog voltages between 0 V and 0.7 V on two parallel 75 Ω resistors ($= 37.5$ Ω). As already mentioned, G can be 0.3 V higher.

VGA male connector
(DB15 male plug)

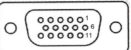
VGA female connector
(DB15 female receptacle)

Pin	Signal	Direction	Simplest setup
1	R (analog red, 0V-0.7V on 37.5Ω)	To monitor	Connected (analog)
2	G (analog green, 0V-0.7V or 0.3V-1V on 37.5Ω)	To monitor	Connected (analog)
3	B (analog blue, 0V-0.7V on 37.5Ω)	To monitor	Connected (analog)
5	GND (general and for +5V)	To monitor	GND
6	GND for R	To monitor	GND
7	GND for G	To monitor	GND
8	GND for B	To monitor	GND
9	No pin or +5V (optional)	To monitor	N/C
10	GND for HSync and Vsync	To monitor	GND
13	Hsync (horizontal sync, 0V/5V waveform)	To monitor	Connected (digital)
14	Vsync (vertical sync, 0V/5V waveform)	To monitor	Connected (digital)

	Old (non-DDC):		DDC1:		DDC2B (and E-DDC):		
4	ID2	From monitor	(ID2	From monitor)	(ID2	From monitor)	N/C
11	ID0	From monitor	(ID0	From monitor)	(ID0	From monitor)	N/C
12	ID1	From monitor	Data	From monitor	SDA	Bidirectional	N/C
15	-----	-----	(ID3	From monitor)	SCL	To monitor	N/C

Figure 15.3
The most common VGA connector, called DB15, with respective pinout. The function of some pins depends on the DDC version. The last column shows the simplest possible connection (just five active wires plus ground).

3) Still in the upper part, two wires are employed for the horizontal and vertical synchronization signals (*Hsync*, *Vsync*), which consist of 0V/5V digital waveforms.

4) The remaining six wires in the upper part of the table are either for ground or for an optional +5V supply voltage.

5) Finally, the lower part of the table shows the signals used for monitor identification. In modern monitors, the DDC standard is used (described later). Note that the pins' functionalities change with the DDC version.

15.3 DDC and EDID

The lower part of the table in figure 15.3 shows three alternatives for monitor identification (parameter passing), identified as non-DDC, DDC1, and DDC2B.

In the case of non-DDC (obsolete), the monitor is identified by means of three wires, called ID0, ID1, and ID2 (ID3 was added later). Each of these wires has a pull-up resistor connected to V_{CC} on the computer side, so the default logic value on these pins is '1'. For a '0' to occur, the display controller must provide a short circuit to GND. A typical monitor identification using ID2-ID1-ID0 (in this order) is the following: "111" → no monitor, "101" → monochrome monitor with resolution under 1024×768, "110" → color monitor with resolution under 1024×768, and "010" → color monitor with the 1024×768 resolution included.

DDC (display data channel) is a standard procedure for computer-monitor communication developed by VESA. With the introduction of DDC, such communication was greatly improved because it allowed much more room for the display to tell its parameters to the graphics controller. Such information is stored in a ROM on the monitor side, and obeys a standard format called EDID (extended display identification data), also defined by VESA. Up to 128 bytes can be stored in the ROM, containing the manufacturer's name, supported resolutions, and other information. Such information is made available every time a pulse occurs in *Vsync* (in other words, *Vsync* acts as the memory-read clock). ID3 and the +5V supply were also included in DDC1, though monitor identification through ID pins became rapidly obsolete, with the EDID structure (accessed through pin 12) preferred instead.

DDC1 was rapidly superseded by DDC2B, which introduced an important change in the DDC channel—it determined that the EDID data should be transmitted using the I^2C bus (section 14.4). I^2C is a robust serial interface consisting of just two wires, called *SDA* (serial data) and *SCL* (serial clock), plus the power supply rails (V_{CC} and GND). Consequently, the EDID information must be stored in a ROM that supports I^2C access. The *SDA* and *SCL* lines are pulled-up to V_{CC} (5 V or 3.3 V) by 15 kΩ resistors installed on the computer side. Since *SDA* is bidirectional, with DDC2B data can now not only be read from the monitor, but can also be sent to it, thus allowing, for example, the graphics controller to change the display's setup. The 128-byte EDID structure was eventually expanded to 256 bytes.

As already mentioned, EDID is a standard data format for storing display-related information in a ROM at the monitor. Like DDC, EDID was also specified by VESA. Currently, the EDID ROM is accessed using the I^2C bus. Indeed, this combination of EDID with I^2C constitutes the monitor-identification part of the DDC2B standard. There are several versions of EDID (1.1, 1.2, 1.3, 2, 3), encompassing 128 and 256 bytes of display information.

15.4   Circuit Diagram

Figure 15.4 shows a block diagram for a VGA system, with the graphics controller on the left and the display controller on the right. The VGA interface is the circuit used in the physical layer that interconnects these two controllers (see figure 15.1b). As shown in figure 15.4, the computer-side circuit can be divided into four sections: *image generator*, *control generator*, *EDID interface*, and finally the DACs (digital to analog converters).

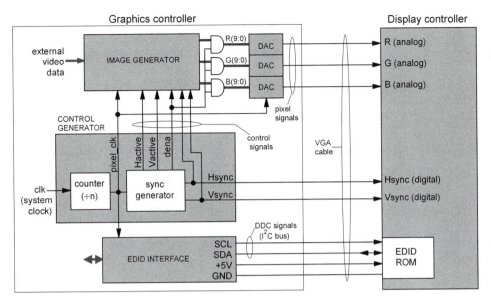

Figure 15.4
VGA interface circuit diagram.

The pixel signals are generated digitally (normally with 10 bits), then converted to analog by the three DACs. These analog voltages constitute the image that will be displayed by the VGA monitor.

Contrary to the pixel signals, the control signals have a fixed constitution (for a given VGA mode, of course). The whole sequence is controlled by *pixel_clk*, whose frequency, for the original (default) VGA, is 25.175 MHz. Five control signals are generated (though in many cases not all are needed): *Hsync* (horizontal synchronism), *Vsync* (vertical synchronism), *Hactive* (portion of *Hsync* during which pixels are displayed), *Vactive* (portion of *Vsync* during which lines of pixels are displayed), and *dena* (display enable).

Hsync and *Vsync* are responsible for determining when a new line or new frame should start, with their timings also determining the VGA mode. *Hactive* and *Vactive* represent the time intervals during which an image is actually being drawn on the screen. Finally, *dena* is responsible for turning the pixel signals OFF during retrace (that is, while the electron beam returns to the beginning of a new line or of a new frame), so it can be obtained by simply ANDing *Hactive* and *Vactive*. Note that only two of the five control signals are transmitted to the monitor.

As already mentioned, the EDID circuit is responsible for retrieving the monitor specifications, stored in the EDID ROM. Note that all signals in the VGA cable are generated by the host (computer) and transmitted to the device (monitor), with the exception of *SDA*, which is bidirectional (reads display parameters and can also adjust them).

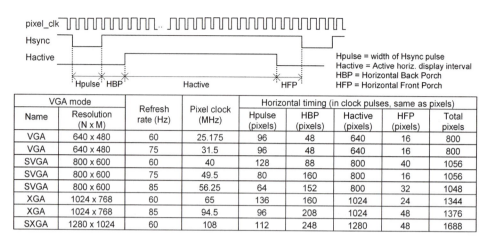

Figure 15.5
Examples of VGA modes and corresponding horizontal timing parameters.

15.5 Control Signals

As we have seen, the control signals are *Hsync, Vsync, Hactive, Vactive*, and *dena*, and their timings define the VGA mode. Several such modes are listed in figure 15.5, where the first line corresponds to the original (default) VGA. Note that in this case the frequency of *pixel_clk* is 25.175 MHz.

Figure 15.5 also shows the waveforms for *Hsync* and *Hactive* (based on *pixel_clk*), which consist of four parts (all measured in number of pixels; i.e., number of clock cycles), called *Hpulse* (width of the horizontal synchronization pulse), *HBP* (horizontal back porch), *Hactive* (active line display interval), and *HFP* (horizontal front porch).

The vertical timing diagram is depicted in figure 15.6, also consisting of four parts (all measured in number of lines or number of *Hsync* cycles), called *Vpulse* (width of the vertical synchronization pulse), *VBP* (vertical back porch), *Vactive* (active column display interval), and *VFP* (vertical front porch).

For example, in the 640 × 480 × 60Hz VGA option, the drawing of one line takes 800 clock cycles (figure 15.5), while one frame requires a time equivalent to 525 lines (figure 15.6). Consequently, to be able to generate 60 frames per second, the clock frequency must be 800 × 525 × 60 = 25.175 MHz, which is the value listed in the figures (indeed, this clock frequency is for a refresh rate of 59.94 Hz, a value inherited from the NTSC television system).

The final control signal is *dena*, which must be low during the blanking intervals (retrace), so it can be easily obtained by ANDing *Hactive* and *Vactive*.

As already mentioned, some monitors support also a form of composite sync called *sync-on-green*, which consists of combining both horizontal and vertical synchronization

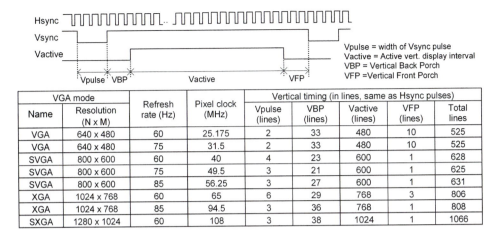

VGA mode		Refresh rate (Hz)	Pixel clock (MHz)	Vertical timing (in lines, same as Hsync pulses)				
Name	Resolution (N x M)			Vpulse (lines)	VBP (lines)	Vactive (lines)	VFP (lines)	Total lines
VGA	640 x 480	60	25.175	2	33	480	10	525
VGA	640 x 480	75	31.5	2	33	480	10	525
SVGA	800 x 600	60	40	4	23	600	1	628
SVGA	800 x 600	75	49.5	3	21	600	1	625
SVGA	800 x 600	85	56.25	3	27	600	1	631
XGA	1024 x 768	60	65	6	29	768	3	806
XGA	1024 x 768	85	94.5	3	36	768	1	808
SXGA	1280 x 1024	60	108	3	38	1024	1	1066

Figure 15.6
Examples of VGA modes and corresponding vertical timing parameters.

pulses with the green signal, thus eliminating the need for the *Hsync* and *Vsync* wires. In this case, the voltage range of *G* is made 0.3 V higher (0.3V-to-1V), so the downward pulses (toward 0 V) of *Hsync* and *Vsync* can be easily detected by the display controller, which can also distinguish *Hsync* from *Vsync* because the latter is much longer.

15.6   Pixel Signals

In the last section we described the control signals, which are application-independent (fixed). In this section we discuss the pixel signals, which vary with the application.

Figure 15.7a shows additional details regarding the DACs (pixel signals). Thanks to the AND gates, *R*, *G*, and *B* are turned OFF when *dena* = '0'. Note that *pixel_clk* is needed to control the data sequence that is applied to the triple DAC. Finally, observe the presence of two new control signals, called *nblank* and *nsync*, which are specifically for the DACs.

An example of DAC for video applications is shown in figure 15.7b, which is the ADV7123 chip from Analog Devices, containing a triple 10-bit converter. The *R* and *B* DACs provide output voltages in the 0V-to-0.7V range, while the *G* DAC includes also a 0.3 V higher (0.3V-to-1V) range, needed when sync-on-green is used. Since the DACs drive dual 75 Ω loads, the output currents from *R* and *B* must be in the 0mA-to-18.7mA range (because $0.7V/(75\Omega//75\Omega) = 18.7$ mA). The additional current source for *G* (which can be turned ON and OFF by *nsync*) is 8 mA (because $0.3V/(75\Omega//75\Omega) = 8$ mA).

The purpose of *nsync* is to cause *G* to operate in the regular (when low) or composite sync (when high) range. The purpose of *nblank* is to blank the screen (when low) by

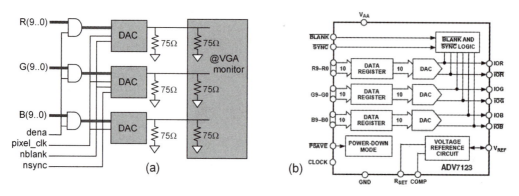

Figure 15.7
(a) Details regarding the DACs; (b) ADV7123 triple DAC.

bringing *R*, *G*, and *B* to their lowest levels: 0 mA (thus 0 V) for *R* and *B* and 0 mA (if operating in the regular range), or 8 mA (hence 0.3 V, if using composite sync) for *G*.

15.7 Setup for the Experiments

Having described all signals involved in the operation of the VGA interface, we present in figure 15.8a a block diagram for the circuit to be used in the experiments (borrowed from figure 15.4). Since the original VGA mode (640 × 480 × 60Hz) will be used, there is no need to read the EDID ROM because support for the original mode is compulsory.

Figure 15.8b repeats the plots of figures 15.5 and 15.6, again for the particular case of 640 × 480 × 60Hz VGA. The physical setup of figure 15.8a and the values of figure 15.8b will be employed in the experiments that follow, in which the host-side VGA interface will be designed using VHDL.

15.8 Comments on VHDL Code for VGA Systems

All remaining sections of this chapter deal with actual VGA circuits designed using VHDL. Based on figure 15.8a, such circuits will be divided into two parts: *control generator*, which is always the same (for a given VGA mode), hence application-independent, and *image generator*, responsible for generating the images used in the examples, hence application-dependent.

The generation/processing of images can be done in a number of ways. For example, images can be built by the VHDL code itself (local hardware, proper only for very simple, geometric images), or can be retrieved from some kind of memory (SRAM, file, etc.), or can be real-time images produced by a video camera. The purpose of the designs presented

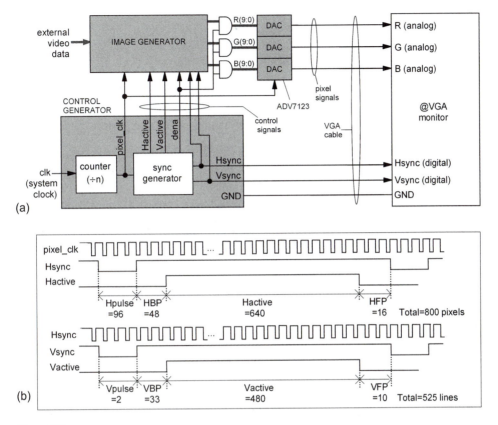

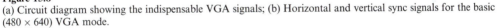

Figure 15.8
(a) Circuit diagram showing the indispensable VGA signals; (b) Horizontal and vertical sync signals for the basic (480 × 640) VGA mode.

in the following sections is to illustrate some of these methods (in other words, the designs are hardware-oriented instead of software-oriented).

15.9   Hardware-Generated Image

The main purpose of this first design is to illustrate how the control signals can be constructed. To keep the focus on that, a very simple image was chosen, so it can be generated by the local hardware (FPGA) without the need for external data (the designs in the next sections will take care of external data). The image consists of just four horizontal stripes of solid colors, with widths 1, 2, 3, and 474 pixels. The first three are red, green, and blue, respectively, while the last one (wide) is determined by three toggle switches, hence allowing eight colors.

red_switch = '0' → no red (*R* = "0000000000" → 0V).

red_switch = '1' → maximum red intensity (*R* = "1111111111" → 0.7V).

green_switch = '0' → no green (*G* = "0000000000" → 0V).

green_switch = '1' → maximum green intensity (*G* = "1111111111" → 0.7V).

blue_switch = '0' → no blue (*B* = "0000000000" → 0V).

blue_switch = '1' → maximum blue intensity (*B* = "1111111111" → 0.7V).

A 50 MHz clock and a triple 10-bit DAC (as in Altera's DE2 board) will be used. The DAC's control signals, *nblank* and *nsync*, must be kept at '1' and '0', respectively.

A VHDL code for this VGA interface is presented below. The control signal parameters of figure 15.8b were entered using GENERIC declarations (lines 7–14), so the code can be easily adjusted to other VGA modes. The signal names (lines 16–21) are from figure 15.8a. Note that because only two (*Hsync*, *Vsync*) of the five control signals are transmitted to the monitor (figure 15.8a), the other three were declared internally (line 25).

The code proper (lines 26–116) was separated into two parts. Part 1 (lines 30–77) implements the *control generator* (figure 15.8a). Note that the code for each signal obeys the timing diagrams of figure 15.8b.

Part 2 of the code (lines 81–115) implements the *image generator*. In lines 85–91, a counter is used to construct a pointer (called *line_counter*) to the image rows. If it points to row 1 (lines 93–96), the color is red. If it points to rows 2–3 (lines 97–100), the color is green. When pointing to rows 4–6 (lines 101–104), it is blue. Finally, when pointing to rows 7–480 (lines 105–108), the color is determined by the three toggle switches. In lines 110–113, *dena* = '0' is used to turn the image OFF during retrace.

This design could obviously have been done using a structural approach, with each of these parts implemented using the COMPONENT construct (such an approach will be illustrated in the next chapter). The reader is invited to compile the code below and observe what happens on the VGA display while playing with the RGB switches.

```
1    ------------------------------------------------------------
2    LIBRARY ieee;
3    USE ieee.std_logic_1164.all;
4    ------------------------------------------------------------
5    ENTITY vga IS
6       GENERIC (
7          Ha: INTEGER := 96;     --Hpulse
8          Hb: INTEGER := 144;    --Hpulse+HBP
9          Hc: INTEGER := 784;    --Hpulse+HBP+Hactive
10         Hd: INTEGER := 800;    --Hpulse+HBP+Hactive+HFP
11         Va: INTEGER := 2;      --Vpulse
12         Vb: INTEGER := 35;     --Vpulse+VBP
13         Vc: INTEGER := 515;    --Vpulse+VBP+Vactive
```

```
14          Vd: INTEGER := 525); --Vpulse+VBP+Vactive+VFP
15      PORT (
16          clk: IN STD_LOGIC; --50MHz in our board
17          red_switch, green_switch, blue_switch: IN STD_LOGIC;
18          pixel_clk: BUFFER STD_LOGIC;
19          Hsync, Vsync: BUFFER STD_LOGIC;
20          R, G, B: OUT STD_LOGIC_VECTOR(9 DOWNTO 0);
21          nblanck, nsync : OUT STD_LOGIC);
22  END vga;
23  ------------------------------------------------------------
24  ARCHITECTURE vga OF vga IS
25      SIGNAL Hactive, Vactive, dena: STD_LOGIC;
26  BEGIN
27      ------------------------------------------------------------
28      --Part 1: CONTROL GENERATOR
29      ------------------------------------------------------------
30      --Static signals for DACs:
31      nblanck <= '1';  --no direct blanking
32      nsync <= '0';    --no sync on green
33      --Create pixel clock (50MHz->25MHz):
34      PROCESS (clk)
35      BEGIN
36          IF (clk'EVENT AND clk='1') THEN
37              pixel_clk <= NOT pixel_clk;
38          END IF;
39      END PROCESS;
40      --Horizontal signals generation:
41      PROCESS (pixel_clk)
42          VARIABLE Hcount: INTEGER RANGE 0 TO Hd;
43      BEGIN
44          IF (pixel_clk'EVENT AND pixel_clk='1') THEN
45              Hcount := Hcount + 1;
46              IF (Hcount=Ha) THEN
47                  Hsync <= '1';
48              ELSIF (Hcount=Hb) THEN
49                  Hactive <= '1';
50              ELSIF (Hcount=Hc) THEN
51                  Hactive <= '0';
52              ELSIF (Hcount=Hd) THEN
53                  Hsync <= '0';
54                  Hcount := 0;
55              END IF;
56          END IF;
```

```
57    END PROCESS;
58    --Vertical signals generation:
59    PROCESS (Hsync)
60       VARIABLE Vcount: INTEGER RANGE 0 TO Vd;
61    BEGIN
62       IF (Hsync'EVENT AND Hsync='0') THEN
63          Vcount := Vcount + 1;
64          IF (Vcount=Va) THEN
65             Vsync <= '1';
66          ELSIF (Vcount=Vb) THEN
67             Vactive <= '1';
68          ELSIF (Vcount=Vc) THEN
69             Vactive <= '0';
70          ELSIF (Vcount=Vd) THEN
71             Vsync <= '0';
72             Vcount := 0;
73          END IF;
74       END IF;
75    END PROCESS;
76    ---Display enable generation:
77    dena <= Hactive AND Vactive;
78    --------------------------------------------------------
79    --Part 2: IMAGE GENERATOR
80    --------------------------------------------------------
81    PROCESS (Hsync, Vsync, Vactive, dena, red_switch,
82       green_switch, blue_switch)
83       VARIABLE line_counter: INTEGER RANGE 0 TO Vc;
84    BEGIN
85       IF (Vsync='0') THEN
86          line_counter := 0;
87       ELSIF (Hsync'EVENT AND Hsync='1') THEN
88          IF (Vactive='1') THEN
89             line_counter := line_counter + 1;
90          END IF;
91       END IF;
92       IF (dena='1') THEN
93          IF (line_counter=1) THEN
94             R <= (OTHERS => '1');
95             G <= (OTHERS => '0');
96             B <= (OTHERS => '0');
97          ELSIF (line_counter>1 AND line_counter<=3) THEN
98             R <= (OTHERS => '0');
99             G <= (OTHERS => '1');
```

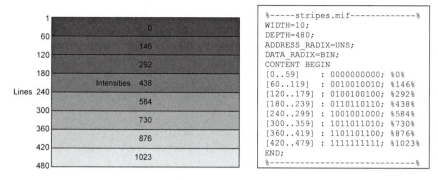

```
%-----stripes.mif-------------%
WIDTH=10;
DEPTH=480;
ADDRESS_RADIX=UNS;
DATA_RADIX=BIN;
CONTENT BEGIN
[0..59]    : 0000000000; %0%
[60..119]  : 0010010010; %146%
[120..179] : 0100100100; %292%
[180..239] : 0110110110; %438%
[240..299] : 1001001000; %584%
[300..359] : 1011011010; %730%
[360..419] : 1101101100; %876%
[420..479] : 1111111111; %1023%
END;
%---------------------------%
```

Figure 15.9
Image to be produced in the design of section 15.10 and corresponding MIF file.

```
100               B <= (OTHERS => '0');
101           ELSIF (line_counter>3 AND line_counter<=6) THEN
102               R <= (OTHERS => '0');
103               G <= (OTHERS => '0');
104               B <= (OTHERS => '1');
105           ELSE
106               R <= (OTHERS => red_switch);
107               G <= (OTHERS => green_switch);
108               B <= (OTHERS => blue_switch);
109           END IF;
110       ELSE
111           R <= (OTHERS => '0');
112           G <= (OTHERS => '0');
113           B <= (OTHERS => '0');
114       END IF;
115    END PROCESS;
116 END vga;
117 ------------------------------------------------------------
```

15.10   Image Generation with a File and On-Chip Memory

In chapter 13 we saw several ways of implementing ROM and RAM memory. It was also described how data from a file can be loaded into such memories. The purpose of the design presented here is to show how an image can be read from a file and loaded into a ROM (or RAM) memory, and from there be sent to a VGA display. A simple image will again be used, which eases the comparison between the actual and the expected result.

Figure 15.9 shows the image to be produced in this design, along with the corresponding data file (of type MIF—see chapter 13). The 480 lines are broken into eight 60-line portions (shown on the left) and the 10-bit intensity (with 10-bit DACs) is broken into eight linearly spaced values (shown in the center). Toggle switches will again be employed for *R*, *G*, and *B*.

A VHDL code for this circuit is presented below. In the library declarations (lines 2–6), note the inclusion of the packages *std_logic_arith*, which contains the function *conv_std_logic_vector* (used in line 60), and *lpm_components*, which specifies the *lpm_rom* cell (used in lines 40–48).

The control signal parameters of figure 15.8b were entered using GENERIC declarations (lines 10–17), so the code can be easily adjusted to other VGA modes. The signal names (lines 19–24) are from figure 15.8a. Because only two (*Hsync*, *Vsync*) of the five control signals are transmitted to the monitor (figure 15.8a), the other three were declared internally (line 28). Two other internal signals were declared in lines 29–30 to deal with the ROM.

The code proper (lines 31–66) was broken into two parts, with part 1 implementing the *control generator* and part 2 implementing the *image generator*. Because the former is exactly the same as that in the previous design, it was omitted in the code below.

The image generator (lines 39–66) is divided into three subsections. The first subsection implements the ROM (lines 40–48). The second (lines 50–61) builds a pointer (called *line_counter*) that is used in line 48 as an address to retrieve data from the ROM, which is assigned to the signal *intensity*. Finally, in the third subsection (lines 63–65), *intensity* is assigned or not to the system colors (*R*, *G*, *B*), depending on the positions of the toggle switches.

As in the previous design, a structural code could have been used, with the control generator in one code and the image generator in another, both then instantiated in the main code by means of the COMPONENT construct.

```
 1   ----------------------------------------------------------------
 2   LIBRARY ieee;
 3   USE ieee.std_logic_1164.all;
 4   USE ieee.std_logic_arith.all;
 5   LIBRARY lpm;
 6   USE lpm.lpm_components.all;
 7   ----------------------------------------------------------------
 8   ENTITY vga IS
 9      GENERIC (
10         Ha: INTEGER := 96;     --Hpulse
11         Hb: INTEGER := 144;    --Hpulse+HBP
12         Hc: INTEGER := 784;    --Hpulse+HBP+Hactive
13         Hd: INTEGER := 800;    --Hpulse+HBP+Hactive+HFP
14         Va: INTEGER := 2;      --Vpulse
15         Vb: INTEGER := 35;     --Vpulse+VBP
```

```
16        Vc: INTEGER := 515;   --Vpulse+VBP+Vactive
17        Vd: INTEGER := 525); --Vpulse+VBP+Vactive+VFP
18     PORT (
19        clk: IN STD_LOGIC; --50MHz in our board
20        red_switch, green_switch, blue_switch: IN STD_LOGIC;
21        pixel_clk: BUFFER STD_LOGIC;
22        Hsync, Vsync: BUFFER STD_LOGIC;
23        R, G, B: OUT STD_LOGIC_VECTOR(9 DOWNTO 0);
24        nblanck, nsync : OUT STD_LOGIC);
25   END vga;
26   ------------------------------------------------------------------------
27   ARCHITECTURE vga OF vga1 IS
28      SIGNAL Hactive, Vactive, dena: STD_LOGIC;
29      SIGNAL address: STD_LOGIC_VECTOR(8 DOWNTO 0);
30      SIGNAL intensity: STD_LOGIC_VECTOR(9 DOWNTO 0);
31   BEGIN
32      ------------------------------------------------
33      --Part 1: CONTROL GENERATOR
34      ------------------------------------------------
35      ... (same as in previous design)
36      ------------------------------------------------
37      --Part 2: IMAGE GENERATOR
38      ------------------------------------------------
39      --ROM instantiation:
40      myrom: lpm_rom
41         GENERIC MAP (
42            lpm_widthad => 9, --address width
43            lpm_outdata => "UNREGISTERED",
44            lpm_address_control => "REGISTERED",
45            lpm_file => "stripes.mif", --data file
46            lpm_width => 10)  --data width
47         PORT MAP (
48            inclock=>NOT pixel_clk, address=>address, q=>intensity);
49      --Create address (row number):
50      PROCESS (Vsync, Hsync)
51         VARIABLE line_counter: INTEGER RANGE 0 TO Vd;
52      BEGIN
53         IF (Vsync='0') THEN
54            line_counter := 0;
55         ELSIF (Hsync'EVENT AND Hsync='1') THEN
56            IF (Vactive='1') THEN
57               line_counter := line_counter + 1;
58            END IF;
```

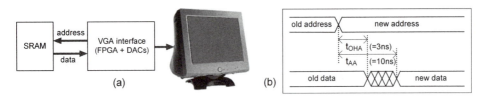

Figure 15.10
(a) Setup for the design of section 15.11 (data read from an external memory); (b) Memory-read timing.

```
59          END IF;
60          address <= conv_std_logic_vector(line_counter, 9);
61      END PROCESS;
62      --Assign color values to R/G/B:
63      R<=intensity WHEN red_switch='1' AND dena='1' ELSE (OTHERS=>'0');
64      G<=intensity WHEN green_switch='1' AND dena='1' ELSE (OTHERS=>'0');
65      B<=intensity WHEN blue_switch='1' AND dena='1' ELSE (OTHERS=>'0');
66  END vga;
67  -------------------------------------------------------------------
```

15.11 Arbitrary Image Generation with a File and Off-Chip Memory

The setup for this experiment is depicted in figure 15.10a, which shows the VGA interface (implemented in the FPGA and followed by DACs) retrieving the video data from an external memory (SRAM, in this example). We will assume that the data, possibly from a bitmap file, has already been stored in the SRAM, so our job is to retrieve the data, process it, and send it to the VGA monitor.

There are several ways of loading data from a file into an SRAM (like the methods seen in chapter 13), but such procedures are normally dependent on the tools being used in the project. For example, if using Altera's DE2 board, a bitmap file containing the target picture can be converted to a raw format using the *ImgConv.exe* program that accompanies the DE2 board (see tutorial in appendix D), which is the method adopted in this design.

A VHDL code that solves this problem is presented below. Since the actual image size must be $640 \times 480 = 307{,}200$ pixels, a color image could not be used because the SRAM available in the DE2 board (ISSI IS61LV25616-10) can only store 256k 16-bit words. To circumvent this limitation, a monochromatic image was employed (obtained with *ImgConv.exe*), with eight bits per pixel, hence allowing 256 shades of gray. To be able to fit this file into the SRAM, two pixels were stored at each address, so the picture occupied 153,600 16-bit words (~60% of the total SRAM space).

The memory-read timing is depicted in figure 15.10b. The SRAM mentioned above is a 10 ns read cycle memory, with t_{OHA} (output hold time) = 3 ns and t_{AA} (address access time) $= 10$ ns. Since our circuit will operate in the basic VGA mode ($640 \times 480 \times 60$Hz), its clock is 25 MHz. Given that each memory address contains two pixels, the memory-read operation only needs to be executed once every two pixel clock cycles—that is, at 12.5 MHz, well under the maximum speed of this memory. Note in the code that the five memory-control signals, *nWE* (write enable, active low), *nCE* (chip enable, active low), *nOE* (output enable, active low), *nLB* (lower byte enable, active low), and *nUB* (upper byte enable, active low) are all held at fixed values (line 39), proper for memory-reading only.

The architecture (lines 26–75) contains again two parts: *control generator* and *image generator*. The former is exactly the same as that in the previous designs, so it was omitted. The latter is in lines 38–75, and was further divided into three subsections. The first of them assigns the static values to the five memory-control pins mentioned above (line 39). The second subsection (lines 41–55) reads the memory and stores the result (16 bits = 2 pixels) locally. Finally, the third subsection (lines 57–74) assigns such bits to the *R*, *G*, and *B* signals.

```
1   -------------------------------------------------------------
2   LIBRARY ieee;
3   USE ieee.std_logic_1164.all;
4   -------------------------------------------------------------
5   ENTITY vga IS
6      GENERIC (
7         Ha: INTEGER := 96;    --Hpulse
8         Hb: INTEGER := 144;   --Hpulse+HBP
9         Hc: INTEGER := 784;   --Hpulse+HBP+Hactive
10        Hd: INTEGER := 800;   --Hpulse+HBP+Hactive+HFP
11        Va: INTEGER := 2;     --Vpulse
12        Vb: INTEGER := 35;    --Vpulse+VBP
13        Vc: INTEGER := 515;   --Vpulse+VBP+Vactive
14        Vd: INTEGER := 525);  --Vpulse+VBP+Vactive+VFP
15     PORT (
16        clk: IN STD_LOGIC;                            --50MHz
17        pixel_clk: BUFFER STD_LOGIC;                  --25MHz
18        Hsync, Vsync: BUFFER STD_LOGIC;               --control
19        R, G, B: OUT STD_LOGIC_VECTOR(9 DOWNTO 0);    --to DACs
20        nblanck, nsync: OUT STD_LOGIC;                --to DACs
21        nWE, nCE, nOE, nLB, nUB: OUT STD_LOGIC;       --to SRAM
22        address: OUT INTEGER RANGE 0 TO 262143;       --to SRAM
23        data: IN STD_LOGIC_VECTOR(15 DOWNTO 0));      --from SRAM
24  END vga;
25  -------------------------------------------------------------
26  ARCHITECTURE vga OF vga IS
```

```
27      SIGNAL Hactive, Vactive, dena: STD_LOGIC;
28      SIGNAL registered_data: STD_LOGIC_VECTOR(15 DOWNTO 0);
29      SIGNAL flag: STD_LOGIC;
30   BEGIN
31      ------------------------------------------------
32      --Part 1: CONTROL GENERATOR
33      ------------------------------------------------
34      ... (same as in previous design)
35      ------------------------------------------------
36      --Part 2: IMAGE GENERATOR
37      ------------------------------------------------
38      --Static signals for SRAM:
39      nWE<='1'; nCE<='0'; nOE<='0'; nLB<='0'; nUB<='0';
40      --Read SRAM and register its data:
41      PROCESS (pixel_clk, Vsync)
42         VARIABLE pixel_counter: INTEGER RANGE 0 TO 262143;
43      BEGIN
44         IF (Vsync='0') THEN
45            pixel_counter := 0;
46            flag <= '0';
47         ELSIF (pixel_clk'EVENT AND pixel_clk='1') THEN
48            IF (dena='1' AND flag='1') THEN
49               registered_data <= data;
50               pixel_counter := pixel_counter + 1;
51            END IF;
52            flag <= NOT flag;
53         END IF;
54         address <= pixel_counter;
55      END PROCESS;
56      --Create image:
57      PROCESS (dena, flag, registered_data)
58      BEGIN
59         IF (dena='1') THEN
60            IF (flag='1') THEN
61               R <= (registered_data(15 DOWNTO 8) & "00");
62               G <= (registered_data(15 DOWNTO 8) & "00");
63               B <= (registered_data(15 DOWNTO 8) & "00");
64            ELSE
65               R <= (registered_data(7 DOWNTO 0) & "00");
66               G <= (registered_data(7 DOWNTO 0) & "00");
67               B <= (registered_data(7 DOWNTO 0) & "00");
68            END IF;
69         ELSE
```

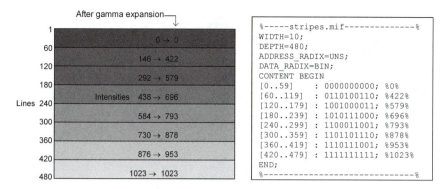

Figure 15.11
Gamma expanded input values for approximately "linear" image reproduction.

```
70              R <= (OTHERS => '0');
71              G <= (OTHERS => '0');
72              B <= (OTHERS => '0');
73          END IF;
74      END PROCESS;
75  END vga;
76  ------------------------------------------------------------
```

15.12   Image Equalization with Gamma Expansion

CRT monitors introduce a nonlinear luminance distortion known as *gamma compression*, because it compresses the input values (x) according to the function x^γ, where $0 \le x \le 1$ and $\gamma \approx 2.2$. Consequently, for a "linear" reproduction, the inputs must be *gamma expanded*; that is, $x^{1/\gamma}$ must be entered instead of x (as in the NTSC television system).

To illustrate this phenomenon, we can display the same image of section 15.10, now with gamma expanded values at the input (see figure 15.11). Note that, because the image is stored in an external (memory initialization) file, nothing in the code of section 15.10 needs to be changed. The reader is invited to compile the code with this new MIF file and compare the resulting image (on a VGA monitor) against that produced in section 15.10.

15.13   Exercises

Note: For exercise solutions, please consult the book website.

Exercise 15.1: 800 × 600 × 75Hz SVGA Interface

a) Say that our monitor supports other modes besides the default VGA mode. What needs to be done for it to change from one mode to the other?

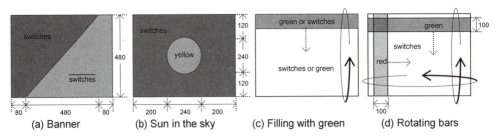

Figure 15.12

b) Say that we want it to operate in the $800 \times 600 \times 75\text{Hz}$ mode. Draw the corresponding horizontal and vertical timing diagrams (as in figure 15.8b).

c) Modify the code in section 15.9 in order to display the same image but operating in the 75 Hz SVGA mode.

Exercise 15.2: Image Generation with Hardware #1 (Banner)

Design a circuit capable of producing the image of figure 15.12a on a VGA monitor. The color on the left half must be determined by three toggle switches (for R, G, B), with the complementary color (R', G', B') automatically assigned to the right half of the banner. The figure must be generated by local hardware (FPGA cells, as in section 15.9).

Exercise 15.3: Image Generation with Hardware #2 (Sun in the Sky)

Design a circuit capable of producing the image of figure 15.12b on a VGA monitor. The color outside the circle must be determined by three toggle switches (for R, G, B). The figure must be generated by local hardware (FPGA cells, as in section 15.9).

Exercise 15.4: Image Generation with Hardware #3 (Filling with Green)

Design a circuit capable of producing the image of figure 15.12c on a VGA monitor, which must be generated by local hardware (FPGA cells, as in section 15.9). It consists of filling the screen with green, from top to bottom, with the base color determined by three toggle switches (for R, G, and B). When the filling is completed, the base color should start filling the screen, also from top to bottom, until green starts filling it again, and so on. The speed of the filling should be one line per frame (with 60 frames/second, a total of $480/60 = 8$ seconds will be needed to fill one screen).

Exercise 15.5: Image Generation with Hardware #4 (Rotating Bar)

Design a circuit capable of producing the image of figure 15.12d on a VGA monitor, which must be generated by local hardware (FPGA cells, as in section 15.9). Each bar must be 100 pixels wide. As soon as the green bar reaches the bottom of the screen it

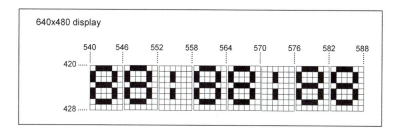

Figure 15.13

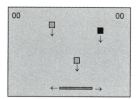

Figure 15.14

must start reentering at the top, the same occurring with the red bar in the horizontal direction. The speed of the green bar must be one line per frame (with 60 frames/second, 8 seconds will be needed for a full bar rotation). Choose for the red bar a speed such that it takes the same time as the green bar to cover one screen.

Exercise 15.6: Image Generation with Hardware #5 (Digital Clock)

Design a circuit that implements the clock of figure 15.13 on a VGA monitor. The positions of the digits on the screen, for the monitor operating in the basic 640 × 480 VGA mode, are given in the figure. The image must be generated by local hardware (FPGA cells, as in section 15.9). (Hint: See a related design in chapter 17.)

Exercise 15.7: Image Generation with Hardware #6 (Arcade Game)

Design a circuit capable of generating the elementary game illustrated in figure 15.14. The player must be able to move the racket horizontally using two pushbuttons. If a clear ball is collected by the racket, the player wins a point; otherwise, if the ball reaches the bottom of the screen, the computer gets a point. A few black balls also fall along with clear balls, which must be avoided by the player. The game ends when a certain score is reached or a black ball hits the racket. The score must be kept on the screen (upper left and right corners). The speed of the balls (four levels of difficulty) must be set by two toggle switches. The image must be generated by local hardware (FPGA cells, as in section 15.9).

Exercise 15.8: Image Generation with a File and On-Chip Memory #1 (Banner)

Following the procedure in section 15.10, design a circuit capable of generating the image of figure 15.12a. The image must first be prepared in a standard text file (MIF, for example), from which it should be read and transferred to the FPGA SRAM memory, then finally displayed on the monitor. The file should contain only '0's (for the left half of the banner) and '1's (for the right half). When reading this file, '0' should be interpreted as green and '1' as red. It is left to the reader to find a way of preparing the text file (with Excel, for example, among several other possibilities).

Exercise 15.9: Image Generation with a File and On-Chip Memory #2 (Sun in the Sky)

Develop a design similar to that in exercise 15.8 for the image of figure 15.12b.

Exercise 15.10: Image Generation with a File and Off-Chip Memory (Arbitrary Picture)

Find a picture that you consider interesting. Following the procedure in section 15.11, design a circuit capable of displaying that picture on a VGA monitor.

16 VHDL Design of DVI Video Interfaces

16.1   Introduction

DVI (digital visual interface) is a digital video interface developed by DDWG (Digital Display Working Group) in 1999 for connecting uncompressed video from computers to LCD monitors. Its core consists of TMDS (transition minimized differential signaling) circuits, studied in chapter 14.

Figure 16.1 illustrates the use of DVI. In (a), the previous technology is shown, which consists of a CRT-based monitor, with which a computer communicates using a VGA (analog) interface (seen in chapter 15). Current technology is depicted in (b), where an LCD monitor is shown, with which a computer communicates using the DVI (digital) interface. Finally, (c) shows the way pixels are counted (from top left to bottom right).

Typical specifications for current LCD monitors are also included in figure 16.1. In this case, the display's *native* (fixed) resolution is 1280 pixels per row, with a total of 1024 rows, refreshed 60 times per second. The required pixel rate then is $1280 \times 1024 \times 60 \approx 79$ Mpps (the actual pixel rate is indeed higher because of the blanking/retrace intervals). This particular resolution (1280×1024) is called SXGA (super extended graphics array). We will see more on display resolutions in section 16.3.

As in chapters 15 and 17, which also deal with video interfaces, the following fundamental aspects will be focused on here:

1) Operation of the DVI interface.

2) How the circuit should be divided to make the design as simple and as standard as possible.

3) How the control signals operate and how they should be generated.

4) How images can be generated (from local hardware, external memory, file, etc.), rather than focusing on the images themselves (software).

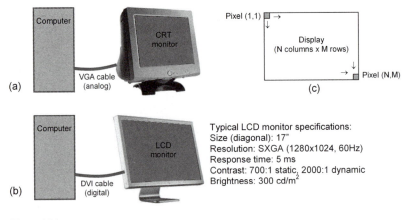

Figure 16.1
(a) VGA interface (analog) between a computer and a CRT monitor; (b) DVI interface (digital) between a computer and an LCD monitor (typical specifications also shown); (c) Pixel count.

16.2   Circuit Diagram

Figure 16.2 shows a general diagram for the DVI circuit. It contains three parts, called TMDS (transition minimized differential signaling), DDC (display data channel), and VGA (video graphics array). TMDS is a line encoder/decoder (studied in chapter 14), DDC is how the monitor tells the computer its characteristics (chapter 15), and VGA (chapter 15) is an optional section whose purpose is to maintain compatibility with analog VGA monitors.

Note that the TMDS part of DVI can contain one or two links, called *link0* (mandatory) and *link1* (optional). Each link consists of three TMDS channels (highlighted in a white box at the top of figure 16.2), which transmit alternately *pixel* data (R, G, and B colors, with eight bits each) or *control* data (two bits per channel, with the horizontal and vertical synchronization signals, *Hsync* and *Vsync*, in the first channel, plus up to five pairs of reserved control bits, $C0$-$C1$,...,$C8$-$C9$, in the other channels). The decision between transmitting pixel or control data is made by a signal called *dena* (display enable), which remains high during the time intervals in which pixels must be effectively written onto the display, and remains low during the blanking/retrace intervals. Note that because a pixel is represented by 8 bits, a total of $256^3 = 16.8$ million distinct colors or shades result.

The clock frequency in DVI systems must lie in the range from 25 MHz (minimum resolution; that is, basic $640 \times 480 \times 60$Hz VGA) up to 165 MHz (default, which allows maximum resolution). A minimum is necessary in order to detect when the DVI cable is idle (22.5 MHz is the actual nominal minimum). Moreover, when a pixel rate above 165 Mpps is required (with blanking included), dual link must be used. However, both links must

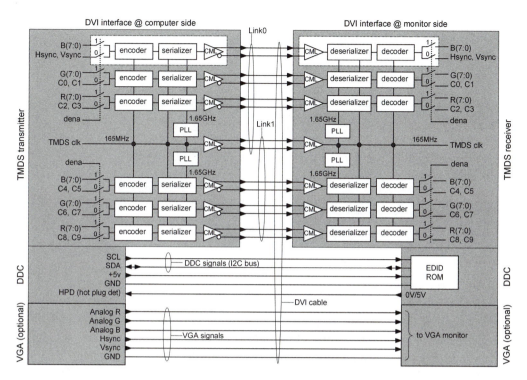

Figure 16.2
DVI circuit, which consists of three parts: TMDS, DDC, and VGA (optional). A complete DVI connector has 29
pins, corresponding to the 25 wires shown above plus 4 grounded shields.

operate at the same speed. Then, if for example 200 Mpps are required, it must not be
165 Mpps in link0 and 35 Mpps in link1, but rather 100 Mpps in each link.

Note that the bit rate through the DVI cable is independent from the type of data
(pixel or control) being transmitted, because in either case 10 bits are produced by the
TMDS encoder. Consequently, when operating at full speed (165 Mpps for single link or
330 Mpps for dual link), a bit rate of 1.65 Gbps is needed through the DVI cable.

Note also that a fast serializer (chapter 14) is needed because the output clock must
be ten times faster than the TMDS clock. This additional 1.65 GHz transmission clock is
obtained with a PLL circuit (Pedroni 2008). Transmissions start with the LSB. For further
details on the TMDS transmitter, see section 14.6.

DVI has also a DDC channel, through which the computer reads the display's features
(supported resolutions, timings, etc.) stored in a ROM on the monitor side (the type of interface used in this channel is called I^{2}C, also studied in chapter 14). The ROM's data format is called EDID (extended display identification data). For example, since any LCD
monitor has a fixed (called *native*) resolution, in order for it to display other formats it

Name	Resolution	Aspect ratio	Number of pixels	Minimum DVI @ refresh rate = 60 Hz [1] and blanking overhead = 10% [2]
VGA	640 x 480	4:3	0.31 M	Single link 25 MHz [3]
SVGA	800 x 600	4:3	0.48 M	Single link 35 MHz
XGA	1024 x 768	4:3	0.79 M	Single link 52 MHz
SXGA	1280 x 1024	5:4	1.31 M	Single link 83 MHz
WXGA+	1440 x 900	16:10	1.30 M	Single link 87 MHz
UXGA	1600 x 1200	4:3	1.92 M	Single link 127 MHz
WUXGA	1920 x 1200	16:10	2.30 M	Single link 152 MHz
QXGA	2048 x 1536	4:3	3.15 M	Dual link 104 MHz [4]
QSXGA	2560 x 2048	5:4	5,24 M	(2x173MHz) Too high for DVI
QUXGA	3200 x 2400	4:3	7.68 M	(2x254MHz) Too high for DVI
WQUXGA	3840 x 2400	16:10	9.22 M	(2x304MHz) Too high for DVI

(1) The refresh rate can also assume other values, like 17Hz, 33Hz, 75Hz, 85Hz, etc.
(2) Illustrative blanking period (can be larger, particularly if GTF is used).
(3) Recall that the DVI clock must be in the 25 to 165 MHz interval.
(4) Recall that when dual-link is used, the links must operate with the same frequency.

Figure 16.3
Examples of computer display resolutions, along with minimum DVI specifications needed to support them. For illustrative purposes, a refresh rate of 60Hz and an overhead of 10% (blanking period) were assumed.

must possess an *image scaler*. Consequently, the native resolution plus those supported by the scaler are the resolutions reported by EDID to the host computer. In case the monitor is not equipped with a scaler, then only the native resolution is reported, but the monitor should still be able to display a legible image for *low-pixel format* (640 × 480 resolution), needed at computer startup. Other details about DDC and EDID can be found in chapter 15.

There is also an HPD (hot plug detect) wire, kept at '1' by the device when plugged in, so the host can know when a monitor is connected to the DVI cable.

Finally, in the lowest part of figure 16.2, which is optional, the analog signals needed to drive a VGA monitor (chapter 15) are shown. The purpose of this analog section, with six wires (*R, G, B, Vsync, Hsynk*, and GND for RGB), is to allow compatibility with the previous monitor generation.

16.3   Display Resolutions

To analyze figure 16.3, recall that a DVI controller can operate with one or two links and with a clock frequency between 25 and 165 MHz.

Examples of standardized computer monitor resolutions are shown in figure 16.3. For illustration purposes, it is assumed that the display is refreshed 60 times per second, with a blanking period that adds just 10% of overhead (other refresh rates also exist, such as 17 Hz, 33 Hz, 75 Hz, 85 Hz, etc.; the same is true for the blanking overhead, which can be larger, particularly if GTF [generalized timing formula, used to calculate the blanking intervals of VGA monitors] is adopted, in which case the overhead is generally over 30%).

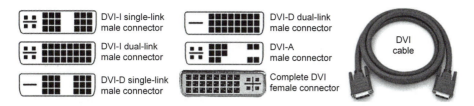

Figure 16.4
Types of DVI connectors and a DVI cable.

16.4   DVI Types and DVI Connectors

Regarding compatibility, there are two types of DVI, called DVI-I and DVI-D.

• DVI-I (I = Integrated; i.e., digital + analog): Contains the actual DVI signals, which are digital, plus the analog VGA signals, thus providing backward compatibility with the previous monitor technology. This is the case in figure 16.2.

• DVI-D (D = digital): This option contains only the true-DVI signals.

Regarding the number of links, there are again two DVI categories, called *single-link* and *dual-link*.

• Single-link DVI: Employs one pair of wires per color (R, G, B), which can transmit up to 165 Mpps to an LCD monitor.

• Dual-link DVI: Employs two pairs of wires per color, being therefore capable of transmitting 330 Mpps to an LCD monitor. This is the case in figure 16.2.

Finally, regarding the DVI connectors, there are five fundamental types, depicted in figure 16.4. The first pair of male connectors is for single- and dual-link DVI-I; the second pair is for single- and dual-link DVI-D; and the last male connector is for the case when just the analog signals of a DVI-I board are indeed communicated to a VGA monitor. A complete DVI female connector and a DVI cable are also shown.

The pinout of a complete DVI connector is shown in figure 16.5, which also shows a female DVI with the respective pin numbers. Observe that the connector contains all 25 wires seen in figure 16.2 plus four pins connected to the seven shielding covers of TMDS signals and TMDS clock. The pins' functions are summarized below.

• There are three pairs of wires, called TMDS0 (pins 18–17, signal B), TMDS1 (pins 10–9, signal G), and TMDS2 (pins 2–1, signal R), to transmit RGB in the first link (*link*0) of the DVI system.

• There are three pairs of wires, called TMDS3 (pins 13–12, signal B), TMDS4 (pins 5–4, signal G), and TMDS5 (pins 21–20, signal R), to transmit RGB in the second link (*link*1) of the DVI system.

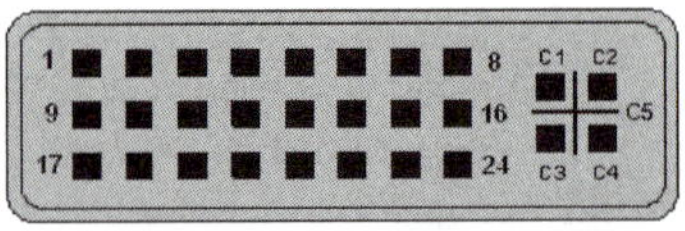

Pin	Signal	Pin	Signal	Pin	Signal	Pin	Signal
1	TMDS2– (digital R)	9	TMDS1– (digital G)	17	TMDS0– (digital B)	C1	Analog R
2	TMDS2+ (digital R)	10	TMDS1+ (digital G)	18	TMDS0+ (digital B)	C2	Analog G
3	Shield TMDS2/4	11	Shield TMDS1/3	19	Shield TMDS0/5	C3	Analog B
4	TMDS4– (digital G)	12	TMDS3– (digital B)	20	TMDS5– (digital R)	C4	Analog Hsync
5	TMDS4+ (digital G)	13	TMDS3+ (digital B)	21	TMDS5+ (digital R)	C5	GND for R, G, B
6	SCL (DDC clock)	14	+5V for DDC	22	Shield TMDS clk	--	--
7	SDA (DDC data)	15	GND for DDC +5V	23	TMDS clk+	--	--
8	Analog Vsync	16	HPD (Hot Plug Det.)	24	TMDS clk–	--	--

Figure 16.5
DVI connector pinout.

• There is one pair of wires, called TMDS clock (pins 23–24), for transmitting the DVI clock from the host to the device.

• Each of the seven pairs of wires above has a shielding cover, connected to pins 3, 11, 19, or 22.

• There are also two DDC wires (SCL, pin 6, and SDA, pin 7), needed for the host to read the device's features, stored in the EDID ROM.

• There is a pair of wires for the power supply (+5V, pin 14, and GND, pin 15) that feeds the EDID ROM when the monitor is off.

• The connector also shows an HPD wire, which allows the host to know when a device is connected to the DVI cable.

• Finally, the remaining six wires (R, G, B, Vsync, Hsynk, and GND for $R/G/B$) are for compatibility with VGA (analog) monitors.

16.5   DVI versus HDMI

While DVI is a video interface for computer-display communication, HDMI (high definition multimedia interface) is a video plus audio interface for communication between consumer electronics products and HDTV (high definition television) sets. Such products include, for example, DVD players, camcorders, cable/satellite boxes, video game consoles, and video projectors.

The transmission/reception portions of HDMI and DVI are similar in the sense that both use a TMDS transmitter/receiver pair and data serializer/deserializer circuits. The data contents, however, exhibit important differences, because HDMI also transmits audio (up to eight channels) and supports YCbCr (luminance plus chrominance) video format. It

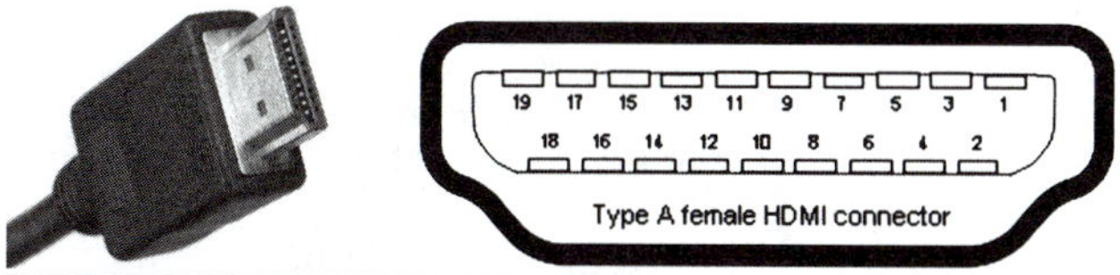

Figure 16.6
HDMI connector and respective pinout.

Pin	Signal	Pin	Signal	Pin	Signal
1	TMDS2+	8	Shield for TMDS0	15	SCL
2	Shield for TMDS2	9	TMDS0–	16	SDA
3	TMDS2–	10	TMDS_clk+	17	GND for CEC/DDC +5V
4	TMDS1+	11	Shield for TMDS_clk	18	+5V
5	Shield for TMDS1	12	TMDS_clk–	19	HPD
6	TMDS1–	13	CEC	--	--
7	TMDS0+	14	Reserved	--	--

also allows more than eight bits per pixel color and a higher clock frequency (340 instead of 165 MHz). However, only one TMDS link is used in HDMI. Dual-link HDMI, compatible with DVI-D, is expected to be used in the future for very high resolution (e.g. 3840×2400) displays.

The HDMI connector is also different from the DVI connector, as can be observed in figure 16.6 (there are indeed three connector sizes/shapes, called types A, B, and C; type A is shown here). Because HDMI uses only one link, it contains 19 wires, against 29 of DVI. As mentioned above, a dual-link version compatible with DVI is expected to be used in the future, then with a DVI-compatible (type B) connector. Because the signals in both systems are electrically compatible, a cable adapter can be used to convert one to another (so a computer can be connected to an HDMI display, for example).

There are so far four HDMI versions, which allow the following maximum display resolutions:

Versions 1.0 and 1.2 (165 MHz single link, 24 bits/pixel, 60 Hz scan): up to 1920×1200

Version 1.3 (340 MHz single link, 48 bits/pixel, 75 Hz scan): up to 2560×1600

Version 1.4 (340 MHz single link, ethernet, 48 bits/pixel, 24 Hz scan): up to 4096×2160.

16.6   Setup for the Experiments

Figure 16.7 shows a complete single-link DVI interface. As in chapter 15, the circuit was broken into subsections, here called *image generator*, *control generator*, and *tmds transmitter*. This division is very important because not all subsections change from one application to another, so the fixed ones only need to be designed once (only the *image generator* portion is application-dependent in figure 16.7).

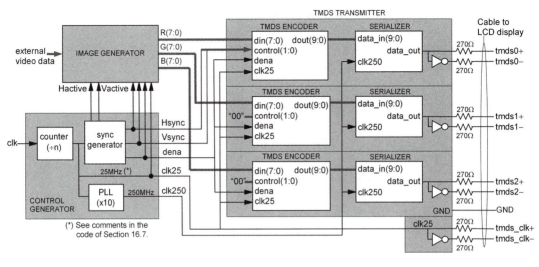

Figure 16.7
Complete single-link DVI interface.

Another peculiarity of figure 16.7 is the pair of 270 ohm resistors at each TMDS output, which allow regular 3.3V LVCMOS pins to emulate CML I/Os (see details in section 14.6).

We will use an FPGA (Cyclone II, available in the Altera DE2 board) whose highest frequency is too low for full-speed DVI operation (1.65 Gbps at the output), making the problem even more interesting from an engineering perspective. Thus we have now two limitations to overcome: the lack of CML I/Os (dealt with in section 14.6) and the speed limit.

To overcome the speed limitation, the system of figure 16.7 will operate in its lowest resolution (basic $640 \times 480 \times 60$Hz VGA mode), thus requiring a 25 MHz clock and producing 250 Mbps in the DVI cable. The system clock is assumed to be 50 MHz, hence called *clk50*, while the TMDS and line clocks, derived from *clk50*, are called *clk25* (25 MHz) and *clk250* (250 MHz), respectively.

16.7   Hardware-Generated Image

We turn now to the design of DVI circuits using VHDL. As in chapter 15, the main purpose of this first design is to illustrate how the control signals, the TMDS encoder, and the serializer (which are the invariant parts of DVI) can be constructed. To keep the focus on that, the same simple image of section 15.9 will be generated here, using local hardware (FPGA), consisting of just four horizontal stripes of solid colors, with widths 1, 2, 3, and 474 pixels. The first three are red, green, and blue, respectively, while the last one (wide) is determined by three toggle switches, hence allowing eight colors.

A structural approach will be adopted in the design, which consists of designing each of the subsections of figure 16.7 separately, then bringing them together in the main code using the COMPONENT construct. Note that the TMDS transmitter is constructed with TMDS encoders plus serializers, both of which were studied in chapter 14 (sections 14.6 and 14.2, respectively).

Part 1: Image Generator

The coder for the image generator is shown below. The inputs are the three toggle switches (line 7) plus the control signals (line 8), while the outputs are the RGB colors (line 9), with eight bits each. The code proper is divided into two subsections; in the first of them (lines 18–24), a pointer (address) to the display rows is implemented, which is then used in the second subsection (lines 26–48) to build the four-stripe image. Recall that there are always five control signals (*Hsync*, *Vsync*, *Hactive*, *Vactive*, *dena*) produced by the control generator, but not all are always needed by the image generator (in this example, *Hactive* was not required).

```
1    -----Image generator:-------------------------------------------------
2    LIBRARY ieee;
3    USE ieee.std_logic_1164.all;
4    ---------------------------------------------------------------------
5    ENTITY image_generator IS
6       PORT (
7          red_switch, green_switch, blue_switch: IN STD_LOGIC;
8          Hsync, Vsync, Vactive, dena: IN STD_LOGIC;
9          R, G, B: OUT STD_LOGIC_VECTOR(7 DOWNTO 0));
10   END image_generator;
11   ---------------------------------------------------------------------
12   ARCHITECTURE image_generator OF image_generator IS
13   BEGIN
14      PROCESS (Hsync, Vsync, dena, red_switch, green_switch, blue_switch)
15         VARIABLE line_counter: INTEGER RANGE 0 TO 480;
16      BEGIN
17         -----Create pointer to LCD rows:-----
18         IF (Vsync='0') THEN
19            line_counter := 0;
20         ELSIF (Hsync'EVENT AND Hsync='1') THEN
21            IF (Vactive='1') THEN
22               line_counter := line_counter + 1;
23            END IF;
24         END IF;
25         -----Create image:-------------------
26         IF (dena='1') THEN
27            IF (line_counter=1) THEN
```

```
28                R <= (OTHERS => '1');
29                G <= (OTHERS => '0');
30                B <= (OTHERS => '0');
31            ELSIF (line_counter=2 OR line_counter=3) THEN
32                R <= (OTHERS => '0');
33                G <= (OTHERS => '1');
34                B <= (OTHERS => '0');
35            ELSIF (line_counter>3 AND line_counter<=6) THEN
36                R <= (OTHERS => '0');
37                G <= (OTHERS => '0');
38                B <= (OTHERS => '1');
39            ELSE
40                R <= (OTHERS => red_switch);
41                G <= (OTHERS => green_switch);
42                B <= (OTHERS => blue_switch);
43            END IF;
44        ELSE
45            R <= (OTHERS => '0');
46            G <= (OTHERS => '0');
47            B <= (OTHERS => '0');
48        END IF;
49    END PROCESS;
50 END image_generator;
51 -------------------------------------------------------------------
```

Part 2: Control Generator

A VHDL code for the control generator is presented below. The parameters for the original 640×480 VGA mode were adopted (from figure 15.8b), entered using the GENERIC attribute (lines 6–14), so they can be easily modified to implement other video modes.

Note that this code is similar to that seen in chapter 15 for the VGA display, with the only particularity that an additional high-frequency clock (*clk250*), for the serializers, is also inferred here (with a PLL). The PLL is declared in lines 27–33, then instantiated in line 36. Other details about the PLL will be given in part 3.

Because we want the circuit to produce 60 frames/second, a 25 MHz clock (*clk25*) is needed (see chapter 15). Hence the system must operate with two clocks: *clk250* (250 MHz, for the serializer) and *clk25* (25 MHz, which is the new system clock). In this kind of situation, and assuming that the system clock might be used to build also other circuits (as parts of the same system), it is important to know the phase relationship between these two clocks. For example, if *clk50* and *clk250* are in phase (every five clock cycles, of course), which is generally the case, then *clk25* can be derived from either one; otherwise, if their relative phases are unknown or too far apart, then it is advisable to divide *clk250* down to get *clk25*. Both cases are included in the code below (see options 1 and 2 in lines 39–44 and 46–56, respectively).

```
1   -----Control generator:-------------------------------------
2   LIBRARY ieee;
3   USE ieee.std_logic_1164.all;
4   -------------------------------------------------------------
5   ENTITY control_generator IS
6      GENERIC (
7         Ha: INTEGER := 96;        --Hpulse
8         Hb: INTEGER := 144;       --Hpulse+HBP
9         Hc: INTEGER := 784;       --Hpulse+HBP+Hactive
10        Hd: INTEGER := 800;       --Hpulse+HBP+Hactive+HFP
11        Va: INTEGER := 2;         --Vpulse
12        Vb: INTEGER := 35;        --Vpulse+VBP
13        Vc: INTEGER := 515;       --Vpulse+VBP+Vactive
14        Vd: INTEGER := 525);      --Vpulse+VBP+Vactive+VFP
15     PORT (
16        clk50: IN STD_LOGIC;            --System clock (50MHz)
17        clk25: BUFFER STD_LOGIC;        --TMDS clock (25MHz)
18        clk250: BUFFER STD_LOGIC;       --Tx clock (250MHz)
19        Hsync: BUFFER STD_LOGIC;        --Horizontal sync
20        Vsync: OUT STD_LOGIC;           --Vertical sync
21        Hactive: BUFFER STD_LOGIC;      --Active portion of Hsync
22        Vactive: BUFFER STD_LOGIC;      --Active portion of Vsync
23        dena: OUT STD_LOGIC);           --Display enable
24  END control_generator;
25  -------------------------------------------------------------
26  ARCHITECTURE control_generator OF control_generator IS
27     COMPONENT altera_pll IS
28        PORT (
29        areset: IN STD_LOGIC;
30        inclk0: IN STD_LOGIC;
31        c0: OUT STD_LOGIC;
32        locked: OUT STD_LOGIC);
33     END COMPONENT;
34  BEGIN
35     -----Generation of clk250:-------------
36     pll: altera_pll PORT MAP ('0', clk50, clk250, OPEN);
37     -----Generation of clk25:--------------
38     ---Option 1: From clk50
39     PROCESS (clk50)
40     BEGIN
41        IF (clk50'EVENT AND clk50='1') THEN
42           clk25 <= NOT clk25;
43        END IF;
44     END PROCESS;
```

```
45    ---Option 2: From clk250
46    --PROCESS (clk250)
47    -- VARIABLE count: INTEGER RANGE 0 TO 5;
48    --BEGIN
49    -- IF (clk250'EVENT AND clk250='1') THEN
50    --    count := count + 1;
51    --    IF (count=5) THEN
52    --       clk25 <= NOT clk25;
53    --       count := 0;
54    --    END IF;
55    -- END IF;
56    --END PROCESS;
57    ---Horizontal signals generation:----
58    PROCESS (clk25)
59       VARIABLE Hcount: INTEGER RANGE 0 TO Hd;
60    BEGIN
61       IF (clk25'EVENT AND clk25='1') THEN
62          Hcount := Hcount + 1;
63          IF (Hcount=Ha) THEN
64             Hsync <= '1';
65          ELSIF (Hcount=Hb) THEN
66             Hactive <= '1';
67          ELSIF (Hcount=Hc) THEN
68             Hactive <= '0';
69          ELSIF (Hcount=Hd) THEN
70             Hsync <= '0';
71             Hcount := 0;
72          END IF;
73       END IF;
74    END PROCESS;
75    -----Vertical signals generation:------
76    PROCESS (Hsync)
77       VARIABLE Vcount: INTEGER RANGE 0 TO Vd;
78    BEGIN
79       IF (Hsync'EVENT AND Hsync='0') THEN
80          Vcount := Vcount + 1;
81          IF (Vcount=Va) THEN
82             Vsync <= '1';
83          ELSIF (Vcount=Vb) THEN
84             Vactive <= '1';
85          ELSIF (Vcount=Vc) THEN
86             Vactive <= '0';
87          ELSIF (Vcount=Vd) THEN
88             Vsync <= '0';
```

```
89                   Vcount := 0;
90              END IF;
91           END IF;
92       END PROCESS;
93       -----Display-enable generation:--------
94       dena <= Hactive AND Vactive;
95   END control_generator;
96   -------------------------------------------------------------
```

Part 3: PLL

If using Altera Quartus II, for example, the file for PLL instantiation is obtained with the MegaWizard Plug-In Manager. Part of such a file is shown below (this portion shows the main parameters, which can even be adjusted directly in the file; notice the divider, multiplier, duty cycle, phase, and input clock period). See other details in appendix F.

```
1    -----PLL:-------------------------------------------------
2    LIBRARY ieee;
3    USE ieee.std_logic_1164.all;
4    -------------------------------------------------------------
5    ENTITY altera_pll IS
6       PORT (areset: IN STD_LOGIC := '0';
7              inclk0: IN STD_LOGIC := '0';
8              c0: OUT STD_LOGIC;
9              locked: OUT STD_LOGIC);
10   END altera_pll;
11   -------------------------------------------------------------
12   ARCHITECTURE SYN OF altera_pll IS
13      ...
14      clk0_divide_by => 1,     --divider
15      clk0_duty_cycle => 50,   --duty cycle
16      clk0_multiply_by => 5,   --multiplier
17      clk0_phase_shift => "0",
18      compensate_clock => "CLK0",
19      gate_lock_signal => "NO",
20      inclk0_input_frequency => 20000, --input clk period in ps
21      intended_device_family => "Cyclone II",
22      ...
23   END ARCHITECTURE;
24   -------------------------------------------------------------
```

Part 4: TMDS Encoder

The TMDS encoder was studied in section 14.6. A corresponding VHDL code is shown below, based directly on the algorithm of figure 14.29 and on the design in section 14.6.

As mentioned there, the flowchart of figure 14.29 is in fact a modified version of that from DDWG, exhibiting a more hardware-oriented flow, which helps organize and optimize the implementation.

```
1   -----TMDS encoder:-------------------------------------------------
2   LIBRARY ieee;
3   USE ieee.std_logic_1164.all;
4   ------------------------------------------------------------------
5   ENTITY tmds_encoder IS
6      PORT (
7         din: IN STD_LOGIC_VECTOR(7 DOWNTO 0);     --pixel data
8         control: IN STD_LOGIC_VECTOR(1 DOWNTO 0);--control data
9         clk25: IN STD_LOGIC;                      --clock
10        dena: IN STD_LOGIC;                       --display enable
11        dout: OUT STD_LOGIC_VECTOR(9 DOWNTO 0)); --output data
12  END tmds_encoder;
13  ------------------------------------------------------------------
14  ARCHITECTURE tmds_encoder OF tmds_encoder IS
15     SIGNAL x: STD_LOGIC_VECTOR(8 DOWNTO 0);   --internal vector
16     SIGNAL onesX: INTEGER RANGE 0 TO 8;       --# of '1's in x
17     SIGNAL onesD: INTEGER RANGE 0 TO 8;       --# of '1's in din
18     SIGNAL disp: INTEGER RANGE -16 TO 15;     --disparity
19  BEGIN
20     -----Computes number of '1's in din:--------
21     PROCESS (din)
22        VARIABLE counterD: INTEGER RANGE 0 TO 8;
23     BEGIN
24        counterD := 0;
25        FOR i IN 0 TO 7 LOOP
26           IF (din(i)='1') THEN
27              counterD := counterD + 1;
28           END IF;
29        END LOOP;
30        onesD <= counterD;
31     END PROCESS;
32     -----Produces the internal vector x:-------
33     PROCESS (din, onesD)
34     BEGIN
35        x(0) <= din(0);
36        IF (onesD>4 OR (onesD=4 AND din(0)='0')) THEN
37           x(1) <= din(1) XNOR x(0);
38           x(2) <= din(2) XNOR x(1);
39           x(3) <= din(3) XNOR x(2);
```

```
40              x(4) <= din(4) XNOR x(3);
41              x(5) <= din(5) XNOR x(4);
42              x(6) <= din(6) XNOR x(5);
43              x(7) <= din(7) XNOR x(6);
44              x(8) <= '0';
45          ELSE
46              x(1) <= din(1) XOR x(0);
47              x(2) <= din(2) XOR x(1);
48              x(3) <= din(3) XOR x(2);
49              x(4) <= din(4) XOR x(3);
50              x(5) <= din(5) XOR x(4);
51              x(6) <= din(6) XOR x(5);
52              x(7) <= din(7) XOR x(6);
53              x(8) <= '1';
54          END IF;
55      END PROCESS;
56      -----Computes the number of '1's in x:-----
57      PROCESS (x)
58          VARIABLE counterX: INTEGER RANGE 0 TO 8;
59      BEGIN
60          counterX := 0;
61          FOR i IN 0 TO 7 LOOP
62              IF (x(i)='1') THEN
63                  counterX := counterX + 1;
64              END IF;
65          END LOOP;
66          onesX <= counterX;
67      END PROCESS;
68      -----Produces output vector and new disparity:--
69      PROCESS (disp, x, onesX, dena, control, clk25)
70          VARIABLE disp_new: INTEGER RANGE -31 TO 31;
71      BEGIN
72          IF (dena='1') THEN
73              dout(8) <= x(8);
74              IF (disp=0 OR onesX=4) THEN
75                  dout(9) <= NOT x(8);
76                  IF (x(8)='0') THEN
77                      dout(7 DOWNTO 0) <= NOT x(7 DOWNTO 0);
78                      disp_new := disp - 2*onesX + 8;
79                  ELSE
80                      dout(7 DOWNTO 0) <= x(7 DOWNTO 0);
81                      disp_new := disp + 2*onesX - 8;
82                  END IF;
```

```
83              ELSE
84                 IF ((disp>0 AND onesX>4) OR (disp<0 AND onesX<4)) THEN
85                    dout(9) <= '1';
86                    dout(7 DOWNTO 0) <= NOT x(7 DOWNTO 0);
87                    IF (x(8)='0') THEN
88                       disp_new := disp - 2*onesX + 8;
89                    ELSE
90                       disp_new := disp - 2*onesX + 10;
91                    END IF;
92                 ELSE
93                    dout(9) <= '0';
94                    dout(7 DOWNTO 0) <= x(7 DOWNTO 0);
95                    IF (x(8)='0') THEN
96                       disp_new := disp + 2*onesX - 10;
97                    ELSE
98                       disp_new := disp + 2*onesX - 8;
99                    END IF;
100                END IF;
101             END IF;
102          ELSE
103             disp_new := 0;
104             IF (control="00") THEN
105                dout <= "1101010100";
106             ELSIF (control="01") THEN
107                dout <= "0010101011";
108             ELSIF (control="10") THEN
109                dout <= "0101010100";
110             ELSE
111                dout <= "1010101011";
112             END IF;
113          END IF;
114          IF (clk25'EVENT AND clk25='1') THEN
115             disp <= disp_new;
116          END IF;
117       END PROCESS;
118    END tmds_encoder;
119    ------------------------------------------------------------------
```

Part 5: Serializer

Serializers were also studied in chapter 14. The code below was borrowed from section 14.2, with the only difference that the PLL is no longer present, because in the current design it was moved to the control generator block.

```
1   -----Serializer:-------------------------------
2   LIBRARY ieee;
3   USE ieee.std_logic_1164.all;
4   -----------------------------------------------
5   ENTITY serializer IS
6      PORT (clk250: IN STD_LOGIC;
7             din: IN STD_LOGIC_VECTOR(9 DOWNTO 0);
8             dout: OUT STD_LOGIC);
9   END serializer;
10  -----------------------------------------------
11  ARCHITECTURE serializer OF serializer IS
12     SIGNAL internal: STD_LOGIC_VECTOR(9 DOWNTO 0);
13  BEGIN
14     PROCESS (clk250)
15        VARIABLE count: INTEGER RANGE 0 TO 10;
16     BEGIN
17        IF (clk250'EVENT AND clk250='1') THEN
18           count := count + 1;
19           IF (count=9) THEN
20              internal <= din;
21           ELSIF (count=10) THEN
22              count := 0;
23           END IF;
24           dout <= internal(count);
25        END IF;
26     END PROCESS;
27  END serializer;
28  -----------------------------------------------
```

Part 6: Main Code

The main code is presented next, under the project name *dvi_stripes* (line 5). Observe in the ENTITY (lines 5–13) that the inputs are *clk50* (50 MHz clock available in our FPGA board) and three toggle switches (for color selection), while the outputs are the TMDS signals that feed the LCD monitor. Each output consists of a pair of wires, identified as *tmds0a* (= TMDS0+) plus *tmds0b* (= TMDS0−), *tmds1a* (= TMDS1+) plus *tmds1b* (= TMDS1−), and so on. As explained in section 14.6, this is due to the fact that FPGAs do not have CML I/Os, so a pair of conventional 3.3V LVTTL or LVCMOS pins, each with a series 270 Ω resistor, can be used to emulate such type of logic (see details in section 14.6 and figure 16.7).

The declarative part of the architecture contains general signal declarations in lines 17–21, followed by five COMPONENT declarations in lines 23–56 (notice that the PLL is not included in this list because it was used by the control generator, so its declaration was already included in that code).

The code proper (lines 58–81) simply assembles the circuit of figure 16.7 using the components just described. It employs one instance of *image_generator* (lines 63–65) and *control_generator* (lines 67–68), plus three instances of *tmds_encoder* (lines 70–72) and *serializer* (lines 73–75). The remaining lines (76–80) are for the complementary TMDS values.

The reader is invited to compile this design and download it to the FPGA board, observing what happens on the screen (of a DVI-driven LCD monitor) while playing with the toggle switches.

```
1    -----Main code:-------------------------------------------------
2    LIBRARY ieee;
3    USE ieee.std_logic_1164.all;
4    ----------------------------------------------------------------
5    ENTITY dvi_stripes IS
6       PORT (
7          clk50: IN STD_LOGIC; --50MHz system clock
8          red_switch, green_switch, blue_switch: IN STD_LOGIC;
9          tmds0a, tmds0b: BUFFER STD_LOGIC;      --TMDS0+, TMDS0-
10         tmds1a, tmds1b: BUFFER STD_LOGIC;      --TMDS1+, TMDS1-
11         tmds2a, tmds2b: BUFFER STD_LOGIC;      --TMDS2+, TMDS2-
12         tmds_clka, tmds_clkb: OUT STD_LOGIC); --TMDS_clk+,TMDS_clk-
13   END dvi_stripes;
14   ----------------------------------------------------------------
15   ARCHITECTURE dvi OF dvi_stripes IS
16      -----Signal declarations:--------------
17      SIGNAL clk25, clk250: STD_LOGIC;
18      SIGNAL Hsync, Vsync, Hactive, Vactive, dena: STD_LOGIC;
19      SIGNAL R, G, B: STD_LOGIC_VECTOR(7 DOWNTO 0);
20      SIGNAL control0, control1, control2: STD_LOGIC_VECTOR(1 DOWNTO 0);
21      SIGNAL data0, data1, data2: STD_LOGIC_VECTOR(9 DOWNTO 0);
22      -----1st component declaration:--------
23      COMPONENT image_generator IS
24         PORT (
25         red_switch, green_switch, blue_switch: IN STD_LOGIC;
26         Hsync, Vsync, Vactive, dena: IN STD_LOGIC;
27         R, G, B: OUT STD_LOGIC_VECTOR(7 DOWNTO 0));
28      END COMPONENT;
29      -----2nd component declaration:--------
30      COMPONENT control_generator IS
31         PORT (
```

```
32      clk50: IN STD_LOGIC;
33      clk25: BUFFER STD_LOGIC;
34      clk250: OUT STD_LOGIC;
35      Hsync: BUFFER STD_LOGIC;
36      Vsync: OUT STD_LOGIC;
37      Hactive: BUFFER STD_LOGIC;
38      Vactive: BUFFER STD_LOGIC;
39      dena: OUT STD_LOGIC);
40  END COMPONENT;
41      -----3rd component declaration:--------
42      COMPONENT tmds_encoder IS
43        PORT (
44        din: IN STD_LOGIC_VECTOR(7 DOWNTO 0);
45        control: IN STD_LOGIC_VECTOR(1 DOWNTO 0);
46        clk25: IN STD_LOGIC;
47        dena: IN STD_LOGIC;
48        dout: OUT STD_LOGIC_VECTOR(9 DOWNTO 0));
49      END COMPONENT;
50      -----4th component declaration:--------
51      COMPONENT serializer IS
52        PORT (
53        clk250: IN STD_LOGIC;
54        din: IN STD_LOGIC_VECTOR(9 DOWNTO 0);
55        dout: OUT STD_LOGIC);
56      END COMPONENT;
57      ---------------------------------------
58  BEGIN
59      control0 <= Vsync & Hsync;
60      control1 <= "00";
61      control2 <= "00";
62      -----Image generator:------------------
63      image_gen: image_generator PORT MAP (
64        red_switch, green_switch, blue_switch, Hsync, Vsync,
65        Vactive, dena, R, G, B);
66      -----Control generator:----------------
67      control_gen: control_generator PORT MAP (
68        clk50, clk25, clk250, Hsync, Vsync, OPEN, Vactive, dena);
69      -----TMDS transmitter:----------------
70      tmds0: tmds_encoder PORT MAP (B, control0, clk25, dena, data0);
71      tmds1: tmds_encoder PORT MAP (G, control1, clk25, dena, data1);
72      tmds2: tmds_encoder PORT MAP (R, control2, clk25, dena, data2);
73      serial0: serializer PORT MAP (clk250, data0, tmds0a);
74      serial1: serializer PORT MAP (clk250, data1, tmds1a);
```

```
75      serial2: serializer PORT MAP (clk250, data2, tmds2a);
76      tmds0b <= NOT tmds0a;
77      tmds1b <= NOT tmds1a;
78      tmds2b <= NOT tmds2a;
79      tmds_clka <= clk25;
80      tmds_clkb <= NOT clk25;
81  END dvi;
82  --------------------------------------------------------------------
```

16.8   Other DVI Designs

Once one knows how to deal with the control signals, TMDS encoder, and serializer, generating images for a DVI-driven LCD monitor becomes essentially the same problem as generating images for a VGA monitor. Consequently, the material seen in chapter 15 applies here too, so the following is recommended:

· For generating regular images with dedicated hardware, follow the procedure of section 15.9 (equivalent to section 16.7).

· For generating regular images with a file and on-chip memory, follow the procedure of section 15.10.

· For generating arbitrary images with a file and off-chip memory, follow the procedure of section 15.11.

For the same reason, all examples and exercises seen there can also be used here.

16.9   Exercises

Note: For exercise solutions, please consult the book website.

Exercise 16.1: TMDS Encoder

a) What are the purposes (benefits) of the TMDS circuit?

b) Why is it called an 8B/10B encoder? Is it the same as the regular 8B/10B?

c) Why is it called a serial data transmission encoder?

d) Why is a PLL generally needed to construct it?

e) What is the nominal maximum *effective* pixel rate with single link?

f) Why was the circuit of figure 14.27b used in the experiments?

Exercise 16.2: Image Generation with Hardware #1 (Banner)

Solve exercise 15.2 for a DVI monitor.

Exercise 16.3: Image Generation with Hardware #2 (Sun in the Sky)

Solve exercise 15.3 for a DVI monitor.

Exercise 16.4: Image Generation with Hardware #3 (Filling with Green)

Solve exercise 15.4 for a DVI monitor.

Exercise 16.5: Image Generation with Hardware #4 (Rotating Bar)

Solve exercise 15.5 for a DVI monitor.

Exercise 16.6: Image Generation with Hardware #5 (Wall Clock)

Solve exercise 15.6 for a DVI monitor.

Exercise 16.7: Image Generation with Hardware #6 (Arcade Game)

Solve exercise 15.7 for a DVI monitor.

Exercise 16.8: Image Generation with a File and On-Chip Memory #1 (Banner)

Solve exercise 15.8 for a DVI monitor.

Exercise 16.9: Image Generation with a File and On-Chip Memory #2 (Sun in the Sky)

Solve exercise 15.9 for a DVI monitor.

Exercise 16.10: Image Generation with a File and Off Chip Memory (Arbitrary Picture)

Solve exercise 15.10 for a DVI monitor.

17 VHDL Design of FPD-Link Video Interfaces

17.1 Introduction

FPD (flat panel display)-Link, along with its extended version, LDI (LVDS Display Interface), are other modern video interfaces, used to drive laptop displays and other small LCDs located at short distances from the driving circuit, particularly in automotive, medical, and industrial applications.

The study of FPD-Link (introduced by National Semiconductor in 1992) in this chapter concludes the series of video interfaces, which included also VGA (chapter 15) and DVI (chapter 16) circuits. In summary, while the TMDS circuit (chapter 16) is used in the physical interface between desktop computers and LCD monitors (by means of the DVI controller) or between computers and other video systems, like video projectors and HDTV (by means of the HDMI controller), FPD-Link and LDI are used in the physical interface between notebooks and their LCD displays and in other applications where the display cable is very short. Other standard interfaces, like *Camera Link*, also employ the FPD-Link encoder.

FPD-Link normally operates with a pixel clock in the 25 to 85 MHz range. The main differences with respect to LDI are that the latter can operate in dual-data-rate mode (active at both clock edges) and also with dual data channels (two 8-bit inputs for each color).

As will be seen, there are also important differences between FPD-Link and TMDS, like the type of I/O and the use or not of DC-balancing techniques. Because FPD-Link is intended only for short cables, a slightly simpler I/O is employed (LVDS, versus CML in TMDS) and DC-balance is omitted, rendering a simpler, lower cost, and lower power video interface.

In the design examples, the 10.4″ LCD of figure 17.1 will be used. Note that the display is connected to two other units: FPD-Link interface (in the FPGA) and a high-voltage generator for the display's backlight. Typical specifications for this type of display follow.

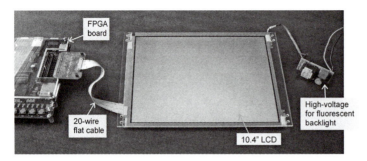

Figure 17.1
10.4″ LCD used in the experiments (flat 20-wire FPD-Link cable and FPGA board shown on the left, high-voltage generator for the backlight shown on the right).

- Diagonal size: 10.4″
- Manufacturers: LG-Philips (model LB104S01), Suntai (model SFA-104A), Samsung, etc.
- Typical native resolution: SVGA (800 × 600).
- Pixel clock: 25 to 85 MHz.
- Color encoding: RGB, six bits per color.
- Monitor cable: Flat 20-wire cable (see figure 17.1).
- Physical interface: FPD-Link.
- Physical interface supply voltage: 3.3 V.
- Backlight type: Dual cold cathode fluorescent lamps (2 × CCFL).
- Backlight supply: ∼500V/6mA RMS, 60 kHz (see figure 17.1).

As in chapters 15 and 16, which also deal with video interfaces, the following fundamental aspects will be focused on here:

1) Operation of the DVI interface.

2) How the circuit should be divided to make the design as simple and as standard as possible.

3) How the control signals operate and how they should be generated.

4) How images can be generated (from local hardware, external memory, file, etc.), rather than focusing on the images themselves (software).

17.2 FPD-Link Encoder

The top-level diagram of an FPD link is shown in figure 17.2. The encoder, shown on the left, transmits typical VGA video signals (seen in chapter 15)—that is, the three fundamen-

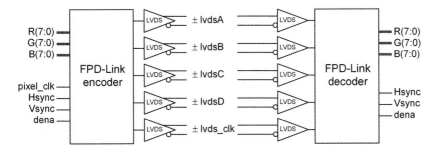

Figure 17.2
FPD-Link interface.

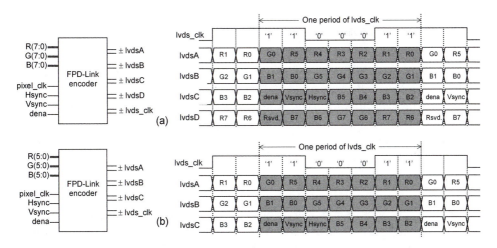

Figure 17.3
(a) Complete FPD-Link encoder (8-bit colors, five LVDS channels) and its timing diagram. (b) Simplified FPD-Link encoder (6-bit colors, four LVDS channels) and corresponding timing diagram.

tal colors (*R*, *G*, and *B*, each represented by eight bits), the pixel clock (*pixel_clk*), and the corresponding control signals (horizontal synchronism, *Hsync*, vertical synchronism, *Vsync*, and display enable, *dena*). The output consists of five LVDS channels, with four of them (*lvdsA* to *lvdsD*) conveying data, while the last one (*lvds_clk*) transmits a modified version of *pixel_clk*. The decoder, shown on the right, recovers the original data.

Figure 17.2 also indicates the type of circuit used in the I/Os, which is LVDS (Pedroni 2008). This is a standard differential circuit that operates with a small voltage difference (0.35 V nominal), resulting in fast, low-power, and low-EMI data transmissions.

A timing diagram is shown in figure 17.3a. Observe that the four serial LVDS outputs contain all three colors (*R*, *G*, *B*) and three control signals (*Hsync*, *Vsync*, *dena*). Note also that the transmission clock (internal to the circuit) is seven times faster than the LVDS

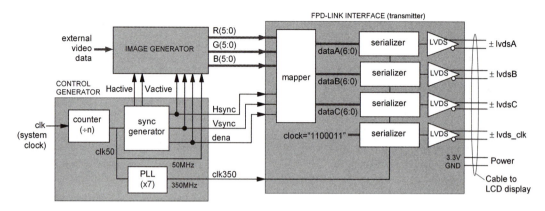

Figure 17.4
6-bit-per-color FPD-Link interface used in the experiments.

clock, with the latter consisting of the "1100011" pattern (see uppermost plot). The transmission starts with the MSB.

A 6-bit-per-color version is depicted in figure 17.3b, which requires only four LVDS channels (three for data, one for the clock). This is a popular implementation in industrial applications, for example.

17.3   Setup for the Experiments

Figure 17.4 shows a complete 6-bit-per-color video circuit to be used in the experiments with FPD-Link. Besides the *FPD-Link interface* block, it contains also a *control generator*, responsible for generating all control signals (SVGA signals *Hsync*, *Vsync*, *Hactive*, *Vactive*, and *dena*, plus the high-frequency clock for the serializer) and an *image generator* (responsible for providing the pixel-related signals, *R*, *G*, and *B*).

A typical resolution for small-display applications based on FPD-Link is SVGA (800 × 600 pixels—see SVGA parameters in figures 15.5 and 15.6), operating with a pixel clock in the 25 to 85 MHz range. With this resolution, $1056 \times 628 = 663,168$ clock cycles are needed to complete one screen, so the refresh rate with 85 MHz is 128 frames/s. To attain 60 frames/s, a pixel clock of 40 MHz is sufficient. We will assume that the clock available in our FPGA board is 50 MHz (hence $n = 1$, or 75 frames/s), so the serializer clock must be 350 MHz. To ease their identifications, *pixel_clk* and *clkTx* (for the serializers) are here called *clk50* and *clk350*, respectively.

In summary, the following parameters will be used in the control generator:

- System clock (*clk50*): 50 MHz.
- Pixel clock (*pixel_clk = clk50*): 50 MHz.

Pin	Signal	Pin	Signal	Pin	Signal	Pin	Signal
1	VDD (3.3V)	6	lvdsA+	11	LvdsC–	16	GND
2	VDD (3.3V)	7	GND	12	LvdsC+	17	GND
3	GND	8	lvdsB–	13	GND	18	GND
4	GND	9	lvdsB+	14	lvds_clk–	19	GND
5	lvdsA–	10	GND	15	lvds_clk+	20	GND

Figure 17.5
Pinout of a 20-pin 6-bit FPD-Link connector.

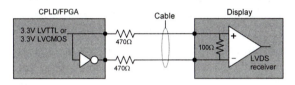

Figure 17.6
LVDS transmitter emulated with 3.3V LVTTL/LVCMOS pads and resistors.

- Clock for serializer receiver ($clkRx = clk50$): 50 MHz.
- Clock for serializer transmitter ($clkTx = clk350$): 350 MHz.
- LVDS clock ($lvds\_clk$): 50 MHz, with "1100011" pattern.
- Horizontal timing (SVGA, figure 15.5): $Hpulse = 128$, $HBP = 88$, $Hactive = 800$, $HFP = 40$ (total = 1056 pixels).
- Vertical timing (SVGA, figure 15.6): $Vpulse = 4$, $VBP = 23$, $Vactive = 600$, $VFP = 1$ (total = 628 lines).

Figure 17.5 shows the pinout of a 20-pin connector used in the 6-bit/color FPD-Link interface (see flat cable on the left of figure 17.1). Note that only the wires relative to the four LVDS pairs are indeed used, with all the other pins at VDD or GND (compare this to the cable on the right of figure 17.4).

A final comment regarding the LVDS I/Os: To make the problem even more interesting, let us assume that our CPLD or FPGA device does not have LVDS pins (or that a different VCCIO was used, such that the 2.5V LVDS standard is no longer available). As described in Pedroni (2008), LVDS receivers must be able to operate with differential voltages as low as 100 mV (350 mV nominal), with a common-mode voltage in the 0.1V-to-2.4V range. Consequently, traditional 3.3V LVTTL/LVCMOS pins can be used. The corresponding arrangement is shown in figure 17.6, which produces a differential voltage of 0.32 V and a common-mode voltage of 1.65 V.

17.4 Hardware-Generated Image

We turn now to the design of actual FPD-Link-driven applications. Similar to what was done in the previous two chapters, the main purpose of this first design is to illustrate how the control signals and the FPD-Link interface can be constructed. To keep the focus on that, a simple image was chosen, which consists of just four horizontal stripes of solid colors with widths 1, 2, 3, and 594 pixels. The first three are red, green, and blue, respectively, while the color of the last one is determined by three toggle switches. SVGA resolution (800×600) and a 50 MHz clock (*clk50*) will be employed.

A corresponding VHDL code is presented below. It is a structural code, which follows figure 17.4 exactly. The circuit is broken into three subcircuits, called *image_generator*, *control_generator*, and *FPDLink_interface*. The first one is application-dependent, whereas the other two exhibit a fixed structure. As indicated in figure 17.4, *FPDLink_interface* is further divided using the *serializer* block. Notice that once the circuit is well understood and then properly divided into blocks (as in figure 17.4), designing it becomes relatively simple.

Part 1: Image Generator

The code below implements the four-stripe image. In the PORT declarations (lines 6–9), all five control signals are included (line 8), making this a kind of generic entity, even though not all five are always needed. The code in the architecture is divided into two parts; the first part (lines 19–25) implements a pointer (address) to the image rows, while the second part (lines 27–49) builds the image proper, which is partially controlled by three toggle switches.

```
 1   -----Image generator:-------------------------------
 2   LIBRARY ieee;
 3   USE ieee.std_logic_1164.all;
 4   ----------------------------------------------------
 5   ENTITY image_generator IS
 6      PORT (
 7         red_switch, green_switch, blue_switch: IN STD_LOGIC;
 8         Hsync, Vsync, Hactive, Vactive, dena: IN STD_LOGIC;
 9         R, G, B: OUT STD_LOGIC_VECTOR(5 DOWNTO 0));
10   END image_generator;
11   ----------------------------------------------------
12   ARCHITECTURE image_generator OF image_generator IS
13   BEGIN
14      PROCESS (Hsync, Vsync, Vactive, dena, red_switch,
15         green_switch, blue_switch)
16         VARIABLE line_counter: INTEGER RANGE 0 TO 600;
17      BEGIN
```

```
18              -----Create pointer to LCD rows:----
19         IF (Vsync='0') THEN
20              line_counter := 0;
21         ELSIF (Hsync'EVENT AND Hsync='1') THEN
22              IF (Vactive='1') THEN
23                   line_counter := line_counter + 1;
24              END IF;
25         END IF;
26         -----Create image:------------------
27         IF (dena='1') THEN
28              IF (line_counter=1) THEN
29                   R <= (OTHERS => '1');
30                   G <= (OTHERS => '0');
31                   B <= (OTHERS => '0');
32              ELSIF (line_counter>1 AND line_counter<=3) THEN
33                   R <= (OTHERS => '0');
34                   G <= (OTHERS => '1');
35                   B <= (OTHERS => '0');
36              ELSIF (line_counter>3 AND line_counter<=6) THEN
37                   R <= (OTHERS => '0');
38                   G <= (OTHERS => '0');
39                   B <= (OTHERS => '1');
40              ELSE
41                   R <= (OTHERS => red_switch);
42                   G <= (OTHERS => green_switch);
43                   B <= (OTHERS => blue_switch);
44              END IF;
45         ELSE
46                   R <= (OTHERS => '0');
47                   G <= (OTHERS => '0');
48                   B <= (OTHERS => '0');
49         END IF;
50    END PROCESS;
51 END image_generator;
52 ----------------------------------------------------------------
```

Part 2: Control Generator

The code below implements the control generator, which produces the signals *Hsync*, *Vsync*, *Hactive*, *Vactive*, and *dena*, plus the fast clock (*clk350*) for the serializers. Recall that this circuit has a fixed structure because it is application-independent.

The SVGA parameters are specified using GENERIC declarations (lines 6–14), so they can be easily changed to any other display resolution. The only input is *clk50* (line 16), while the outputs are the signals just mentioned (lines 17–22).

The architecture contains, in its declarative part, a component declaration (a PLL, lines 26–31), which is then instantiated in line 72 to create *clk350*. The process in lines 34–50 creates the horizontal control signals (in accordance with figure 15.5), while that in lines 52–68 creates the vertical control signals (in accordance with figure 15.6). Note that, except for the fast clock, this code is similar to that for the control generator in chapter 15.

Still regarding the PLL, the file that is automatically generated during its instantiation must be included among the project files. For example, if using Altera's Quartus II to synthesize the circuit, then the MegaWizard Plug-In Manager tool can be employed to attain the PLL (that is the case in the example below, where the name chosen for the generated file is *altera_pll.vhd*). Further details can be seen in appendix G.

```
1   -----Control generator:-------------------------------------
2   LIBRARY ieee;
3   USE ieee.std_logic_1164.all;
4   -----------------------------------------------------------
5   ENTITY control_generator IS
6      GENERIC (                 --SVGA parameters
7         Ha: INTEGER := 128;    --Hpulse
8         Hb: INTEGER := 216;    --Hpulse+HBP
9         Hc: INTEGER := 1016;   --Hpulse+HBP+Hactive
10        Hd: INTEGER := 1056;   --Hpulse+HBP+Hactive+HFP
11        Va: INTEGER := 4;      --Vpulse
12        Vb: INTEGER := 27;     --Vpulse+VBP
13        Vc: INTEGER := 627;    --Vpulse+VBP+Vactive
14        Vd: INTEGER := 628);   --Vpulse+VBP+Vactive+VFP
15     PORT (
16        clk50: IN STD_LOGIC;          --50MHz system clock
17        clk350: OUT STD_LOGIC;        --350MHz serializer clock
18        Hsync: BUFFER STD_LOGIC;      --Horizontal sync
19        Vsync:  OUT STD_LOGIC;        --Vertical sync
20        Hactive: BUFFER STD_LOGIC;    --Horiz. display interval
21        Vactive: BUFFER STD_LOGIC;    --Vert. display interval
22        dena: OUT STD_LOGIC);         --Display enable
23  END control_generator;
24  -----------------------------------------------------------
25  ARCHITECTURE control_generator OF control_generator IS
26     COMPONENT altera_pll IS
27        PORT (areset: IN STD_LOGIC;
28              inclk0: IN STD_LOGIC;
29              c0: OUT STD_LOGIC;
30              locked: OUT STD_LOGIC);
31     END COMPONENT;
```

```
32  BEGIN
33      ---Horizontal signals:------------
34      PROCESS (clk50)
35          VARIABLE Hcount: INTEGER RANGE 0 TO Hd;
36      BEGIN
37          IF (clk50'EVENT AND clk50='1') THEN
38              Hcount := Hcount + 1;
39              IF (Hcount=Ha) THEN
40                  Hsync <= '1';
41              ELSIF (Hcount=Hb) THEN
42                  Hactive <= '1';
43              ELSIF (Hcount=Hc) THEN
44                  Hactive <= '0';
45              ELSIF (Hcount=Hd) THEN
46                  Hsync <= '0';
47                  Hcount := 0;
48              END IF;
49          END IF;
50      END PROCESS;
51      ---Vertical signals:-------------
52      PROCESS (Hsync)
53          VARIABLE Vcount: INTEGER RANGE 0 TO Vd;
54      BEGIN
55          IF (Hsync'EVENT AND Hsync='0') THEN
56              Vcount := Vcount + 1;
57              IF (Vcount=Va) THEN
58                  Vsync <= '1';
59              ELSIF (Vcount=Vb) THEN
60                  Vactive <= '1';
61              ELSIF (Vcount=Vc) THEN
62                  Vactive <= '0';
63              ELSIF (Vcount=Vd) THEN
64                  Vsync <= '0';
65                  Vcount := 0;
66              END IF;
67          END IF;
68      END PROCESS;
69      ---Display enable:----------------
70      dena <= Hactive AND Vactive;
71      ---Serializer clock (350MHz):-----
72      pll: altera_pll PORT MAP ('0', clk50, clk350, OPEN);
73  END control_generator;
74  ------------------------------------------------------------
```

Part 3: Serializer

Serializers were studied and also designed in section 14.2, based on which the code below was assembled. The only difference is that here the PLL, which multiplies the system clock (50 MHz) by 7 (resulting a 350 MHz clock for the serializers), is implemented in another block (*control_generator*).

```
1    -----Serializer:--------------------------------
2    LIBRARY ieee;
3    USE ieee.std_logic_1164.all;
4    ------------------------------------------------
5    ENTITY serializer IS
6       PORT (clk350: IN STD_LOGIC;
7             din: IN STD_LOGIC_VECTOR(6 DOWNTO 0);
8             dout: OUT STD_LOGIC);
9    END serializer;
10   ------------------------------------------------
11   ARCHITECTURE serializer OF serializer IS
12      SIGNAL internal: STD_LOGIC_VECTOR(6 DOWNTO 0);
13   BEGIN
14      PROCESS (clk350)
15         VARIABLE count: INTEGER RANGE 0 TO 7 := 0;
16      BEGIN
17         IF (clk350'EVENT AND clk350='1') THEN
18            count := count + 1;
19            IF (count=6) THEN
20               internal <= din;
21            ELSIF (count=7) THEN
22               count := 0;
23            END IF;
24            dout <= internal(6-count); --MSB first
25         END IF;
26      END PROCESS;
27   END serializer;
28   ------------------------------------------------
```

Part 4: FPD-Link Interface

The code below is a direct implementation of the FPD-Link interface circuit shown in figure 17.4. It contains only two types of blocks: *mapper* and *serializer*. The former is just a wiring, while the latter was designed in the code above, so it enters here as a subcircuit (COMPONENT).

```
 1   ----FPDlink interface:-------------------------------------------------
 2   LIBRARY ieee;
 3   USE ieee.std_logic_1164.all;
 4   -----------------------------------------------------------------------
 5   ENTITY FPDlink_interface IS
 6      PORT (
 7         clk350: IN STD_LOGIC; --350MHz serializer clock
 8         Hsync, Vsync, dena: IN STD_LOGIC;
 9         R, G, B: IN STD_LOGIC_VECTOR(5 DOWNTO 0);
10         lvdsA: OUT STD_LOGIC;
11         lvdsB: OUT STD_LOGIC;
12         lvdsC: OUT STD_LOGIC;
13         lvds_clk: OUT STD_LOGIC);
14   END FPDlink_interface;
15   -----------------------------------------------------------------------
16   ARCHITECTURE FPDlink_interface OF FPDlink_interface IS
17      SIGNAL dataA, dataB, dataC, data_clk: STD_LOGIC_VECTOR(6 DOWNTO 0);
18      COMPONENT serializer IS
19         PORT (clk350: IN STD_LOGIC;
20               din: IN STD_LOGIC_VECTOR(6 DOWNTO 0);
21               dout: OUT STD_LOGIC);
22      END COMPONENT;
23   BEGIN
24      -----Mapper:----------------------------
25      dataA <= G(0) & R(5 DOWNTO 0);
26      dataB <= B(1 DOWNTO 0) & G(5 DOWNTO 1);
27      dataC <= dena & Vsync & Hsync & B(5 DOWNTO 2);
28      data_clk <= "1100011";
29      -----Serializers:----------------------
30      serialA: serializer PORT MAP (clk350, dataA, lvdsA);
31      serialB: serializer PORT MAP (clk350, dataB, lvdsB);
32      serialC: serializer PORT MAP (clk350, dataC, lvdsC);
33      serial_clk: serializer PORT MAP (clk350, data_clk, lvds_clk);
34   END FPDlink_interface;
35   -----------------------------------------------------------------------
```

Part 5: Main Code

Finally, the main code is shown below, under the project name *FPDLink_stripes* (line 5). It simply joins the subcircuits designed above to create the complete FPD-Link driver and the intended image. Note in lines 77–80 that complementary signals are created in order to be able to use the arrangement of figure 17.6, which emulates LVDS pads. The reader is invited to compile this code and test it in the FPGA board, checking what happens on the LCD screen while playing with the *R*, *G*, and *B* switches.

```
 1   -----Main code:-----------------------------------------------------------
 2   LIBRARY ieee;
 3   USE ieee.std_logic_1164.all;
 4   ---------------------------------------------------------------------------
 5   ENTITY FPDlink_stripes IS
 6      PORT (
 7         clk50: IN STD_LOGIC; --50MHz system clock
 8         red_switch, green_switch, blue_switch: IN STD_LOGIC;
 9         lvdsA1: BUFFER STD_LOGIC;
10         lvdsA2: OUT STD_LOGIC;
11         lvdsB1: BUFFER STD_LOGIC;
12         lvdsB2: OUT STD_LOGIC;
13         lvdsC1: BUFFER STD_LOGIC;
14         lvdsC2: OUT STD_LOGIC;
15         lvds_clk1: BUFFER STD_LOGIC;
16         lvds_clk2: OUT STD_LOGIC);
17   END FPDlink_stripes;
18   ---------------------------------------------------------------------------
19   ARCHITECTURE FPDlink_stripes OF FPDlink_stripes IS
20      ----Signal declarations:--------------
21      SIGNAL clk350: STD_LOGIC;
22      SIGNAL Hsync, Vsync, Hactive, Vactive, dena: STD_LOGIC;
23      SIGNAL R, G, B: STD_LOGIC_VECTOR(5 DOWNTO 0);
24      SIGNAL dataA, dataB, dataC, data_clk: STD_LOGIC_VECTOR(6 DOWNTO 0);
25      ----1st component declaration:--------
26      COMPONENT image_generator IS
27         PORT (
28            red_switch, green_switch, blue_switch: IN STD_LOGIC;
29            Hsync, Vsync, Hactive, Vactive, dena: IN STD_LOGIC;
30            R, G, B: OUT STD_LOGIC_VECTOR(5 DOWNTO 0));
31      END COMPONENT;
32      ----2nd component declaration:--------
33      COMPONENT control_generator IS
34         GENERIC (
35            Ha: INTEGER := 128;
36            Hb: INTEGER := 216;
37            Hc: INTEGER := 1016;
38            Hd: INTEGER := 1056;
39            Va: INTEGER := 4;
40            Vb: INTEGER := 27;
41            Vc: INTEGER := 627;
42            Vd: INTEGER := 628);
43         PORT (
44            clk50: IN STD_LOGIC;
```

```
45              clk350: OUT STD_LOGIC;
46              Hsync: BUFFER STD_LOGIC;
47              Vsync: OUT STD_LOGIC;
48              Hactive: OUT STD_LOGIC;
49              Vactive: OUT STD_LOGIC;
50              dena: OUT STD_LOGIC);
51      END COMPONENT;
52      ----3rd component declaration:--------
53      COMPONENT FPDlink_interface IS
54        PORT (
55              clk350: IN STD_LOGIC;
56              Hsync, Vsync, dena: IN STD_LOGIC;
57              R, G, B: IN STD_LOGIC_VECTOR(5 DOWNTO 0);
58              lvdsA: OUT STD_LOGIC;
59              lvdsB: OUT STD_LOGIC;
60              lvdsC: OUT STD_LOGIC;
61              lvds_clk: OUT STD_LOGIC);
62      END COMPONENT;
63      -------------------------------------
64  BEGIN
65      ----Image_generator:----------------
66      stripes: image_generator PORT MAP (
67          red_switch, green_switch, blue_switch,
68          Hsync, Vsync, Hactive, Vactive, dena, R, G, B);
69      ----Control signals:----------------
70      control: control_generator PORT MAP (
71          clk50, clk350, Hsync, Vsync, Hactive, Vactive, dena);
72      ----FPDlink interface:--------------
73      fpd_link: FPDlink_interface PORT MAP (
74          clk350, Hsync, Vsync, dena, R, G, B,
75          lvdsA1, lvdsB1, lvdsC1, lvds_clk1);
76      ---------------------------------
77      lvdsA2 <= NOT lvdsA1;
78      lvdsB2 <= NOT lvdsB1;
79      lvdsC2 <= NOT lvdsC1;
80      lvds_clk2 <= NOT lvds_clk1;
81  END FPDlink_stripes;
82  -----------------------------------------------------------------------
```

17.5   Hardware-Generated Image with Characters

This design is for a circuit that implements a digital clock to be displayed on an FPD-Link
driven LCD (a similar design was presented in section 12.5, with SSDs). As shown in

Figure 17.7
Clock to be designed in section 17.5.

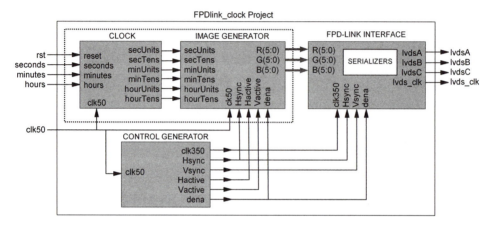

Figure 17.8
Circuit for the wall clock in section 17.5.

figure 17.7, it must exhibit hours, minutes, and seconds, separated by colons, and must have four control buttons that provide the features below.

Reset: When asserted, must zero the display, with precedence over any other button.

Seconds: When asserted, must increase the speed of the counter by a factor of 8 (fast adjustment of seconds).

Minutes: When asserted, must increase the speed of the counter by a factor of 252 (fast adjustment of minutes).

Hours: When asserted, must increase the speed of the counter by a factor of 8,192 (fast adjustment of hours).

As before, powers of two were used in the speed-up factors above to minimize the amount of hardware (these dividers are just shifters).

As in the previous section, a structural design will be developed. A circuit diagram is suggested in figure 17.8, where the overall system is broken into four subcircuits, called *clock*, *image_generator*, *control_generator*, and *FPDlink_interface*. As before, the last

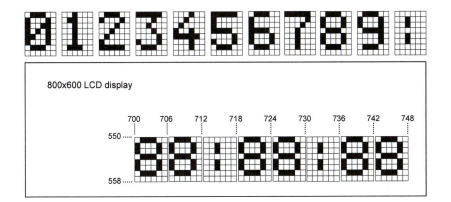

Figure 17.9
Digit representations and digit positions on the screen for the design in section 17.5.

two exhibit a fixed structure, while the first two, which together are responsible for generating the pixels, vary according to the application. As indicated in the figure, the *FPDlink_interface* subcircuit is further divided using the *serializer* block.

Observe also in figure 17.8 the circuit's global inputs and outputs. The inputs consist of *clk50* (50 MHz system clock) plus four pushbuttons to control the clock. The outputs are four LVDS channels, directly connected to the cable that goes to the display. As already seen, each LVDS output consists of two wires, which in our case are obtained with the setup of figure 17.6, so for each LVDS output a pair of signals are indeed needed.

As in the previous design, only the image generator needs to be changed. To ease its construction, such a part was assembled with two subcircuits (*clock* and *image_generator*) in the present design. Therefore, all that is needed is to replace *image_generator* in the previous design with *clock* plus the new *image_generator* in the present design (followed, of course, by the respective adjustments in the SIGNAL and COMPONENT declarations in the main code).

The representation used for the digits in the image generator is depicted in figure 17.9, which also shows the positions of the digits on the screen. The ":" (colon) symbol was employed to separate the pairs of digits.

Part 1: Clock

A VHDL code for the clock is shown below. It does not contain pixel signals yet (converting the signals below into image signals is left to the *image_generator*). The inputs are *clk50* (system clock) plus the four control pushbuttons, from which the electrical signals for the clock digits are provided. The code is divided into two parts; the first part (lines 25–28) defines the speed-up factors for clock adjustment, while the second part (lines 30–80) builds the clock proper.

```
1    -----Clock:-------------------------------------------------------
2    LIBRARY ieee;
3    USE ieee.std_logic_1164.all;
4    ------------------------------------------------------------------
5    ENTITY clock IS
6       GENERIC (fclk: INTEGER := 50_000_000); --system clock freq.
7       PORT (
8          clk50: IN STD_LOGIC;     --system clock
9          rst: IN STD_LOGIC;       --clock reset
10         seconds: IN STD_LOGIC;  --fast adjustment for seconds
11         minutes: IN STD_LOGIC;  --fast adjustment for minutes
12         hours: IN STD_LOGIC;     --fast adjustment for hours
13         secUnits: OUT NATURAL RANGE 0 TO 9;    --units of seconds
14         secTens: OUT NATURAL RANGE 0 TO 5;     --tens of seconds
15         minUnits: OUT NATURAL RANGE 0 TO 9;    --units of minutes
16         minTens: OUT NATURAL RANGE 0 TO 5;     --tens of minutes
17         hourUnits: OUT NATURAL RANGE 0 TO 9;   --units of hours
18         hourTens: OUT NATURAL RANGE 0 TO 2);   --tens of hours
19   END clock;
20   ------------------------------------------------------------------
21   ARCHITECTURE clock OF clock IS
22      SIGNAL limit: INTEGER RANGE 0 TO fclk;
23   BEGIN
24      ---Define speed-up factors:-----------
25      limit <= fclk/8192 WHEN hours='1' ELSE
26                fclk/252 WHEN minutes='1' ELSE
27                fclk/8 WHEN seconds='1' ELSE
28                fclk;
29      ---Design the clock:------------------
30      PROCESS (clk50, rst)
31         VARIABLE one_second: NATURAL RANGE 0 TO fclk;
32         VARIABLE secU: NATURAL RANGE 0 TO 10;
33         VARIABLE secT: NATURAL RANGE 0 TO 6;
34         VARIABLE minU: NATURAL RANGE 0 TO 10;
35         VARIABLE minT: NATURAL RANGE 0 TO 6;
36         VARIABLE hourU: NATURAL RANGE 0 TO 10;
37         VARIABLE hourT: NATURAL RANGE 0 TO 3;
38      BEGIN
39         IF (rst='1') THEN
40            one_second := 0; secU := 0; secT := 0;
41            minU := 0; minT := 0; hourU := 0; hourT := 0;
42         ELSIF (clk50'EVENT AND clk50='1') THEN
43            one_second := one_second + 1;
```

```
44              IF (one_second=limit) THEN
45                 one_second := 0;
46                secU := secU + 1;
47                IF (secU=10) THEN
48                   secU := 0;
49                   secT := secT + 1;
50                   IF (secT=6) THEN
51                      secT := 0;
52                      minU := minU + 1;
53                      IF (minU=10) THEN
54                         minU := 0;
55                         minT := minT + 1;
56                         IF (minT=6) THEN
57                            minT := 0;
58                            hourU := hourU + 1;
59                            IF ((hourT/=2 AND hourU=10) OR
60                                (hourT=2 AND hourU=4)) THEN
61                               hourU := 0;
62                               hourT := hourT + 1;
63                               IF (hourT=3) THEN
64                                  hourT := 0;
65                               END IF;
66                            END IF;
67                         END IF;
68                      END IF;
69                   END IF;
70                END IF;
71             END IF;
72          END IF;
73          ---Pass values to output:---------
74          secUnits <= secU;
75          secTens <= secT;
76          minUnits <= minU;
77          minTens <= minT;
78          hourUnits <= hourU;
79          hourTens <= hourT;
80       END PROCESS;
81    END clock;
82    ----------------------------------------------------------------
```

Part 2: Image Generator

A VHDL code for the image generator is presented below. The digits are represented by means of constants (specified in lines 23–77). Note that a FUNCTION, called *bcd_to_digit6x8*, was created in lines 79–93 to make the data conversion from BCD (binary coded decimal) format to the proper display format (defined by the constants). This function is then called six times in lines 101–106. In lines 108–118, horizontal (*x*) and vertical (*y*) pointers (addresses) are created, which are then used in lines 120–140 to build the image.

```
1    -----Image generator:-------------------------------------------------
2    LIBRARY ieee;
3    USE ieee.std_logic_1164.all;
4    -------------------------------------------------------------------------
5    ENTITY image_generator IS
6       PORT (
7          clk50, Hsync, Hactive, Vactive, dena: IN STD_LOGIC;
8          secUnits: IN NATURAL RANGE 0 TO 9;
9          secTens: IN NATURAL RANGE 0 TO 5;
10         minUnits: IN NATURAL RANGE 0 TO 9;
11         minTens: IN NATURAL RANGE 0 TO 5;
12         hourUnits: IN NATURAL RANGE 0 TO 9;
13         hourTens: IN NATURAL RANGE 0 TO 2;
14         R, G, B: OUT STD_LOGIC_VECTOR(5 DOWNTO 0));
15   END image_generator;
16   -------------------------------------------------------------------------
17   ARCHITECTURE image_generator OF image_generator IS
18      ---Type/signal/constant declarations:---
19      TYPE digit6x8 IS ARRAY (1 TO 8, 1 TO 6) OF STD_LOGIC;
20      SIGNAL digit_secU, digit_secT: digit6x8;
21      SIGNAL digit_minU, digit_minT: digit6x8;
22      SIGNAL digit_hourU, digit_hourT: digit6x8;
23      CONSTANT zero: digit6x8 := (
24         ('0','1','1','1','0','0'), ('1','0','0','0','1','0'),
25         ('1','0','0','1','1','0'), ('1','0','1','0','1','0'),
26         ('1','1','0','0','1','0'), ('1','0','0','0','1','0'),
27         ('0','1','1','1','0','0'), ('0','0','0','0','0','0'));
28      CONSTANT one: digit6x8 := (
29         ('0','0','1','0','0','0'), ('0','1','1','0','0','0'),
30         ('0','0','1','0','0','0'), ('0','0','1','0','0','0'),
31         ('0','0','1','0','0','0'), ('0','0','1','0','0','0'),
32         ('0','1','1','1','0','0'), ('0','0','0','0','0','0'));
33      CONSTANT two: digit6x8 := (
34         ('0','1','1','1','0','0'), ('1','0','0','0','1','0'),
35         ('0','0','0','1','0','0'), ('0','0','1','0','0','0'),
36         ('0','1','0','0','0','0'), ('1','0','0','0','0','0'),
37         ('1','1','1','1','1','0'), ('0','0','0','0','0','0'));
```

```
38    CONSTANT three: digit6x8 := (
39        ('1','1','1','1','1','0'), ('0','0','0','1','0','0'),
40        ('0','0','1','0','0','0'), ('0','0','0','1','0','0'),
41        ('0','0','0','0','1','0'), ('1','0','0','0','1','0'),
42        ('0','1','1','1','0','0'), ('0','0','0','0','0','0'));
43    CONSTANT four: digit6x8 := (
44        ('0','0','0','1','0','0'), ('0','0','1','0','0','0'),
45        ('0','1','0','0','0','0'), ('1','0','0','1','0','0'),
46        ('1','1','1','1','1','0'), ('0','0','0','1','0','0'),
47        ('0','0','0','1','0','0'), ('0','0','0','0','0','0'));
48    CONSTANT five: digit6x8 := (
49        ('1','1','1','1','1','0'), ('1','0','0','0','0','0'),
50        ('1','0','0','0','0','0'), ('1','1','1','1','0','0'),
51        ('0','0','0','0','1','0'), ('1','0','0','0','1','0'),
52        ('0','1','1','1','0','0'), ('0','0','0','0','0','0'));
53    CONSTANT six: digit6x8 := (
54        ('0','1','1','1','0','0'), ('1','0','0','0','0','0'),
55        ('1','0','0','0','0','0'), ('1','1','1','1','0','0'),
56        ('1','0','0','0','1','0'), ('1','0','0','0','1','0'),
57        ('0','1','1','1','0','0'), ('0','0','0','0','0','0'));
58    CONSTANT seven: digit6x8 := (
59        ('1','1','1','1','1','0'), ('0','0','0','0','1','0'),
60        ('0','0','0','1','0','0'), ('0','0','1','0','0','0'),
61        ('0','0','1','0','0','0'), ('0','0','1','0','0','0'),
62        ('0','0','1','0','0','0'), ('0','0','0','0','0','0'));
63    CONSTANT eight: digit6x8 := (
64        ('0','1','1','1','0','0'), ('1','0','0','0','1','0'),
65        ('1','0','0','0','1','0'), ('0','1','1','1','0','0'),
66        ('1','0','0','0','1','0'), ('1','0','0','0','1','0'),
67        ('0','1','1','1','0','0'), ('0','0','0','0','0','0'));
68    CONSTANT nine: digit6x8 := (
69        ('0','1','1','1','0','0'), ('1','0','0','0','1','0'),
70        ('1','0','0','0','1','0'), ('0','1','1','1','1','0'),
71        ('0','0','0','0','1','0'), ('0','0','0','0','1','0'),
72        ('0','1','1','1','0','0'), ('0','0','0','0','0','0'));
73    CONSTANT colon: digit6x8 := (
74        ('0','0','0','0','0','0'), ('0','0','1','0','0','0'),
75        ('0','0','1','0','0','0'), ('0','0','0','0','0','0'),
76        ('0','0','1','0','0','0'), ('0','0','1','0','0','0'),
77        ('0','0','0','0','0','0'), ('0','0','0','0','0','0'));
78    ---Function construction:----------------
79    FUNCTION bcd_to_digit6x8 (SIGNAL input: INTEGER) RETURN digit6x8 IS
80    BEGIN
```

```
81        CASE input IS
82           WHEN 0 => return zero;
83           WHEN 1 => return one;
84           WHEN 2 => return two;
85           WHEN 3 => return three;
86           WHEN 4 => return four;
87           WHEN 5 => return five;
88           WHEN 6 => return six;
89           WHEN 7 => return seven;
90           WHEN 8 => return eight;
91           WHEN OTHERS => return nine;
92        END CASE;
93     END bcd_to_digit6x8;
94     -------------------------------------------
95  BEGIN
96     PROCESS (clk50, Hsync, dena)
97        VARIABLE x: INTEGER RANGE 0 TO 800; --horiz. coordinate
98        VARIABLE y: INTEGER RANGE 0 TO 600; --vert. coordinate
99     BEGIN
100       ---Make BCD to digit6x8 conversion:----
101       digit_secU <= bcd_to_digit6x8(secUnits);
102       digit_secT <= bcd_to_digit6x8(secTens);
103       digit_minU <= bcd_to_digit6x8(minUnits);
104       digit_minT <= bcd_to_digit6x8(minTens);
105       digit_hourU <= bcd_to_digit6x8(hourUnits);
106       digit_hourT <= bcd_to_digit6x8(hourTens);
107       ---Create horizontal coordinate:-------
108       IF (clk50'EVENT AND clk50='1') THEN
109          IF (Hactive='1') THEN x := x + 1;
110          ELSE x := 0;
111          END IF;
112       END IF;
113       ---Create vertical coordinate:---------
114       IF (Hsync'EVENT AND Hsync='1') THEN
115          IF (Vactive='1') THEN y := y + 1;
116          ELSE y := 0;
117          END IF;
118       END IF;
119       ---Create image (w/ digits in red):----
120       G <= (OTHERS => '0');
121       B <= (OTHERS => '0');
122       IF (x>=700 AND x<706)AND (y>=550 AND y<558) THEN
123          R <= (OTHERS => digit_hourT(y-550, x-700));
```

```
124         ELSIF (x>=706 AND x<712)AND (y>=550 AND y<558) THEN
125            R <= (OTHERS => digit_hourU(y-550, x-706));
126         ELSIF (x>=712 AND x<718)AND (y>=550 AND y<558) THEN
127            R <= (OTHERS => colon(y-550, x-712));
128         ELSIF (x>=718 AND x<724)AND (y>=550 AND y<558) THEN
129            R <= (OTHERS => digit_minT(y-550, x-718));
130         ELSIF (x>=724 AND x<730)AND (y>=550 AND y<558) THEN
131            R <= (OTHERS => digit_minU(y-550, x-724));
132         ELSIF (x>=730 AND x<736)AND (y>=550 AND y<558) THEN
133            R <= (OTHERS => colon(y-550, x-730));
134         ELSIF (x>=736 AND x<742)AND (y>=550 AND y<558) THEN
135            R <= (OTHERS => digit_secT(y-550, x-736));
136         ELSIF (x>=742 AND x<748)AND (y>=550 AND y<558) THEN
137            R <= (OTHERS => digit_secU(y-550, x-742));
138         ELSE
139            R <= (OTHERS => '0');
140         END IF;
141      END PROCESS;
142   END image_generator;
143   ------------------------------------------------------------------
```

Part 3: Control Generator

Same as part 2 of the previous design (section 17.4).

Part 4: Serializer

Same as part 3 of the previous design (section 17.4).

Part 5: FPD-Link Interface

Same as part 4 of the previous design (section 17.4).

Part 6: Main Code

Finally, the main code is shown below, under the project name *fpdlink_clock* (line 5). It simply combines the subcircuits designed above to create the complete FPD-Link driver plus the intended image. The components are declared in lines 34–91, then instantiated in lines 95–108. Note in lines 110–113 that complementary signals are again created in order to be able to use the arrangement of figure 17.6 to emulate LVDS pins.

```
1   ------------------------------------------------------------------
2   LIBRARY ieee;
3   USE ieee.std_logic_1164.all;
4   ------------------------------------------------------------------
5   ENTITY fpdlink_clock IS
```

```
6    PORT (
7      clk50: IN STD_LOGIC;
8      rst: IN STD_LOGIC;
9      seconds: IN STD_LOGIC;
10     minutes: IN STD_LOGIC;
11     hours: IN STD_LOGIC;
12     lvdsA1: BUFFER STD_LOGIC;
13     lvdsA2: OUT STD_LOGIC;
14     lvdsB1: BUFFER STD_LOGIC;
15     lvdsB2: OUT STD_LOGIC;
16     lvdsC1: BUFFER STD_LOGIC;
17     lvdsC2: OUT STD_LOGIC;
18     lvds_clk1: BUFFER STD_LOGIC;
19     lvds_clk2: OUT STD_LOGIC);
20   END fpdlink_clock;
21   ----------------------------------------------------------------
22   ARCHITECTURE fpdlink_clock OF fpdlink_clock IS
23     ----Signal declarations:-------------
24     SIGNAL clk350: STD_LOGIC;
25     SIGNAL Hsync, Vsync, Hactive, Vactive, dena: STD_LOGIC;
26     SIGNAL R, G, B: STD_LOGIC_VECTOR(5 DOWNTO 0);
27     SIGNAL secUnits: NATURAL RANGE 0 TO 9;
28     SIGNAL secTens: NATURAL RANGE 0 TO 5;
29     SIGNAL minUnits: NATURAL RANGE 0 TO 9;
30     SIGNAL minTens: NATURAL RANGE 0 TO 5;
31     SIGNAL hourUnits: NATURAL RANGE 0 TO 9;
32     SIGNAL hourTens: NATURAL RANGE 0 TO 2;
33     ----1st component declaration:-------
34     COMPONENT clock IS
35       GENERIC (fclk: INTEGER := 50_000_000);
36       PORT (
37         clk50: IN STD_LOGIC;
38         rst: IN STD_LOGIC;
39         seconds: IN STD_LOGIC;
40         minutes: IN STD_LOGIC;
41         hours: IN STD_LOGIC;
42         secUnits: OUT NATURAL RANGE 0 TO 9;
43         secTens: OUT NATURAL RANGE 0 TO 5;
44         minUnits: OUT NATURAL RANGE 0 TO 9;
45         minTens: OUT NATURAL RANGE 0 TO 5;
46         hourUnits: OUT NATURAL RANGE 0 TO 9;
47         hourTens: OUT NATURAL RANGE 0 TO 2);
```

```
48    END COMPONENT;
49    ----2nd component declaration:-------
50    COMPONENT image_generator IS
51       PORT (
52          clk50, Hsync, Hactive, Vactive, dena: IN STD_LOGIC;
53          secUnits: IN NATURAL RANGE 0 TO 9;
54          secTens: IN NATURAL RANGE 0 TO 5;
55          minUnits: IN NATURAL RANGE 0 TO 9;
56          minTens: IN NATURAL RANGE 0 TO 5;
57          hourUnits: IN NATURAL RANGE 0 TO 9;
58          hourTens: IN NATURAL RANGE 0 TO 2;
59          R, G, B: OUT STD_LOGIC_VECTOR(5 DOWNTO 0));
60    END COMPONENT;
61    ----3rd component declaration:-------
62    COMPONENT control_generator IS
63       GENERIC (
64          Ha: INTEGER := 128;
65          Hb: INTEGER := 216;
66          Hc: INTEGER := 1016;
67          Hd: INTEGER := 1056;
68          Va: INTEGER := 4;
69          Vb: INTEGER := 27;
70          Vc: INTEGER := 627;
71          Vd: INTEGER := 628);
72       PORT (
73          clk50: IN STD_LOGIC;
74          clk350: OUT STD_LOGIC;
75          Hsync: BUFFER STD_LOGIC;
76          Vsync: OUT STD_LOGIC;
77          Hactive: BUFFER STD_LOGIC;
78          Vactive: BUFFER STD_LOGIC;
79          dena: OUT STD_LOGIC);
80    END COMPONENT;
81    ----4th component declaration:-------
82    COMPONENT FPDlink_interface IS
83       PORT (
84          clk350: IN STD_LOGIC;
85          Hsync, Vsync, dena: IN STD_LOGIC;
86          R, G, B: IN STD_LOGIC_VECTOR(5 DOWNTO 0);
87          lvdsA: OUT STD_LOGIC;
88          lvdsB: OUT STD_LOGIC;
89          lvdsC: OUT STD_LOGIC;
```

```
90              lvds_clk: OUT STD_LOGIC);
91      END COMPONENT;
92      -------------------------------------
93   BEGIN
94      ----Clock:--------------------------
95      timing: clock PORT MAP (
96          clk50, rst, seconds, minutes, hours,
97          secUnits, secTens, minUnits, minTens, hourUnits, hourTens);
98      ----Image_generator:-----------------
99      image: image_generator PORT MAP (
100         clk50, Hsync, Hactive, Vactive, dena, secUnits, secTens,
101         minUnits, minTens, hourUnits, hourTens, R, G, B);
102     ----Control signals:-----------------
103     control: control_generator PORT MAP (
104         clk50, clk350, Hsync, Vsync, Hactive, Vactive, dena);
105     ----FPDlink interface:--------------
106     fpd_link: FPDlink_interface PORT MAP (
107         clk350, Hsync, Vsync, dena, R, G, B,
108         lvdsA1, lvdsB1, lvdsC1, lvds_clk1);
109     -------------------------------------
110     lvdsA2 <= NOT lvdsA1;
111     lvdsB2 <= NOT lvdsB1;
112     lvdsC2 <= NOT lvdsC1;
113     lvds_clk2 <= NOT lvds_clk1;
114  END fpdlink_clock;
115  -----------------------------------------------------------------
```

17.6 Other Designs

Once one knows how to deal with the control signals, FPD-Link encoder, and serializer, generating images for an FPD-Link-driven LCD monitor becomes essentially the same problem as generating images for a VGA monitor. Consequently, the material seen in chapter 15 applies here too, so the following is recommended:

• For generating regular images with dedicated hardware, follow the procedure of section 15.9.

• For generating regular images with a file and on-chip memory, follow the procedure of section 15.10.

• For generating arbitrary images with a file and off-chip memory, follow the procedure of section 15.11.

For the same reason, all examples and exercises seen there can also be used here.

17.7   Exercises

Note: For exercise solutions, please consult the book website.

Exercise 17.1: FPD-Link Encoder

a) What are the purposes (benefits) of the FPD-Link circuit?

b) Does it have redundant bits (like 8B/10B in TMDS)?

c) Why is it called a serial data transmission encoder?

d) Why is a PLL generally needed to construct it?

e) What is the nominal maximum pixel rate?

f) Why was the circuit of figure 17.6 used in the experiments?

Exercise 17.2: Image Generation with Hardware #1 (Banner)

Solve exercise 15.2 for an FDP-Link-driven display operating with SVGA resolution. Adjust the horizontal values to 100, 600, 100.

Exercise 17.3: Image Generation with Hardware #2 (Sun in the Sky)

Solve exercise 15.3 for an FDP-Link-driven display operating with SVGA resolution. Adjust the horizontal values to 250, 300, 250, and the vertical values to 150, 300, 150.

Exercise 17.4: Image Generation with Hardware #3 (Filling with Green)

Solve exercise 15.4 for an FDP-Link-driven display operating with SVGA resolution.

Exercise 17.5: Image Generation with Hardware #4 (Rotating Bar)

Solve exercise 15.5 for an FDP-Link-driven display operating with SVGA resolution.

Exercise 17.6: Image Generation with Hardware #5 (Digital Clock)

Compile the code for the clock in section 17.5 and physically test it in your FPGA board using an FDP-Link-driven LCD display.

Exercise 17.7: Image Generation with Hardware #6 (Arcade Game)

Solve exercise 15.7 for an FDP-Link-driven display operating with SVGA resolution.

Exercise 17.8: Image Generation with a File and On-Chip Memory #1 (Banner)

Solve exercise 15.8 for an FDP-Link-driven display operating with SVGA resolution. Adjust the image as in exercise 17.2.

Exercise 17.9: Image Generation with a File and On-Chip Memory #2 (Sun in the Sky)

Solve exercise 15.9 for an FDP-Link-driven display operating with SVGA resolution. Adjust the image as in exercise 17.3.

Exercise 17.10: Image Generation with a File and Off-Chip Memory (Arbitrary Picture)

Solve exercise 15.10 for an FDP-Link-driven display operating with SVGA resolution.

APPENDICES

A Programmable Logic Devices

A.1 Introduction

The purpose of programmable logic devices (PLDs) is to attain integrated circuits whose *hardware* is programmable. Therefore, different from microcontrollers, whose *tasks* are programmable but their hardware is fixed, the hardware itself is programmable in a PLD, so with the same device a huge selection of different circuits—including microprocessors—can be implemented.

The first PLDs were called PLA (programmable logic array) and PAL (programmable array logic), introduced by Signetics and Monolithic Memories, respectively, in the mid-1970s. A major limitation of these first PLDs was the fact that they employed only traditional logic gates (no flip-flops), so they were adequate only for the implementation of *combinational* circuits.

A major advancement occurred in the early 1980s, when Lattice introduced GAL (generic array logic), which included at each device output what they called a *macrocell*. Besides containing a flip-flop, each macrocell had also several multiplexers to allow data to the routed to the output, to a neighboring cell, or back to the programmable array itself, thus conferring the device not only the capability of implementing sequential circuits (due to the flip-flops), but also a much greater flexibility (due to data routing).

These devices (PLA, PAL, GAL, and other variants) are now collectively referred to as SPLDs (simple PLDs), of which GAL is the only one still manufactured in a stand-alone package.

In the mid-1980s, several GAL devices were fabricated in the same chip, using a more sophisticated routing scheme, more advanced silicon technology, and several additional features (like JTAG support and interface to several logic standards). Such approach became known as CPLD (complex PLD). CPLDs are currently very popular due to their high density, high performance, low cost (CPLDs can be found for less than a dollar), and more recently even relatively low power consumption (for example, Altera Max IIZ and Xilinx CoolRunner II series).

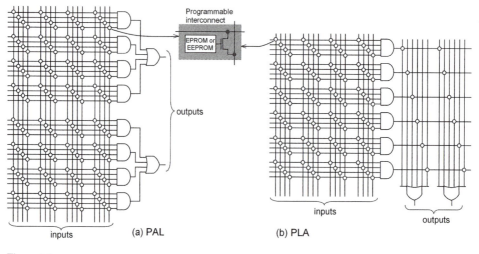

Figure A.1

Still in the mid-1980s, FPGAs (field programmable gate arrays) were introduced. FPGAs differ from CPLDs in architecture, technology, built-in features, and cost. They aim mainly at the implementation of large, complex, high-performance systems.

A last comment regards the programming scheme of PLDs. In SPLDs and CPLDs, nonvolatile memory (EEPROM in older devices, Flash in newer ones) is employed, so the configuration is not lost when the power is turned off. FPGAs, on the other hand, normally employ volatile memory (SRAM), so they must be reprogrammed every time the power is turned back on. In order to do so, special nonvolatile configuration memories are sold by FPGA companies to hold the configuration data, which can be automatically retrieved and loaded by the FPGA when the power is turned on.

A.2   PAL and PLA Devices

Figure A.1 shows the general architecture employed in PAL and PLA devices. In (a), programmable AND gate arrays are connected to nonprogrammable OR gate arrays, while in (b) both arrays are programmable (the little circles indicate programmable connections).

These implementations are based on the fact that any Boolean function can be expressed as an SOP (sum-of-products), that is, if $a_1, a_2, \ldots, a_N$ are the logic inputs, then an output x can be computed as $x = m_1 + m_2 + \cdots + m_M$, where $m_i = f_i(a_1, a_2, \ldots, a_N)$ are the minterms of the function x. For example, $x = a_1 \bar{a}_2 + a_2 a_3 \bar{a}_4 + \bar{a}_1 \bar{a}_2 a_3 a_4 \bar{a}_5$. Consequently, the products (minterms) are computed by the AND gates, while the sums are computed by the OR gates.

Due to the programmable OR array, with the same number of gates the PLA circuit can compute more Boolean functions than the PAL circuit. These programmable cells, however, raised the capacitive load of the lines departing from the AND gates, making these gates slower; they also caused the circuit to occupy a larger silicon space, increasing its cost. For these reasons, the PAL architecture was more successful.

The early technology employed in the fabrication of these devices was bipolar, with 5 V supply and current consumption (with open outputs) around 200 mA. The maximum frequency was on the order of 100 MHz, and the programmable cells were generally of EPROM type, later improved to EEPROM and MOS transistors (see inset in figure A.1).

The PAL architecture was employed in the fabrication of GALs, while PLAs became essentially obsolete. However, in spite of the GAL (hence PAL) architecture being used in most CPLDs, the PLA architecture reappeared in the Xilinx CoolRunner II family of low-power CPLDs. Thus, indirectly, both PAL and PLA are still present in current devices.

A.3   GAL Devices

GAL caused a major advancement in the acceptance of PLDs. It employed the PAL architecture, to which a macrocell was added at each output. This new arrangement provided two major advantages: the inclusion of flip-flops allowed the construction of sequential circuits and the inclusion of programmable routing, by means of several multiplexers, allowed the output signal to be registered or unregistered, to be sent to a neighboring cell, and also to be fed back to the programmable AND array, enormously enhancing the device's functionalities.

A then-popular GAL device, called GAL16V8, is depicted in figure A.2. It has 16 inputs and 8 outputs in a 20-pin package (note that eight pins are actually bidirectional). Note that a macrocell is present at each output. Even though GAL devices are still manufactured, their main use is as building blocks in CPLDs.

A.4   CPLD Devices

The basic approach to the construction of CPLDs is illustrated in figure A.3. It consists of several PLDs (in general of GAL type) fabricated on a single chip, with a sophisticated switching array used to interconnect them and to the I/O pins. Moreover, CPLDs normally exhibit a few additional features, like JTAG support and interface to other logic standards (for example, LVTTL/LVCMOS [Pedroni 2008]).

Figure A.4 shows the main CPLDs from Altera and Xilinx. Observe the building block, which is GAL in two families and PLA in another; in Altera's Max II series, a "simplified" FPGA architecture is indeed used, so if that becomes the tendency, the time of true CPLDs might be nearing its end.

Also inspect (and compare) the other parameters in Figure A.4, such as the CMOS technology, the core and I/O voltages, the type of memory used to store the configuration, the number of flip-flops available, and so on.

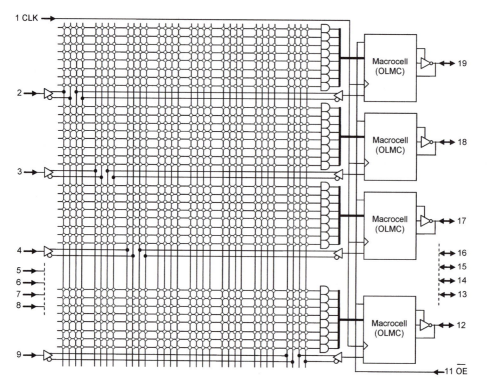

Figure A.2

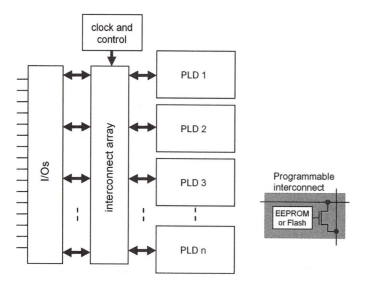

Figure A.3

Feature	Altera CPLDs		Xilinx CPLDs	
	Max 3000	Max II (-, G, Z)	XC9500	CoolRunner II
Core voltage	3.3V	3.3V, 2.5V, 1.8V	3.3V	1.8V
I/O voltages	5V, 3.3V, 2.5V	3.3V, 2.5V, 1.8V, 1.5V	5V, 3.3V, 2.5V	3.3V, 2.5V, 1.8V
Building block	GAL	LAB (10 LEs)	GAL	PLA
Number of macrocells	32 – 512	-----	36 – 288	32 – 512
Number of logic elements	-----	240 – 2210	-----	-----
Number of flip-flops	1 per macrocell	1 per logic element	1 per macrocell	1 per macrocell (DE)
Number of user I/O pins	34 – 208	80 – 272	34 – 192	33 – 270
Supported I/Os	TTL, LVTTL, LVCMOS	LVTTL, LVCMOS, PCI	LVTTL, LVCMOS	LVTTL, LVCMOS, SSTL, HSTL
Schmitt trigger at I/Os	No	Yes	Yes	Yes
Approx. max. frequency	227MHz	304MHz	222MHz	323MHz
Standby current	10mA – 350mA	29uA – 12mA	12mA – 500mA	20uA
Technology	CMOS 0.3um EEPROM	CMOS 0.18um Flash	CMOS 0.35um Flash	CMOS 0.18um Flash
User Flash	No	8 kbits	No	No

Figure A.4

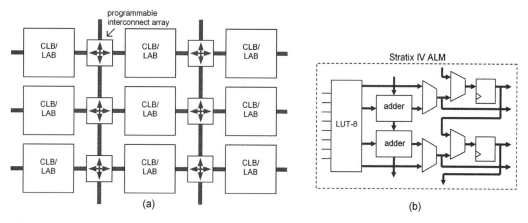

(a)   (b)

Figure A.5

A.5   FPGA Devices

The general approach to the construction of FPGAs is depicted in figure A.5a. Note that it consists of a matrix of special cells, called CLBs (configurable logic blocks) by Xilinx or LABs (logic array blocks) by Altera. The top-performance FPGAs from these two companies (at the time of this writing) are Virtex 6 and Stratix IV, both fabricated with state-of-the-art 40 nm CMOS technology.

Just to illustrate what goes inside these blocks (CLB, LAB), the contents of an ALM (adaptive logic module) are depicted in figure A.5b (each CLB consists of 2 slices, while

Feature	Stratix IV E	Stratix IV GX	Stratix IV GT
Number of LABs	4,224 – 27,244	2,904 – 21,248	9,120 – 21,248
Number of ALMs	10 per LAB	10 per LAB	10 per LAB
Number of flip-flops	2 per ALM	2 per ALM	2 per ALM
User SRAM bits	9,564 – 31,491	7,370 – 27,376	17,133 – 27,376
Max. frequency	600 MHz	600 MHz	600 MHz
18x18 multipliers in MACs	512 – 1,360	384 – 1,288	832 – 1,288
Max. frequency	540 MHz	540 MHz	540 MHz
PLLs	4 – 12	3 – 12	8 – 12
Number of transceivers	-----	8 – 48	36 – 48
Speed	-----	0.6 to 6.5 or 8.5 Gbps	2.5 to 6.5 or 8.5 or 10.3 Gbps
Core voltage	0.9V – 1V		
Technology	CMOS 40nm with SRAM configuration		
User I/O pins	288 – 1,104		
External memory interfaces	SDRAM, DDR2 SDRAM, DDR3 SDRAM, RLDRAM, QDRII SRAM, QDRII+ SRAM		
Max. frequency	533 MHz		
Additional support	LVCMOS, LVDS, SSTL, HSTL, PCI Express, Ethernet, RapidIO, SONET, SPI, etc.		

Figure A.6

Feature	Virtex 6 LXT	Virtex 6 SXT
Number of CLBs	5,820 – 59,280	24,600 – 37,200
Number of Slices	2 per LAB	2 per LAB
Number of flip-flops	8 per Slice	8 per Slice
User SRAM bits	5,616 – 25,920	25,344 – 38,304
Max. frequency		
25x18 multipliers in MACs	288 – 864	1,344 – 2,016
Number of transceivers	0 – 36	24 – 36
Speed	0.15 to 6.5 Gbps	0.15 to 6.5 Gbps
Core voltage	0.9V – 1V	
Technology	CMOS 40nm with SRAM configuration	
User I/O pins	360 – 1,200	
External memory interfaces	SDRAM, DDR2 SDRAM, RLDRAM, QDRII SRAM	
Additional support	LVCMOS, LVDS, SSTL, HSTL, PCI Express, Ethernet, SPI, etc.	

Figure A.7

each LAB consists of 10 ALMs). Note the existence of an 8-input LUT (look-up table), responsible for the combinational logic (the LUT replaces the AND + OR arrangement of PAL or PLA), plus two dedicated adders, two flip-flops, and finally some data routers (multiplexers).

The top FPGAs from Altera and Xilinx are briefly described in figure A.6 and A.7, respectively. Compare them with each other and also with the CPLDs of figure A.4. Observe, among other things, the following in FPGAs:

- Their core voltage is much lower than that of CPLDs.
- The technology is state-of-the-art (40 nm).
- The traditional resources (logic cells and flip-flops) are much more abundant.
- There are also extra resources, like user SRAM, multipliers, and PLLs.
- They support a much wider, much more complex set of I/Os.
- Transceivers are among the most modern I/O additions.
- The configuration memory is SRAM instead of EEPROM or Flash.

B Altera Quartus II Tutorial

This tutorial briefly describes Quartus II, from Altera, a design software for CPLD/FPGA-based circuits. The description is based on Quartus II 9.0 sp1 Web Edition, available free of charge at www.altera.com. The registered multiplexer of figure B.1 will be used as an example. Only VHDL input will be considered.

The tutorial is divided into eight parts:

B.1 Introduction

B.2 Starting a New Project

B.3 Synthesizing the Design

B.4 Inspecting Synthesis Results

B.5 Simulating the Circuit

B.6 Making Pin Assignments

B.7 Physically Implementing the Circuit

B.8 Interpreting the Fitter Equations

B.1 Introduction

Quartus II 9.0 allows integrated synthesis and simulation as follows.

Synthesis: Synthesis with the Quartus II synthesizer.

Simulation: Manual graphical simulation with the Quartus II simulator (VHDL testbenches not allowed).

Note: Unfortunately, Altera has decided to no longer offer its simulator after version 9.1 of Quartus II (~2011).

Because Quartus II does not support automated verification (simulation with VHDL testbenches), an Altera edition of ModelSim is provided, which is described in appendix D.

The registered multiplexer studied in section 10.5 (repeated in figure B.1a, without the unregistered output) will be used in this tutorial. The corresponding design file

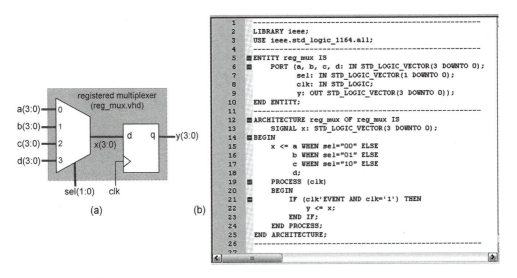

Figure B.1

(*reg_mux.vhd*) is shown in figure B.1b. For the simulations, the same stimuli used in section 10.5 (figure 10.6) will be employed.

B.2   Starting a New Project

a) Launch Quartus II.

b) Create a new project by selecting File > New Project Wizard. The dialog of figure B.2 will be opened.

c) In the working directory field of figure B.2, select the directory where all project files should be located (*reg_mux*, in this tutorial). If the directory does not exist yet, just type in the desired name (with the proper path, of course) and Quartus II will create it for you. Preferably, use the same name for the directory, the project, and the main VHDL entity.

d) In the project name field of figure B.2, enter the desired project name (*reg_mux*). Note that the entity name field is automatically filled with the same name. Click Finish until the Project Navigator (figure B.3a) is displayed.

e) Now enter the VHDL code. If the code of figure B.1b was already typed and saved in the work directory, proceed to section B.3. Otherwise, open the VHDL editor by clicking ⬜ or selecting File > New, which calls up the dialog box of figure B.3b. Select VHDL File and click OK. A blank page will be presented. Type the VHDL code and save it as *reg_mux.vhd*.

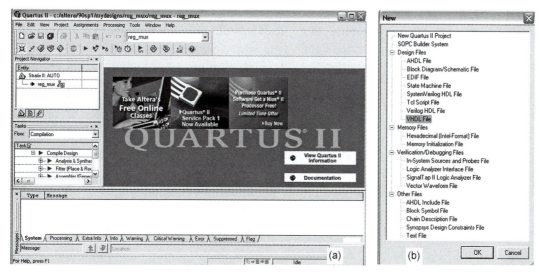

Figure B.2

Figure B.3

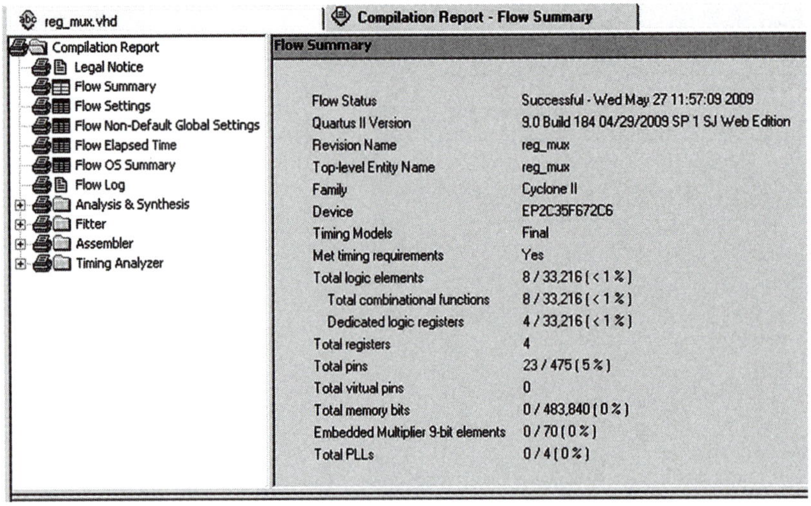

Figure B.4

B.3 Synthesizing the Design

a) Define the device in which the circuit should be implemented by selecting Assignments > Device, then in the Family field select Cyclone II, and in the Target Device field either select Auto device selected by the Fitter (in this case the compiler will pick a device from the Cyclone II family for you) or mark Specific device selected in 'Available devices' list and select, for example, the chip used in the DE2 board (EP2C35F672C6).

b) Compile the design by clicking ▶ or selecting Processing > Start Compilation.

For a faster compilation, particularly useful while the VHDL code is still being debugged, click ♦ or select Processing > Start > Start Analysis and Synthesis. In this case, no timing information is recorded. After the code is working properly, full compilation (▶) should then be performed.

c) When the compilation ends, the Compilation Report of figure B.4 is exhibited, which contains several pieces of valuable information, some of which are described below.

B.4 Inspecting Synthesis Results

This section describes some of the results produced by the compiler.

a) Device type and number of pins: Check in figure B.4 if the device type is the intended one (Cyclone II EP2C35F672C6, in this example). Check also if the total number of pins is as expected ($4 \times 4 + 2 + 1 = 19$ inputs $+ 4$ outputs $= 23$ pins).

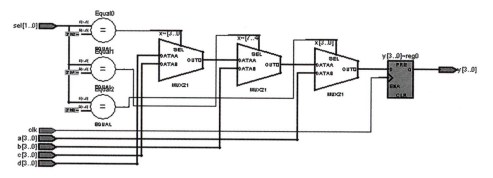

Figure B.5

b) Number of logic elements: Figure B.4 also shows the amount of logic needed to implement the circuit. In this case, eight logic elements were needed, out of ~33k LEs available in the chosen device.

c) Number of registers: Since y is a 4-bit signal (figure B.1a), four flip-flops are expected to be inferred by the synthesizer. This too can also be checked in the flow summary of figure B.4, which shows a total of four registers.

d) RTL View: This tool shows how the code was interpreted by the compiler (before optimization and fitting). Select Tools > Netlist Viewers > RTL Viewer, which exhibits the circuit of figure B.5 (or similar). Observe that because 2-input multiplexers are available in the device, a total of three units were used to implement the 4-input mux (the other three blocks at the input are for processing *sel*). Note also the 4-bit register at the output.

e) Equations: They represent the actual circuit implemented by the compiler. In the compilation report, select Fitter > Equations (if this option is not available, go to Tools > Options > General > Processing and mark Automatically generate equation file during compilation, then recompile the code). To interpret the equations, see first section B.8. For example, "A1L41Q = DFFEAS(A1L60, GLOBAL(A1L2), ...);" means that the internal signal A1L41Q is produced by a D-type flip-flop (DFF) whose data input comes from A1L60 and whose clock input is fed by the global signal A1L2. Confirm in the equations that the total number of DFFs is indeed four.

f) Timing analysis: In the compilation report, select Timing Analyzer > Summary and check the worst-case values for the three time delays below (these parameters are used, for example, to determine the circuit's maximum clock frequency).

tco (clock to output delay): Time necessary to obtain a valid output after a clock transition.

tsu (clock setup time): Time during which the data and/or enable inputs must be stable before a clock transition occurs.

Input Pins

	Name	Pin #	I/O Bank	X coordinate	Y coordinate	Cell number
1	a[0]	AE9	8	20	0	0
2	a[1]	H11	3	18	36	3
3	a[2]	G21	5	65	33	0
4	a[3]	A9	3	20	36	2
5	b[0]	C13	3	31	36	2
6	b[1]	G22	5	65	33	1
7	b[2]	AC10	8	20	0	1
8	b[3]	AC9	8	20	0	2
9	c[0]	H12	3	18	36	2
10	c[1]	B10	3	22	36	3
11	c[2]	C10	3	20	36	1
12	c[3]	AD10	8	22	0	1
13	clk	P2	1	0	18	2
14	d[0]	F6	2	0	33	0
15	d[1]	D11	3	22	36	0
16	d[2]	B9	3	20	36	3
17	d[3]	A10	3	22	36	2
18	sel[0]	D13	3	31	36	3
19	sel[1]	G11	3	22	36	1

Output Pins

	Name	Pin #	I/O Bank	X coordinate	Y coordinate	Cell number
1	y[0]	D10	3	20	36	0
2	y[1]	E10	3	18	36	0
3	y[2]	E8	3	7	36	3
4	y[3]	F11	3	18	36	1

Figure B.6

th (clock hold time): Time during which the data and/or enable inputs must remain stable after a clock transition has occurred.

g) Pin assignments: In section B.6, it will be shown how to make or change pin assignments. For now, we want to simply check the assignments made automatically by the compiler. In the Compilation Report, select Fitter > Resource Section > Input Pins. This leads to the table on the left of figure B.6. Next, select Fitter > Resource Section > Output Pins to see the output pins, shown on the right of figure B.6.

B.5 Simulating the Circuit

a) To perform manual graphical simulation, we need first to create (draw) the input waveforms, based on which the simulator will calculate and plot the output waveform. Click or select File > New, which will lead again to the dialog of figure B.3b.

b) Select Vector Waveform File and click OK. The wave pane of figure B.7a will then be exhibited.

b) Select View > Fit in Window (or press Ctrl+W) to have the complete plot exhibited in the waveforms window.

c) The time axis in figure B.7 goes from 0 to 1 us. If a different end time is needed, select Edit > End Time and enter the desired value. Press Ctrl+W again to see the whole time axis.

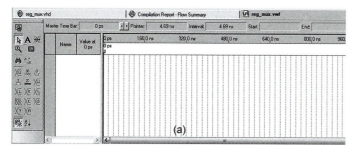

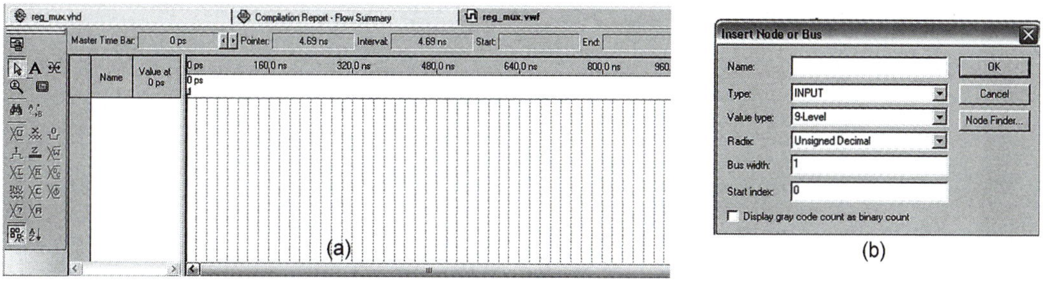

Figure B.7

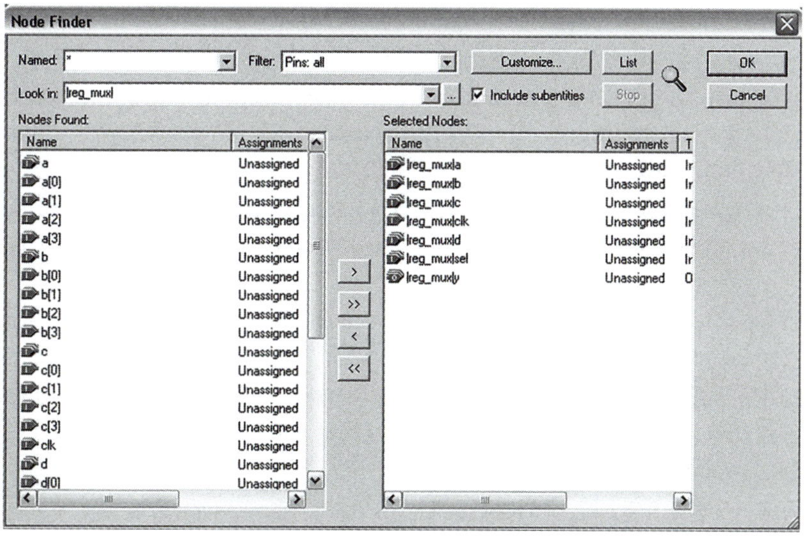

Figure B.8

d) The grid can also be adjusted as needed. Select Edit > Grid Size and enter 40 ns.

e) Now add the signals to the waveform editor. Press the right mouse button in the white area under Name and select Insert > Insert Node or Bus, which leads to the dialog of figure B.7b.

f) In the Radix field select Unsigned Decimal, then click Node Finder. The dialog box of figure B.8 (but empty) will be presented.

g) In the Filter field, select Pins: All (another useful option is Pins: All & Registers: Post-Fitting) then click List. The window will be filled with the signals (partially) shown in the left column of figure B.8. Select the desired signals, which can be copied to the column on the right with [>] (individually), with [»] (all), or by simply double-clicking the left mouse

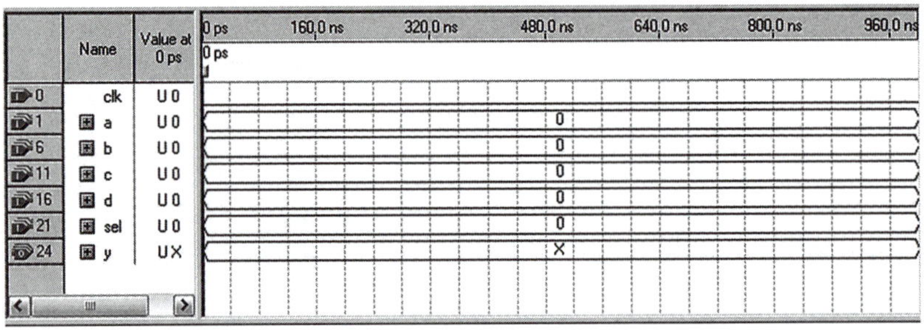

Figure B.9

button on the name of the desired signal. Click OK twice, which will fill the waveforms window with the signals of figure B.9.

h) To change the position (order) of a signal in figure B.9, just select its name, then press and hold the mouse button on it, dragging it to the desired position. Note in figure B.9 that all input signals come first, with the clock at the top.

i) Now we need to draw the input waveforms (*clk*, *a*, *b*, *c*, *d*, *sel*), after which the simulator will compute and draw the output waveform (*y*). The stimuli of figure B.10 will be adopted.

1) Draw the waveform for *clk*: Select line *clk* and click the clock icon X₆. Enter 80 ns for the period and '0' for the initial value. Click OK.

2) Draw the waveform for *a*: Highlight line *a* from 0 to 80 ns, click the arbitrary value icon X?, enter 2, and click OK. Repeat the operation entering 3 for the interval from 80 ns up to the end of the simulation (1 us).

3) Draw the waveforms for for *b*, *c*, *d*: Repeat the process above, entering the values shown in figure B.10.

4) Draw the waveform for *sel*: Select line *sel* and click the counter icon X₆. Enter Start value = 0, Increment = 1, and Count every = 160 ns. Click OK.

5) Save the file with the same name as the entity's and with extension *.vwf* (vector waveform file)—that is, *reg_mux.vwf*. Recall that the last waveform (*y*) will be filled by the simulator.

j) Now we must choose between *functional* or *timing* (default) simulation. The former checks only the design functionalities, while the latter also includes the device's internal propagation delays, thus representing the actual circuit. For functional simulation, continue to step (k) below. For timing simulation, go to step (l).

k) Functional simulation: Select Processing > Generate Functional Simulation Netlist. After the program finishes producing the netlist, select Assignments > Settings > Simulator Set-

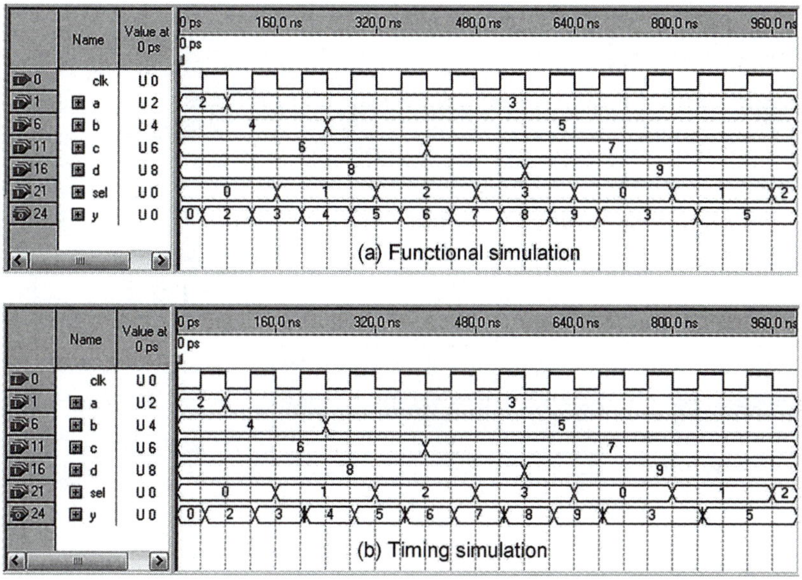

Figure B.10

tings. In the Simulation Mode list, choose Functional. Finally, in the Simulation input field, enter the name of the waveform file (*reg_mux.vwf*) and click OK. When done, go to step (m).

l) Timing simulation (default settings): Select Assignments > Settings > Simulator Settings. In the Simulation Mode list, choose Timing. In the Simulation input field, enter the name of the waveform file (*reg_mux.vwf*) and click OK.

m) Recompile the code in case the setup in (k) or (l) was modified, then proceed to step (n). Otherwise, go directly to (n).

n) Run the simulation by clicking ![icon] or selecting Processing > Start Simulation. The simulator will draw the waveform for *y* in figure B.10a if it is a functional simulation (note that there are no delays between the transitions of *clk* and *y*) or in figure B.10b if it is a timing simulation (note the propagation delay between positive clock transitions and the settling of *y*).

o) Examine the results: When the simulation ends, Quartus II displays the results in a separate window. Instead of examining that window, click the tab related to the original waveforms window (*vwf* plots), to which the new results will then be automatically copied. If the results are not copied to the original waveforms window, select Assignments > Settings > Simulator Settings > Simulation Output Files and mark Overwrite Simulation Input File with Simulation Results, then rerun the simulation.

	To △	Location	I/O Bank	I/O Standard	General Function	Special Function
1	a			3.3-V LVTTL		
2	a[0]			3.3-V LVTTL		
3	a[1]			3.3-V LVTTL		
4	a[2]			3.3-V LVTTL		
5	a[3]			3.3-V LVTTL		
6	b			3.3-V LVTTL		
7	b[0]			3.3-V LVTTL		
8	b[1]			3.3-V LVTTL		
9	b[2]			3.3-V LVTTL		
10	b[3]			3.3-V LVTTL		
11	c			3.3-V LVTTL		
12	c[0]			3.3-V LVTTL		
13	c[1]			3.3-V LVTTL		
14	c[2]			3.3-V LVTTL		
15	c[3]			3.3-V LVTTL		
16	clk	PIN_N2	2	3.3-V LVTTL	Dedicated Clock	CLK0, LVDSCLK0p,
17	d			3.3-V LVTTL		
18	d[0]			3.3-V LVTTL		

Figure B.11

B.6   Making Pin Assignments

Below are three ways of making or changing pin assignments:

1) With Pin Planner (click or select Assignments > Pins or Assignments > Pin Planner)

2) With Assignment Editor (click 🖉 or select Assignments > Assignment Editor)

3) With a CSV (comma separated value) file.

The first two are manual and equivalent, while the third is an automated (imported) assignment. The second and third methods are described below. To delete pin assignments, select Assignments > Remove Assignments.

Manual Pin Assignments with Assignment Editor

a) Click 🖉 or select Assignments > Assignment Editor, which opens the window of figure B.11. In the Category field, select Pin. If the pin names are not shown, click 🖳 (Show All Known Pin Names).

b) As an example, the pin for *clk* is assigned in figure B.11. Double-click the left mouse button in the white area under Location. A pin number can be selected from the pull-down menu, or its name can be typed in (for example, for PIN_N2, only N2 or n2 needs to be typed). Since the clock is a special signal, a specific type of pin should be assigned to it, identified as Dedicated Clock in the pin list. Select, for example, N2.

c) Repeat this process for each of the circuit pins.

d) Finally, recompile the design.

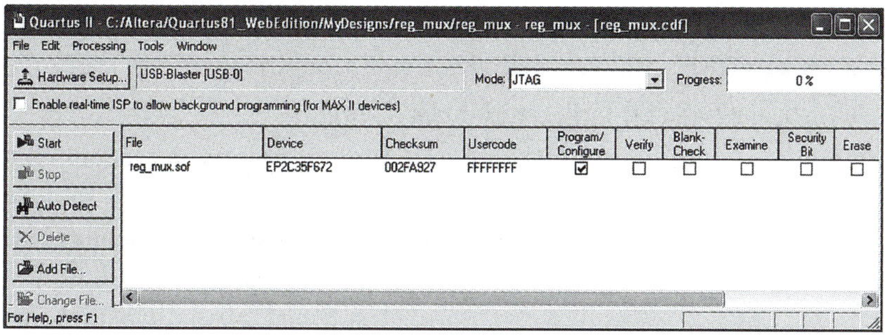

Figure B.12

Importing Pin Assignments with a CSV File

Pin assignments can be exported to and imported from a file, saved with the extension *.csv*. The initial pin assignments, made automatically by the compiler, cannot be exported.

a) Select Assignments > Assignment Editor.

b) Select Assignments > Import Assignments. As an example, try to import the pin assignments for the DE2 board, available in a file called *DE2_pin_assignments.csv*.

c) Recompile the code.

d) To export a pin assignment, select Assignments > Pin Planner, then File > Export, and save the file with the extension *.csv*.

B.7 Physically Implementing the Circuit

a) Connect the board containing the CPLD or FPGA device to an USB port of your computer and turn the power on the board on. If this is the first time that you are using a board that interfaces using an USB port, execute section E.2 of appendix E: Installing the USB-Blaster Driver (this is needed only once).

b) In Quartus II, click the Programmer icon 🔲 or select Tools > Programmer. Figure B.12 will be displayed. Note the following in the figure: the programmer file is *reg_mux.sof*; the Program/Configure box is checked; the driver is USB-Blaster; finally, the mode is JTAG.

c) Click Start, and the device will be programmed.

B.8 Interpreting the Fitter Equations

Below are the main symbols used in the Fitter equations.

a) Logic operators: ! (NOT), & (AND), # (OR), $ (XOR)

b) Flip-flops:

DFF (D, CLK, CLRN, PRN) (DFF with reset and preset, both active low)

DFFE (D, CLK, CLRN, PRN, ENA) (DFF above plus enable input)

DFFEA (D, CLK, CLRN, PRN, ENA, ADATA, ALOAD) (DFF above plus asynchronous data load)

DFFEAS (D, CLK, CLRN, PRN, ENA, ADATA, ALOAD) (DFF above with synchronous clear)

TFFE (T, CLK, CLRN, PRN, ENA) (TFF with reset, preset, and enable)

Note: Recall that in the context of this book an output-zeroing command is called *reset* when it is *asynchronous* or *clear* if it is *synchronous*.

C Xilinx ISE Tutorial

This tutorial briefly describes the ISE synthesis/simulation suite from Xilinx, a design software for CPLD/FPGA-based circuits. The description is based on ISE 11.1 WebPack, available free of charge at www.xilinx.com. The registered multiplexer of figure C.1 will be used as an example. Only VHDL input will be considered.

The tutorial is divided into eight parts:

C.1 Introduction

C.2 Starting a New Project

C.3 Synthesizing the Design

C.4 Inspecting Synthesis Results

C.5 Simulating the Circuit with ISim

C.6 Simulating the Circuit with ModelSim

C.7 Making Pin Assignments

C.8 Physically Implementing the Design

C.1 Introduction

ISE 11.1 allows integrated synthesis and simulation as follows.

Synthesis: With XST (Xilinx Synthesis Technology), Precision RTL (from Mentor Graphics), or Synplify (from Synopsys).

Simulation: With ISim (ISE Simulator, from Xilinx), ModelSim (from Mentor Graphics), NC-Sim (from Cadence), or VCS (from Synopsys)

Note: Unfortunately, Xilinx has decided, starting in version 11.1 of ISE to remove the waveform generator from its simulator. This means that manual graphical inputs (which can be very helpful in the classroom, particularly in the beginning of VHDL/Verilog courses) are no longer possible (only stimuli from testbench files can now be entered into the simulator).

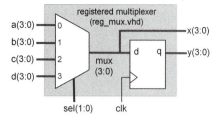

Figure C.1

To compensate for the limitation above, we have included in the ModelSim tutorial (appendix D) some Tcl commands to construct script files so the user can still employ that kind of simulation (though it would be much simpler with a properly designed GUI, as in Quartus II, up to version 9.1).

The registered multiplexer studied in section 10.5 (repeated in figure C.1) will be used in this tutorial. Its design and test files are the following:

Design file: reg_mux.vhd, seen in section 10.5.

Testbench file: reg_mux_tb.vhd, with one option (without automated verification) seen in example 10.4 and another (with automated verification) seen in example 10.6. The former will be employed here.

C.2 Starting a New Project

a) Create a directory where all the design files should be located (work library). Copy the files *reg_mux.vhd* and *reg_mux_tb.vhd* mentioned earlier to that directory. If they were not typed yet, the ISE text editor can be used (step (f) ahead).

b) Launch ISE.

c) Select File > New Project, which will open the Create New Project dialog of figure C.2a. Enter the project name and location (note that the project name is automatically copied to the project location). Click Next, which opens the dialog of figure C.2b.

d) In the Device Properties dialog (figure C.2b), select the device, the synthesizer (XST), the simulator (ISim—we will deal with ModelSim later), and the language (VHDL). Click Next and Finish until the Project Navigator (figure C.3) is opened.

e) Observe the Project Navigator's four windows: Sources, Processes, Transcript, and Workspace. Make sure that Sources is set to Implementation.

f) Open the design (*reg_mux.vhd*) and testbench (*reg_mux_tb.vhd*) files (they will be displayed in the Workspace window). If they were not typed yet, click ⃞ or select File > New > Text File to enter each file.

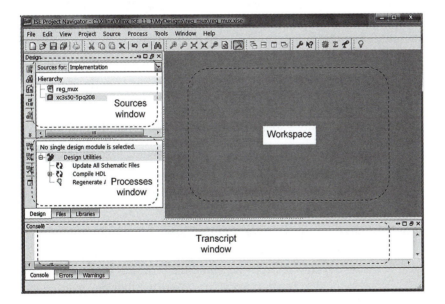

Figure C.2

Figure C.3

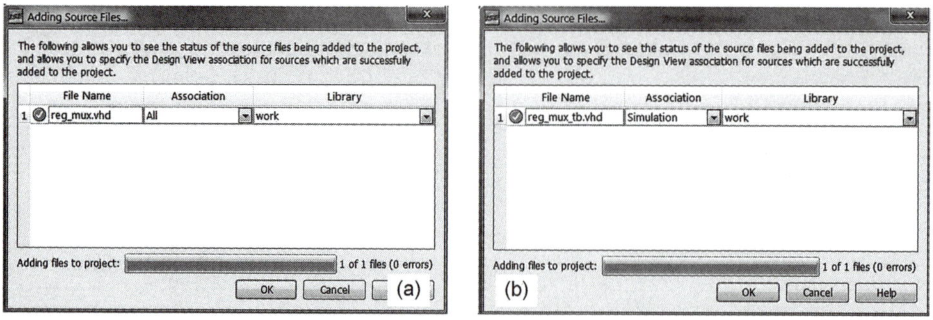

Figure C.4

C.3   Synthesizing the Design

a) First, we must define the project "sources," which are the two files prepared above. Select Project > Add Source, highlight *reg_mux.vhd* and click Open. This opens the dialog of figure C.4a. In the Association field, select All. Do the same for *reg_mux_tb.vhd*, now selecting Simulation in the Association field, as shown in figure C.4b.

Note: The first of the files above is the source for synthesis, while the second is the source for simulation. Consequently, if one prefers, the latter can be included later (in section C.5 or C.6).

Note: If at any point you need to remove sources from the project, proceed as follows.

To remove a design file: In the Source for field of the Sources window, select Implementation. Right-click the file name and select Remove.

To remove a testbench file: In the Source for field of the Sources window, select Behavioral Simulation. Right-click the file name and select Remove.

b) Next, set the synthesis effort. Select Project > Design Goals & Strategies and choose Balanced.

c) The design is now ready to be synthesized. At this point, the Processes window will look like that in figure C.5. Note that under the synthesis process there are four possible actions. For example, if the input files are still being debugged, one might prefer to run just the Check Syntax module. To synthesize the circuit, double-click Synthesize—XST. Some of the results from the synthesis process will be examined in the next section.

d) At any point you can restart the whole process by selecting Project > Cleanup Project Files, which will delete the files created by ISE (but not the design and testbench files).

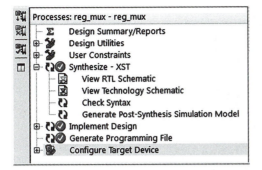

Figure C.5

reg_mux Project Status (05/28/2009 - 19:08:05)			
Project File:	reg_mux.ise	Implementation State:	Synthesized
Module Name:	reg_mux	• Errors:	No Errors
Target Device:	xc3s50-5pq208	• Warnings:	No Warnings
Product Version:	ISE 11.1	• Routing Results:	
Design Goal:	Balanced	• Timing Constraints:	
Design Strategy:	Xilinx Default (unlocked)	• Final Timing Score:	

Device Utilization Summary (estimated values)			[-]
Logic Utilization	Used	Available	Utilization
Number of Slices	4	768	0%
Number of Slice Flip Flops	4	1536	0%
Number of 4 input LUTs	8	1536	0%
Number of bonded IOBs	27	124	21%
Number of GCLKs	1	8	12%

Design Overview — Summary — IOB Properties — Module Level Utilization — Timing Constraints — Pinout Report — Clock Report — Static Timing; Errors and Warnings — Synthesis Messages — Translation Messages — Map Messages — Place and Route Messages — Timing Messages — Bitgen Messages — All Current Messages; Detailed Reports — Synthesis Report; Design Properties

Figure C.6

C.4 Inspecting Synthesis Results

The synthesis reports contain several pieces of valuable information, some of which are examined below.

a) Design parameters: On the upper left corner of the Workspace window select Design Overview > Summary. The result is partially shown in figure C.6. Observe in the Project Status table the name of the design, the device used, and the design goal (Balanced).

b) Device utilization: In the Device Utilization Summary table of figure C.6 observe that four slices were used (out of 768 available in the chosen device), four flip-flops were inferred (as expected), and that the circuit requires 27 pins (expected: $4 \times 4 + 2 + 1$ inputs $+ 2 \times 4$ outputs $= 27$).

c) RTL view: Double-click View RTL Schematic in figure C.5 or select Tools > Schematic Viewer > RTL. The Create RTL Schematic dialog will be opened. Click the "+" icon next

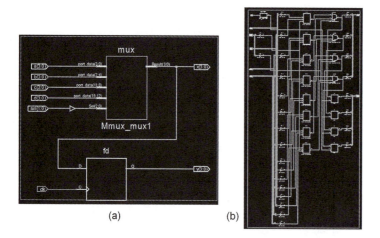

Figure C.7

to Signal, select all signals in the list, then click Add to copy the signals to the column on the right. Finally, click Create Schematic. The result is shown in figure C.7a. Note that this circuit coincides with that in figure C.1. This is how our circuit was understood by the synthesizer (after optimization and place & route it might look a little different, but obviously still has the same functionalities).

d) Technology view: Double-click View Technology Schematic in figure C.5 or select Tools > Schematic Viewer > Technology, then proceed as in step (c) above, which will cause the final circuit to be exhibited (figure C.7b).

e) Timing analysis: Select Tools > Timing Analyzer > Post-Place & Route. After the process is concluded, check in the Report Navigation the time values in Setup/Hold to clock clk and in Pad to pad (a few ns are generally reported), which help give an initial rough idea about the design's maximum speed.

C.5   Simulating the Circuit with ISim

a) If not done yet, enter the source file for simulation (in this tutorial, it was done in step (a) of sections C.2 and C.3.

b) We must choose between *functional* or *timing* simulation. The former checks only the design functionalities, while the latter includes also the device's internal propagation delays, thus representing the actual circuit.

• For functional simulation, continue in step (c) below.

• For timing simulation, go to step (d).

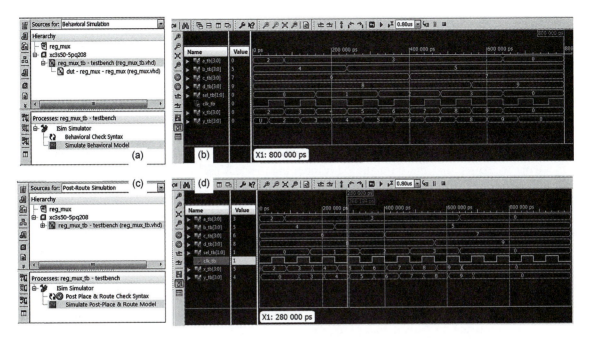

Figure C.8

c) Functional simulation: In the Sources window of the Project Navigator, select Behavioral Simulation in the Sources for field (as shown in the upper part of figure C.8a). Then highlight the testbench file, which will cause the ISim menu to be exhibited in the Processes window (lower part of figure C.8a). Double-click Simulate Behavioral Model in the Processes window. This will start ISim, which displays the waveforms window of figure C.8b. When done, go to step (e).

d) Timing simulation: In the Sources window of the Project Navigator, select Post-Route Simulation in the Sources for field (as shown in the upper part of figure C.8c). Then highlight the testbench file, which will cause the ISim menu to be exhibited in the Processes window (lower part of figure C.8c). Double-click Simulate Post-Place & Route Model in the Processes window. This will start ISim, which displays the waveform window of figure C.8d.

e) In figure C.8b or C.8d (depending on the simulation type), select all signals but *clk* and change their radix to unsigned decimal (right-click a signal name and select Radix > Unsigned Decimal).

f) Click the Zoom to Full View icon ⊠ to see the whole plot. Note that the simulation time interval (0.8 us in this example) can be changed to any other value.

g) Examine the results in figure C.8. Note particularly the following:

• They coincide with the results in example 10.4.

• If it is a functional simulation (figure C.8b), there are no time delays between input and output transitions.

• If it is a timing simulation (figure C.8d), then there are time delays between input and output transitions. For example, the marker and cursor in the figure show a 6.2 ns time delay between the clock transition and the change in y.

h) To get acquainted with ISim, practice with the Run and Zoom controllers. For example, click the Run for the Time Specified icon ⬛, followed by the Zoom to Full View icon ⬛, to see that the simulation advances another 0.8 us. Now click the Restart icon ⬛, followed by ⬛ and ⬛.

i) Also practice with:

• Previous/Next Transition icons ⬛ ⬛ (for example, select x and then click one of these icons and observe the corresponding time value at the cursor foot).

• Go to Time 0 and Go to the Latest Time icons ⬛ ⬛.

• Cursor and time markers ⬛ ⬛ ⬛.

C.6   Simulating the Circuit with ModelSim

a) If not done yet, enter the source file for simulation (in this tutorial, it was done in step (a) of sections C.2 and C.3.

b) In the simulation above, ISim was used. To change to ModelSim, right-click anywhere in the Sources window and select Design Properties, which opens a dialog similar to that in figure C.2b. In the Simulator field, select the ModelSim version available.

c) Now proceed as in section C.5 to run either functional or timing simulation. When ModelSim is started, follow the ModelSim tutorial of appendix D.

Note: If ISE is unable to find the ModelSim executable file, select Edit > Preferences and enter the full path to the executable file in the Model Tech Simulator field.

C.7   Making Pin Assignments

a) Start the PlanAhead component of ISE by selecting Tools > PlanAhead > I/O Pin Planning (PlanAhead) Post-Synthesis, which opens the PlanAhead navigator of figure C.9.

b) To have the I/O pins placed automatically, select Tools > Auto-Place I/O Ports. Click Next and Finish. Observe that device pins are assigned to all 27 circuit ports. To remove all pin assignments, select Tools > Clear Placement Constraints and just follow the dialogs.

c) In the I/O Ports window of the PlanAhead navigator (figure C.9), click any port name and observe that it is highlighted in both the Device and the Package floorplan windows.

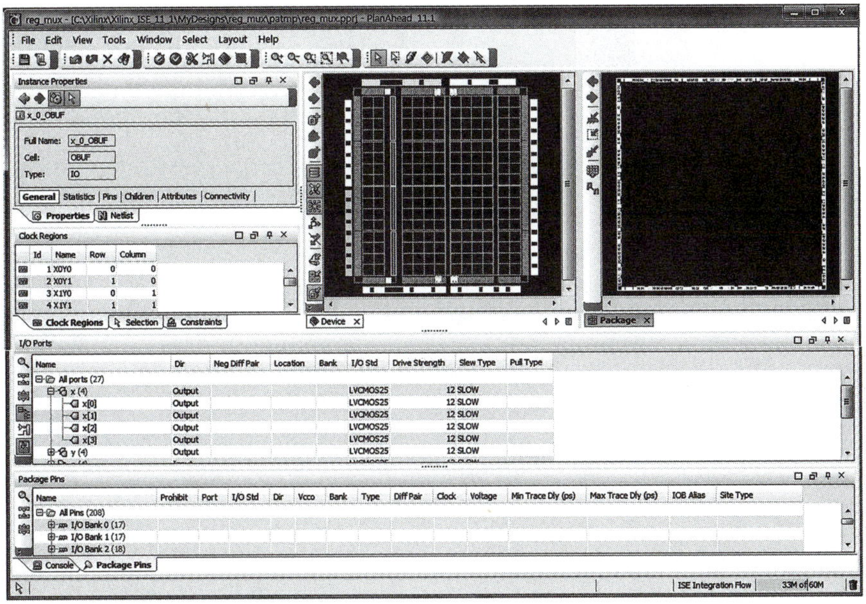

Figure C.9

d) To change pin assignments, again use the I/O Ports window. Double-click the port name (say, a[0]) or any other position in that line, which will cause that specific port to be displayed in the I/O Port Properties window. In the Site field, enter the desired pin name, then click Apply. If that pin is available, the assignment will be accepted.

e) When done making manual pin assignments, run Tools > DRC to check (fix) any inconsistencies, then recompile the design.

C.8 Physically Implementing the Design

a) Connect the development board containing the target CPLD or FPGA device to your computer and turn the power on the board on.

b) In the Sources for field of the Sources window select Implementation.

c) Still in the Sources window, highlight the design file (*reg_mux.vhd*), which will cause the menu shown in the Processes window of figure C.5 to be displayed.

d) In the Processes window, double-click Implement Design.

e) Next, double-click Generate Programming File.

f) Finally, double-click Configure Target Device, which starts the iMPACT tool responsible for configuring the device (iMPACT requires some setups, like the chain type—Boundary Scan, for example; details about iMPACT are available at the Xilinx website).

D ModelSim Tutorial

This tutorial briefly describes ModelSim, from Mentor Graphics, a simulator for VHDL-based (and other) designs. This tutorial is a complement to chapter 10, which deals exclusively with simulation, and is based on ModelSim 6.3g (web edition for Altera devices available free of charge at www.altera.com).

The tutorial is divided into six parts:

D.1 Introduction

The circuit used in this tutorial is shown in figure D.1a, which is a four-stage single-bit shift register. Simulation results obtained with the Quartus II simulator are included in figure D.1b. In the tutorial, the same input waveforms will be generated, so the same output values are expected.

To perform the simulations, two files must be created by the user: a *design* file (here called *mydesign.vhd*) and a *test* file containing the testbench (here called *mydesign_tb.vhd*). Both are shown in figure D.2. Note that the stimuli created by the latter are based on figure D.1b.

For *functional* simulations, only these two files are needed. However, for *timing* simulations, two additional files are required, both generated by the synthesizer when under the

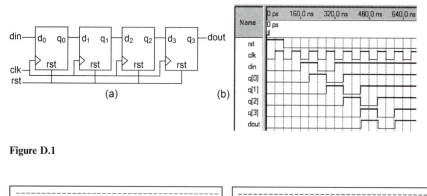

Figure D.1

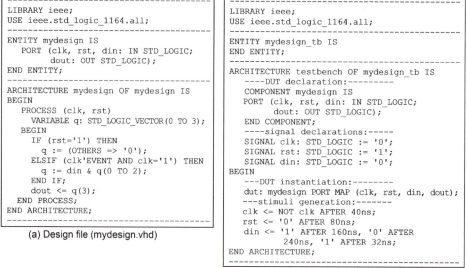

(a) Design file (mydesign.vhd)

(b) Testbench file (mydesign_tb.vhd)

Figure D.2

proper setup. One is a post synthesis file, while the other is a file with annotated propagation delays in SDF format. Both types of simulations are described in this appendix.

D.2   Preparing the Simulation Environment

a) Create a directory where all files should be located.

b) Copy both files of figure D.2 to that directory. If the files were not typed yet, the ModelSim editor can be used (see step f below).

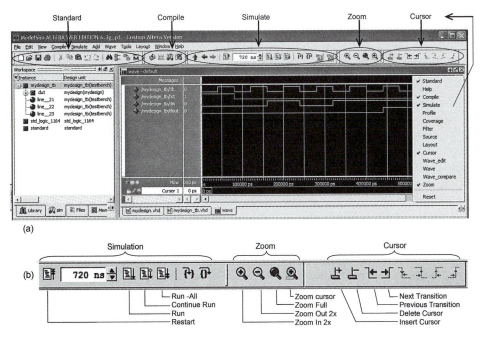

Figure D.3

c) Launch ModelSim and adopt the following setup:

For the windows: In View, mark Workspace and unmark any other options.

For the tools bar: Click the right mouse button on any empty space in the tools bar, which will cause the pull-down menu shown on the right of figure D.3a to be exhibited. Only five tools need to be selected: Standard, Compile, Simulate, Cursor, and Zoom (the corresponding tool menus can be dragged to any desired position). Details regarding three of these tools are shown in figure D.3b.

d) Select File > Change Directory and change to the directory created in step a.

e) If the files *mydesign.vhd* and *mydesign_tb.vhd* already exist, open them. Otherwise, type them following the procedure below.

f) To type the files, select File > New > Source > VHDL, which opens the VHDL editor. Type the files and save them in the directory created in step a.

g) Create now the work library by selecting File > New > Library. This opens the dialog of figure D.4. If the word "work" was not entered automatically in both fields (as in the figure), type it in. Click OK. This concludes the preparation for the simulation.

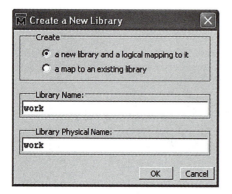

Figure D.4

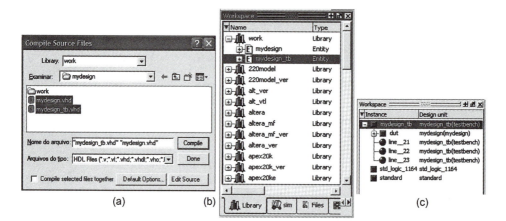

(a)   (b)   (c)

Figure D.5

D.3 Running a Functional Simulation

For *functional* simulation, proceed below. For *timing* simulation, go to section D.4.

a) Compile the files by clicking or by selecting Compile > Compile. This opens the Compile Source Files dialog of figure D.5a. Select both files (as in the figure) and click Compile. When finished, click Done. If you are debugging a file, you can compile it separately until the problems are fixed.

b) To simulate the design, start by left double-clicking *mydesign_tb* in the Workspace (see figure D.5b). Alternatively, you can select Simulate > Start Simulation, which opens

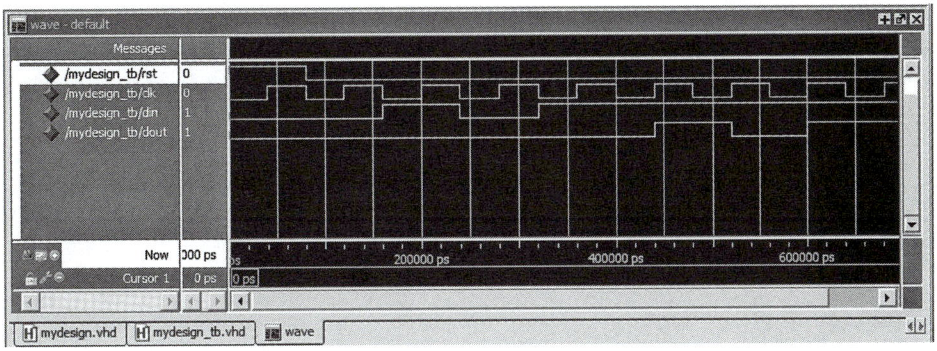

Figure D.6

the Start Simulation dialog. In it, click the "+" icon to expand the work library, select *mydesign_tb* and click OK. (For *timing* simulation, the latter option must be employed.)

c) When the process ends, figure D.5c is displayed in the *sim* tab of the Workspace. Right click *mydesign_tb* and select Add > Add to Wave. The wave pane of figure D.6 will then be exhibited (but without the waveforms). Any signal in the wave list can be dragged up or down (normally reset and clock are wanted at the top). To do so, press and hold the left mouse button on the signal's name and move it to the desired position.

d) We must now display the waveforms. First, set the simulation time interval by selecting Simulate > Runtime Options. Enter 720 ns.

e) To run the simulation, click ⬛. Alternatively, you can select Simulate > Run > Run 100. The waveforms of figure D.6 will then be exhibited.

f) Click the Zoom Full icon 🔍 (see figure D.3b) to have the complete plot displayed in the window.

g) Repeat steps e–f a few times and observe that the plot grows 720 ns each time.

h) Clean the waveforms window by clicking the Restart icon ⬛ (see figure D.3b), then repeat steps e–f. Finally, inspect the results (note that they coincide with those in figure D.1b).

D.4   Running a Timing Simulation

To run a *timing* simulation, two additional files are needed: post synthesis and SDF. These files are generated by the synthesizer when the proper setup is in place (the synthesizer must know which simulator will be employed and that it is a timing simulation). For example, in the case of Quartus II, these files are saved with the extensions *.vho* (VHDL output) and *_vhd.sdo* (SDF output). To illustrate how it is done, the case of Quartus II is described below.

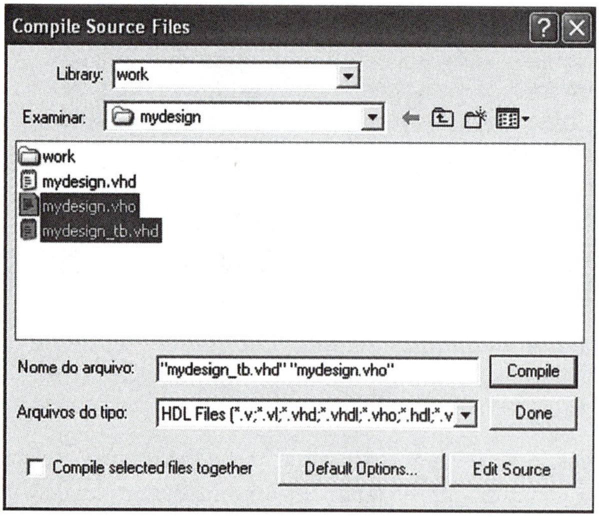

Figure D.7

a) Get the additional files

• In Quartus II, select Assignments > Settings > Simulator Settings and choose Timing in the Simulation Mode list. Click OK.

• Select Assignments > EDA Tools Settings > Simulator and choose ModelSim-Altera in the Tool Name list. Click OK.

• Compile the design (*mydesign.vhd*). In the work directory, a *simulation/modelsim* (default name) subdirectory is automatically created, containing the files *mydesign.vho* and *mydesign_vhd.sdo*.

• Copy these two files to the ModelSim work directory and return to the ModelSim software.

b) Back to ModelSim, compile the design by clicking [icon] or by selecting Compile > Compile. This opens the Compile Source Files dialog of figure D.7. Select the postsynthesis file (*mydesign.vho*) and the test file (*mydesign_tb.vhd*), as shown in the figure, and click Compile. When finished, click Done.

c) Now we can simulate the design. Select Simulate > Start Simulation, which opens the Start Simulation dialog of figure D.8a. In it, select *mydesign_tb* (as in the figure) and click the SDF tab, leading to the dialog of figure D.8b. In the SDF File field, enter the address to the SDF file. In the Apply to Region field, type */mydesign_tb/DUT*). Click OK in both dialogs, which will cause the simulation to start.

d) When the process is finished, go to step c of section D.3 and continue from there. Observe that the waveforms obtained in the timing simulation are similar to those in figure

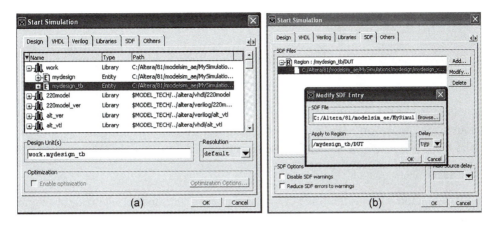

Figure D.8

D.6, but now with time delays included (note, for example, that *dout* does not change immediately when a clock transition occurs).

D.5 Running Manual Graphical Simulations

In the simulations shown in the previous sections, the input waveforms were provided by a VHDL code (testbench file) and the simulator was controlled using ModelSim's GUI.

ModelSim also allows manual graphical simulation (that is, with the input waveforms drawn by the user, as in Quartus II). In this case, only the *design* file (*mydesign.vhd*) is required, as summarized below.

For *functional* simulation: Proceed as in section D.3, but use only the *design* file (*mydesign.vhd*) in the compilation step.

For *timing* simulation: Proceed as in section D.4, but use only the *post synthesis* file (*mydesign.vho*) in the compilation step.

Once the Add > Add to Wave step is reached, the waveforms must be drawn. This can be done using either the GUI or the command line (Tcl commands, described below). Because the options in the former are very limited, the latter is a better choice. Such a procedure is described in the next section.

D.6 Running ModelSim with Tcl Commands and DO File

Tcl (tool command language) is a scripting language. Like any other EDA software, ModelSim too can be run using Tcl commands (this is called *command-line mode*, as opposed to *GUI mode*, employed in the previous sections). The main commands for the present purpose are described below.

1) *Add wave* command: Adds a new wave to the wave pane. For example,

```
add wave clk
```

2) *Run* command: Below are popular options for the run command. The first run command causes the simulation to be run until the time reaches the time limit set in the simulator. The other two cause the simulation to advance 300ns (ps is the default time unit).

```
run, run 300ns, run 300000
```

3) *Restart* command: Causes the waveforms to be cleared, with the simulator returning to time zero.

```
restart
```

4) *Force* command: Allows the construction of waveforms. Two popular options are shown here.

The syntax below is for generating clocks (-r stands for "repeat" and -freeze is the default value, so it can be omitted):

```
force -freeze <signal_name> <value> <time>, <value> <time> -r <time>
```

Example: Below is a clock with initial value '0', 150ns period, and 50% duty cycle.

```
force clk 0 0ns, 1 75ns -r 150ns
```

The next syntax is for generating arbitrary waveforms:

```
force -freeze <signal_name> <value> [<time>] [, <value>] [<time>] ...
```

Example: Below is a reset with initial value '1' during 50ns, then '0' forever. When leaving a space between the time value and the time unit, enclose them with braces (for example, 50ns = {50 ns}).

```
force rst 1 run 50ns force rst 0 run
```

5) *Do* command: Runs a DO file. For example,

```
do mydesign.do
```

A DO file is a script file consisting of Tcl commands. It constitutes a very effective way of documenting and reusing test sequences.

An example of DO file is shown in figure D.9b, which produces the signals *clk*, *rst*, and *din* depicted in figure D.9a. If we run any of the previous simulations up to the Add > Add

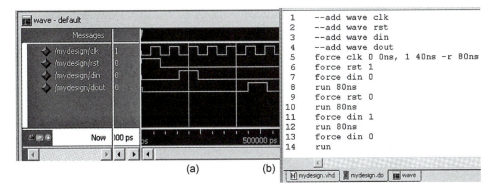

(a)   (b)

Figure D.9

to Wave step, the rest can be performed with this file. Save it with the extension .*do* and run it with the command "do mydesign.do". The resulting waveform (*dout*) is shown in the last plot of figure D.9a; note that it coincides with that in figure D.1b.

If the add wave commands that are commented out in the DO file are included, then the step Add > Add to Wave does not need to be executed.

D.7   Using Breakpoints

Breakpoints are useful for code analysis and debugging. In the description below, we will assume that functional simulation has just been executed.

a) Include the Objects window in the design navigator by selecting View > Objects. The result is in figure D.10 (dark area).

b) Open the design file (*mydesign.vhd*) in the main window (figure D.10).

c) The red lines accept breakpoints. Click, for example, the BP column in the direction of lines 18 and 20, which introduces breakpoints (red balls, figure D.10) for *q* and *dout* in those lines.

d) Click on a red ball to deactivate it (black ball), then click it again to reactivate it.

e) Click (Restart) to clean the wave pane.

f) Perform the simulation by clicking ▣↓ (Run-All), which causes the simulator to proceed until a breakpoint is found (a blue pointer will indicate the present simulation point).

g) Observe that the signal values are updated in the Objects window. Another way of verifying a signal's value is by selecting it, then right-clicking the mouse and selecting Examine.

h) Click ▣↓ several other times to see how the values of *q* and *dout* progress.

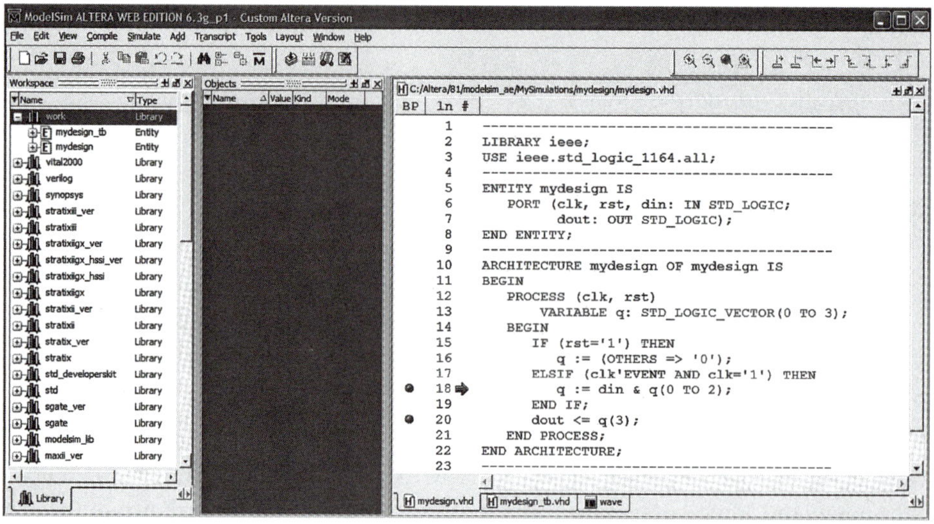

Figure D.10

i) Remove the breakpoints by clicking the right mouse button on them and selecting
Remove Breakpoint.

j) Finally, sweep the code one step at a time by clicking [↯]. This causes the simulator to
go from one red line to the next, successively.

D.8   Creating a Project

In the simulations above, the files were entered directly into the simulation environment. In
this section, we show how a project can be created first. A project saves the simulation
status, easing its continuation. It also helps organize and document the simulations.

Preparing the Project

a) Create a directory for the Project.

b) Start ModelSim (adopt the setup suggested in section D.2).

c) Select File > New > Project, which opens the Create Project dialog of figure D.11a.
Choose a name for the project, provide the directory location, leave *work* as the default
library, and click OK. This opens the Add Items to the Project dialog of figure D.11b.

d) In the dialog of figure D.11b, click Add Existing File, which leads to the Add File to
Project dialog of figure D.11c. Enter each file name, mark Copy to Project Directory, and
click OK.

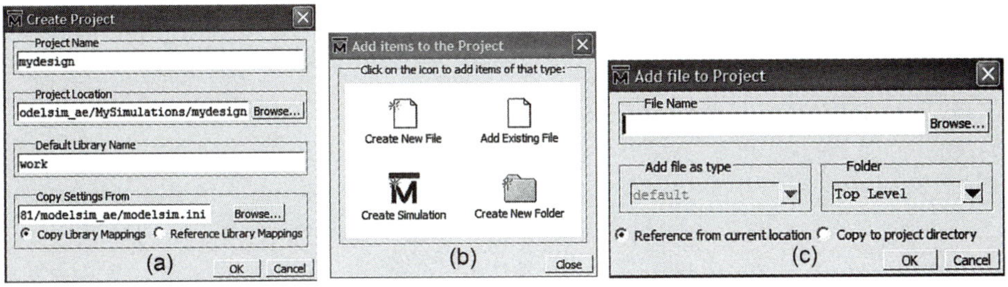

Figure D.11

Compilation

e) Select Compile > Compile Order, which opens the Compile Order dialog.

f) The design file must come first. If it does not, select it and use the arrows on the right-hand side to move it to the desired position. Click Auto Generate. When finished, click OK twice.

g) In the Workspace, select both files and click ⊻ (or select Compile > Compile All) to compile the project.

Simulation

h) From this point on the procedure is the same as before. Therefore, go to section D.3 for functional simulation or D.4 for timing simulation.

E Altera DE2 Board Tutorial

Due to its wide set of modern features, the DE2 board is very helpful in the teaching of digital design and VHDL. In this tutorial, only its most fundamental aspects are presented, which include also the *DE2 Control Panel*. Further details can be seen at www.altera.com. It is important to mention that several other helpful boards are available in the market, from Altera, Xilinx, and other CPLD/FPGA companies.

The tutorial is divided into six parts:

E.1 Board Features

E.2 Installing the USB-Blaster Driver

E.3 The DE2 Control Panel (DE2-CP)

E.4 Configuring the FPGA

E.5 Starting the DE2-CP

E.6 Introductory Exercises with the DE2 Board and its Control Panel

E.1 Board Features

The DE2 board is shown in figure E.1a and its features are summarized in figure E.1b. It includes several types of switches, displays, memories, serial ports, and video and audio circuits, plus two clock generators, the USB-Blaster driver, configuration memory and extension headers.

As indicated in figure E.1b, the USB-Blaster driver is used to program the FPGA, with two options available.

1) Directly from Quartus II: In this case, the switch must be in the RUN position, so the configuration data goes directly to the FPGA's internal SRAM configuration memory. The type of file used in this case is called SOF (SRAM object file). Since the SRAM is volatile, the configuration is lost when the board is powered off.

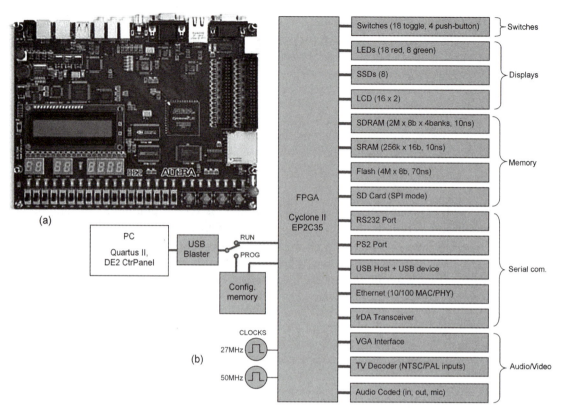

Figure E.1

2) With an external configuration memory: In this case, the switch must be in the PROG position, so the configuration data is sent to an external nonvolatile memory. The power must then be turned off and back on to cause the configuration data to be automatically retrieved from that memory by the FPGA controller. The type of file used in this case is called POF (programmer object file).

E.2   Installing the USB-Blaster Driver

Contrary to previous Altera boards, such as UP1 and UP2, which were connected to a PC through its parallel port and were configured using the Byte-Blaster driver, the DE2 board connects via an USB port and is configured using the USB-Blaster driver. Hence, before using the board, that driver must be installed.

The first time you plug the DE2 board to an USB port on your computer, the New Hardware Wizard will be opened. Just follow the dialogs and respond as follows.

First dialog: Can Windows connect to Windows Update to search for the software? No, not this time.

Second dialog: What do you want the wizard to do? Install from a list or specific location (advanced).

Third dialog: Now select the *usb-blaster* folder in the Quartus directory and just proceed until the installation is completed.

E.3 The DE2 Control Panel (DE2-CP)

The DE2 board is shipped with a software called Control Panel, which allows the user to interact, via a computer, with several of its parts, as follows.

- 26 LEDs (18 red, 8 green)
- 8 SSDs (seven-segment displays)
- 1 LCD (liquid crystal display, 16×2 alphanumeric)
- 1 PS2 keyboard
- 1 SRAM memory chip (ISSI IS61LV25616 $\rightarrow$ 256k $\times$ 16b, 10ns)
- 1 SDRAM memory chip (PSC A2V64S4 $\rightarrow$ 2M $\times$ 8b $\times$ 4banks, 10ns)
- 1 Flash memory chip (Spansion S29AL032D $\rightarrow$ 4M $\times$ 8b, 70ns)
- 1 VGA video interface, with 10-bit DACs

The following sections describe how to install and use the DE2-CP.

E.4 Configuring the FPGA

In this part, the SOF file named *DE2_USB_API.sof* must be downloaded to the FPGA. Its purpose is to configure the FPGA as needed to interface with the several chips available on the DE2 board, after which the board can be controlled by the DE2-CP.

a) Prepare the DE2 board (connect the power and USB cables, turn the power on, and place the RUN/PROG switch in RUN).

b) Start Quartus II. No project needs to be created here.

c) Open the Programmer (click 🔲 or select Tools > Programmer), which will open the window of figure E.2. In case Quartus II was already open, delete any files that might appear in the programmer, then click Add File to add the file *DE2_USB_API.sof* (as in the figure).

d) Make sure that the selected hardware is USB-Blaster (upper left corner of figure E.2).

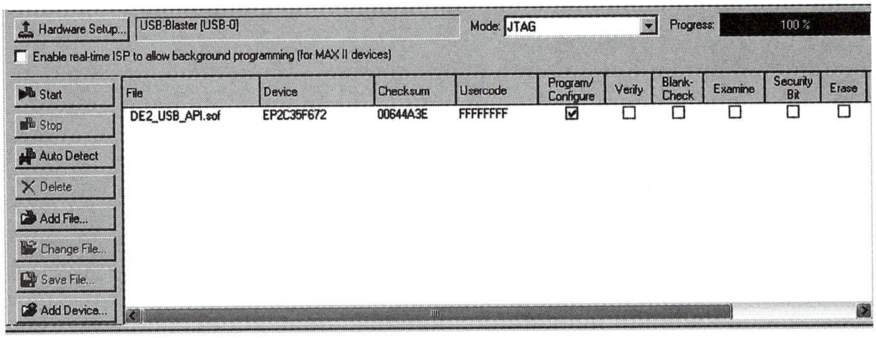

Figure E.2

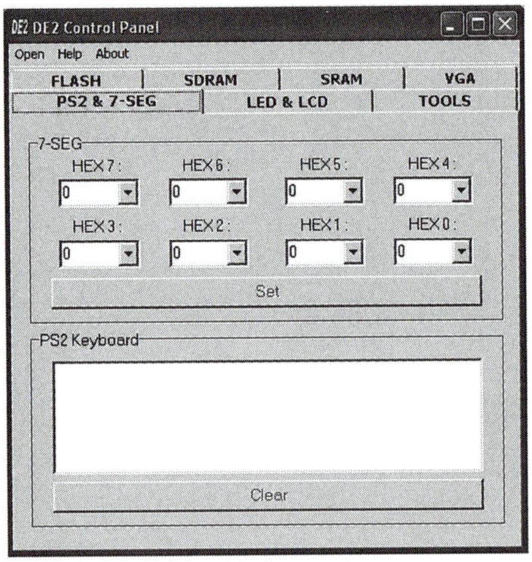

Figure E.3

e) Mark the Program/Configure box and click Start. During programming, only two blue LEDs will remain lit on the board.

E.5   Starting the DE2-CP

Once the FPGA has been configured, its utilization can commence.

a) Start the DE2-CP program by running *DE2_Control_Panel.exe*. The user interface of figure E.3 will be displayed.

b) Check in the DE2-CP (figure E.3) the presence of all peripherals listed in section E.3.

c) Open the USB connection by selecting Open > Open USB Port 0.

d) Select the TOOLS tab. In all three pull-down menus select Host USB Port, then click Configure.

e) Now the DE2-CP is ready to be used. Some exercises are presented next, but first observe the note below.

Note: When done, close the USB connection by selecting Open > Close USB Port. If you intend to return to Quartus II for some other project, close the control panel of figure E.3.

E.6   Introductory Exercises with the DE2 Board and its Control Panel

Exercise E.1: LEDs and LCD

a) Make sure that sections E.4–E.5 were executed.

b) Select the LED & LCD tab.

c) Mark some of the LEDs and click Set. Observe that the corresponding LEDs are lit.

d) Now write some text in the LCD box and click Set. Observe that the text is displayed in the LCD.

e) When done, proceed to exercise E.2 or close the DE2-CP (see Note in section E.5).

Exercise E.2: SSDs

a) Make sure that sections E.4–E.5 were executed.

b) Select the PS2 & 7-SEG tab.

c) Write any hexadecimal value in the 0-to-F range (it can be picked from the pull down menu) in any of the SSDs.

d) Click Set and observe that the digit is displayed in the corresponding SSD.

e) When done, proceed to exercise E.3 or close the DE2-CP (see Note in section E.5).

Exercise E.3: Loading Data from HEX and TXT Files into the SDRAM Chip

a) Make sure that sections E.4–E.5 were executed.

b) Select the SDRAM tab.

c) In the Random Access field, enter a value in wDATA and click Write, then click Read to see in rDATA whether the value was indeed stored in that memory address. Repeat the procedure for other addresses.

d) Using a text editor, type a file containing only hexadecimal symbols (for example, 0123456789ABCDEF0000) and save it with the extension *.hex* (not to be confused with the Intel HEX format seen in chapter 13).

e) In the Sequential Write box, mark File Length and then click Write a File to SDRAM.

f) Now in the Sequential Read part, copy to the Length field the same address obtained in the Length box of the Sequential Write part.

g) Click Load SDRAM Content to a File. Enter the desired file name and save it with the extension *.hex*.

h) Open the file in a text editor and check whether it matches that loaded into the SDRAM.

i) Repeat the procedure above using a file with a different extension (*.txt*, for example), and include in it nonhexadecimal symbols (other ASCII characters). What happens in this case?

j) Repeat the procedure once again, using the same file (with nonhexadecimal symbols) but saved with the extension *.hex*. What happens now?

Exercise E.4: Loading Data from a BMP File into the SRAM Chip

Before doing this exercise, read appendix F: BMP-to-RAW File Converter Tutorial.

a) Make sure that sections E.4–E.5 were executed.

b) Locate in the DE2 material a bitmap file called *picture.bmp* (see figure E.4), which is a 640×480 color image. Suppose that we want to load it into the SRAM chip available on the DE2 board for subsequent display on a VGA monitor.

c) Using the file converter *ImgConv.exe* (appendix F), make the conversion of *picture.bmp*. Choose the following conversion parameters: *color = red, output file name = picture*. Four files will then be generated: *picture_BW.txt, picture_BW.dat, picture_GRAY .dat*, and *picture_RGB.dat*. The file *picture_GRAY.dat* will be used in this exercise. It contains $640 \times 480 = 307{,}200$ pixels of 8 bits each (256 shades of gray).

d) As mentioned in the introduction, the SRAM is organized as 256 kwords of 16 bits each. Therefore, for the 307,200 bytes that represent the image to fit in the SRAM, they have to be loaded as 153,600 16-bit words (two pixels/word).

e) Observe also that this SRAM's access time is 10 ns, so the access speed when reading this memory must stay below 100 MHz.

f) In the DE2-CP (figure E.3), select the SRAM tab. In Sequential Write, mark File Length, then click Write a File to SRAM.

g) You can now test whether some kind of data has actually been stored in the SRAM by entering an address in Random Access and then clicking Read. Observe the result in rDATA, which is expected to be (in general) different from 0000.

h) Next, view the data stored in the SRAM with a VGA video monitor. Plug the monitor to the DE2 VGA connector.

Figure E.4

i) Select the TOOLS tab in the DE2-CP. In SRAM Multiplexer, select Asynchronous 1 and click Configure. This causes the SRAM chip to be directly connected to the VGA controller implemented by the file *DE2_USB_API.sof* loaded in section E.4.

j) Finally, select the VGA tab and unmark Default Image. The new image, stored in the SRAM, should be displayed by the monitor.

F BMP-to-RAW File Converter Tutorial

ImgConv.exe is another software program provided by Altera along with the DE2 board. Its purpose is to convert picture files of type *bitmap* into a *raw data format* (nonstandardized format, consisting of a listing of pixel values) that can be understood by the DE2 Control Panel software, for subsequent storage in the external memory chips available on that board.

F.1 The File Converter (*ImgConv.exe*)

Figure F.1 shows the window that is opened when *ImgConv.exe* is run. The screen exhibits the bmp image to be converted, whose size must be 640 × 480 (generally with 256 colors). The conversion controls (along with explanations) are shown on the right.

Say that the input file is called *test.bmp*. *ImgConv.exe* will then produce the following four files:

a) *test_BW.txt:* Text file with black-and-white data (only 00 = black and FF = white values), helpful when MIF and HEX files are subsequently wanted. The threshold and the reference color for this conversion are established in fields (1) and (2) of figure F.1. If the color intensity is above the threshold, FF is produced; otherwise, the result is 00.

b) *test_BW.dat:* Raw data file for the black-and-white picture.

c) *test_GRAY.dat:* Raw data file for the gray image (each pixel is represented by a single 8-bit value, thus allowing 256 shades of gray). The reference color for this conversion is established in field (2) of figure F.1. Note that this is a very simplified converter, because it takes into account the intensity of only one of the image colors. For example, if red is chosen, only the intensity of the red component will matter.

d) *test_RGB.dat:* Raw data file for the color image.

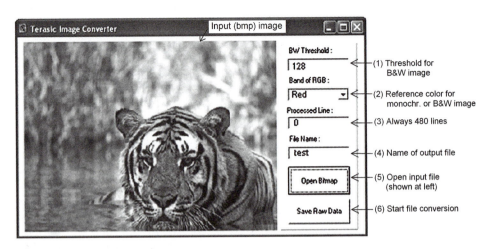

Figure F.1

F.2   File Preparation

a) The input file can be bmp, jpg, or similar, while the output file must be 640 × 480 bmp. If the file's type and size are bmp and 640 × 480, respectively, proceed to section F.3. Otherwise, continue below.

b) Open your file in a picture editor (say, Microsoft Paint).

c) Select Image > Attributes and enter Width = 640, Height = 480.

d) Save the resized image as a 256-color bmp picture.

F.3   File Conversion (input must be 640 × 480 bmp)

a) Start *ImgConv.exe*.

b) Click Open Bitmap (button (5) in figure F.1) to load the file to be converted.

c) Choose the conversion parameters (fields (1) and (2) in figure F.1).

d) Choose a name for the output file (field (4) in figure F.1).

e) Click Save Raw Data (button (6) in figure F.1) to run the conversion.

G Using Macrofunctions

Microfunctions include LPM (library of parameterized modules) units and IP (intellectual property) units among others. They are special I/O drivers, PLLs, memory blocks, and so on. The purpose of this tutorial is to show how such units can be instantiated in a VHDL code. Quartus II will be used as an example, but equivalent procedures exist for other VHDL compilers.

G.1 Preparing the Design

The circuit of figure G.1 will be used as an example. It contains a DFF, which can obviously be inferred by the code, and also a PLL (phase locked loop), which is a semi-analog unit, and so cannot be implemented directly by the code. In this example, the PLL is employed to multiply a 50 MHz clock by 2.5, producing a 125 MHz clock.

There are several application examples in the book in which clock multiplication is required (see, for example, chapters 16 and 17). This, of course, can only be done if the target device contains user PLLs.

A code for the circuit of figure G.1 is shown below, without the PLL. The project name is *clock_multiplier* (line 5). It produces the upper part (DFF) of figure G.1. The lower part (PLL) will be provided in section G.2.

```
1   -------------------------------------------------
2   LIBRARY ieee;
3   USE ieee.std_logic_1164.all;
4   -------------------------------------------------
5   ENTITY clock_multiplier IS
6      PORT (clk, din: IN STD_LOGIC;
7             clkout, dout: OUT STD_LOGIC);
8   END ENTITY;
9   -------------------------------------------------
10  ARCHITECTURE clock_multiplier OF clock_multiplier IS
11  BEGIN
```

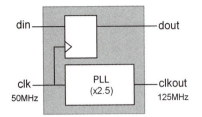

Figure G.1

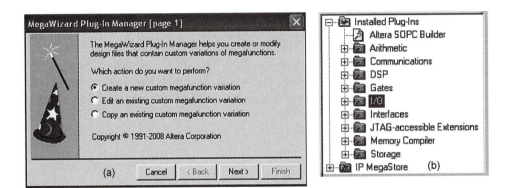

Figure G.2

```
12      PROCESS (clk)
13      BEGIN
14          IF (clk'EVENT AND clk='1') THEN
15              dout <= din;
16          END IF;
17      END PROCESS;
18  END ARCHITECTURE;
19  ----------------------------------------------------
```

G.2   Instantiating the PLL

The next step is to instantiate the PLL into the *clock_multiplier* code seen in section G.1. If one knows the PLL code well, then the file can be copied from the corresponding Altera library and edited directly. A more usual approach is to use the MegaWizard Plug-In Manager to make the instantiation.

a) With the project open in Quartus II, select Tools > MegaWizard Plug-In Manager, which will open the dialog of figure G.2a. Mark the option as shown and click Next, which will open the menu of figure G.2b.

Figure G.3

b) Expand the I/O list in figure G.2b and select ALTPLL. This will open the dialog of figure G.3a. Enter the speed grade of your device and the frequency of the input clock.

c) In the dialog of figure G.3b and a few more that follow, enter 5 and 2 for the clock multiplication and division factors, *c0* for the output, and *altera_pll.vhd* as the name for the resulting file (could be any other name). When finished, the file *altera_pll.vhd* is automatically included in the project's folder (work library). A declaration file is also created to ease the instantiation of the PLL in the main code, but that file will not be used here.

d) Open the *altera_pll.vhd* file and copy its entity. Take it to the main code, where it must be entered as a COMPONENT. The result is shown in lines 12–17 of the new code below. Finally, create an instantiation for that component, as in line 27 of the code below (under the label *mypll*). Now the project is complete, ready to be compiled and simulated.

```
1    ---------------------------------------------------------
2    LIBRARY ieee;
3    USE ieee.std_logic_1164.all;
4    ---------------------------------------------------------
5    ENTITY clock_multiplier IS
6        PORT (clk, din: IN STD_LOGIC;
7             clkout, dout: OUT STD_LOGIC);
8    END ENTITY;
9    ---------------------------------------------------------
10   ARCHITECTURE clock_multiplier OF clock_multiplier IS
11       ---PLL declaration:---------
12       COMPONENT altera_pll IS
13          PORT (areset: IN STD_LOGIC;
14               inclk0: IN STD_LOGIC;
15               c0: OUT STD_LOGIC;
16               locked: OUT STD_LOGIC);
```

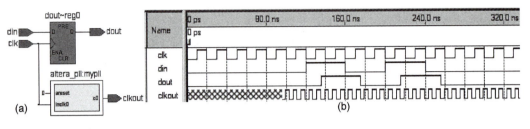

(a)   (b)

Figure G.4

```
17      END COMPONENT;
18   BEGIN
19      ---Flip-flop:--------------
20      PROCESS (clk)
21      BEGIN
22         IF (clk'EVENT AND clk='1') THEN
23            dout <= din;
24         END IF;
25      END PROCESS;
26      ---PLL instantiation:-------
27      mypll: altera_pll PORT MAP ('0', clk, clkout, OPEN);
28   END ARCHITECTURE;
29   -------------------------------------------------------
```

e) Compile the code.

f) Before proceeding to the simulation, select Tools > Netlist Viewers > RLT Viewer. The circuit of figure G.4a will be exhibited. Note that it matches the intended design of figure G.1.

g) Finally, simulate the design. With the stimuli *clk* and *din* of figure G.4b applied to the circuit, the *dout* and *clkout* responses are produced. Note that *dout* is as expected. Likewise, after the locking period, *clkout* is a 125 MHz signal.

H Package *standard* (2002 and 2008)

The VHDL 2008 version of the package *standard* is specified in section 16.3 of the *IEEE 1076-2008 Standard VHDL Language Reference Manual*, and is presented below. The new features (to be implemented) are explicitly identified in the code. The order of the text was rearranged to show related types near each other.

```
-------------------------------------------------------------------------
PACKAGE standard IS

  TYPE BIT IS ('0', '1');
  -- "and", "or", "nand", "nor", "xor", "xnor", "not"
  -- "=", "/=", "<", "<=", ">", ">="
  -- "?=", "?/=", "?<", "?<=", "?>", "?>=" (VHDL 2008)
  -- "??", MINIMUM, MAXIMUM, RISING_EDGE, FALLING_EDGE, TO_STRING (VHDL 2008)

  TYPE BIT_VECTOR IS ARRAY (NATURAL RANGE <>) OF BIT;
  -- "and", "or", "nand", "nor", "xor", "xnor", "not"
  -- "=", "/=", "<", "<=", ">", ">="
  -- "sll", "srl", "sla", "sra", "rol", "ror"
  -- "&"
  -- "?=", "?/=" (VHDL 2008)
  -- MINIMUM, MAXIMUM, TO_STRING, TO_OSTRING, TO_HSTRING (VHDL 2008)

  TYPE BOOLEAN IS (FALSE, TRUE);
  -- "and", "or", "nand", "nor", "xor", "xnor", "not"
  -- "=", "/=", "<", "<=", ">", ">="
  -- MINIMUM, MAXIMUM, RISING_EDGE, FALLING_EDGE, TO_STRING (VHDL 2008)

  TYPE BOOLEAN_VECTOR IS ARRAY (NATURAL RANGE <>) OF BOOLEAN; --VHDL 2008
  -- "and", "or", "nand", "nor", "xor", "xnor", "not"
  -- "=", "/=", "<", "<=", ">", ">="
  -- "sll", "srl", "sla", "sra", "rol", "ror"
  -- "&"
  -- "?=", "?/="
  --MINIMUM, MAXIMUM
```

```
TYPE INTEGER IS RANGE implementation_defined;
-- "+", "-", "*", "/", "**", "abs", "rem", "mod"
-- "=", "/=", "<", "<=", ">", ">="
-- MINIMUM, MAXIMUM, TO_STRING (VHDL 2008)

-- Default range -2147483647 TO 2147483647

--TYPE UNIVERSAL_INTEGER IS RANGE implementation_defined;
-- "+", "-", "*", "/", "abs", "rem", "mod"
-- "=", "/=", "<", "<=", ">", ">="
-- MINIMUM, MAXIMUM, TO_STRING (VHDL 2008)

SUBTYPE NATURAL IS INTEGER RANGE 0 TO INTEGER'HIGH;
-- Same operators as INTEGER

SUBTYPE POSITIVE IS INTEGER RANGE 1 TO INTEGER'HIGH;
-- Same operators as INTEGER

TYPE INTEGER_VECTOR IS ARRAY (NATURAL RANGE <>) OF INTEGER; --VHDL 2008
-- "=", "/=", "<", "<=", ">", ">="
-- "&"
-- MINIMUM, MAXIMUM

TYPE CHARACTER IS (
   nul,  soh,  stx,  etx,  eot,  enq,  ack,  bel,
   bs,   ht,   lf,   vt,   ff,   cr,   so,   si,
   dle,  dc1,  dc2,  dc3,  dc4,  nak,  syn,  etb,
   can,  em,   sub,  esc,  fsp,  gsp,  rsp,  usp,
   ' ',  '!',  '"',  '#',  '$',  '%',  '&',  ''',
   '(',  ')',  '*',  '+',  ',',  '-',  '.',  '/',
   '0',  '1',  '2',  '3',  '4',  '5',  '6',  '7',
   '8',  '9',  ':',  ';',  '<',  '=',  '>',  '?',
   '@',  'A',  'B',  'C',  'D',  'E',  'F',  'G',
   'H',  'I',  'J',  'K',  'L',  'M',  'N',  'O',
   'P',  'Q',  'R',  'S',  'T',  'U',  'V',  'W',
   'X',  'Y',  'Z',  '[',  '\',  ']',  '^',  '_',
   ''',  'a',  'b',  'c',  'd',  'e',  'f',  'g',
   'h',  'i',  'j',  'k',  'l',  'm',  'n',  'o',
   'p',  'q',  'r',  's',  't',  'u',  'v',  'w',
   'x',  'y',  'z',  '{',  '|',  '}',  '~',  del,
   c128, c129, c130, c131, c132, c133, c134, c135,
   c136, c137, c138, c139, c140, c141, c142, c143,
   c144, c145, c146, c147, c148, c149, c150, c151,
   c152, c153, c154, c155, c156, c157, c158, c159,
   ' ',  '¡',  '¢',  '£',  '¤',  '¥',  '|',  '§',
```

```
     '¨',    '©',    'ª',    '«',    '¬',    '-',    '®',    '‾',
     '°',    '±',    '²',    '³',    '´',    'µ',    '¶',    '·',
     '¸',    '¹',    '⁰',    '»',    '¼',    '½',    '¾',    '¿',
     'À',    'Á',    'Â',    'Ã',    'Ä',    'Å',    'Æ',    'Ç',
     'È',    'É',    'Ê',    'Ë',    'Ì',    'Í',    'Î',    'Ï',
     'Ð',    'Ñ',    'Ò',    'Ó',    'Ô',    'Õ',    'Ö',    '×',
     'Ø',    'Ù',    'Ú',    'Û',    'Ü',    'Ý',    'Þ',    'ß',
     'à',    'á',    'â',    'ã',    'ä',    'å',    'æ',    'ç',
     'è',    'é',    'ê',    'ë',    'ì',    'í',    'î',    'ï',
     'ð',    'ñ',    'ò',    'ó',    'ô',    'õ',    'ö',    '÷',
     'ø',    'ù',    'ú',    'û',    'ü',    'ý',    'þ',    'ÿ');
  -- "=", "/=", "<", "<=", ">", ">="
  -- MINIMUM, MAXIMUM, TO_STRING (VHDL 2008)


TYPE STRING IS ARRAY (positive RANGE <>) OF CHARACTER;
-- "=", "/=", "<", "<=", ">", ">="
-- "&"
-- MINIMUM, MAXIMUM (VHDL 2008)


TYPE REAL IS RANGE implementation_defined;
-- "=", "/=", "<", "<=", ">", ">="
-- "+", "-", "*", "/", "**", "abs"
-- MINIMUM, MAXIMUM, TO_STRING (VHDL 2008)


--TYPE UNIVERSAL_REAL IS RANGE implementation_defined;
-- "=", "/=", "<", "<=", ">", ">="
-- "+", "-", "*", "/", "abs"
-- MINIMUM, MAXIMUM, TO_STRING (VHDL 2008)


TYPE REAL_VECTOR IS ARRAY (NATURAL RANGE <>) OF REAL; --VHDL 2008
-- "=", "/="
-- "&"
-- MINIMUM, MAXIMUM


TYPE TIME IS RANGE implementation_defined;
  UNITS
    fs;
    ps = 1000 fs;
    ns = 1000 ps;
    us = 1000 ns;
    ms = 1000 us;
    sec = 1000 ms;
    min = 60 sec;
    hr = 60 min;
  END UNITS;
```

```
-- "=", "/=", "<", "<=", ">", ">="
-- "+", "-", "*", "/", "rem", "mod"
-- MINIMUM, MAXIMUM, TO_STRING (VHDL 2008)

SUBTYPE DELAY_LENGTH IS TIME RANGE 0 fs TO TIME'HIGH;

TYPE TIME_VECTOR IS ARRAY (NATURAL RANGE <>) OF TIME; --VHDL 2008
-- "=", "/="
-- "&"
-- MINIMUM, MAXIMUM

IMPURE FUNCTION NOW RETURN DELAY_LENGTH;

TYPE SEVERITY_LEVEL IS (NOTE, WARNING, ERROR, FAILURE);
-- "=", "/=", "<", "<=", ">", ">="
-- MINIMUM, MAXIMUM, TO_STRING (VHDL 2008)

TYPE FILE_OPEN_KIND IS (READ_MODE, WRITE_MODE, APPEND_MODE);
-- "=", "/=", "<", "<=", ">", ">="
-- MINIMUM, MAXIMUM, TO_STRING (VHDL 2008)

TYPE FILE_OPEN_STATUS IS (OPEN_OK, STATUS_ERROR, NAME_ERROR, MODE_ERROR);
-- "=", "/=", "<", "<=", ">", ">="
-- MINIMUM, MAXIMUM, TO_STRING (VHDL 2008)

ATTRIBUTE FOREIGN: STRING;
-- TO_STRING (VHDL 2008)

END standard;
----------------------------------------------------------------------------
```

I Package *std_logic_1164* (1993 and 2008)

This appendix is divided into two parts.

Part I: Contains the first (active) version of the *std_logic_1164* package, specified in the IEEE 1164 standard of 1993, used as part of all VHDL versions since then, previous to VHDL 2008.

Part II: Contains the expanded version (new features still to be implemented) of *std_logic_1164*, which is part of VHDL 2008.

The order of the text was rearranged slightly to improve readability.

Part I: Package *std_logic_1164* in VHDL 93

Note the following particularities in the code below:

1) The main types and subtypes are STD_ULOGIC, STD_ULOGIC_VECTOR, STD_LOGIC, and STD_LOGIC_VECTOR (the last two are industry standards).

2) Only logical operators are defined for them (there are no arithmetic, comparison, or shift operators).

3) Type-conversion and edge-detection functions are also included.

```
--------------------------------------------------------------------------
PACKAGE std_logic_1164 IS

  -----Types and subtypes:--------------------------
  TYPE STD_ULOGIC IS (
    'U', -- Uninitialized
    'X', -- Forcing Unknown
    '0', -- Forcing 0
    '1', -- Forcing 1
    'Z', -- High Impedance
    'W', -- Weak Unknown
```

```
  'L', -- Weak 0
  'H', -- Weak 1
  '-'  -- Don't care
  );
TYPE STD_ULOGIC_VECTOR IS ARRAY (NATURAL RANGE <>) OF STD_ULOGIC;
FUNCTION resolved (s : STD_ULOGIC_VECTOR) RETURN STD_ULOGIC;
SUBTYPE STD_LOGIC IS resolved STD_ULOGIC;
TYPE STD_LOGIC_VECTOR IS ARRAY (NATURAL RANGE <>) OF STD_LOGIC;
SUBTYPE X01 IS resolved STD_ULOGIC RANGE 'X' TO '1'; --('X','0','1')
SUBTYPE X01Z IS resolved STD_ULOGIC RANGE 'X' TO 'Z'; --('X','0','1','Z')
SUBTYPE UX01 IS resolved STD_ULOGIC RANGE 'U' TO '1'; --('U','X','0','1')
SUBTYPE UX01Z IS resolved STD_ULOGIC RANGE 'U' TO 'Z'; --('U','X','0','1','Z')

-----Logical operators:--------------------------
FUNCTION "and" (l : STD_ULOGIC; r : STD_ULOGIC) RETURN UX01;
FUNCTION "and" (l, r : STD_LOGIC_VECTOR) RETURN STD_LOGIC_VECTOR;
FUNCTION "and" (l, r : STD_ULOGIC_VECTOR) RETURN STD_ULOGIC_VECTOR;
FUNCTION "nand" (l : STD_ULOGIC; r : STD_ULOGIC) RETURN UX01;
FUNCTION "nand" (l, r : STD_LOGIC_VECTOR) RETURN STD_LOGIC_VECTOR;
FUNCTION "nand" (l, r : STD_ULOGIC_VECTOR) RETURN STD_ULOGIC_VECTOR;
FUNCTION "or" (l : STD_ULOGIC; r : STD_ULOGIC) RETURN UX01;
FUNCTION "or" (l, r : STD_LOGIC_VECTOR) RETURN STD_LOGIC_VECTOR;
FUNCTION "or" (l, r : STD_ULOGIC_VECTOR) RETURN STD_ULOGIC_VECTOR;
FUNCTION "nor" (l : STD_ULOGIC; r : STD_ULOGIC) RETURN UX01;
FUNCTION "nor" (l, r : STD_LOGIC_VECTOR) RETURN STD_LOGIC_VECTOR;
FUNCTION "nor" (l, r : STD_ULOGIC_VECTOR) RETURN STD_ULOGIC_VECTOR;
FUNCTION "xor" (l : STD_ULOGIC; r : STD_ULOGIC) RETURN UX01;
FUNCTION "xor" (l, r : STD_LOGIC_VECTOR) RETURN STD_LOGIC_VECTOR;
FUNCTION "xor" (l, r : STD_ULOGIC_VECTOR) RETURN STD_ULOGIC_VECTOR;
FUNCTION "not" (l : STD_ULOGIC) RETURN UX01;
FUNCTION "not" (l : STD_LOGIC_VECTOR) RETURN STD_LOGIC_VECTOR;
FUNCTION "not" (l : STD_ULOGIC_VECTOR) RETURN STD_ULOGIC_VECTOR;
--FUNCTION "xnor" (l : STD_ULOGIC; r : STD_ULOGIC) RETURN UX01;
--FUNCTION "xnor" (l, r : STD_LOGIC_VECTOR) RETURN STD_LOGIC_VECTOR;
--FUNCTION "xnor" (l, r : STD_ULOGIC_VECTOR) RETURN STD_ULOGIC_VECTOR;

-----Type conversion and strength strippers:---------
FUNCTION TO_BIT (s : STD_ULOGIC; xmap : BIT := '0') RETURN BIT;
FUNCTION TO_BITVECTOR (s : STD_LOGIC_VECTOR ; xmap : BIT := '0') RETURN
BIT_VECTOR;
FUNCTION TO_BITVECTOR (s : STD_ULOGIC_VECTOR; xmap : BIT := '0') RETURN
BIT_VECTOR;
FUNCTION TO_STDULOGIC (b : BIT ) RETURN STD_ULOGIC;
FUNCTION TO_STDLOGICVECTOR (b : BIT_VECTOR) RETURN STD_LOGIC_VECTOR;
```

```
FUNCTION TO_STDLOGICVECTOR (s : STD_ULOGIC_VECTOR) RETURN STD_LOGIC_VECTOR;
FUNCTION TO_STDULOGICVECTOR (b : BIT_VECTOR) RETURN STD_ULOGIC_VECTOR;
FUNCTION TO_STDULOGICVECTOR (s : STD_LOGIC_VECTOR) RETURN STD_ULOGIC_VECTOR;
FUNCTION TO_X01 (s : STD_LOGIC_VECTOR) RETURN STD_LOGIC_VECTOR;
FUNCTION TO_X01 (s : STD_ULOGIC_VECTOR) RETURN STD_ULOGIC_VECTOR;
FUNCTION TO_X01 (s : STD_ULOGIC) RETURN X01;
FUNCTION TO_X01 (b : BIT_VECTOR) RETURN STD_LOGIC_VECTOR;
FUNCTION TO_X01 (b : BIT_VECTOR) RETURN STD_ULOGIC_VECTOR;
FUNCTION TO_X01 (b : BIT) RETURN X01;
FUNCTION TO_X01Z (s : STD_LOGIC_VECTOR) RETURN STD_LOGIC_VECTOR;
FUNCTION TO_X01Z (s : STD_ULOGIC_VECTOR) RETURN STD_ULOGIC_VECTOR;
FUNCTION TO_X01Z (s : STD_ULOGIC) RETURN X01Z;
FUNCTION TO_X01Z (b : BIT_VECTOR) RETURN STD_LOGIC_VECTOR;
FUNCTION TO_X01Z (b : BIT_VECTOR) RETURN STD_ULOGIC_VECTOR;
FUNCTION TO_X01Z (b : BIT) RETURN X01Z;
FUNCTION TO_UX01 (s : STD_LOGIC_VECTOR) RETURN STD_LOGIC_VECTOR;
FUNCTION TO_UX01 (s : STD_ULOGIC_VECTOR) RETURN STD_ULOGIC_VECTOR;
FUNCTION TO_UX01 (s : STD_ULOGIC) RETURN UX01;
FUNCTION TO_UX01 (b : BIT_VECTOR) RETURN STD_LOGIC_VECTOR;
FUNCTION TO_UX01 (b : BIT_VECTOR) RETURN STD_ULOGIC_VECTOR;
FUNCTION TO_UX01 (b : BIT) RETURN UX01;

-----Object contains an unknown:-------------------
FUNCTION IS_X (s : STD_ULOGIC_VECTOR) RETURN BOOLEAN;
FUNCTION IS_X (s : STD_LOGIC_VECTOR) RETURN BOOLEAN;
FUNCTION IS_X (s : STD_ULOGIC) RETURN BOOLEAN;

-----Edge detection:-----------------------------
FUNCTION RISING_EDGE (SIGNAL  s : STD_ULOGIC) RETURN BOOLEAN;
FUNCTION FALLING_EDGE (SIGNAL s : STD_ULOGIC) RETURN BOOLEAN;

END std_logic_1164;
-------------------------------------------------------------------------------
```

Part II: Package *std_logic_1164* in VHDL 2008

Note the following particularities in the code below:

1) STD_LOGIC_VECTOR is now a subtype of STD_ULOGIC_VECTOR, hence operators defined for the latter are automatically overloaded to the former.

2) More logical operator options.

3) The XNOR operator was uncommented.

4) Inclusion of matching operators ("?=", "?/=", "?<", "?<=", "?>", "?>=", "??").

5) Some shift operators were also included.

6) Inclusion of string-conversion, read, and write operations.

7) Plus a number of aliases for type-conversion and other functions.

```
-------------------------------------------------------------------------------
USE std.textio.all;

PACKAGE std_logic_1164 IS

  -----Types and SUBTYPEs:-------------------------
  TYPE STD_ULOGIC IS (
      'U', -- Uninitialized
      'X', -- Forcing Unknown
      '0', -- Forcing 0
      '1', -- Forcing 1
      'Z', -- High Impedance
      'W', -- Weak Unknown
      'L', -- Weak 0
      'H', -- Weak 1
      '-'  -- Don't care
      )
  TYPE STD_ULOGIC_VECTOR IS array (NATURAL RANGE <>) of STD_ULOGIC;
  FUNCTION resolved (s : STD_ULOGIC_VECTOR) RETURN STD_ULOGIC;
  SUBTYPE STD_LOGIC IS resolved STD_ULOGIC;
  SUBTYPE STD_LOGIC_VECTOR IS (resolved) STD_ULOGIC_VECTOR;
  SUBTYPE X01 IS resolved STD_ULOGIC RANGE 'X' TO '1'; -- ('X','0','1')
  SUBTYPE X01Z IS resolved STD_ULOGIC RANGE 'X' TO 'Z'; -- ('X','0','1','Z')
  SUBTYPE UX01 IS resolved STD_ULOGIC RANGE 'U' TO '1'; -- ('U','X','0','1')
  SUBTYPE UX01Z IS resolved STD_ULOGIC RANGE 'U' TO 'Z'; -- ('U','X','0','1','Z')

  -----Logical operators:-------------------------
  FUNCTION "and" (l, r : STD_ULOGIC_VECTOR) RETURN STD_ULOGIC_VECTOR;
  FUNCTION "and" (l : STD_ULOGIC; r : STD_ULOGIC) RETURN UX01;
  FUNCTION "and" (l : STD_ULOGIC_VECTOR; r : STD_ULOGIC) RETURN
  STD_ULOGIC_VECTOR;
  FUNCTION "and" (l : STD_ULOGIC; r : STD_ULOGIC_VECTOR) RETURN
  STD_ULOGIC_VECTOR;
  FUNCTION "and" (l : STD_ULOGIC_VECTOR) RETURN STD_ULOGIC;
  FUNCTION "nand" (l : STD_ULOGIC; r : STD_ULOGIC) RETURN UX01;
  FUNCTION "nand" (l, r : STD_ULOGIC_VECTOR) RETURN STD_ULOGIC_VECTOR;
  FUNCTION "nand" (l : STD_ULOGIC_VECTOR; r : STD_ULOGIC) RETURN
  STD_ULOGIC_VECTOR;
  FUNCTION "nand" (l : STD_ULOGIC; r : STD_ULOGIC_VECTOR) RETURN
  STD_ULOGIC_VECTOR;
```

```
FUNCTION "nand" (l : STD_ULOGIC_VECTOR) RETURN STD_ULOGIC;
FUNCTION "or" (l : STD_ULOGIC; r : STD_ULOGIC) RETURN UX01;
FUNCTION "or" (l, r : STD_ULOGIC_VECTOR) RETURN STD_ULOGIC_VECTOR;
FUNCTION "or" (l : STD_ULOGIC_VECTOR; r : STD_ULOGIC) RETURN
STD_ULOGIC_VECTOR;
FUNCTION "or" (l : STD_ULOGIC; r : STD_ULOGIC_VECTOR) RETURN
STD_ULOGIC_VECTOR;
FUNCTION "or" (l : STD_ULOGIC_VECTOR) RETURN STD_ULOGIC;
FUNCTION "nor" (l : STD_ULOGIC; r : STD_ULOGIC) RETURN UX01;
FUNCTION "nor" (l, r : STD_ULOGIC_VECTOR) RETURN STD_ULOGIC_VECTOR;
FUNCTION "nor" (l : STD_ULOGIC_VECTOR; r : STD_ULOGIC) RETURN
STD_ULOGIC_VECTOR;
FUNCTION "nor" (l : STD_ULOGIC; r : STD_ULOGIC_VECTOR) RETURN
STD_ULOGIC_VECTOR;
FUNCTION "nor" (l : STD_ULOGIC_VECTOR) RETURN STD_ULOGIC;
FUNCTION "xor" (l : STD_ULOGIC; r : STD_ULOGIC) RETURN UX01;
FUNCTION "xor" (l, r : STD_ULOGIC_VECTOR) RETURN STD_ULOGIC_VECTOR;
FUNCTION "xor" (l : STD_ULOGIC_VECTOR; r : STD_ULOGIC) RETURN
STD_ULOGIC_VECTOR;
FUNCTION "xor" (l : STD_ULOGIC; r : STD_ULOGIC_VECTOR) RETURN
STD_ULOGIC_VECTOR;
FUNCTION "xor" (l : STD_ULOGIC_VECTOR) RETURN STD_ULOGIC;
FUNCTION "xnor" (l : STD_ULOGIC; r : STD_ULOGIC) RETURN UX01;
FUNCTION "xnor" (l, r : STD_ULOGIC_VECTOR) RETURN STD_ULOGIC_VECTOR;
FUNCTION "xnor" (l : STD_ULOGIC_VECTOR; r : STD_ULOGIC) RETURN
STD_ULOGIC_VECTOR;
FUNCTION "xnor" (l : STD_ULOGIC; r : STD_ULOGIC_VECTOR) RETURN
STD_ULOGIC_VECTOR;
FUNCTION "xnor" (l : STD_ULOGIC_VECTOR) RETURN STD_ULOGIC;
FUNCTION "not" (l : STD_ULOGIC) RETURN UX01;
FUNCTION "not" (l     : STD_ULOGIC_VECTOR) RETURN STD_ULOGIC_VECTOR;

-----Shift operators:---------------------------
FUNCTION "sll" (l : STD_ULOGIC_VECTOR; r : INTEGER) RETURN STD_ULOGIC_VECTOR;
FUNCTION "srl" (l : STD_ULOGIC_VECTOR; r : INTEGER) RETURN STD_ULOGIC_VECTOR;
FUNCTION "rol" (l : STD_ULOGIC_VECTOR; r : INTEGER) RETURN STD_ULOGIC_VECTOR;
FUNCTION "ror" (l : STD_ULOGIC_VECTOR; r : INTEGER) RETURN STD_ULOGIC_VECTOR;

--------Matching comparison operators:------------
--The following operations are predefined:
--FUNCTION "?=" (l, r : STD_ULOGIC) RETURN STD_ULOGIC;
--FUNCTION "?=" (l, r : STD_ULOGIC_VECTOR) RETURN STD_ULOGIC;
--FUNCTION "?/=" (l, r : STD_ULOGIC) RETURN STD_ULOGIC;
--FUNCTION "?/=" (l, r : STD_ULOGIC_VECTOR) RETURN STD_ULOGIC;
```

```
--FUNCTION "?<"  (l, r : STD_ULOGIC) RETURN STD_ULOGIC;
--FUNCTION "?<=" (l, r : STD_ULOGIC) RETURN STD_ULOGIC;
--FUNCTION "?>"  (l, r : STD_ULOGIC) RETURN STD_ULOGIC;
--FUNCTION "?>=" (l, r : STD_ULOGIC) RETURN STD_ULOGIC;
FUNCTION "??" (l : STD_ULOGIC) RETURN BOOLEAN;

-----Type conversion:---------------------------
FUNCTION TO_BIT (s: STD_ULOGIC; xmap : BIT := '0') RETURN BIT;
FUNCTION TO_BITVECTOR (s: STD_ULOGIC_VECTOR; xmap: BIT := '0') RETURN
BIT_VECTOR;
FUNCTION TO_STDULOGIC (b: BIT) RETURN STD_ULOGIC;
FUNCTION TO_STDLOGICVECTOR (b: BIT_VECTOR) RETURN STD_LOGIC_VECTOR;
FUNCTION TO_STDLOGICVECTOR (s: STD_ULOGIC_VECTOR) RETURN STD_LOGIC_VECTOR;
FUNCTION TO_STDULOGICVECTOR (b: BIT_VECTOR) RETURN STD_ULOGIC_VECTOR;
FUNCTION TO_STDULOGICVECTOR (s: STD_LOGIC_VECTOR) RETURN STD_ULOGIC_VECTOR;
ALIAS TO_BIT_VECTOR IS TO_BITVECTOr[STD_ULOGIC_VECTOR, BIT RETURN
BIT_VECTOR];
ALIAS TO_BV IS TO_BITVECTOR[STD_ULOGIC_VECTOR, BIT RETURN BIT_VECTOR];
ALIAS TO_STD_LOGIC_VECTOR IS TO_STDLOGICVECTOR[BIT_VECTOR RETURN
STD_LOGIC_VECTOR];
ALIAS TO_SLV IS TO_STDLOGICVECTOr[BIT_VECTOR RETURN STD_LOGIC_VECTOR];
ALIAS TO_STD_LOGIC_VECTOR IS TO_STDLOGICVECTOR[STD_ULOGIC_VECTOR RETURN
STD_LOGIC_VECTOR];
ALIAS TO_SLV IS TO_STDLOGICVECTOR[STD_ULOGIC_VECTOR RETURN STD_LOGIC_VECTOR];
ALIAS TO_STD_ULOGIC_VECTOR IS TO_STDULOGICVECTOR[BIT_VECTOR RETURN
STD_ULOGIC_VECTOR];
ALIAS TO_SULV IS TO_STDULOGICVECTOR[BIT_VECTOR RETURN STD_ULOGIC_VECTOR];
ALIAS TO_STD_ULOGIC_VECTOR IS TO_STDULogicVector[STD_LOGIC_VECTOR RETURN
STD_ULOGIC_VECTOR];
ALIAS TO_SULV IS TO_STDULOGICVECTOR[STD_LOGIC_VECTOR RETURN
STD_ULOGIC_VECTOR];
FUNCTION TO_01 (s : STD_ULOGIC_VECTOR; xmap : STD_ULOGIC := '0') RETURN
STD_ULOGIC_VECTOR;
FUNCTION TO_01 (s : STD_ULOGIC; xmap : STD_ULOGIC := '0') RETURN STD_ULOGIC;
FUNCTION TO_01 (s : BIT_VECTOR; xmap : STD_ULOGIC := '0') RETURN
STD_ULOGIC_VECTOR;
FUNCTION TO_01 (s : BIT; xmap : STD_ULOGIC := '0') RETURN STD_ULOGIC;
FUNCTION TO_X01 (s : STD_ULOGIC_VECTOR) RETURN STD_ULOGIC_VECTOR;
FUNCTION TO_X01 (s : STD_ULOGIC) RETURN X01;
FUNCTION TO_X01 (b : BIT_VECTOR) RETURN STD_ULOGIC_VECTOR;
FUNCTION TO_X01 (b : BIT) RETURN X01;
FUNCTION TO_X01Z (s : STD_ULOGIC_VECTOR) RETURN STD_ULOGIC_VECTOR;
FUNCTION TO_X01Z (s : STD_ULOGIC) RETURN X01Z;
FUNCTION TO_X01Z (b : BIT_VECTOR) RETURN STD_ULOGIC_VECTOR;
```

```
FUNCTION TO_X01Z (b : BIT) RETURN X01Z;
FUNCTION TO_UX01 (s : STD_ULOGIC_VECTOR) RETURN STD_ULOGIC_VECTOR;
FUNCTION TO_UX01 (s : STD_ULOGIC) RETURN UX01;
FUNCTION TO_UX01 (b : BIT_VECTOR) RETURN STD_ULOGIC_VECTOR;
FUNCTION TO_UX01 (b : BIT) RETURN UX01;

-----String conversion:-------------------------
--Predefined: FUNCTION TO_STRING (value : STD_ULOGIC) RETURN STRING;
--Predefined: FUNCTION TO_ STRING (value : STD_ULOGIC_VECTOR) RETURN STRING;
FUNCTION TO_OSTRING (value : STD_ULOGIC_VECTOR) RETURN STRING;
FUNCTION TO_HSTRING (value : STD_ULOGIC_VECTOR) RETURN STRING;
ALIAS TO_BSTRING IS TO_STRING [STD_ULOGIC_VECTOR RETURN STRING];
ALIAS TO_BINARY_STRING IS TO_STRING [STD_ULOGIC_VECTOR RETURN STRING];
ALIAS TO_OCTAL_STRING IS TO_OSTRING [STD_ULOGIC_VECTOR RETURN STRING];
ALIAS TO_HEX_STRING IS TO_HSTRING [STD_ULOGIC_VECTOR RETURN STRING];

-----Read operations:-------------------------
PROCEDURE READ (l : INOUT LINE; value : OUT STD_ULOGIC; GOOD : OUT BOOLEAN);
PROCEDURE READ (l : INOUT LINE; value : OUT STD_ULOGIC);
PROCEDURE READ (l : INOUT LINE; value : OUT STD_ULOGIC_VECTOR; GOOD : OUT
BOOLEAN);
PROCEDURE READ (l : INOUT LINE; value : OUT STD_ULOGIC_VECTOR);
PROCEDURE OREAD (l : INOUT LINE; value : OUT STD_ULOGIC_VECTOR; GOOD : OUT
BOOLEAN);
PROCEDURE OREAD (l : INOUT LINE; value : OUT STD_ULOGIC_VECTOR);
PROCEDURE HREAD (l : INOUT LINE; value : OUT STD_ULOGIC_VECTOR; GOOD: OUT
BOOLEAN);
PROCEDURE HREAD (l : INOUT LINE; value : OUT STD_ULOGIC_VECTOR);
ALIAS BREAD IS READ [LINE, STD_ULOGIC_VECTOR, BOOLEAN];
ALIAS BREAD IS READ [LINE, STD_ULOGIC_VECTOR];
ALIAS BINARY_READ IS READ [LINE, STD_ULOGIC_VECTOR, BOOLEAN];
ALIAS BINARY_READ IS READ [LINE, STD_ULOGIC_VECTOR];
ALIAS OCTAL_READ IS OREAD [LINE, STD_ULOGIC_VECTOR, BOOLEAN];
ALIAS OCTAL_READ IS OREAD [LINE, STD_ULOGIC_VECTOR];
ALIAS HEX_READ IS HREAD [LINE, STD_ULOGIC_VECTOR, BOOLEAN];
ALIAS HEX_READ IS HREAD [LINE, STD_ULOGIC_VECTOR];

-----Write operations:-------------------------
PROCEDURE WRITE (l : INOUT LINE; value: IN STD_ULOGIC;
  JUSTIFIED: IN SIDE := RIGHT; FIELD : IN WIDTH := 0);
PROCEDURE WRITE (l : INOUT LINE; value: IN STD_ULOGIC_VECTOR;
  JUSTIFIED : IN SIDE := RIGHT; FIELD : IN WIDTH := 0);
PROCEDURE OWRITE (l : INOUT LINE; value : IN STD_ULOGIC_VECTOR;
  JUSTIFIED : IN SIDE := RIGHT; FIELD : IN WIDTH := 0);
```

```
PROCEDURE HWRITE (l : INOUT LINE; value : IN STD_ULOGIC_VECTOR;
   JUSTIFIED : IN SIDE := RIGHT; FIELD : IN WIDTH := 0);
ALIAS BWRITE IS WRITE [LINE, STD_ULOGIC_VECTOR, SIDE, WIDTH];
ALIAS BINARY_WRITE IS WRITE [LINE, STD_ULOGIC_VECTOR, SIDE, WIDTH];
ALIAS OCTAL_WRITE IS OWRITE [LINE, STD_ULOGIC_VECTOR, SIDE, WIDTH];
ALIAS HEX_WRITE IS HWRITE [LINE, STD_ULOGIC_VECTOR, SIDE, WIDTH];

-----Object contains an unknown:-----------------
FUNCTION IS_X (s : STD_ULOGIC_VECTOR) RETURN BOOLEAN;
FUNCTION IS_X (s : STD_ULOGIC) RETURN BOOLEAN;

-----Edge detection:----------------------------
FUNCTION RISING_EDGE (SIGNAL s : STD_ULOGIC) RETURN BOOLEAN;
FUNCTION FALLING_EDGE (SIGNAL s : STD_ULOGIC) RETURN BOOLEAN;

END std_logic_1164;
```

J ┃ Package *numeric_std* (1997 and 2008)

The previous version of this package was specified in the *IEEE 1076.3-1997 Standard VHDL Synthesis Packages* document, which includes also another package, called *numeric_bit*. The difference between *numeric_std* and *numeric_bit* is that the base type in the former is STD_LOGIC, while in the latter it is BIT.

This appendix is divided into two parts, as follows:

Part I: Contains the 1997 version of the *numeric_std* package.

Part II: Contains the expanded version (to be implemented) of *numeric_std*, which is part of VHDL 2008.

The order of the text was rearranged slightly to improve readability.

Part I: Package *numeric_std* in the 1997 Specification

Note the following particularities in the code below:

1) It defines the types UNSIGNED and SIGNED, which have STD_LOGIC as their base type (subtype).

2) Large operator sets are specified, including arithmetic, logical, comparison, and shift.

3) The shift operators do not include arithmetic shift (SLA, SRA).

4) Several type-conversion and other functions are also included in the package.

```
-----------------------------------------------------------------
LIBRARY ieee;
USE ieee.std_logic_1164.all;

PACKAGE numeric_std IS

    --------Types:-------------------------------------
    TYPE UNSIGNED IS ARRAY (NATURAl range <>) of STD_LOGIC;
    TYPE SIGNED IS ARRAY (NATURAl range <>) of STD_LOGIC;
```

```
--------Arithmetic operators:--------------------
FUNCTION "+" (l, r: UNSIGNED) RETURN UNSIGNED;
FUNCTION "+" (l, r: SIGNED) RETURN SIGNED;
FUNCTION "+" (l: UNSIGNED; r: NATURAL) RETURN UNSIGNED;
FUNCTION "+" (l: NATURAL; r: UNSIGNED) RETURN UNSIGNED;
FUNCTION "+" (l: INTEGER; r: SIGNED) RETURN SIGNED;
FUNCTION "+" (l: SIGNED; r: INTEGER) RETURN SIGNED;
FUNCTION "-" (arg: SIGNED) RETURN SIGNED; --minus operation
FUNCTION "-" (l, r: UNSIGNED) RETURN UNSIGNED;
FUNCTION "-" (l, r: SIGNED) RETURN SIGNED;
FUNCTION "-" (l: UNSIGNED;r: NATURAL) RETURN UNSIGNED;
FUNCTION "-" (l: NATURAL; r: UNSIGNED) RETURN UNSIGNED;
FUNCTION "-" (l: SIGNED; r: INTEGER) RETURN SIGNED;
FUNCTION "-" (l: INTEGER; r: SIGNED) RETURN SIGNED;
FUNCTION "*" (l, r: UNSIGNED) RETURN UNSIGNED;
FUNCTION "*" (l, r: SIGNED) RETURN SIGNED;
FUNCTION "*" (l: UNSIGNED; r: NATURAL) RETURN UNSIGNED;
FUNCTION "*" (l: NATURAL; r: UNSIGNED) RETURN UNSIGNED;
FUNCTION "*" (l: SIGNED; r: INTEGER) RETURN SIGNED;
FUNCTION "*" (l: INTEGER; r: SIGNED) RETURN SIGNED;
FUNCTION "/" (l, r: UNSIGNED) RETURN UNSIGNED;
FUNCTION "/" (l, r: SIGNED) RETURN SIGNED;
FUNCTION "/" (l: UNSIGNED; r: NATURAL) RETURN UNSIGNED;
FUNCTION "/" (l: NATURAL; r: UNSIGNED) RETURN UNSIGNED;
FUNCTION "/" (l: SIGNED; r: INTEGER) RETURN SIGNED;
FUNCTION "/" (l: INTEGER; r: SIGNED) RETURN SIGNED;
FUNCTION "abs" (arg: SIGNED) RETURN SIGNED;
FUNCTION "rem" (l, r: UNSIGNED) RETURN UNSIGNED;
FUNCTION "rem" (l, r: SIGNED) RETURN SIGNED;
FUNCTION "rem" (l: UNSIGNED; r: NATURAL) RETURN UNSIGNED;
FUNCTION "rem" (l: NATURAL; r: UNSIGNED) RETURN UNSIGNED;
FUNCTION "rem" (l: SIGNED; r: INTEGER) RETURN SIGNED;
FUNCTION "rem" (l: INTEGER; r: SIGNED) RETURN SIGNED;
FUNCTION "mod" (l, r: UNSIGNED) RETURN UNSIGNED;
FUNCTION "mod" (l, r: SIGNED) RETURN SIGNED;
FUNCTION "mod" (l: UNSIGNED; r: NATURAL) RETURN UNSIGNED;
FUNCTION "mod" (l: NATURAL; r: UNSIGNED) RETURN UNSIGNED;
FUNCTION "mod" (l: SIGNED; r: INTEGER) RETURN SIGNED;
FUNCTION "mod" (l: INTEGER; r: SIGNED) RETURN SIGNED;

--------Logical operators:-----------------------
FUNCTION "not" (l: UNSIGNED) RETURN UNSIGNED;
FUNCTION "and" (l, r: UNSIGNED) RETURN UNSIGNED;
FUNCTION "or" (l, r: UNSIGNED) RETURN UNSIGNED;
```

```
FUNCTION "nand" (l, r: UNSIGNED) RETURN UNSIGNED;
FUNCTION "nor" (l, r: UNSIGNED) RETURN UNSIGNED;
FUNCTION "xor" (l, r: UNSIGNED) RETURN UNSIGNED;
FUNCTION "xnor" (l, r: UNSIGNED) RETURN UNSIGNED;
FUNCTION "not" (l: SIGNED) RETURN SIGNED;
FUNCTION "and" (l, r: SIGNED) RETURN SIGNED;
FUNCTION "or" (l, r: SIGNED) RETURN SIGNED;
FUNCTION "nand" (l, r: SIGNED) RETURN SIGNED;
FUNCTION "nor" (l, r: SIGNED) RETURN SIGNED;
FUNCTION "xor" (l, r: SIGNED) RETURN SIGNED;
FUNCTION "xnor" (l, r: SIGNED) RETURN SIGNED;

-------Comparison operators:---------------------
FUNCTION ">" (l, r: UNSIGNED) RETURN BOOLEAN;
FUNCTION ">" (l, r: SIGNED) RETURN BOOLEAN;
FUNCTION ">" (l: NATURAL; r: UNSIGNED) RETURN BOOLEAN;
FUNCTION ">" (l: INTEGER; r: SIGNED) RETURN BOOLEAN;
FUNCTION ">" (l: UNSIGNED; r: NATURAL) RETURN BOOLEAN;
FUNCTION ">" (l: SIGNED; r: INTEGER) RETURN BOOLEAN;
FUNCTION "<" (l, r: UNSIGNED) RETURN BOOLEAN;
FUNCTION "<" (l, r: SIGNED) RETURN BOOLEAN;
FUNCTION "<" (l: NATURAL; r: UNSIGNED) RETURN BOOLEAN;
FUNCTION "<" (l: INTEGER; r: SIGNED) RETURN BOOLEAN;
FUNCTION "<" (l: UNSIGNED; r: NATURAL) RETURN BOOLEAN;
FUNCTION "<" (l: SIGNED; r: INTEGER) RETURN BOOLEAN;
FUNCTION "<=" (l, r: UNSIGNED) RETURN BOOLEAN;
FUNCTION "<=" (l, r: SIGNED) RETURN BOOLEAN;
FUNCTION "<=" (l: NATURAL; r: UNSIGNED) RETURN BOOLEAN;
FUNCTION "<=" (l: INTEGER; r: SIGNED) RETURN BOOLEAN;
FUNCTION "<=" (l: UNSIGNED; r: NATURAL) RETURN BOOLEAN;
FUNCTION "<=" (l: SIGNED; r: INTEGER) RETURN BOOLEAN;
FUNCTION ">=" (l, r: UNSIGNED) RETURN BOOLEAN;
FUNCTION ">=" (l, r: SIGNED) RETURN BOOLEAN;
FUNCTION ">=" (l: NATURAL; r: UNSIGNED) RETURN BOOLEAN;
FUNCTION ">=" (l: INTEGER; r: SIGNED) RETURN BOOLEAN;
FUNCTION ">=" (l: UNSIGNED; r: NATURAL) RETURN BOOLEAN;
FUNCTION ">=" (l: SIGNED; r: INTEGER) RETURN BOOLEAN;
FUNCTION "=" (l, r: UNSIGNED) RETURN BOOLEAN;
FUNCTION "=" (l, r: SIGNED) RETURN BOOLEAN;
FUNCTION "=" (l: NATURAL; r: UNSIGNED) RETURN BOOLEAN;
FUNCTION "=" (l: INTEGER; r: SIGNED) RETURN BOOLEAN;
FUNCTION "=" (l: UNSIGNED; r: NATURAL) RETURN BOOLEAN;
FUNCTION "=" (l: SIGNED; r: INTEGER) RETURN BOOLEAN;
FUNCTION "/=" (l, r: UNSIGNED) RETURN BOOLEAN;
```

```
FUNCTION "/=" (l, r: SIGNED) RETURN BOOLEAN;
FUNCTION "/=" (l: NATURAL; r: UNSIGNED) RETURN BOOLEAN;
FUNCTION "/=" (l: INTEGER; r: SIGNED) RETURN BOOLEAN;
FUNCTION "/=" (l: UNSIGNED; r: NATURAL) RETURN BOOLEAN;
FUNCTION "/=" (l: SIGNED; r: INTEGER) RETURN BOOLEAN;

--------Shift operators:-------------------------
FUNCTION "sll" (arg: UNSIGNED; count: INTEGER) RETURN UNSIGNED;
FUNCTION "sll" (arg: SIGNED; count: INTEGER) RETURN SIGNED;
FUNCTION "srl" (arg: UNSIGNED; count: INTEGER) RETURN UNSIGNED;
FUNCTION "srl" (arg: SIGNED; count: INTEGER) RETURN SIGNED;
FUNCTION "rol" (arg: UNSIGNED; count: INTEGER) RETURN UNSIGNED;
FUNCTION "rol" (arg: SIGNED; count: INTEGER) RETURN SIGNED;
FUNCTION "ror" (arg: UNSIGNED; count: INTEGER) RETURN UNSIGNED;
FUNCTION "ror" (arg: SIGNED; count: INTEGER) RETURN SIGNED;
FUNCTION SHIFT_LEFT (arg: UNSIGNED; count: NATURAL) RETURN UNSIGNED;
FUNCTION SHIFT_RIGHT (arg: UNSIGNED; count: NATURAL) RETURN UNSIGNED;
FUNCTION SHIFT_LEFT (arg: SIGNED; count: NATURAL) RETURN SIGNED;
FUNCTION SHIFT_RIGHT (arg: SIGNED; count: NATURAL) RETURN SIGNED;
FUNCTION ROTATE_LEFT (arg: UNSIGNED; count: NATURAL) RETURN UNSIGNED;
FUNCTION ROTATE_RIGHT (arg: UNSIGNED; count: NATURAL) RETURN UNSIGNED;
FUNCTION ROTATE_LEFT (arg: SIGNED; count: NATURAL) RETURN SIGNED;
FUNCTION ROTATE_RIGHT (arg: SIGNED; count: NATURAL) RETURN SIGNED;

--------Type conversion:-------------------------
FUNCTION TO_INTEGER (arg: UNSIGNED) RETURN NATURAL;
FUNCTION TO_INTEGER (arg: SIGNED) RETURN INTEGER;
FUNCTION TO_UNSIGNED (arg: INTEGER; size: NATURAL) RETURN UNSIGNED;
FUNCTION TO_SIGNED (arg: INTEGER; size: NATURAL) RETURN SIGNED;
FUNCTION TO_01 (S: UNSIGNED; xmap: STD_LOGIC := '0') RETURN UNSIGNED;
FUNCTION TO_01 (S: SIGNED; xmap: STD_LOGIC := '0') RETURN SIGNED;

--------Resize functions:-------------------------
FUNCTION RESIZE (arg: SIGNED; new_size: NATURAL) RETURN SIGNED;
FUNCTION RESIZE (arg: UNSIGNED; new_size: NATURAL) RETURN UNSIGNED;

--------Match functions:-------------------------
FUNCTION STD_MATCH (l, r: STD_ULOGIC) RETURN BOOLEAN;
FUNCTION STD_MATCH (l, r: UNSIGNED) RETURN BOOLEAN;
FUNCTION STD_MATCH (l, r: SIGNED) RETURN BOOLEAN;
FUNCTION STD_MATCH (l, r: STD_LOGIC_VECTOR) RETURN BOOLEAN;
FUNCTION STD_MATCH (l, r: STD_ULOGIC_VECTOR) RETURN BOOLEAN;

END numeric_std;
--------------------------------------------------------------------------
```

Part II: Package *numeric_std* in VHDL 2008

Note the following particularities in the code below (with respect to the previous version):

1) The definitions of UNSIGNED and SIGNED are slightly different.

2) All operator sets were expanded as well as the type-conversion functions list.

3) Several new functions were included, for matching operators, string conversion, read, write, and so on.

```
--------------------------------------------------------------------------------
USE std.textio.all;
LIBRARY ieee;
USE ieee.std_logic_1164.all;

PACKAGE numeric_std IS

  CONSTANT CopyRightNotice: STRING := "Copyright © 2008 IEEE. All rights
  reserved.";

  -----Types and subtypes:----------------------------
  TYPE UNRESOLVED_UNSIGNED IS ARRAY (NATURAl range <>) of STD_ULOGIC;
  TYPE UNRESOLVED_SIGNED IS ARRAY (NATURAl range <>) of STD_ULOGIC;
  SUBTYPE UNSIGNED IS (resolved) UNRESOLVED_UNSIGNED;
  SUBTYPE SIGNED IS (resolved) UNRESOLVED_SIGNED;
  ALIAS U_UNSIGNED IS UNRESOLVED_UNSIGNED;
  ALIAS U_SIGNED IS UNRESOLVED_SIGNED;

  -----Arithmetic operators:--------------------------
  FUNCTION "+" (l, r : UNRESOLVED_UNSIGNED) RETURN UNRESOLVED_UNSIGNED;
  FUNCTION "+"(l : UNRESOLVED_UNSIGNED; r : STD_ULOGIC) RETURN
  UNRESOLVED_UNSIGNED;
  FUNCTION "+"(l : STD_ULOGIC; r : UNRESOLVED_UNSIGNED) RETURN
  UNRESOLVED_UNSIGNED;
  FUNCTION "+" (l, r : UNRESOLVED_SIGNED) RETURN UNRESOLVED_SIGNED;
  FUNCTION "+"(l : UNRESOLVED_SIGNED; r : STD_ULOGIC) RETURN UNRESOLVED_SIGNED;
  FUNCTION "+"(l : STD_ULOGIC; r : UNRESOLVED_SIGNED) RETURN UNRESOLVED_SIGNED;
  FUNCTION "+" (l : UNRESOLVED_UNSIGNED; r : NATURAL) RETURN
  UNRESOLVED_UNSIGNED;
  FUNCTION "+" (l : NATURAL; r : UNRESOLVED_UNSIGNED) RETURN
  UNRESOLVED_UNSIGNED;
  FUNCTION "+" (l : INTEGER; r : UNRESOLVED_SIGNED) RETURN UNRESOLVED_SIGNED;
  FUNCTION "+" (l : UNRESOLVED_SIGNED; r : INTEGER) RETURN UNRESOLVED_SIGNED;
  FUNCTION "-" (arg : UNRESOLVED_SIGNED) RETURN UNRESOLVED_SIGNED;
  FUNCTION "-" (l, r : UNRESOLVED_UNSIGNED) RETURN UNRESOLVED_UNSIGNED;
```

```
FUNCTION "-"(l : UNRESOLVED_UNSIGNED; r : STD_ULOGIC) RETURN
UNRESOLVED_UNSIGNED;
FUNCTION "-"(l : STD_ULOGIC; r : UNRESOLVED_UNSIGNED) RETURN
UNRESOLVED_UNSIGNED;
FUNCTION "-" (l, r : UNRESOLVED_SIGNED) RETURN UNRESOLVED_SIGNED;
FUNCTION "-"(l : UNRESOLVED_SIGNED; r : STD_ULOGIC) RETURN UNRESOLVED_SIGNED;
FUNCTION "-"(l : STD_ULOGIC; r : UNRESOLVED_SIGNED) RETURN UNRESOLVED_SIGNED;
FUNCTION "-" (l : UNRESOLVED_UNSIGNED; r : NATURAL) RETURN
UNRESOLVED_UNSIGNED;
FUNCTION "-" (l : NATURAL; r : UNRESOLVED_UNSIGNED) RETURN
UNRESOLVED_UNSIGNED;
FUNCTION "-" (l : UNRESOLVED_SIGNED; r : INTEGER) RETURN UNRESOLVED_SIGNED;
FUNCTION "-" (l : INTEGER; r : UNRESOLVED_SIGNED) RETURN UNRESOLVED_SIGNED;
FUNCTION "*" (l, r : UNRESOLVED_UNSIGNED) RETURN UNRESOLVED_UNSIGNED;
FUNCTION "*" (l, r : UNRESOLVED_SIGNED) RETURN UNRESOLVED_SIGNED;
FUNCTION "*" (l : UNRESOLVED_UNSIGNED; r : NATURAL) RETURN
UNRESOLVED_UNSIGNED;
FUNCTION "*" (l : NATURAL; r : UNRESOLVED_UNSIGNED) RETURN
UNRESOLVED_UNSIGNED;
FUNCTION "*" (l : UNRESOLVED_SIGNED; r : INTEGER) RETURN UNRESOLVED_SIGNED;
FUNCTION "*" (l : INTEGER; r : UNRESOLVED_SIGNED) RETURN UNRESOLVED_SIGNED;
FUNCTION "/" (l, r : UNRESOLVED_UNSIGNED) RETURN UNRESOLVED_UNSIGNED;
FUNCTION "/" (l, r : UNRESOLVED_SIGNED) RETURN UNRESOLVED_SIGNED;
FUNCTION "/" (l : UNRESOLVED_UNSIGNED; r : NATURAL) RETURN
UNRESOLVED_UNSIGNED;
FUNCTION "/" (l : NATURAL; r : UNRESOLVED_UNSIGNED) RETURN
UNRESOLVED_UNSIGNED;
FUNCTION "/" (l : UNRESOLVED_SIGNED; r : INTEGER) RETURN UNRESOLVED_SIGNED;
FUNCTION "/" (l : INTEGER; r : UNRESOLVED_SIGNED) RETURN UNRESOLVED_SIGNED;
FUNCTION "abs" (arg : UNRESOLVED_SIGNED) RETURN UNRESOLVED_SIGNED;
FUNCTION "rem" (l, r : UNRESOLVED_UNSIGNED) RETURN UNRESOLVED_UNSIGNED;
FUNCTION "rem" (l, r : UNRESOLVED_SIGNED) RETURN UNRESOLVED_SIGNED;
FUNCTION "rem" (l : UNRESOLVED_UNSIGNED; r : NATURAL) RETURN
UNRESOLVED_UNSIGNED;
FUNCTION "rem" (l : NATURAL; r : UNRESOLVED_UNSIGNED) RETURN
UNRESOLVED_UNSIGNED;
FUNCTION "rem" (l : UNRESOLVED_SIGNED; r : INTEGER) RETURN UNRESOLVED_SIGNED;
FUNCTION "rem" (l : INTEGER; r : UNRESOLVED_SIGNED) RETURN UNRESOLVED_SIGNED;
FUNCTION "mod" (l, r : UNRESOLVED_UNSIGNED) RETURN UNRESOLVED_UNSIGNED;
FUNCTION "mod" (l, r : UNRESOLVED_SIGNED) RETURN UNRESOLVED_SIGNED;
FUNCTION "mod" (l : UNRESOLVED_UNSIGNED; r : NATURAL) RETURN
UNRESOLVED_UNSIGNED;
FUNCTION "mod" (l : NATURAL; r : UNRESOLVED_UNSIGNED) RETURN
UNRESOLVED_UNSIGNED;
```

```
FUNCTION "mod" (l : UNRESOLVED_SIGNED; r : INTEGER) RETURN UNRESOLVED_SIGNED;
FUNCTION "mod" (l : INTEGER; r : UNRESOLVED_SIGNED) RETURN UNRESOLVED_SIGNED;
FUNCTION FIND_LEFTMOST (arg : UNRESOLVED_UNSIGNED; y : STD_ULOGIC) RETURN
INTEGER;
FUNCTION FIND_LEFTMOST (arg : UNRESOLVED_SIGNED; y : STD_ULOGIC) RETURN
INTEGER;
FUNCTION FIND_RIGHTMOST (arg : UNRESOLVED_UNSIGNED; y : STD_ULOGIC) RETURN
INTEGER;
FUNCTION FIND_RIGHTMOST (arg : UNRESOLVED_SIGNED; y : STD_ULOGIC) RETURN
INTEGER;

-----Logical operators:----------------------------
FUNCTION "not" (l : UNRESOLVED_UNSIGNED) RETURN UNRESOLVED_UNSIGNED;
FUNCTION "not" (l : UNRESOLVED_SIGNED) RETURN UNRESOLVED_SIGNED;
FUNCTION "and" (l, r : UNRESOLVED_UNSIGNED) RETURN UNRESOLVED_UNSIGNED;
FUNCTION "and" (l, r : UNRESOLVED_SIGNED) RETURN UNRESOLVED_SIGNED;
FUNCTION "and" (l : STD_ULOGIC; r : UNRESOLVED_UNSIGNED) RETURN
UNRESOLVED_UNSIGNED;
FUNCTION "and" (l : UNRESOLVED_UNSIGNED; r : STD_ULOGIC) RETURN
UNRESOLVED_UNSIGNED;
FUNCTION "and" (l : STD_ULOGIC; r : UNRESOLVED_SIGNED) RETURN
UNRESOLVED_SIGNED;
FUNCTION "and" (l : UNRESOLVED_SIGNED; r : STD_ULOGIC) RETURN
UNRESOLVED_SIGNED;
FUNCTION "and" (l : UNRESOLVED_SIGNED) RETURN STD_ULOGIC;
FUNCTION "and" (l : UNRESOLVED_UNSIGNED) RETURN STD_ULOGIC;
FUNCTION "or" (l, r : UNRESOLVED_UNSIGNED) RETURN UNRESOLVED_UNSIGNED;
FUNCTION "or" (l, r : UNRESOLVED_SIGNED) RETURN UNRESOLVED_SIGNED;
FUNCTION "or" (l : STD_ULOGIC; r : UNRESOLVED_UNSIGNED) RETURN
UNRESOLVED_UNSIGNED;
FUNCTION "or" (l : UNRESOLVED_UNSIGNED; r : STD_ULOGIC) RETURN
UNRESOLVED_UNSIGNED;
FUNCTION "or" (l : STD_ULOGIC; r : UNRESOLVED_SIGNED) RETURN
UNRESOLVED_SIGNED;
FUNCTION "or" (l : UNRESOLVED_SIGNED; r : STD_ULOGIC) RETURN
UNRESOLVED_SIGNED;
FUNCTION "or" (l : UNRESOLVED_SIGNED) RETURN STD_ULOGIC;
FUNCTION "or" (l : UNRESOLVED_UNSIGNED) RETURN STD_ULOGIC;
FUNCTION "nand" (l, r : UNRESOLVED_UNSIGNED) RETURN UNRESOLVED_UNSIGNED;
FUNCTION "nand" (l, r : UNRESOLVED_SIGNED) RETURN UNRESOLVED_SIGNED;
FUNCTION "nand" (l : STD_ULOGIC; r : UNRESOLVED_UNSIGNED) RETURN
UNRESOLVED_UNSIGNED;
FUNCTION "nand" (l : UNRESOLVED_UNSIGNED; r : STD_ULOGIC) RETURN
UNRESOLVED_UNSIGNED;
```

```
FUNCTION "nand" (l : STD_ULOGIC; r : UNRESOLVED_SIGNED) RETURN
UNRESOLVED_SIGNED;
FUNCTION "nand" (l : UNRESOLVED_SIGNED; r : STD_ULOGIC) RETURN
UNRESOLVED_SIGNED;
FUNCTION "nand" (l : UNRESOLVED_SIGNED) RETURN STD_ULOGIC;
FUNCTION "nand" (l : UNRESOLVED_UNSIGNED) RETURN STD_ULOGIC;
FUNCTION "nor" (l, r : UNRESOLVED_UNSIGNED) RETURN UNRESOLVED_UNSIGNED;
FUNCTION "nor" (l, r : UNRESOLVED_SIGNED) RETURN UNRESOLVED_SIGNED;
FUNCTION "nor" (l : STD_ULOGIC; r : UNRESOLVED_UNSIGNED) RETURN
UNRESOLVED_UNSIGNED;
FUNCTION "nor" (l : UNRESOLVED_UNSIGNED; r : STD_ULOGIC) RETURN
UNRESOLVED_UNSIGNED;
FUNCTION "nor" (l : STD_ULOGIC; r : UNRESOLVED_SIGNED) RETURN
UNRESOLVED_SIGNED;
FUNCTION "nor" (l : UNRESOLVED_SIGNED; r : STD_ULOGIC) RETURN
UNRESOLVED_SIGNED;
FUNCTION "nor" (l : UNRESOLVED_SIGNED) RETURN STD_ULOGIC;
FUNCTION "nor" (l : UNRESOLVED_UNSIGNED) RETURN STD_ULOGIC;
FUNCTION "xor" (l : UNRESOLVED_SIGNED) RETURN STD_ULOGIC;
FUNCTION "xor" (l, r : UNRESOLVED_UNSIGNED) RETURN UNRESOLVED_UNSIGNED;
FUNCTION "xor" (l, r : UNRESOLVED_SIGNED) RETURN UNRESOLVED_SIGNED;
FUNCTION "xor" (l : STD_ULOGIC; r : UNRESOLVED_UNSIGNED) RETURN
UNRESOLVED_UNSIGNED;
FUNCTION "xor" (l : UNRESOLVED_UNSIGNED; r : STD_ULOGIC) RETURN
UNRESOLVED_UNSIGNED;
FUNCTION "xor" (l : STD_ULOGIC; r : UNRESOLVED_SIGNED) RETURN
UNRESOLVED_SIGNED;
FUNCTION "xor" (l : UNRESOLVED_SIGNED; r : STD_ULOGIC) RETURN
UNRESOLVED_SIGNED;
FUNCTION "xor" (l : UNRESOLVED_UNSIGNED) RETURN STD_ULOGIC;
FUNCTION "xnor" (l, r : UNRESOLVED_UNSIGNED) RETURN UNRESOLVED_UNSIGNED;
FUNCTION "xnor" (l, r : UNRESOLVED_SIGNED) RETURN UNRESOLVED_SIGNED;
FUNCTION "xnor" (l : STD_ULOGIC; r : UNRESOLVED_UNSIGNED) RETURN
UNRESOLVED_UNSIGNED;
FUNCTION "xnor" (l : UNRESOLVED_UNSIGNED; r : STD_ULOGIC) RETURN
UNRESOLVED_UNSIGNED;
FUNCTION "xnor" (l : STD_ULOGIC; r : UNRESOLVED_SIGNED) RETURN
UNRESOLVED_SIGNED;
FUNCTION "xnor" (l : UNRESOLVED_SIGNED; r : STD_ULOGIC) RETURN
UNRESOLVED_SIGNED;
FUNCTION "xnor" (l : UNRESOLVED_SIGNED) RETURN STD_ULOGIC;
FUNCTION "xnor" (l : UNRESOLVED_UNSIGNED) RETURN STD_ULOGIC;
```

```
-----Comparison operators:--------------------------
FUNCTION ">" (l, r : UNRESOLVED_UNSIGNED) RETURN BOOLEAN;
FUNCTION ">" (l, r : UNRESOLVED_SIGNED) RETURN BOOLEAN;
FUNCTION ">" (l : NATURAL; r : UNRESOLVED_UNSIGNED) RETURN BOOLEAN;
FUNCTION ">" (l : INTEGER; r : UNRESOLVED_SIGNED) RETURN BOOLEAN;
FUNCTION ">" (l : UNRESOLVED_UNSIGNED; r : NATURAL) RETURN BOOLEAN;
FUNCTION ">" (l : UNRESOLVED_SIGNED; r : INTEGER) RETURN BOOLEAN;
FUNCTION "<" (l, r : UNRESOLVED_UNSIGNED) RETURN BOOLEAN;
FUNCTION "<" (l, r : UNRESOLVED_SIGNED) RETURN BOOLEAN;
FUNCTION "<" (l : NATURAL; r : UNRESOLVED_UNSIGNED) RETURN BOOLEAN;
FUNCTION "<" (l : INTEGER; r : UNRESOLVED_SIGNED) RETURN BOOLEAN;
FUNCTION "<" (l : UNRESOLVED_UNSIGNED; r : NATURAL) RETURN BOOLEAN;
FUNCTION "<" (l : UNRESOLVED_SIGNED; r : INTEGER) RETURN BOOLEAN;
FUNCTION "<=" (l, r : UNRESOLVED_UNSIGNED) RETURN BOOLEAN;
FUNCTION "<=" (l, r : UNRESOLVED_SIGNED) RETURN BOOLEAN;
FUNCTION "<=" (l : NATURAL; r : UNRESOLVED_UNSIGNED) RETURN BOOLEAN;
FUNCTION "<=" (l : INTEGER; r : UNRESOLVED_SIGNED) RETURN BOOLEAN;
FUNCTION "<=" (l : UNRESOLVED_UNSIGNED; r : NATURAL) RETURN BOOLEAN;
FUNCTION "<=" (l : UNRESOLVED_SIGNED; r : INTEGER) RETURN BOOLEAN;
FUNCTION ">=" (l, r : UNRESOLVED_UNSIGNED) RETURN BOOLEAN;
FUNCTION ">=" (l, r : UNRESOLVED_SIGNED) RETURN BOOLEAN;
FUNCTION ">=" (l : NATURAL; r : UNRESOLVED_UNSIGNED) RETURN BOOLEAN;
FUNCTION ">=" (l : INTEGER; r : UNRESOLVED_SIGNED) RETURN BOOLEAN;
FUNCTION ">=" (l : UNRESOLVED_UNSIGNED; r : NATURAL) RETURN BOOLEAN;
FUNCTION ">=" (l : UNRESOLVED_SIGNED; r : INTEGER) RETURN BOOLEAN;
FUNCTION "=" (l, r : UNRESOLVED_UNSIGNED) RETURN BOOLEAN;
FUNCTION "=" (l, r : UNRESOLVED_SIGNED) RETURN BOOLEAN;
FUNCTION "=" (l : NATURAL; r : UNRESOLVED_UNSIGNED) RETURN BOOLEAN;
FUNCTION "=" (l : INTEGER; r : UNRESOLVED_SIGNED) RETURN BOOLEAN;
FUNCTION "=" (l : UNRESOLVED_UNSIGNED; r : NATURAL) RETURN BOOLEAN;
FUNCTION "=" (l : UNRESOLVED_SIGNED; r : INTEGER) RETURN BOOLEAN;
FUNCTION "/=" (l, r : UNRESOLVED_UNSIGNED) RETURN BOOLEAN;
FUNCTION "/=" (l, r : UNRESOLVED_SIGNED) RETURN BOOLEAN;
FUNCTION "/=" (l : NATURAL; r : UNRESOLVED_UNSIGNED) RETURN BOOLEAN;
FUNCTION "/=" (l : INTEGER; r : UNRESOLVED_SIGNED) RETURN BOOLEAN;
FUNCTION "/=" (l : UNRESOLVED_UNSIGNED; r : NATURAL) RETURN BOOLEAN;
FUNCTION "/=" (l : UNRESOLVED_SIGNED; r : INTEGER) RETURN BOOLEAN;
FUNCTION MINIMUM (l, r : UNRESOLVED_UNSIGNED) RETURN UNRESOLVED_UNSIGNED;
FUNCTION MINIMUM (l, r : UNRESOLVED_SIGNED) RETURN UNRESOLVED_SIGNED;
FUNCTION MINIMUM (l : NATURAL; r : UNRESOLVED_UNSIGNED) RETURN
UNRESOLVED_UNSIGNED;
FUNCTION MINIMUM (l : INTEGER; r : UNRESOLVED_SIGNED) RETURN
UNRESOLVED_SIGNED;
```

```
FUNCTION MINIMUM (l : UNRESOLVED_UNSIGNED; r : NATURAL) RETURN
UNRESOLVED_UNSIGNED;
FUNCTION MINIMUM (l : UNRESOLVED_SIGNED; r : INTEGER) RETURN
UNRESOLVED_SIGNED;
FUNCTION MAXIMUM (l, r : UNRESOLVED_UNSIGNED) RETURN UNRESOLVED_UNSIGNED;
FUNCTION MAXIMUM (l, r : UNRESOLVED_SIGNED) RETURN UNRESOLVED_SIGNED;
FUNCTION MAXIMUM (l : NATURAL; r : UNRESOLVED_UNSIGNED) RETURN
UNRESOLVED_UNSIGNED;
FUNCTION MAXIMUM (l : INTEGER; r : UNRESOLVED_SIGNED) RETURN
UNRESOLVED_SIGNED;
FUNCTION MAXIMUM (l : UNRESOLVED_UNSIGNED; r : NATURAL) RETURN
UNRESOLVED_UNSIGNED;
FUNCTION MAXIMUM (l : UNRESOLVED_SIGNED; r : INTEGER) RETURN
UNRESOLVED_SIGNED;

-----Comparison matching operators:-----------------
FUNCTION "?>" (l, r : UNRESOLVED_UNSIGNED) RETURN STD_ULOGIC;
FUNCTION "?>" (l, r : UNRESOLVED_SIGNED) RETURN STD_ULOGIC;
FUNCTION "?>" (l : NATURAL; r : UNRESOLVED_UNSIGNED) RETURN STD_ULOGIC;
FUNCTION "?>" (l : INTEGER; r : UNRESOLVED_SIGNED) RETURN STD_ULOGIC;
FUNCTION "?>" (l : UNRESOLVED_UNSIGNED; r : NATURAL) RETURN STD_ULOGIC;
FUNCTION "?>" (l : UNRESOLVED_SIGNED; r : INTEGER) RETURN STD_ULOGIC;
FUNCTION "?<" (l, r : UNRESOLVED_UNSIGNED) RETURN STD_ULOGIC;
FUNCTION "?<" (l, r : UNRESOLVED_SIGNED) RETURN STD_ULOGIC;
FUNCTION "?<" (l : NATURAL; r : UNRESOLVED_UNSIGNED) RETURN STD_ULOGIC;
FUNCTION "?<" (l : INTEGER; r : UNRESOLVED_SIGNED) RETURN STD_ULOGIC;
FUNCTION "?<" (l : UNRESOLVED_UNSIGNED; r : NATURAL) RETURN STD_ULOGIC;
FUNCTION "?<" (l : UNRESOLVED_SIGNED; r : INTEGER) RETURN STD_ULOGIC;
FUNCTION "?<=" (l, r : UNRESOLVED_UNSIGNED) RETURN STD_ULOGIC;
FUNCTION "?<=" (l, r : UNRESOLVED_SIGNED) RETURN STD_ULOGIC;
FUNCTION "?<=" (l : NATURAL; r : UNRESOLVED_UNSIGNED) RETURN STD_ULOGIC;
FUNCTION "?<=" (l : INTEGER; r : UNRESOLVED_SIGNED) RETURN STD_ULOGIC;
FUNCTION "?<=" (l : UNRESOLVED_UNSIGNED; r : NATURAL) RETURN STD_ULOGIC;
FUNCTION "?<=" (l : UNRESOLVED_SIGNED; r : INTEGER) RETURN STD_ULOGIC;
FUNCTION "?>=" (l, r : UNRESOLVED_UNSIGNED) RETURN STD_ULOGIC;
FUNCTION "?>=" (l, r : UNRESOLVED_SIGNED) RETURN STD_ULOGIC;
FUNCTION "?>=" (l : NATURAL; r : UNRESOLVED_UNSIGNED) RETURN STD_ULOGIC;
FUNCTION "?>=" (l : INTEGER; r : UNRESOLVED_SIGNED) RETURN STD_ULOGIC;
FUNCTION "?>=" (l : UNRESOLVED_UNSIGNED; r : NATURAL) RETURN STD_ULOGIC;
FUNCTION "?>=" (l : UNRESOLVED_SIGNED; r : INTEGER) RETURN STD_ULOGIC;
FUNCTION "?=" (l, r : UNRESOLVED_UNSIGNED) RETURN STD_ULOGIC;
FUNCTION "?=" (l, r : UNRESOLVED_SIGNED) RETURN STD_ULOGIC;
FUNCTION "?=" (l : NATURAL; r : UNRESOLVED_UNSIGNED) RETURN STD_ULOGIC;
FUNCTION "?=" (l : INTEGER; r : UNRESOLVED_SIGNED) RETURN STD_ULOGIC;
```

```
FUNCTION "?=" (l : UNRESOLVED_UNSIGNED; r : NATURAL) RETURN STD_ULOGIC;
FUNCTION "?=" (l : UNRESOLVED_SIGNED; r : INTEGER) RETURN STD_ULOGIC;
FUNCTION "?/=" (l, r : UNRESOLVED_UNSIGNED) RETURN STD_ULOGIC;
FUNCTION "?/=" (l, r : UNRESOLVED_SIGNED) RETURN STD_ULOGIC;
FUNCTION "?/=" (l : NATURAL; r : UNRESOLVED_UNSIGNED) RETURN STD_ULOGIC;
FUNCTION "?/=" (l : INTEGER; r : UNRESOLVED_SIGNED) RETURN STD_ULOGIC;
FUNCTION "?/=" (l : UNRESOLVED_UNSIGNED; r : NATURAL) RETURN STD_ULOGIC;
FUNCTION "?/=" (l : UNRESOLVED_SIGNED; r : INTEGER) RETURN STD_ULOGIC;

-----Shift operators:------------------------------
FUNCTION "sll" (arg : UNRESOLVED_UNSIGNED; count : INTEGER) RETURN
UNRESOLVED_UNSIGNED;
FUNCTION "sll" (arg : UNRESOLVED_SIGNED; count : INTEGER) RETURN
UNRESOLVED_SIGNED;
FUNCTION "srl" (arg : UNRESOLVED_UNSIGNED; count : INTEGER) RETURN
UNRESOLVED_UNSIGNED;
FUNCTION "srl" (arg : UNRESOLVED_SIGNED; count : INTEGER) RETURN
UNRESOLVED_SIGNED;
FUNCTION "rol" (arg : UNRESOLVED_UNSIGNED; count : INTEGER) RETURN
UNRESOLVED_UNSIGNED;
FUNCTION "rol" (arg : UNRESOLVED_SIGNED; count : INTEGER) RETURN
UNRESOLVED_SIGNED;
FUNCTION "ror" (arg : UNRESOLVED_UNSIGNED; count : INTEGER) RETURN
UNRESOLVED_UNSIGNED;
FUNCTION "ror" (arg : UNRESOLVED_SIGNED; count : INTEGER) RETURN
UNRESOLVED_SIGNED;
FUNCTION "sla" (arg : UNRESOLVED_UNSIGNED; count : INTEGER) RETURN
UNRESOLVED_UNSIGNED;
FUNCTION "sla" (arg : UNRESOLVED_SIGNED; count : INTEGER) RETURN
UNRESOLVED_SIGNED;
FUNCTION "sra" (arg : UNRESOLVED_UNSIGNED; count : INTEGER) RETURN
UNRESOLVED_UNSIGNED;
FUNCTION "sra" (arg : UNRESOLVED_SIGNED; count : INTEGER) RETURN
UNRESOLVED_SIGNED;
FUNCTION SHIFT_LEFT (arg : UNRESOLVED_UNSIGNED; count : NATURAL) RETURN
UNRESOLVED_UNSIGNED;
FUNCTION SHIFT_RIGHT (arg : UNRESOLVED_UNSIGNED; count : NATURAL) RETURN
UNRESOLVED_UNSIGNED;
FUNCTION SHIFT_LEFT (arg : UNRESOLVED_SIGNED; count : NATURAL) RETURN
UNRESOLVED_SIGNED;
FUNCTION SHIFT_RIGHT (arg : UNRESOLVED_SIGNED; count : NATURAL) RETURN
UNRESOLVED_SIGNED;
FUNCTION ROTATE_LEFT (arg : UNRESOLVED_UNSIGNED; count : NATURAL) RETURN
UNRESOLVED_UNSIGNED;
```

```
FUNCTION ROTATE_RIGHT (arg : UNRESOLVED_UNSIGNED; count : NATURAL) RETURN
UNRESOLVED_UNSIGNED;
FUNCTION ROTATE_LEFT (arg : UNRESOLVED_SIGNED; count : NATURAL) RETURN
UNRESOLVED_SIGNED;
FUNCTION ROTATE_RIGHT (arg : UNRESOLVED_SIGNED; count : NATURAL) RETURN
UNRESOLVED_SIGNED;

-----Type conversion:----------------------------
FUNCTION TO_INTEGER (arg : UNRESOLVED_UNSIGNED) RETURN NATURAL;
FUNCTION TO_INTEGER (arg : UNRESOLVED_SIGNED) RETURN INTEGER;
FUNCTION TO_UNSIGNED (arg, size : NATURAL) RETURN UNRESOLVED_UNSIGNED;
FUNCTION TO_SIGNED (arg : INTEGER; size : NATURAL) RETURN UNRESOLVED_SIGNED;
FUNCTION TO_UNSIGNED (arg : NATURAL; size_res : UNRESOLVED_UNSIGNED) RETURN
UNRESOLVED_UNSIGNED;
FUNCTION TO_SIGNED (arg : INTEGER; size_res : UNRESOLVED_SIGNED) RETURN
UNRESOLVED_SIGNED;
FUNCTION TO_01 (s : UNRESOLVED_UNSIGNED; xmap : STD_ULOGIC := '0') RETURN
UNRESOLVED_UNSIGNED;
FUNCTION TO_01 (s : UNRESOLVED_SIGNED; xmap : STD_ULOGIC := '0') RETURN
UNRESOLVED_SIGNED;
FUNCTION TO_X01 (s : UNRESOLVED_UNSIGNED) RETURN UNRESOLVED_UNSIGNED;
FUNCTION TO_X01 (s : UNRESOLVED_SIGNED) RETURN UNRESOLVED_SIGNED;
FUNCTION TO_X01Z (s : UNRESOLVED_UNSIGNED) RETURN UNRESOLVED_UNSIGNED;
FUNCTION TO_X01Z (s : UNRESOLVED_SIGNED) RETURN UNRESOLVED_SIGNED;
FUNCTION TO_UX01 (s : UNRESOLVED_UNSIGNED) RETURN UNRESOLVED_UNSIGNED;
FUNCTION TO_UX01 (s : UNRESOLVED_SIGNED) RETURN UNRESOLVED_SIGNED;
FUNCTION IS_X (s : UNRESOLVED_UNSIGNED) RETURN BOOLEAN;
FUNCTION IS_X (s : UNRESOLVED_SIGNED) RETURN BOOLEAN;

-----Resize functions:----------------------------
FUNCTION RESIZE (arg : UNRESOLVED_SIGNED; new_size : NATURAL) RETURN
UNRESOLVED_SIGNED;
FUNCTION RESIZE (arg : UNRESOLVED_UNSIGNED; new_size : NATURAL) RETURN
UNRESOLVED_UNSIGNED;
FUNCTION RESIZE (arg, size_res : UNRESOLVED_UNSIGNED) RETURN
UNRESOLVED_UNSIGNED;
FUNCTION RESIZE (arg, size_res : UNRESOLVED_SIGNED) RETURN UNRESOLVED_SIGNED;

-----Match functions:----------------------------
FUNCTION STD_MATCH (l, r : STD_ULOGIC) RETURN BOOLEAN;
FUNCTION STD_MATCH (l, r : UNRESOLVED_UNSIGNED) RETURN BOOLEAN;
FUNCTION STD_MATCH (l, r : UNRESOLVED_SIGNED) RETURN BOOLEAN;
FUNCTION STD_MATCH (l, r : STD_ULOGIC_VECTOR) RETURN BOOLEAN;
```

```
-----String conversion:-------------------------
--Predefined: FUNCTION to_string (value : UNRESOLVED_UNSIGNED) RETURN STRING;
--Predefined: FUNCTION to_string (value : UNRESOLVED_SIGNED) RETURN STRING;
FUNCTION TO_OSTRING (value : UNRESOLVED_UNSIGNED) RETURN STRING;
FUNCTION TO_OSTRING (value : UNRESOLVED_SIGNED) RETURN STRING;
FUNCTION TO_HSTRING (value : UNRESOLVED_UNSIGNED) RETURN STRING;
FUNCTION TO_HSTRING (value : UNRESOLVED_SIGNED) RETURN STRING;
ALIAS TO_BSTRING IS TO_STRING [UNRESOLVED_UNSIGNED RETURN STRING];
ALIAS TO_BSTRING IS TO_STRING [UNRESOLVED_SIGNED RETURN STRING];
ALIAS TO_BINARY_STRING IS TO_STRING [UNRESOLVED_UNSIGNED RETURN STRING];
ALIAS TO_BINARY_STRING IS TO_STRING [UNRESOLVED_SIGNED RETURN STRING];
ALIAS TO_OCTAL_STRING IS TO_OSTRING [UNRESOLVED_UNSIGNED RETURN STRING];
ALIAS TO_OCTAL_STRING IS TO_OSTRING [UNRESOLVED_SIGNED RETURN STRING];
ALIAS TO_HEX_STRING IS TO_HSTRING [UNRESOLVED_UNSIGNED RETURN STRING];
ALIAS TO_HEX_STRING IS TO_HSTRING [UNRESOLVED_SIGNED RETURN STRING];

-----Read operations:------------------------
PROCEDURE READ(l : INOUT LINE; value : OUT UNRESOLVED_UNSIGNED; GOOD : OUT
BOOLEAN);
PROCEDURE READ(l : INOUT LINE; value : OUT UNRESOLVED_UNSIGNED);
PROCEDURE READ(l : INOUT LINE; value : OUT UNRESOLVED_SIGNED; GOOD : OUT
BOOLEAN);
PROCEDURE READ(l : INOUT LINE; value : OUT UNRESOLVED_SIGNED);
PROCEDURE OREAD (l : INOUT LINE; value : OUT UNRESOLVED_UNSIGNED; GOOD : OUT
BOOLEAN);
PROCEDURE OREAD (l : INOUT LINE; value : OUT UNRESOLVED_SIGNED; GOOD : OUT
BOOLEAN);
PROCEDURE OREAD (l : INOUT LINE; value : OUT UNRESOLVED_UNSIGNED);
PROCEDURE OREAD (l : INOUT LINE; value : OUT UNRESOLVED_SIGNED);
PROCEDURE HREAD (l : INOUT LINE; value : OUT UNRESOLVED_UNSIGNED; GOOD : OUT
BOOLEAN);
PROCEDURE HREAD (l : INOUT LINE; value : OUT UNRESOLVED_SIGNED; GOOD : OUT
BOOLEAN);
PROCEDURE HREAD (l : INOUT LINE; value : OUT UNRESOLVED_UNSIGNED);
PROCEDURE HREAD (l : INOUT LINE; value : OUT UNRESOLVED_SIGNED);
ALIAS BREAD IS READ [LINE, UNRESOLVED_UNSIGNED, BOOLEAN];
ALIAS BREAD IS READ [LINE, UNRESOLVED_SIGNED, BOOLEAN];
ALIAS BREAD IS READ [LINE, UNRESOLVED_UNSIGNED];
ALIAS BREAD IS READ [LINE, UNRESOLVED_SIGNED];
ALIAS BINARY_READ IS READ [LINE, UNRESOLVED_UNSIGNED, BOOLEAN];
ALIAS BINARY_READ IS READ [LINE, UNRESOLVED_SIGNED, BOOLEAN];
ALIAS BINARY_READ IS READ [LINE, UNRESOLVED_UNSIGNED];
ALIAS BINARY_READ IS READ [LINE, UNRESOLVED_SIGNED];
```

```
ALIAS OCTAL_READ IS OREAD [LINE, UNRESOLVED_UNSIGNED, BOOLEAN];
ALIAS OCTAL_READ IS OREAD [LINE, UNRESOLVED_SIGNED, BOOLEAN];
ALIAS OCTAL_READ IS OREAD [LINE, UNRESOLVED_UNSIGNED];
ALIAS OCTAL_READ IS OREAD [LINE, UNRESOLVED_SIGNED];
ALIAS HEX_READ IS HREAD [LINE, UNRESOLVED_UNSIGNED, BOOLEAN];
ALIAS HEX_READ IS HREAD [LINE, UNRESOLVED_SIGNED, BOOLEAN];
ALIAS HEX_READ IS HREAD [LINE, UNRESOLVED_UNSIGNED];
ALIAS HEX_READ IS HREAD [LINE, UNRESOLVED_SIGNED];

-----Write operations:--------------------------
PROCEDURE WRITE (l : INOUT LINE; value : IN UNRESOLVED_UNSIGNED;
   JUSTIFIED : IN SIDE := RIGHT; field : IN WIDTH := 0);
PROCEDURE WRITE (l : INOUT LINE; value : IN UNRESOLVED_SIGNED;
   JUSTIFIED : IN SIDE := RIGHT; field : IN WIDTH := 0);
PROCEDURE OWRITE (l : INOUT LINE; value : IN UNRESOLVED_UNSIGNED;
   JUSTIFIED : IN SIDE := RIGHT; field : IN WIDTH := 0);
PROCEDURE OWRITE (l : INOUT LINE; value : IN UNRESOLVED_SIGNED;
   JUSTIFIED : IN SIDE := RIGHT; field : IN WIDTH := 0);
PROCEDURE HWRITE (l : INOUT LINE; value : IN UNRESOLVED_UNSIGNED;
   JUSTIFIED : IN SIDE := RIGHT; field : IN WIDTH := 0);
PROCEDURE HWRITE (l : INOUT LINE; value : IN UNRESOLVED_SIGNED;
   JUSTIFIED : IN SIDE := RIGHT; field : IN WIDTH := 0);
ALIAS BWRITE IS WRITE [LINE, UNRESOLVED_UNSIGNED, SIDE, WIDTH];
ALIAS BWRITE IS WRITE [LINE, UNRESOLVED_SIGNED, SIDE, WIDTH];
ALIAS BINARY_WRITE IS WRITE [LINE, UNRESOLVED_UNSIGNED, SIDE, WIDTH];
ALIAS BINARY_WRITE IS WRITE [LINE, UNRESOLVED_SIGNED, SIDE, WIDTH];
ALIAS OCTAL_WRITE IS OWRITE [LINE, UNRESOLVED_UNSIGNED, SIDE, WIDTH];
ALIAS OCTAL_WRITE IS OWRITE [LINE, UNRESOLVED_SIGNED, SIDE, WIDTH];
ALIAS HEX_WRITE IS HWRITE [LINE, UNRESOLVED_UNSIGNED, SIDE, WIDTH];
ALIAS HEX_WRITE IS HWRITE [LINE, UNRESOLVED_SIGNED, SIDE, WIDTH];

END PACKAGE numeric_std;
---------------------------------------------------------------------------
```

K Package *std_logic_arith*

This package, introduced by Synopsys, has the same purposes as *numeric_std*, to which it is *partially* equivalent.

Observe the following particularities in the code below:

1) The main types are UNSIGNED and SIGNED, which have STD_LOGIC as the base type (subtype).

2) There are no logical operators.

3) The arithmetic operators do not include /, **, REM, MOD.

4) The shift operators do not include arithmetic shift and rotation.

5) Several type-conversion functions are included in the package.

```
-------------------------------------------------------------------------
LIBRARY ieee;
USE ieee.std_logic_1164.all;

PACKAGE std_logic_arith IS

    --------Types and subtype:------------------------
    TYPE UNSIGNED IS ARRAY (NATURAL RANGE <>) OF STD_LOGIC;
    TYPE SIGNED IS ARRAY (NATURAL RANGE <>) OF STD_LOGIC;
    SUBTYPE SMALL_INT IS INTEGER RANGE 0 TO 1;

    --------Arithmetic operators:---------------------
    FUNCTION "+"(l: UNSIGNED; r: UNSIGNED) RETURN UNSIGNED;
    FUNCTION "+"(l: SIGNED; r: SIGNED) RETURN SIGNED;
    FUNCTION "+"(l: UNSIGNED; r: SIGNED) RETURN SIGNED;
    FUNCTION "+"(l: SIGNED; r: UNSIGNED) RETURN SIGNED;
    FUNCTION "+"(l: UNSIGNED; r: INTEGER) RETURN UNSIGNED;
    FUNCTION "+"(l: INTEGER; r: UNSIGNED) RETURN UNSIGNED;
    FUNCTION "+"(l: SIGNED; r: INTEGER) RETURN SIGNED;
    FUNCTION "+"(l: INTEGER; r: SIGNED) RETURN SIGNED;
```

```
FUNCTION "+"(l: UNSIGNED; r: STD_ULOGIC) RETURN UNSIGNED;
FUNCTION "+"(l: STD_ULOGIC; r: UNSIGNED) RETURN UNSIGNED;
FUNCTION "+"(l: SIGNED; r: STD_ULOGIC) RETURN SIGNED;
FUNCTION "+"(l: STD_ULOGIC; r: SIGNED) RETURN SIGNED;
FUNCTION "+"(l: UNSIGNED; r: UNSIGNED) RETURN STD_LOGIC_VECTOR;
FUNCTION "+"(l: SIGNED; r: SIGNED) RETURN STD_LOGIC_VECTOR;
FUNCTION "+"(l: UNSIGNED; r: SIGNED) RETURN STD_LOGIC_VECTOR;
FUNCTION "+"(l: SIGNED; r: UNSIGNED) RETURN STD_LOGIC_VECTOR;
FUNCTION "+"(l: UNSIGNED; r: INTEGER) RETURN STD_LOGIC_VECTOR;
FUNCTION "+"(l: INTEGER; r: UNSIGNED) RETURN STD_LOGIC_VECTOR;
FUNCTION "+"(l: SIGNED; r: INTEGER) RETURN STD_LOGIC_VECTOR;
FUNCTION "+"(l: INTEGER; r: SIGNED) RETURN STD_LOGIC_VECTOR;
FUNCTION "+"(l: UNSIGNED; r: STD_ULOGIC) RETURN STD_LOGIC_VECTOR;
FUNCTION "+"(l: STD_ULOGIC; r: UNSIGNED) RETURN STD_LOGIC_VECTOR;
FUNCTION "+"(l: SIGNED; r: STD_ULOGIC) RETURN STD_LOGIC_VECTOR;
FUNCTION "+"(l: STD_ULOGIC; r: SIGNED) RETURN STD_LOGIC_VECTOR;
FUNCTION "-"(l: UNSIGNED; r: UNSIGNED) RETURN UNSIGNED;
FUNCTION "+"(l: UNSIGNED) RETURN UNSIGNED;
FUNCTION "+"(l: SIGNED) RETURN SIGNED;
FUNCTION "+"(l: UNSIGNED) RETURN STD_LOGIC_VECTOR;
FUNCTION "+"(l: SIGNED) RETURN STD_LOGIC_VECTOR;
FUNCTION "-"(l: SIGNED; r: SIGNED) RETURN SIGNED;
FUNCTION "-"(l: UNSIGNED; r: SIGNED) RETURN SIGNED;
FUNCTION "-"(l: SIGNED; r: UNSIGNED) RETURN SIGNED;
FUNCTION "-"(l: UNSIGNED; r: INTEGER) RETURN UNSIGNED;
FUNCTION "-"(l: INTEGER; r: UNSIGNED) RETURN UNSIGNED;
FUNCTION "-"(l: SIGNED; r: INTEGER) RETURN SIGNED;
FUNCTION "-"(l: INTEGER; r: SIGNED) RETURN SIGNED;
FUNCTION "-"(l: UNSIGNED; r: STD_ULOGIC) RETURN UNSIGNED;
FUNCTION "-"(l: STD_ULOGIC; r: UNSIGNED) RETURN UNSIGNED;
FUNCTION "-"(l: SIGNED; r: STD_ULOGIC) RETURN SIGNED;
FUNCTION "-"(l: STD_ULOGIC; r: SIGNED) RETURN SIGNED;
FUNCTION "-"(l: UNSIGNED; r: UNSIGNED) RETURN STD_LOGIC_VECTOR;
FUNCTION "-"(l: SIGNED; r: SIGNED) RETURN STD_LOGIC_VECTOR;
FUNCTION "-"(l: UNSIGNED; r: SIGNED) RETURN STD_LOGIC_VECTOR;
FUNCTION "-"(l: SIGNED; r: UNSIGNED) RETURN STD_LOGIC_VECTOR;
FUNCTION "-"(l: UNSIGNED; r: INTEGER) RETURN STD_LOGIC_VECTOR;
FUNCTION "-"(l: INTEGER; r: UNSIGNED) RETURN STD_LOGIC_VECTOR;
FUNCTION "-"(l: SIGNED; r: INTEGER) RETURN STD_LOGIC_VECTOR;
FUNCTION "-"(l: INTEGER; r: SIGNED) RETURN STD_LOGIC_VECTOR;
FUNCTION "-"(l: UNSIGNED; r: STD_ULOGIC) RETURN STD_LOGIC_VECTOR;
FUNCTION "-"(l: STD_ULOGIC; r: UNSIGNED) RETURN STD_LOGIC_VECTOR;
FUNCTION "-"(l: SIGNED; r: STD_ULOGIC) RETURN STD_LOGIC_VECTOR;
```

```
FUNCTION "-"(l: STD_ULOGIC; r: SIGNED) RETURN STD_LOGIC_VECTOR;
FUNCTION "-"(l: SIGNED) RETURN STD_LOGIC_VECTOR;
FUNCTION "*"(l: UNSIGNED; r: UNSIGNED) RETURN UNSIGNED;
FUNCTION "*"(l: SIGNED; r: SIGNED) RETURN SIGNED;
FUNCTION "*"(l: SIGNED; r: UNSIGNED) RETURN SIGNED;
FUNCTION "*"(l: UNSIGNED; r: SIGNED) RETURN SIGNED;
FUNCTION "*"(l: UNSIGNED; r: UNSIGNED) RETURN STD_LOGIC_VECTOR;
FUNCTION "*"(l: SIGNED; r: SIGNED) RETURN STD_LOGIC_VECTOR;
FUNCTION "*"(l: SIGNED; r: UNSIGNED) RETURN STD_LOGIC_VECTOR;
FUNCTION "*"(l: UNSIGNED; r: SIGNED) RETURN STD_LOGIC_VECTOR;
FUNCTION "ABS"(l: SIGNED) RETURN SIGNED;
FUNCTION "ABS"(l: SIGNED) RETURN STD_LOGIC_VECTOR;

--------Comparison operators:---------------------
FUNCTION "<"(l: UNSIGNED; r: UNSIGNED) RETURN BOOLEAN;
FUNCTION "<"(l: SIGNED; r: SIGNED) RETURN BOOLEAN;
FUNCTION "<"(l: UNSIGNED; r: SIGNED) RETURN BOOLEAN;
FUNCTION "<"(l: SIGNED; r: UNSIGNED) RETURN BOOLEAN;
FUNCTION "<"(l: UNSIGNED; r: INTEGER) RETURN BOOLEAN;
FUNCTION "<"(l: INTEGER; r: UNSIGNED) RETURN BOOLEAN;
FUNCTION "<"(l: SIGNED; r: INTEGER) RETURN BOOLEAN;
FUNCTION "<"(l: INTEGER; r: SIGNED) RETURN BOOLEAN;
FUNCTION "<="(l: UNSIGNED; r: UNSIGNED) RETURN BOOLEAN;
FUNCTION "<="(l: SIGNED; r: SIGNED) RETURN BOOLEAN;
FUNCTION "<="(l: UNSIGNED; r: SIGNED) RETURN BOOLEAN;
FUNCTION "<="(l: SIGNED; r: UNSIGNED) RETURN BOOLEAN;
FUNCTION "<="(l: UNSIGNED; r: INTEGER) RETURN BOOLEAN;
FUNCTION "<="(l: INTEGER; r: UNSIGNED) RETURN BOOLEAN;
FUNCTION "<="(l: SIGNED; r: INTEGER) RETURN BOOLEAN;
FUNCTION "<="(l: INTEGER; r: SIGNED) RETURN BOOLEAN;
FUNCTION ">"(l: UNSIGNED; r: UNSIGNED) RETURN BOOLEAN;
FUNCTION ">"(l: SIGNED; r: SIGNED) RETURN BOOLEAN;
FUNCTION ">"(l: UNSIGNED; r: SIGNED) RETURN BOOLEAN;
FUNCTION ">"(l: SIGNED; r: UNSIGNED) RETURN BOOLEAN;
FUNCTION ">"(l: UNSIGNED; r: INTEGER) RETURN BOOLEAN;
FUNCTION ">"(l: INTEGER; r: UNSIGNED) RETURN BOOLEAN;
FUNCTION ">"(l: SIGNED; r: INTEGER) RETURN BOOLEAN;
FUNCTION ">"(l: INTEGER; r: SIGNED) RETURN BOOLEAN;
FUNCTION ">="(l: UNSIGNED; r: UNSIGNED) RETURN BOOLEAN;
FUNCTION ">="(l: SIGNED; r: SIGNED) RETURN BOOLEAN;
FUNCTION ">="(l: UNSIGNED; r: SIGNED) RETURN BOOLEAN;
FUNCTION ">="(l: SIGNED; r: UNSIGNED) RETURN BOOLEAN;
FUNCTION ">="(l: UNSIGNED; r: INTEGER) RETURN BOOLEAN;
FUNCTION ">="(l: INTEGER; r: UNSIGNED) RETURN BOOLEAN;
```

```
FUNCTION ">="(l: SIGNED; r: INTEGER) RETURN BOOLEAN;
FUNCTION ">="(l: INTEGER; r: SIGNED) RETURN BOOLEAN;
FUNCTION "="(l: UNSIGNED; r: UNSIGNED) RETURN BOOLEAN;
FUNCTION "="(l: SIGNED; r: SIGNED) RETURN BOOLEAN;
FUNCTION "="(l: UNSIGNED; r: SIGNED) RETURN BOOLEAN;
FUNCTION "="(l: SIGNED; r: UNSIGNED) RETURN BOOLEAN;
FUNCTION "="(l: UNSIGNED; r: INTEGER) RETURN BOOLEAN;
FUNCTION "="(l: INTEGER; r: UNSIGNED) RETURN BOOLEAN;
FUNCTION "="(l: SIGNED; r: INTEGER) RETURN BOOLEAN;
FUNCTION "="(l: INTEGER; r: SIGNED) RETURN BOOLEAN;
FUNCTION "/="(l: UNSIGNED; r: UNSIGNED) RETURN BOOLEAN;
FUNCTION "/="(l: SIGNED; r: SIGNED) RETURN BOOLEAN;
FUNCTION "/="(l: UNSIGNED; r: SIGNED) RETURN BOOLEAN;
FUNCTION "/="(l: SIGNED; r: UNSIGNED) RETURN BOOLEAN;
FUNCTION "/="(l: UNSIGNED; r: INTEGER) RETURN BOOLEAN;
FUNCTION "/="(l: INTEGER; r: UNSIGNED) RETURN BOOLEAN;
FUNCTION "/="(l: SIGNED; r: INTEGER) RETURN BOOLEAN;
FUNCTION "/="(l: INTEGER; r: SIGNED) RETURN BOOLEAN;

--------Shift operators:-------------------------
FUNCTION SHL(arg: UNSIGNED; count: UNSIGNED) RETURN UNSIGNED;
FUNCTION SHL(arg: SIGNED; count: UNSIGNED) RETURN SIGNED;
FUNCTION SHR(arg: UNSIGNED; count: UNSIGNED) RETURN UNSIGNED;
FUNCTION SHR(arg: SIGNED; count: UNSIGNED) RETURN SIGNED;

--------Type conversion:-------------------------
FUNCTION CONV_INTEGER(arg: INTEGER) RETURN INTEGER;
FUNCTION CONV_INTEGER(arg: UNSIGNED) RETURN INTEGER;
FUNCTION CONV_INTEGER(arg: SIGNED) RETURN INTEGER;
FUNCTION CONV_INTEGER(arg: STD_ULOGIC) RETURN SMALL_INT;
FUNCTION CONV_UNSIGNED(arg: INTEGER; size: INTEGER) RETURN UNSIGNED;
FUNCTION CONV_UNSIGNED(arg: UNSIGNED; size: INTEGER) RETURN UNSIGNED;
FUNCTION CONV_UNSIGNED(arg: SIGNED; size: INTEGER) RETURN UNSIGNED;
FUNCTION CONV_UNSIGNED(arg: STD_ULOGIC; size: INTEGER) RETURN UNSIGNED;
FUNCTION CONV_SIGNED(arg: INTEGER; size: INTEGER) RETURN SIGNED;
FUNCTION CONV_SIGNED(arg: UNSIGNED; size: INTEGER) RETURN SIGNED;
FUNCTION CONV_SIGNED(arg: SIGNED; size: INTEGER) RETURN SIGNED;
FUNCTION CONV_SIGNED(arg: STD_ULOGIC; size: INTEGER) RETURN SIGNED;
FUNCTION CONV_STD_LOGIC_VECTOR(arg: INTEGER; size: INTEGER) RETURN
STD_LOGIC_VECTOR;
FUNCTION CONV_STD_LOGIC_VECTOR(arg: UNSIGNED; size: INTEGER) RETURN
STD_LOGIC_VECTOR;
FUNCTION CONV_STD_LOGIC_VECTOR(arg: SIGNED; size: INTEGER) RETURN
STD_LOGIC_VECTOR;
```

```
    FUNCTION CONV_STD_LOGIC_VECTOR(arg: STD_ULOGIC; size: INTEGER) RETURN
    STD_LOGIC_VECTOR;

    --------Resize functions:-------------------------
    FUNCTION EXT(arg: STD_LOGIC_VECTOR; size: INTEGER) RETURN STD_LOGIC_VECTOR;
    FUNCTION SXT(arg: STD_LOGIC_VECTOR; size: INTEGER) RETURN STD_LOGIC_VECTOR;

END std_logic_arith;
--------------------------------------------------------------------------------
```

L Package *std_logic_signed*

This package was introduced by Synopsys with the purpose of defining *signed operators* (arithmetic, comparison, and some shift) for the type STD_LOGIC_VECTOR. Note that it does not specify any new data types. A similar package, called *std_logic_unsigned*, also exists, which defines *unsigned* operators (arithmetic, comparison, and some shift) for the same data type (STD_LOGIC_VECTOR).

```
----------------------------------------------------------------------
LIBRARY ieee;
USE ieee.std_logic_1164.all;
USE ieee.std_logic_arith.all;

PACKAGE std_logic_signed IS

    --------Arithmetic operators:--------------------
    FUNCTION "+"(l: STD_LOGIC_VECTOR; r: STD_LOGIC_VECTOR) RETURN
    STD_LOGIC_VECTOR;
    FUNCTION "+"(l: STD_LOGIC_VECTOR; r: INTEGER) RETURN STD_LOGIC_VECTOR;
    FUNCTION "+"(l: INTEGER; r: STD_LOGIC_VECTOR) RETURN STD_LOGIC_VECTOR;
    FUNCTION "+"(l: STD_LOGIC_VECTOR; r: STD_LOGIC) RETURN STD_LOGIC_VECTOR;
    FUNCTION "+"(l: STD_LOGIC; r: STD_LOGIC_VECTOR) RETURN STD_LOGIC_VECTOR;
    FUNCTION "+"(l: STD_LOGIC_VECTOR) RETURN STD_LOGIC_VECTOR;
    FUNCTION "-"(l: STD_LOGIC_VECTOR; r: STD_LOGIC_VECTOR) RETURN
    STD_LOGIC_VECTOR;
    FUNCTION "-"(l: STD_LOGIC_VECTOR; r: INTEGER) RETURN STD_LOGIC_VECTOR;
    FUNCTION "-"(l: INTEGER; r: STD_LOGIC_VECTOR) RETURN STD_LOGIC_VECTOR;
    FUNCTION "-"(l: STD_LOGIC_VECTOR; r: STD_LOGIC) RETURN STD_LOGIC_VECTOR;
    FUNCTION "-"(l: STD_LOGIC; r: STD_LOGIC_VECTOR) RETURN STD_LOGIC_VECTOR;
    FUNCTION "-"(l: STD_LOGIC_VECTOR) RETURN STD_LOGIC_VECTOR;
    FUNCTION "*"(l: STD_LOGIC_VECTOR; r: STD_LOGIC_VECTOR) RETURN
    STD_LOGIC_VECTOR;
    FUNCTION "abs"(l: STD_LOGIC_VECTOR) RETURN STD_LOGIC_VECTOR;
```

```
     -------Comparison operators: ---------------------
     FUNCTION "<"(l: STD_LOGIC_VECTOR; r: STD_LOGIC_VECTOR) RETURN BOOLEAN;
     FUNCTION "<"(l: STD_LOGIC_VECTOR; r: INTEGER) RETURN BOOLEAN;
     FUNCTION "<"(l: INTEGER; r: STD_LOGIC_VECTOR) RETURN BOOLEAN;
     FUNCTION "<="(l: STD_LOGIC_VECTOR; r: STD_LOGIC_VECTOR) RETURN BOOLEAN;
     FUNCTION "<="(l: STD_LOGIC_VECTOR; r: INTEGER) RETURN BOOLEAN;
     FUNCTION "<="(l: INTEGER; r: STD_LOGIC_VECTOR) RETURN BOOLEAN;
     FUNCTION ">"(l: STD_LOGIC_VECTOR; r: STD_LOGIC_VECTOR) RETURN BOOLEAN;
     FUNCTION ">"(l: STD_LOGIC_VECTOR; r: INTEGER) RETURN BOOLEAN;
     FUNCTION ">"(l: INTEGER; r: STD_LOGIC_VECTOR) RETURN BOOLEAN;
     FUNCTION ">="(l: STD_LOGIC_VECTOR; r: STD_LOGIC_VECTOR) RETURN BOOLEAN;
     FUNCTION ">="(l: STD_LOGIC_VECTOR; r: INTEGER) RETURN BOOLEAN;
     FUNCTION ">="(l: INTEGER; r: STD_LOGIC_VECTOR) RETURN BOOLEAN;
     FUNCTION "="(l: STD_LOGIC_VECTOR; r: STD_LOGIC_VECTOR) RETURN BOOLEAN;
     FUNCTION "="(l: STD_LOGIC_VECTOR; r: INTEGER) RETURN BOOLEAN;
     FUNCTION "="(l: INTEGER; r: STD_LOGIC_VECTOR) RETURN BOOLEAN;
     FUNCTION "/="(l: STD_LOGIC_VECTOR; r: STD_LOGIC_VECTOR) RETURN BOOLEAN;
     FUNCTION "/="(l: STD_LOGIC_VECTOR; r: INTEGER) RETURN BOOLEAN;
     FUNCTION "/="(l: INTEGER; r: STD_LOGIC_VECTOR) RETURN BOOLEAN;

     --------Shift operators:--------------------------
     FUNCTION SHL(arg:STD_LOGIC_VECTOR; count:STD_LOGIC_VECTOR) RETURN
     STD_LOGIC_VECTOR;
     FUNCTION SHR(arg:STD_LOGIC_VECTOR; count:STD_LOGIC_VECTOR) RETURN
     STD_LOGIC_VECTOR;

     --------Type conversion:--------------------------
     FUNCTION CONV_INTEGER(arg: STD_LOGIC_VECTOR) RETURN INTEGER;
     --remove this since it is already in STD_LOGIC_arith:
     --FUNCTION CONV_STD_LOGIC_VECTOR(arg:INTEGER; size:INTEGER) RETURN
     STD_LOGIC_VECTOR;

END std_logic_signed;
---------------------------------------------------------------------------
```

M Package *textio* (2002 and 2008)

The package *textio* is specified in section 16.4 of the *IEEE 1076-2008 Standard VHDL Language Reference Manual*. The expansion introduced in VHDL 2008 includes *flush*, *minimum*, *maximum*, *to_string*, *justify*, *tee*, and additional *read* and *write* operations, plus a number of aliases.

```
------------------------------------------------------------------------------
PACKAGE textio IS

  ------Type definitions for text I/O:--------------
  TYPE LINE IS ACCESS STRING;
  --"=", "/=", deallocate

  TYPE TEXT IS FILE OF STRING;
  --file_open, file_close, read, write, endfile, flush

  TYPE SIDE IS (RIGHT, LEFT);
  --"=", "/=", "<", "<=", ">", ">=", minimum, maximum, to_string

  SUBTYPE WIDTH IS NATURAL;

  FUNCTION JUSTIFY (value:STRING; JUSTIFIED:SIDE:=RIGHT; field:WIDTH:=0) RETURN
  STRING;

  ------Standard text files:----------------------
  FILE INPUT : TEXT OPEN READ_MODE IS "STD_INPUT";
  FILE OUTPUT : TEXT OPEN WRITE_MODE IS "STD_OUTPUT";

  ------Input routines for standard types:-----------
  PROCEDURE READLINE(FILE f: TEXT; l: OUT LINE);
  PROCEDURE READ(l: INOUT LINE; value: OUT BIT; good: OUT BOOLEAN);
  PROCEDURE READ(l: INOUT LINE; value: OUT BIT);
  PROCEDURE READ(l: INOUT LINE; value: OUT BIT_VECTOR; good: OUT BOOLEAN);
  PROCEDURE READ(l: INOUT LINE; value: OUT BIT_VECTOR);
  PROCEDURE READ(l: INOUT LINE; value: OUT BOOLEAN; good: OUT BOOLEAN);
```

```
PROCEDURE READ(l: INOUT LINE; value: OUT BOOLEAN);
PROCEDURE READ(l: INOUT LINE; value: OUT CHARACTER; good: OUT BOOLEAN);
PROCEDURE READ(l: INOUT LINE; value: OUT CHARACTER);
PROCEDURE READ(l: INOUT LINE; value: OUT INTEGER; good: OUT BOOLEAN);
PROCEDURE READ(l: INOUT LINE; value: OUT INTEGER);
PROCEDURE READ(l: INOUT LINE; value: OUT REAL; good: OUT BOOLEAN);
PROCEDURE READ(l: INOUT LINE; value: OUT REAL);
PROCEDURE READ(l: INOUT LINE; value: OUT STRING; good: OUT BOOLEAN);
PROCEDURE READ(l: INOUT LINE; value: OUT STRING);
PROCEDURE READ(l: INOUT LINE; value: OUT TIME; good: OUT BOOLEAN);
PROCEDURE READ(l: INOUT LINE; value: OUT TIME);
PROCEDURE SREAD(l: INOUT LINE; value: OUT STRING; STRLEN: OUT NATURAL);
PROCEDURE OREAD (l: INOUT LINE; value: OUT BIT_VECTOR; good: OUT BOOLEAN);
PROCEDURE OREAD (l: INOUT LINE; value: OUT BIT_VECTOR);
PROCEDURE HREAD (l: INOUT LINE; value: OUT BIT_VECTOR; good: OUT BOOLEAN);
PROCEDURE HREAD (l: INOUT LINE; value: OUT BIT_VECTOR);
ALIAS STRING_READ IS SREAD [LINE, STRING, NATURAL];
ALIAS BREAD IS READ [LINE, BIT_VECTOR, BOOLEAN];
ALIAS BREAD IS READ [LINE, BIT_VECTOR];
ALIAS BINARY_READ IS READ [LINE, BIT_VECTOR, BOOLEAN];
ALIAS BINARY_READ IS READ [LINE, BIT_VECTOR];
ALIAS OCTAL_READ IS OREAD [LINE, BIT_VECTOR, BOOLEAN];
ALIAS OCTAL_READ IS OREAD [LINE, BIT_VECTOR];
ALIAS HEX_READ IS HREAD [LINE, BIT_VECTOR, BOOLEAN];
ALIAS HEX_READ IS HREAD [LINE, BIT_VECTOR];

------Output routines for standard types:----------
PROCEDURE WRITELINE(FILE F : TEXT; L : INOUT LINE);
PROCEDURE TEE (FILE f: TEXT; l: INOUT LINE);
PROCEDURE WRITE(l: INOUT LINE; value: IN BIT;
  JUSTIFIED: IN SIDE := RIGHT; field: IN WIDTH := 0);
PROCEDURE WRITE(l: INOUT LINE; value : IN BIT_VECTOR;
  JUSTIFIED: IN SIDE := RIGHT; field: IN WIDTH := 0);
PROCEDURE WRITE(l: INOUT LINE; value: IN BOOLEAN;
  JUSTIFIED: IN SIDE := RIGHT; field: IN WIDTH := 0);
PROCEDURE WRITE(l: INOUT LINE; value: IN CHARACTER;
  JUSTIFIED: IN SIDE := RIGHT; field: IN WIDTH := 0);
PROCEDURE WRITE(l: INOUT LINE; value: IN INTEGER;
  JUSTIFIED: IN SIDE := RIGHT; field: IN WIDTH := 0);
PROCEDURE WRITE(l: INOUT LINE; value: IN REAL;
  JUSTIFIED: IN SIDE := RIGHT; field: IN WIDTH := 0; DIGITS: IN NATURAL := 0);
PROCEDURE WRITE(l: INOUT LINE; value: IN REAL; FORMAT: IN STRING;
PROCEDURE WRITE(l: INOUT LINE; value: IN STRING;
  JUSTIFIED: IN SIDE := RIGHT; field: IN WIDTH := 0);
```

```
PROCEDURE WRITE(l: INOUT LINE; value: IN TIME;
  JUSTIFIED: IN SIDE := RIGHT; field: IN WIDTH := 0; UNIT: IN TIME := ns);
PROCEDURE OWRITE (l: INOUT LINE; value: IN BIT_VECTOR;
  JUSTIFIED: IN SIDE := RIGHT; field: IN WIDTH := 0);
PROCEDURE HWRITE (l: INOUT LINE; value: IN BIT_VECTOR;
  JUSTIFIED:IN SIDE := RIGHT; field: IN WIDTH := 0);
ALIAS SWRITE IS WRITE [LINE, STRING, SIDE, WIDTH];
ALIAS STRING_WRITE IS WRITE [LINE, STRING, SIDE, WIDTH];
ALIAS BWRITE IS WRITE [LINE, BIT_VECTOR, SIDE, WIDTH];
ALIAS BINARY_WRITE IS WRITE [LINE, BIT_VECTOR, SIDE, WIDTH];
ALIAS OCTAL_WRITE IS OWRITE [LINE, BIT_VECTOR, SIDE, WIDTH];
ALIAS HEX_WRITE IS HWRITE [LINE, BIT_VECTOR, SIDE, WIDTH];

END textio;
-------------------------------------------------------------------------------
```

N Package *numeric_std_unsigned* (2008)

This package was introduced in VHDL 2008. It is expected to replace *std_logic_unsigned* in the future. A similar package, called *numeric_bit_unsigned*, also exists, which defines almost the same operators for the data types BIT and BIT_VECTOR instead of STD_LOGIC and STD_LOGIC_VECTOR. It too was introduced in VHDL 2008.

```
-----------------------------------------------------------------------------
LIBRARY ieee;
USE ieee.std_logic_1164.all;

PACKAGE numeric_std_unsigned is

  CONSTANT CopyRightNotice : STRING := "Copyright 2008 IEEE. All rights
  reserved.";

  -----Arithmetic operators:-----------------------
  FUNCTION "+" (l, r: STD_ULOGIC_VECTOR) RETURN STD_ULOGIC_VECTOR;
  FUNCTION "+" (l: STD_ULOGIC_VECTOR; r: STD_ULOGIC) RETURN STD_ULOGIC_VECTOR;
  FUNCTION "+" (l: STD_ULOGIC; r: STD_ULOGIC_VECTOR) RETURN STD_ULOGIC_VECTOR;
  FUNCTION "+" (l: STD_ULOGIC_VECTOR; r: NATURAL) RETURN STD_ULOGIC_VECTOR;
  FUNCTION "+" (l: NATURAL; r: STD_ULOGIC_VECTOR) RETURN STD_ULOGIC_VECTOR;
  FUNCTION "-" (l, r: STD_ULOGIC_VECTOR) RETURN STD_ULOGIC_VECTOR;
  FUNCTION "-" (l: STD_ULOGIC_VECTOR; r: STD_ULOGIC) RETURN STD_ULOGIC_VECTOR;
  FUNCTION "-" (l: STD_ULOGIC; r: STD_ULOGIC_VECTOR) RETURN STD_ULOGIC_VECTOR;
  FUNCTION "-" (l: STD_ULOGIC_VECTOR; r: NATURAL) RETURN STD_ULOGIC_VECTOR;
  FUNCTION "-" (l: NATURAL; r: STD_ULOGIC_VECTOR) RETURN STD_ULOGIC_VECTOR;
  FUNCTION "*" (l, r: STD_ULOGIC_VECTOR) RETURN STD_ULOGIC_VECTOR;
  FUNCTION "*" (l: STD_ULOGIC_VECTOR; r: NATURAL) RETURN STD_ULOGIC_VECTOR;
  FUNCTION "*" (l: NATURAL; r: STD_ULOGIC_VECTOR) RETURN STD_ULOGIC_VECTOR;
  FUNCTION "/" (l, r: STD_ULOGIC_VECTOR) RETURN STD_ULOGIC_VECTOR;
  FUNCTION "/" (l: STD_ULOGIC_VECTOR; r: NATURAL) RETURN STD_ULOGIC_VECTOR;
  FUNCTION "/" (l: NATURAL; r: STD_ULOGIC_VECTOR) RETURN STD_ULOGIC_VECTOR;
  FUNCTION "rem" (l, r: STD_ULOGIC_VECTOR) RETURN STD_ULOGIC_VECTOR;
```

```
FUNCTION "rem" (l: STD_ULOGIC_VECTOR; r: NATURAL) RETURN STD_ULOGIC_VECTOR;
FUNCTION "rem" (l: NATURAL; r: STD_ULOGIC_VECTOR) RETURN STD_ULOGIC_VECTOR;
FUNCTION "mod" (l, r: STD_ULOGIC_VECTOR) RETURN STD_ULOGIC_VECTOR;
FUNCTION "mod" (l: STD_ULOGIC_VECTOR; r: NATURAL) RETURN STD_ULOGIC_VECTOR;
FUNCTION "mod" (l: NATURAL; r: STD_ULOGIC_VECTOR) RETURN STD_ULOGIC_VECTOR;
FUNCTION find_leftmost (arg: STD_ULOGIC_VECTOR; y: STD_ULOGIC) RETURN
INTEGER;
FUNCTION find_rightmost (arg: STD_ULOGIC_VECTOR; y: STD_ULOGIC) RETURN
INTEGER;

-----Comparison operators:----------------------
FUNCTION ">" (l, r: STD_ULOGIC_VECTOR) RETURN BOOLEAN;
FUNCTION ">" (l: NATURAL; r: STD_ULOGIC_VECTOR) RETURN BOOLEAN;
FUNCTION ">" (l: STD_ULOGIC_VECTOR; r: NATURAL) RETURN BOOLEAN;
FUNCTION "<" (l, r: STD_ULOGIC_VECTOR) RETURN BOOLEAN;
FUNCTION "<" (l: NATURAL; r: STD_ULOGIC_VECTOR) RETURN BOOLEAN;
FUNCTION "<" (l: STD_ULOGIC_VECTOR; r: NATURAL) RETURN BOOLEAN;
FUNCTION "<=" (l, r: STD_ULOGIC_VECTOR) RETURN BOOLEAN;
FUNCTION "<=" (l: NATURAL; r: STD_ULOGIC_VECTOR) RETURN BOOLEAN;
FUNCTION "<=" (l: STD_ULOGIC_VECTOR; r: NATURAL) RETURN BOOLEAN;
FUNCTION ">=" (l, r: STD_ULOGIC_VECTOR) RETURN BOOLEAN;
FUNCTION ">=" (l: NATURAL; r: STD_ULOGIC_VECTOR) RETURN BOOLEAN;
FUNCTION ">=" (l: STD_ULOGIC_VECTOR; r: NATURAL) RETURN BOOLEAN;
FUNCTION "=" (l, r: STD_ULOGIC_VECTOR) RETURN BOOLEAN;
FUNCTION "=" (l: NATURAL; r: STD_ULOGIC_VECTOR) RETURN BOOLEAN;
FUNCTION "=" (l: STD_ULOGIC_VECTOR; r: NATURAL) RETURN BOOLEAN;
FUNCTION "/=" (l, r: STD_ULOGIC_VECTOR) RETURN BOOLEAN;
FUNCTION "/=" (l: NATURAL; r: STD_ULOGIC_VECTOR) RETURN BOOLEAN;
FUNCTION "/=" (l: STD_ULOGIC_VECTOR; r: NATURAL) RETURN BOOLEAN;
FUNCTION MINIMUM (l, r: STD_ULOGIC_VECTOR) RETURN STD_ULOGIC_VECTOR;
FUNCTION MINIMUM (l: NATURAL; r: STD_ULOGIC_VECTOR) RETURN STD_ULOGIC_VECTOR;
FUNCTION MINIMUM (l: STD_ULOGIC_VECTOR; r: NATURAL) RETURN STD_ULOGIC_VECTOR;
FUNCTION MAXIMUM (l, r: STD_ULOGIC_VECTOR) RETURN STD_ULOGIC_VECTOR;
FUNCTION MAXIMUM (l: NATURAL; r: STD_ULOGIC_VECTOR) RETURN STD_ULOGIC_VECTOR;
FUNCTION MAXIMUM (l: STD_ULOGIC_VECTOR; r: NATURAL) RETURN STD_ULOGIC_VECTOR;

-----Matching comparison operators:--------------
FUNCTION "?>" (l, r: STD_ULOGIC_VECTOR) RETURN STD_ULOGIC;
FUNCTION "?>" (l: NATURAL; r: STD_ULOGIC_VECTOR) RETURN STD_ULOGIC;
FUNCTION "?>" (l: STD_ULOGIC_VECTOR; r: NATURAL) RETURN STD_ULOGIC;
FUNCTION "?<" (l, r: STD_ULOGIC_VECTOR) RETURN STD_ULOGIC;
FUNCTION "?<" (l: NATURAL; r: STD_ULOGIC_VECTOR) RETURN STD_ULOGIC;
FUNCTION "?<" (l: STD_ULOGIC_VECTOR; r: NATURAL) RETURN STD_ULOGIC;
FUNCTION "?<=" (l, r: STD_ULOGIC_VECTOR) RETURN STD_ULOGIC;
```

```
FUNCTION "?<=" (l: NATURAL; r: STD_ULOGIC_VECTOR) RETURN STD_ULOGIC;
FUNCTION "?<=" (l: STD_ULOGIC_VECTOR; r: NATURAL) RETURN STD_ULOGIC;
FUNCTION "?>=" (l, r: STD_ULOGIC_VECTOR) RETURN STD_ULOGIC;
FUNCTION "?>=" (l: NATURAL; r: STD_ULOGIC_VECTOR) RETURN STD_ULOGIC;
FUNCTION "?>=" (l: STD_ULOGIC_VECTOR; r: NATURAL) RETURN STD_ULOGIC;
FUNCTION "?=" (l, r: STD_ULOGIC_VECTOR) RETURN STD_ULOGIC;
FUNCTION "?=" (l: NATURAL; r: STD_ULOGIC_VECTOR) RETURN STD_ULOGIC;
FUNCTION "?=" (l: STD_ULOGIC_VECTOR; r: NATURAL) RETURN STD_ULOGIC;
FUNCTION "?/=" (l, r: STD_ULOGIC_VECTOR) RETURN STD_ULOGIC;
FUNCTION "?/=" (l : NATURAL; r: STD_ULOGIC_VECTOR) RETURN STD_ULOGIC;
FUNCTION "?/=" (l : STD_ULOGIC_VECTOR; r: NATURAL) RETURN STD_ULOGIC;

-----Shift operators:----------------------------
FUNCTION "sla" (arg: STD_ULOGIC_VECTOR; count: INTEGER) RETURN
STD_ULOGIC_VECTOR;
FUNCTION "sra" (arg: STD_ULOGIC_VECTOR; count: INTEGER) RETURN
STD_ULOGIC_VECTOR;
FUNCTION SHIFT_LEFT (arg: STD_ULOGIC_VECTOR; count: NATURAL) RETURN
STD_ULOGIC_VECTOR;
FUNCTION SHIFT_RIGHT (arg: STD_ULOGIC_VECTOR; count: NATURAL) RETURN
STD_ULOGIC_VECTOR;
FUNCTION ROTATE_LEFT (arg: STD_ULOGIC_VECTOR; count: NATURAL) RETURN
STD_ULOGIC_VECTOR;
FUNCTION ROTATE_RIGHT (arg: STD_ULOGIC_VECTOR; count: NATURAL) RETURN
STD_ULOGIC_VECTOR;

-----Resize functions:---------------------------
FUNCTION RESIZE (arg: STD_ULOGIC_VECTOR; new_size: NATURAL) RETURN
STD_ULOGIC_VECTOR;
FUNCTION RESIZE (arg, size_res: STD_ULOGIC_VECTOR) RETURN STD_ULOGIC_VECTOR;

-----Type conversion:----------------------------
FUNCTION TO_INTEGER (arg: STD_ULOGIC_VECTOR) RETURN NATURAL;
FUNCTION TO_STDLOGICVECTOR (arg, size: NATURAL) RETURN STD_LOGIC_VECTOR;
FUNCTION TO_STDLOGICVECTOR (arg: NATURAL; size_res: STD_ULOGIC_VECTOR) RETURN
STD_LOGIC_VECTOR;
FUNCTION TO_STDULOGICVECTOR (arg, size: NATURAL) RETURN STD_ULOGIC_VECTOR;
FUNCTION TO_STDULOGICVECTOR (arg: NATURAL; size_res: STD_ULOGIC_VECTOR)
RETURN STD_ULOGIC_VECTOR;
ALIAS TO_STD_LOGIC_VECTOR IS TO_STDLOGICVECTOR[NATURAL, NATURAL RETURN
STD_LOGIC_VECTOR];
ALIAS TO_SLV IS TO_STDLOGICVECTOR[NATURAL, NATURAL RETURN STD_LOGIC_VECTOR];
ALIAS TO_STD_LOGIC_VECTOR IS TO_STDLOGICVECTOR[NATURAL, STD_ULOGIC_VECTOR
RETURN STD_LOGIC_VECTOR];
```

```
    ALIAS TO_SLV IS TO_STDLOGICVECTOR[NATURAL, STD_ULOGIC_VECTOR RETURN
    STD_LOGIC_VECTOR];
    ALIAS TO_STD_ULOGIC_VECTOR IS TO_STDULOGICVECTOR[NATURAL, NATURAL RETURN
    STD_ULOGIC_VECTOR];
    ALIAS TO_SULV IS TO_STDULOGICVECTOR[NATURAL, NATURAL RETURN
    STD_ULOGIC_VECTOR];
    ALIAS TO_STD_ULOGIC_VECTOR IS TO_STDULOGICVECTOR[NATURAL, STD_ULOGIC_VECTOR
    RETURN STD_ULOGIC_VECTOR];
    ALIAS TO_SULV IS TO_STDULOGICVECTOR[NATURAL, STD_ULOGIC_VECTOR RETURN
    STD_ULOGIC_VECTOR];

END PACKAGE numeric_std_unsigned;
-------------------------------------------------------------------------------
```

O Reserved Words in VHDL 2008

ABS	ENTITY	NULL	SEQUENCE
ACCESS	EXIT	OF	SEVERITY
AFTER	FAIRNESS	ON	SIGNAL
ALIAS	FILE	OPEN	SHARED
ALL	FOR	OR	SLA
AND	FORCE	OTHERS	SLL
ARCHITECTURE	FUNCTION	OUT	SRA
ARRAY	GENERATE	PACKAGE	SRL
ASSERT	GENERIC	PARAMETER	STRONG
ASSUME	GROUP	PORT	SUBTYPE
ASSUME_GUARANTEE	GUARDED	POSTPONED	THEN
ATTRIBUTE	IF	PROCEDURE	TO
BEGIN	IMPURE	PROCESS	TRANSPORT
BLOCK	IN	PROPERTY	TYPE
BODY	INERTIAL	PROTECTED	UNAFFECTED
BUFFER	INOUT	PURE	UNITS
BUS	IS	RANGE	UNTIL
CASE	LABEL	RECORD	USE
COMPONENT	LIBRARY	REGISTER	VARIABLE
CONFIGURATION	LINKAGE	REJECT	VMODE
CONSTANT	LITERAL	RELEASE	VPROP
CONTEXT	LOOP	REM	VUNIT
COVER	MAP	REPORT	WAIT
DEFAULT	MOD	RESTRICT	WHEN
DISCONNECT	NAND	RESTRICT_GUARANTEE	WHILE
DOWNTO	NEW	RETURN	WITH
ELSE	NEXT	ROL	XNOR
ELSIF	NOR	ROR	XOR
END	NOT	SELECT	

Bibliography

Ashenden, P. J. *The Designer's Guide to VHDL*, 3rd edition, San Francisco, CA: Morgan Kaufmann, 2008.

Bishop, D. *Fixed-Point Package User's Guide*, www.eda.org/fphdl, 2006.

Bishop, D. *Floating-Point Package User's Guide*, www.eda.org/fphdl, 2006.

Fixed- and floating-point standard packages for VHDL 2008 and for compatibility with previous VHDL versions, www.eda.org/fphdl, 2010.

IEEE 1076-1987 Standard VHDL Language Reference Manual, 1987.

IEEE 1076-1993 Standard VHDL Language Reference Manual, 1993.

IEEE 1164-1993 Standard Multivalue Logic System for VHDL Model Interoperability, 1993.

IEEE 1076.2-1996 Standard VHDL Mathematical Packages, 1996.

IEEE 1076.3-1997 Standard VHDL Synthesis Packages, 1997.

IEEE 1076-2000 Standard VHDL Language Reference Manual, 2000.

IEEE 1076-2002 Standard VHDL Language Reference Manual, 2002.

IEEE 1076.6-2004 Standard for VHDL Register Transfer Level (RTL) Synthesis, 2004.

IEEE 1076-2008 Standard VHDL Language Reference Manual, 2008.

Pedroni, V. A. *Circuit Design with VHDL*, 1st edition, Cambridge, MA: MIT Press, 2004.

Pedroni, V. A. *Digital Electronics and Design with VHDL*, Burlington, MA: Morgan Kaufmann, 2008.

Index